Metals in Medicine

Metals in Medicine

JAMES C. DABROWIAK

Department of Chemistry, Syracuse University, New York, USA

A John Wiley and Sons, Ltd., Publication

This edition first published 2009
© 2009 John Wiley & Sons, Ltd

Registered office

John Wiley & Sons Ltd, The Atrium, Southern Gate, Chichester, West Sussex, PO19 8SQ, United Kingdom

For details of our global editorial offices, for customer services and for information about how to apply for permission to reuse the copyright material in this book please see our website at www.wiley.com.

Library of Congress Cataloging-in-Publication Data

Dabrowiak, James C.
  Metals in medicine / James C. Dabrowiak.
     p. cm.
  Includes bibliographical references and index.
  ISBN 978-0-470-68196-1 (cloth) – ISBN 978-0-470-68197-8 (pbk.) 1. Metals in medicine.
2. Metals–Therapeutic use. I. Title.
  RM666.M513D33 2009
  615'.231–dc22
                                                                  2009028763

ISBN 978-0-470-68196-1 (H/B)   ISBN 978-0-470-68197-8 (P/B)

A catalogue record for this book is available from the British Library.

Typeset in 10/12pt Times by Thomson Digital, Noida, India.
Printed and bound in Great Britain by CPI Antony Rowe, Chippenham, Wiltshire.

To my wife, Tatiana, 'Tati'
without whose love, support and understanding
this book would not have been possible

Barnett (Barney) Rosenberg, the discoverer of cisplatin, was a remarkable scientist. He transformed a laboratory discovery into one of the most important drugs for treating cancer and he did it in an era with no previous success of a metal-based anticancer drug. The story of cisplatin will always be inspirational to the budding scientist, sustaining to those doing research in the field and motivational to all pursuing the unknown. Barney Rosenberg passed away on August 8, 2009. He will be greatly missed.

Contents

Feature Boxes

Preface

Metals in Medicine is a textbook for undergraduate and graduate students in chemistry, biochemistry, biology and the related areas of biophysics, pharmacology and bioengineering. The first chapter of the book presents basic bonding concepts in inorganic chemistry and provides a brief overview of the physical and chemical properties of metal complexes using concepts and ideas presented in general chemistry. The more demanding concept of quantum mechanics, although not generally discussed in beginning level chemistry courses, is also briefly covered at an easy-to-understand level in Chapter 1. Chapter 2 emphasizes the nature and structure of biological targets, the reactivity of metal complexes in the biological milieu, and methods for measuring the efficacy and toxicity of agents. The steps from drug discovery to marketplace are also briefly outlined and discussed in this chapter.

The remaining six chapters of *Metals in Medicine* focus on individual metallo-drugs, drug candidates and metal-containing agents used to treat and diagnose disease, their synthesis, structures, formulations, pharmacokinetics and known mechanisms of action, and important physical and chemical principles that apply, while the last chapter addresses the role of inorganic chemistry in the emerging and exciting field of nanomedicine.

No attempt was made to cover all of the metal-containing compounds that are actively being used or being considered for use in medicine, but rather select topics were focused upon, in order to present a brief overview of the area, stressing important chemical, physical and biological principles, and pointing out where the area might be headed. It was felt that this was the best way to prevent 'saturation' and leave room for the motivated student to discover more on their own.

Chapter 3 of the text covers cisplatin, which is arguably the most important metal-containing agent used in medicine and without which this book would not be possible. In this chapter, the student is introduced to how cisplatin was discovered, its physical and chemical properties, and possible methods by which it may kill cancer cells, covering mechanisms that involve DNA as well as protein targets in the cell. Chapter 4 discusses the later-generation cisplatin analogues, carboplatin and oxaliplatin, which are in worldwide use for treating cancer, as well as other platinum drugs that have gained regional approval and a select number of platinum drug candidates that are in clinical trails. The huge success of cisplatin prompted the search for other metal complexes with antitumor properties and Chapter 5 addresses compounds of ruthenium, titanium and gallium that exhibit anticancer activity. In order to emphasize the breadth of application of inorganic chemistry in medicine, the use of gold complexes for treating arthritis, cancer and other diseases is presented in Chapter 6, while the use of vanadium for treating diabetes, copper in Wilson's, Menkes and Alzheimer's diseases, and zinc-bicyclam as a stem-cell mobilizing agent and for potentially treating AIDS are covered in Chapter 7.

Although metal-containing agents have had a major impact on treating disease, they are also widely used for disease detection. Chapter 8 outlines the importance of radioactive technetium complexes in diagnostic nuclear medicine and discusses the use of paramagnetic gadolinium compounds as contrast enhancing agents in magnetic resonance imaging, MRI. This chapter also briefly discusses the use of radioactive agents for palliative care and cancer treatment in radioimmunotherapy.

Chapter 9 covers the design and construction of nano-size structures for biomedical applications in nanomedicine. Since nanomedicine is one of the most dynamic areas of science, presenting all of the developments in the field with an inorganic theme proved impossible, so a limited number of examples, chosen largely because they extend the information in previous chapters, are presented and discussed. Although nanomedicine has enormous potential, the fact that nanomaterials are totally alien to the biological system has raised serious questions about the health risks that they may pose to humans. This topic is also discussed in Chapter 9.

Throughout the book are Feature Boxes that expand important concepts in metals in medicine, including relevant physical techniques, structures of biological targets and transport molecules, the discovery of cisplatin, synthesis of compounds, special assays and principles behind medical techniques. Although the Feature Boxes were not intended to be comprehensive, they provide sufficient information to show 'how things work', with additional information being found in the extensive list of references at the end of each chapter. Following each chapter are specifically designed problems, with solutions, that allow the student to apply the laws of thermodynamics and the principles of equilibrium and kinetics to problem solving in the topic being addressed.

While this textbook is designed for teaching a one-semester course on the role of metal complexes in medicine, it could also be used to teach basic coordination chemistry against the exciting backdrop of metals in medicine. It has been the author's experience that students with no previous background in inorganic chemistry and with sights on careers in medicine will easily accept learning some of the most challenging aspects of inorganic chemistry as long as the ultimate goal is learning how metal-containing agents are used in medicine. It is also clear that those students who are committed to chemistry and its related disciplines find the subject matter totally intoxicating and easily acclimatize to the many biochemical and medical aspects that metals in medicine involves.

In reading the extensive volume of literature needed to write this book, the author was impressed, indeed humbled, by the huge body and quality of work produced by investigators in the field. Clearly, many decisions needed to be made about what or what not to include, but in the end the guiding principle was on what to tell students wishing to gain an overview of an exciting area of science and, most importantly, to let them see how they might fit into the area and ultimately how they could help to move it forward. Since selection was of course carried out by the author, he accepts full responsibility for the emphasis of the book and any omissions and inaccuracies that it may contain.

James C. Dabrowiak
May 2009

Web Site

PowerPoint slides of all figures from this book, along with the solutions to the problems, can be found at http://www.wiley.com/go/dabrowiak.

Acknowledgments

It is difficult to write any book without the help, encouragement and support of many people. Special thanks to a professional colleague and personal friend, Jerry Goodisman, who read and commented on various parts of the manuscript for its substance and technical accuracy, and thanks also to many professional colleagues and support staff at Syracuse University for their insight and suggestions. Dr Matthew D. Hall at the National Institutes of Health suggested many important concepts and ideas in the development stage of the book that were ultimately incorporated into the finished manuscript. Through the years a number of outstanding graduate students, because they often focused thinking in unexpected directions, have helped shape the views of the author, and thus their influence is also in this work. Gratitude is also expressed to the graduate and undergraduate students in the 2009 edition of the course 'Metals in Medicine', who helped to make many sections of the book stronger as the manuscript was being written. Spending countless hours reading and writing is a strain on a relationship and my wife, Tatiana, deserves enormous credit for understanding what needed to be done and, most importantly, for 'being there' when things were not going as well as they might.

1

Inorganic Chemistry Basics

No description of the metal-containing compounds that have found their way into medicine would be useful without first providing basic information on the bonding in metal complexes, their spectral and magnetic properties and, most importantly, the manner in which they react with water and biological targets in the cell. The approach taken in this chapter assumes background knowledge of general and organic chemistry with no previous exposure to inorganic chemistry, as would occur in a junior- or senior-level course at most universities. The concepts presented are for the most part intuitive, requiring basic knowledge of chemistry and physics, but sometimes more abstract issues like quantum mechanics – which explains the spectral properties of metal complexes – will also need to be covered. The overall goal of this chapter is to bring all readers to a common level, providing them with the 'core' of information needed to understand how and why, from the chemical perspective, metal complexes play important roles in medicine.

1.1 Crystal field theory

The bonding that exists in metal complexes, their spectral and magnetic properties and their chemical reactivity are not easily explained using a single theory. However, one approach that is often used in a basic presentation of bonding concepts in transition metal chemistry is crystal field CF theory, which because it is based on simple electrostatic arguments, is relatively easy to understand. In CF theory and MO theory the interactions between the metal ion (M) and the groups attached to it (called ligands and denoted by L) are considered to be electrostatic in nature and the bonding in the compound is described as being salt-like in character. The metal ion, a *cation*, electrostatically interacts with a series of surrounding ligands, which are usually negatively charged or, if they are uncharged, have the negative end of a dipole directed toward the metal ion. Barring any serious steric interactions between the ligands, the arrangements about the metal ion generally have high-symmetry geometries. For example, a 6-coordinte complex – that is, a compound with six ligands attached to the metal ion – has an octahedral arrangement of ligands, while five-coordinate complexes have square or trigonal bipyramidal arrangements, four-coordinate structures are tetrahedral and square planar, and so on. These geometries, along with compounds and intermediates commonly encountered in metal complexes used in medicine, are shown in Figure 1.1.

Metals in Medicine   James C. Dabrowiak
© 2009 John Wiley & Sons, Ltd

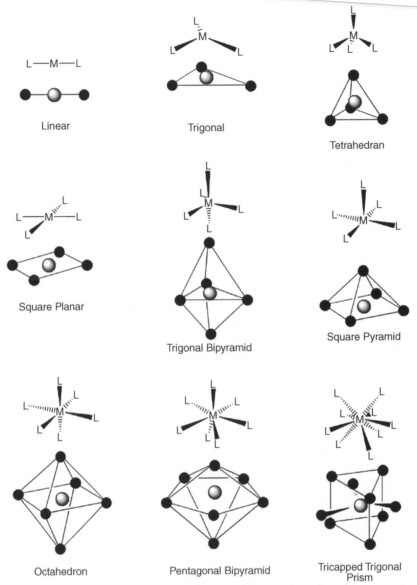

Figure 1.1   *Common geometries of metal complexes and intermediates found in inorganic chemistry*

1.1.1   Octahedral crystal field

The first-row transition metal series, which begins with scandium, Sc, fills the 3*d* level of the atom, while the second- and third-row transition metal series, which begin with yttrium, Y, and lanthanum, La, respectively, fill the 4*d* (second row) and 5*d* (third row) orbitals of the atom. The transition metal ions and the electronic configurations of common oxidation states are shown in Figure 1.2. Since ions of these elements have electron occupancies in the *d* level, which is considered the 'valence' level of the ion, CF theory focuses on the change in energy of the *d-orbital*s when charges representing the ligands approach the metal ion and form salt-like bonds.

	3 3B	4 3B	5 5B	6 6B	7 7B	8 8B	9 8B	10 8B	11 1B	12 2B
Ions	$+2, d^1$ $+3, d^0$	$+2, d^2$ $+3, d^1$ $+4, d^0$	$+2, d^3$ $+3, d^2$ $+4, d^1$ $+5, d^0$	$+2, d^4$ $+3, d^3$ $+4, d^2$ $+5, d^1$ $+6, d^0$	$+1, d^6$ $+2, d^5$ $+3, d^4$ $+4, d^3$ $+5, d^2$ $+6, d^1$ $+7, d^0$	$+2, d^6$ $+3, d^5$ $+4, d^4$	$+2, d^7$ $+3, d^6$ $+4, d^5$	$+2, d^8$ $+3, d^7$ $+4, d^6$	$+1, d^{10}$ $+2, d^9$ $+3, d^8$	$+2, d^{10}$
1st Row $3d^n$	^{21}Sc	^{22}Ti	^{23}V	^{24}Cr	^{25}Mn	^{26}Fe	^{27}Co	^{28}Ni	^{29}Cu	^{30}Zn
2nd Row $4d^n$	^{39}Y	^{40}Zr	^{41}Nb	^{42}Mo	^{43}Tc	^{44}Ru	^{45}Rh	^{46}Pd	^{47}Pd	^{48}Cd
3rd Row $5d^n$	^{57}La	^{72}Hf	^{73}Ta	^{74}W	^{75}Re	^{76}Os	^{77}Ir	^{78}Pt	^{79}Au	^{80}Hg

Figure 1.2 *Transition metal ions and their electronic configurations for various oxidation states*

The spatial arrangements of the five *d*-orbitals on a Cartesian coordinate system are shown in Figure 1.3. The shapes shown represent the probability of finding an electron in a volume of space about the nucleus of the metal ion. If the metal ion has no bonded ligands – this is referred to as a *free ion* – the energies of all five *d*-orbitals will be the same and are said to be *five-fold degenerate* in energy. This situation is shown on the left side of Figure 1.4. Let's suppose that instead of existing as a free ion, the metal ion is part of a stable complex consisting of six negatively-charged ligands bound to the metal ion in an octahedral array. The way that crystal field theory approaches this situation is to consider what happens to the five *d*-orbitals in the

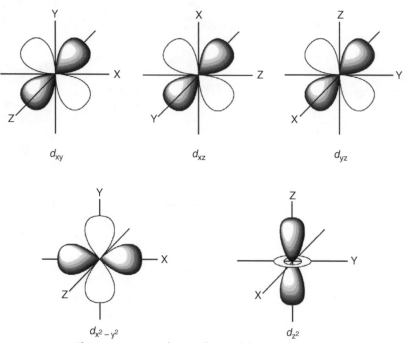

Figure 1.3 *Boundary surfaces of the five d-orbitals*

electrostatic field set that is up by the ligands. The first thing that the theory does is to consider a situation in which the total negative charge of the ligands is 'smeared' equally over the surface of a sphere with a radius equal to the metal–ligand bond distance and with the metal ion at its center. Since the d-orbitals have electrons in them and the surface of the sphere is negatively charged, the energies of the d-orbitals will be raised; that is, they will become less stable relative to the free ion, due to electrostatic repulsion between the d-electrons and the negatively-charged surface of the sphere. Since the charge on the sphere has no 'directionality' – that is, the negative charges are equally distributed over the entire surface of the sphere – all five d-orbitals must experience the same electrostatic perturbation from the sphere and move as a group to a new energy, E_o (see Figure 1.4). The next step is to redistribute the charge on the surface of the sphere and concentrate it at the six points where the axes penetrate the sphere. If the charge at each of the six points is identical, this will produce a perfect octahedral crystal field about the central metal ion and simulate what the d-orbitals experience in an octahedral metal complex. It should be evident that since $d_{x^2-y^2}$ and d_{z^2} are pointed directly at the charges (ligands), they must experience a different perturbation than the three orbitals, d_{xz}, d_{yz}, d_{xy}, that are directed between the charges. While it may not be obvious that both $d_{x^2-y^2}$ and d_{z^2} should experience an identical perturbation from the octahedral field, quantum mechanics shows that d_{z^2}, which has a ring of electron density in the xy plane (Figure 1.3), is actually a composite of two orbitals that are identical to $d_{x^2-y^2}$ except that they lie in the yz and xz planes. Thus, since d_{z^2} is a composite of two orbitals that look like $d_{x^2-y^2}$, it makes sense that the crystal field will affect d_{z^2} and $d_{x^2-y^2}$ identically, as shown in Figure 1.4. It should also be evident that since these orbitals are pointed directly at the ligands, they feel the electrostatic repulsion directly, and thus their energies are *raised* relative to the energy of the spherical field, E_o. It is possible to show that if the total charge on the sphere is simply rearranged or 'localized' to certain positions on the sphere, the energy of the system cannot change; that is, E_o for the sphere and the octahedral field must be the same. This is the *center of gravity* rule, which applies to electrostatic models of this type. The consequences of this is that if two orbitals,

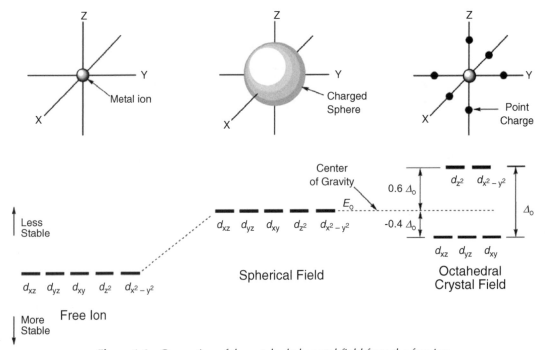

Figure 1.4   *Generation of the octahedral crystal field from the free ion*

$d_{x^2-y^2}$ and d_{z^2}, are raised by a certain amount, the remaining three, d_{xz}, d_{yz}, d_{xy}, must be *lowered* by a certain amount. Inspection of the shapes and orientations of d_{xz}, d_{yz}, d_{xy} shows that since these orbitals are directed 45° to the axes of the system, and each is related to the others by a simple rotation, all must experience exactly the same perturbation from the charges which are on the axes of the system. This set of orbitals, which are 'triply degenerate', is often referred to as the 't_{2g}' set due to its symmetry properties. In a similar fashion, the orbitals, $d_{x^2-y^2}$ and d_{z^2} which are 'doubly degenerate' are referred to as the 'e_g' set. The labels t_{2g} and e_g are products of the application of *group theory*, a mathematical tool for characterizing the symmetry properties of molecules.

Simple electrostatic arguments show that the spacing between the t_{2g} and e_g levels depends on the distance that the charge is from the origin of the system and the magnitude of the charge. If the distance is decreased or if the magnitude of the negative charge is increased, the splitting between t_{2g} and e_g will increase. As we will see, metal complexes can be made with a wide variety of attached ligands, some of which are negatively charged, for example, Cl^-, CN^- and so on, and some of which are electrically neutral, for example, H_2O, NH_3 and so on. However, one thing that all ligands have in common is that they direct electrons, usually a lone pair, toward the metal ion, and these electrons become the 'point charges' in the crystal field model describing the electronic structure of the complex. Since the ability of different ligands to perturb the d-orbitals varies considerably, the spitting between the t_{2g} and e_g sets of orbitals can be quite different for different complexes. In order to address this, crystal field theory denotes the splitting between the t_{2g} and e_g sets as Δ_o, which is the *crystal field splitting parameter*. The subscript 'o' in Δ_o indicates that a crystal field of octahedral symmetry is being addressed. If there are no attached ligands – that is, in the free ion case – there can be no crystal field and Δ_o is zero. Since the splitting between the levels is different for different metal complexes, Δ_o, which carries units of energy usually expressed in wavenumbers (cm^{-1}), varies over a wide range. However, the relative displacement of the t_{2g} and e_g levels *in terms of* Δ_o from the center of gravity, E_o, is the same for all octahedral complexes with the e_g level at $0.6\,\Delta_o$ and the t_{2g} level at $-0.4\,\Delta_o$. These values arise because (2 orbitals) × (0.6 Δ_o) + (3 orbitals) × ($-0.4\,\Delta_o$) $\equiv 0$, which satisfies the center of gravity rule. It should be evident that E_o is the *average crystal field*.

1.1.2   Other crystal fields

Numerous anticancer drugs containing Pt^{+2} have a square planar geometry in which four ligands at the corners of a square are bonded to the metal ion (Figure 1.1). The best way to generate the square planar crystal field splitting pattern for the d-orbitals is to first consider an intermediate field called the *tetragonal crystal field*. Suppose that the charge on each of the two point charges on the plus and minus z-axis of the octahedral crystal field is slightly reduced in magnitude relative to the other four charges in the plane or, the equivalent situation, wherein the magnitude of the charges on the plus and minus z-axis remain unchanged but the charges are moved to greater distance from the metal ion than the four charges in the plane. In this case, the electrostatic field on the z-axis is *less than* the field seen by the metal ion on the x- and y-axes of the system. As a consequence of this *asymmetry* or non-equivalence in the field, all orbitals with z-components – that is, d_{xz}, d_{yz} and d_{z^2} – will have their energies *lowered*; that is, they will become *more stable* in the applied field (Figure 1.5). Since d_{z^2} is pointed directly at the weaker charges on the z-axis, it must experience greater stabilization – that is, more lowering – than d_{xz}, d_{yz}, which are directed away from the point charges. As a consequence of the center of gravity rule, if some levels go down in energy, others – that is, $d_{x^2-y^2}$ and d_{xy} – must become *less stable* and their energies must be raised.

The limiting case of the tetragonal distortion is the *square planar geometry* in which the two charges on the z-axis have been reduced to zero; that is, there are only four charges in the plane of the system. The removal of the axial charges causes a significant stabilization in d_{z^2}, which moves downward in the energy diagram and passes below (becomes more stable than) the d_{xy} orbital. Since d_{xz}, d_{yz} also have z-components, they are also stabilized by the loss of the axial charges, but to a lesser extent than d_{z^2}. The resulting crystal field splitting diagram, sometime called the *square planar limit*, is shown in Figure 1.5.

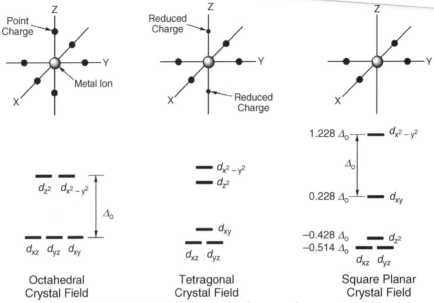

Octahedral   Tetragonal   Square Planar
Crystal Field   Crystal Field   Crystal Field

Figure 1.5 *Octahedral, tetragonal and square planar crystal field*

A second very common structure for metal complexes with four groups bonded to the central metal ion is the *tetrahedral* geometry (Figure 1.1). Compared to the previous examples, rationalizing the *d*-orbital splitting pattern for the tetrahedral geometry is less straightforward. Figure 1.6 shows a Cartesian coordinate system in the center of a cube. Placement of charges at opposite corners of opposite faces of the cube and hypothetically connecting them to the metal ion in the center of the cube generates the tetrahedral geometry; that is, all charge–metal–charge angles are 109.5°. It should be evident from the figure that none of the *d*-orbitals point directly at the charges, and although other relative arrangements of the cube on the *d*-orbital coordinate system are possible, all lead to the conclusion given for the splitting pattern shown in Figure 1.6. The tetrahedral crystal field has a doubly degenerate set of orbitals, the d_{z^2} and $d_{x^2-y^2}$, termed for symmetry reasons the '*e*' set, which is *lowest* in energy, and a triply degenerate set, d_{xz}, d_{yz} and d_{xy}, called the 't_2' set, which is *highest* in energy. While this pattern is exactly the opposite of the octahedral case, the labels *e* and t_2, which also come from group theory, are missing the subscript 'g'. This is because the octahedron has a symmetry element called the *center of inversion* (*i*), which is associated with a mathematical operation in which each point charge of the

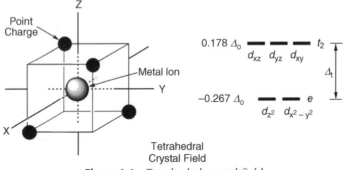

Tetrahedral
Crystal Field

Figure 1.6 *Tetrahedral crystal field*

Table 1.1 Relative energies of the d-orbital in various crystal fields[a]

CN	Structure	d_{z^2}	$d_{x^2-y^2}$	d_{xy}	d_{xz}	d_{yz}
2	Linear[b]	1.028	−0.628	−0.628	0.114	0.114
3	Trigonal[c]	−0.321	0.546	0.546	−0.386	−0.386
4	Tetrahedral	−0.267	−0.267	0.178	0.178	0.178
4	Square Planar[c]	−0.428	1.228	0.228	−0.514	−0.514
5	Trigonal Bipyramid[d]	0.707	−0.082	−0.082	−0.272	−0.272
5	Square Pyramid[d]	0.086	0.914	−0.086	−0.457	−0.457
6	Octahedron	0.600	0.600	−0.400	−0.400	−0.400
7	Pentagonal Bipyramid[d]	0.493	0.282	0.282	−0.528	−0.528
9	Tricapped Trigonal Prism	−0.225	−0.038	−0.038	0.151	0.151

[a] Values given are in units of Δ_o.
[b] Ligands or charges are along the z-axis.
[c] Ligands or charges are in the xy plane.
[d] Pyramid base in xy plane. From Table 9.14, p. 412 of Huheey, J.E. (1983) *Inorganic Chemistry: Principles of Structure and Reactivity*, 3rd edn, Harper & Row Publisher, New York.

structure can be passed along a straight line through the central metal ion to reach an identical point charge (Figure 1.5). Since i is not present in the tetrahedron, the subscript 'g' is missing from the labels. Although the tetrahedral pattern is the exact opposite of the splitting pattern for the octahedron, the magnitude of the splitting between the e and t_2 levels for the tetrahedral geometry, denoted as Δ_t, is only 4/9 the value of the splitting between t_{2g} and e_g of the octahedron; that is, $\Delta_t = 4/9\ \Delta_o$ or 0.445 Δ_o. Thus, for the tetrahedron, the t_2 orbital set is at 0.178 Δ_o and the e orbital set is at −0.267 Δ_o.

Table 1.1 gives the energies of the five d-orbitals for common geometries in terms of the octahedral crystal field splitting parameter, Δ_o. The values in the table, which were calculated using a point charge crystal field model, can be used to determine the orbital energy diagrams for geometries other than the octahedral, tetrahedral and square-planar geometries discussed above. The entries in the table assume that if there is more than one electron in the pattern, which is almost always the case, there is no interaction between the electrons, which is never the case. The energies given are the so-called *one electron energies* for the various orbitals in the different crystal fields. If there is more than one electron in the pattern, the electrons can be in the same or different orbitals and will 'see' each other through what are called *configuration interactions*, and the energies of levels given in Table 1.1 will be adjusted to new values. Since determining the new energies of the orbitals is beyond the scope of our work, and changes are in most cases small, the entries in Table 1.1 are reasonable approximations for all multiple-electron systems encountered in this text.

1.1.3 Factors affecting the crystal field splitting parameter, Δ

1.1.3.1 Spectrochemical series

As was earlier pointed out, the point charges used to generate the splitting patterns for various geometries simulate the electrostatic effects of ligands that are bonded to the metal ion. Extensive spectral and magnetic studies on a large number of transition-metal complexes showed that the electronic effect exerted by a specific ligand on the d-orbitals of the metal ion is essentially a property of that ligand and independent of the geometry of the complex, the nature of the metal ion or its oxidation state. This characteristic allowed ranking of common ligands in terms of their 'd-orbital splitting power', to produce a series called the *spectrochemical*

Table 1.2   *Factors affecting the crystal field splitting parameter,* Δ

The Ligand[a]	*Spectrochemical Series, Increasing* Δ
	$I^- < Br^- < S^{-2} < NC\underline{S}^- < Cl^- < NO_3^- < N_3^- < F^- < OH^- < C_2O_4^{-2} \approx H_2O < \underline{N}CS^-$
	$< CH_3C\underline{N} < NH_3 < en < bipy < phen < \underline{N}O_2^- < \underline{P}Ph_3 < \underline{C}N^- < \underline{C}O$
The Metal Ion	*Principal Quantum Number, n*
	First-row transition metal ion, 3*d* level, Δ^{3d}
	Second-row transition metal ion, 4*d* level, $\Delta^{4d} \sim 1.5\, \Delta^{3d}$
	Third-row transition metal ion, 5*d* level, $\Delta^{5d} \sim 1.75\, \Delta^{3d}$
The Metal Ion[*]	*Oxidation State, Increasing* Δ
	$M^+ < M^{+2} < M^{+3} < M^{+4} < M^{+5}$

[a] The underscored atom is the donor atom to the metal ion. *en*, ethylenediamine, 1, 2 diaminoethane; *bipy*, 2, 2' bipyridine; *phen*, 1, 10 phenanthroline.

series (Table 1.2). Ligands on the left of the series, which are referred to as *weak field ligands*, for example I^- and Br^-, cause a small splitting in the *d*-orbitals, while ligands on the right of the series, for example CN^- (cyanide), CO (carbon monoxide) – *strong field ligands* – cause a large splitting in the orbitals. While there is little doubt that the order of the ligands in the series is correct (the order is obtained from experiment), the series does not seem to follow our intuitive feeling about which ligands should be high in the series and which should be low. For example, CO, which is uncharged, is highest in the series but iodide, I^-, which is negatively charged, is lowest in the series. Based on the electrostatic arguments put forth in connection with the crystal field this makes little sense: I^- should have a *greater* perturbation on the *d*-orbitals than uncharged CO. Clearly, factors other than simple electrostatic effects must influence Δ_o. While the crystal field model works well for most of the cases encountered in this text, complexes which have considerable overlap between the orbitals on the metal and ligand – that is, when covalent bonding is present – cause the theory to 'bend' but not completely break down. How basic crystal field theory needs to be modified to accommodate this will be addressed in a later section.

1.1.3.2   *Principal quantum number, n*

While the spectrochemical series rank orders the experimentally-measured effects of ligands on the splitting of the *d*-orbitals, it is also possible to make some general statements concerning the effects of the metal ion on the magnitude of Δ. If one moves down a given column in the periodic chart, the quantum number *n*, which is called the *principal quantum number*, increases. For example, the first-row transition metal series elements have electrons in the 3*d* ($n = 3$) level, the second-row in the 4*d* ($n = 4$) level and the third-row in the 5*d* ($n = 5$) level of the atom. Experimentally, it has been found that the magnitude of the crystal field splitting parameter Δ increases in the order $3d < 4d < 5d$, with $\Delta\, 4d \sim 1.5\, (\Delta\, 3d)$ and $\Delta\, 5d \sim (1.75\, \Delta\, 3d)$ (Table 1.2). The effects of this increase with *n* can easily be seen for the series $[Co(NH_3)_6]^{3+}$, $3d^6$, $[Rh(NH_3)_6]^{3+}$, $4d^6$ and $[Ir(NH_3)_6]^{3+}$, $5d^6$, which have identical geometries (octahedral), ligands (ammonia) and metal ion oxidation states (+3), and belong to the same family (column) of the periodic chart. The values of Δ_o for these complexes are $\sim 22\,000\,cm^{-1}$, $\sim 34\,000\,cm^{-1}$ and $\sim 41\,000\,cm^{-1}$, respectively, which shows that moving down a given column in the periodic chart does indeed cause the values of the crystal field splitting parameter to *increase* by the approximate amounts given. Since atoms, and ions, become larger with atomic number, M–L bond lengths increase in moving from the first to the second and third rows of the transition metal series. Simple point-charge arguments would predict that if the M–L distance were increased, the magnitude of Δ would *decrease*, not increase as observed. The fact that the opposite is found is further proof that the simple point-charge model cannot be entirely correct and that other factors are important in determining the magnitude of Δ.

1.1.3.3 *Metal ion oxidation state*

Experimentally, it can be shown that *increasing* the charge on the metal ion – that is, increasing its oxidation state – causes the *d-d* absorption bands of the complex to shift toward the UV region of the spectrum, which means that Δ has *increased* (Table 1.2). Since the ionic radius of any ion decreases with an increase in the net positive charge on the ion, the *distance* between the metal ion and its bonded ligands must decrease when oxidation state is increased. Since decreased distance would lead to greater electrostatic repulsions between electrons on the metal ion and the ligands, the observed trends in Δ (with changes in oxidation state on the metal ion) *are* predicted by simple crystal field arguments.

1.1.4 High- and low-spin complexes

When considering the ways in which electrons can occupy energy levels of an atom, ion or molecule, *Hund's rule* states that the electronic configuration with the lowest overall energy is one for which the spins for the electrons are unpaired, even if it means placing electrons in a nearby less-stable orbital (level) in order to do so. For the free ion, the five *d*-orbitals are degenerate in energy and electrons are added to the orbitals by *maximizing the number of unpaired spins*. If, for example, there are four electrons in the *d*-level of a free ion, it is possible to place the electrons in the level in a number of different ways, some of which are shown in Figure 1.7. Experimentally, Figure 1.7a, which has the maximum amount of spin unpairing, is known to be the lowest-energy (most stable) configuration. When describing the electron spin of any system, it is best to use the value of the magnetic spin quantum number, m_s, associated with the spin angular momentum of the electron. Each electron has spin angular momentum of $\pm^1\!/_2$ in units of $h/2\pi$. With n electrons, the maximum possible value of the *total magnetic spin quantum number S* is n/2 (all electrons unpaired) and the minimum possible value of S is zero (if n is even) or $^1\!/_2$ (if n is odd). For simplicity, the term $h/2\pi$, where h is Planck's constant, is usually dropped. Thus, for the configuration shown in Figure 1.7a, $S = (4)(1/2) = 2$, while S for Figure 1.7b is $(+1/2 -1/2 + 1/2 + 1/2) = 1$ and for Figure 1.7c is $(+1/2 -1/2 + 1/2 -/2) = 0$. Two factors associated with electronic configurations, *coulombic interactions* and *spin correlations*, form the basis for Hund's rule. Since placement of two electrons in the same orbital forces them to occupy the same regions of space, the coulombic repulsion between the electrons will be high, thus destabilizing the system. This obvious electrostatic repulsion makes it easy to see why maximum spin unpairing, maximum S, is desirable. While columbic considerations are important, the ability to exchange one electron with another in a given configuration *without changing S* is even more important. This aspect of Hund's rule, which is a product of quantum mechanics, in called *spin correlation* or *exchange energy*. Both of these factors – coulombic (electrostatic) and spin correlation (exchange energy) – drive the system to obtain maximum spin unpairing,

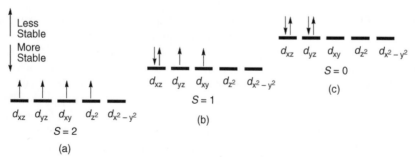

Figure 1.7 *Some possible electronic configurations for the d⁴ free ion and their respective values of S*

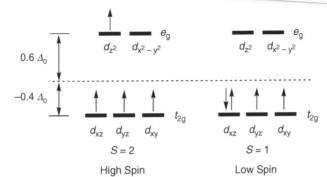

Figure 1.8   *High and low spin possibilities for d^4 in an octahedral crystal field*

and for the free ion case, where all of the d-orbitals have the same energy, *the configuration with the largest value of S always has the lowest, most negative, energy.*

In the presence of an octahedral crystal field the five d-orbitals are not degenerate in energy, and a decision needs to be made on how to place the electrons into the d-orbital pattern to create the lowest energy configuration. For example, d^4 in an octahedral crystal field can have two possible electronic configurations, which are shown in Figure 1.8. One situation, called the *high-spin* case, has one electron in each of the three t_{2g} orbitals and the fourth electron in one of the d-orbitals of the e_g level, giving $S = 2$. The energy gained by the system due to the presence of the crystal field is called the *crystal field stabilization energy* or *CFSE*, which in this case is $CFSE_{hs} = (3 \text{ electrons})(-0.4\,\Delta_o) + (1 \text{ electron})(0.6\,\Delta_o) = -0.6\,\Delta_o$. This possibility can be written as $t_{2g}^3 e_g^1$. An alternative possibility for arranging the electrons in the levels is also shown in Figure 1.8. In this case all four electrons are in the t_{2g} set of orbitals and since *Pauli's principle* must be obeyed, two of the electrons in the same orbital must have their spin oppositely aligned. This gives $S = 1$ for the configuration, which is called the *low-spin* case. The crystal field crystallization energy for this configuration is $CFSE_{ls} = (4 \text{ electrons})(-0.4\,\Delta_o) = -1.6\,\Delta_o + P$, where P is the energy required for pairing two of the electrons in one of the orbitals. This possibility can be written as t_{2g}^4. Since P is the energy *lost* due to coulombic and exchange effects, its sign is *positive*, meaning that it *destabilizes* the system. Which possibility is found – high-spin or low-spin – clearly depends on which $CFSE - CFSE_{ls}$ or $CFSE_{hs}$ – has the larger *negative* value. This can be found by equating $CFSE_{ls}$ to $CFSE_{hs}$, which gives $-1.6\,\Delta_o + P = -0.6\,\Delta_o$ or $P = \Delta_o$. If the ligands bonded to the metal ion produce a splitting in the d-orbitals with $\Delta_o > P$, the *low-spin* possibility, $S = 1$, will be more stable. If the ligands produce a splitting with $\Delta_o < P$, the *high-spin* situation, $S = 2$, will be more stable.

The electronic configuration, value of S and CFSE for weak and strong octahedral fields are given in Table 1.3. A point to make concerning the entries in Table 1.3 is that the value of CFSE given is the stabilization energy that the system accrues as a result of the presence of the crystal field. This means that the 'reference point' for determining the CFSE is the free-ion case. For example, for the strong-field configuration, t_{2g}^6, which has $S = 0$, the value of CFSE in Table 1.3 of $-2.4\,\Delta_o + 2P$ was determined by first writing the electronic configuration for the free-ion case and determining its energy in terms of P, the pairing energy. With six d-electrons and five orbitals, there must be one paired set of electrons or one unit of P for the free ion. Next, the energy of the system due to the presence of the crystal field, in terms of Δ_o and P, was written, which in this case is $-2.4\,\Delta_o + 3P$. There are three pairs of electrons in the t_{2g} set of orbitals, hence $3P$, and the extra energy in terms of Δ_o due to the crystal field is $-2.4\,\Delta_o$. Taking the difference between the energy in the presence of the field and the energy of the free ion gives $CFSE = -2.4\,\Delta_o + 3P - P$ or $-2.4\,\Delta_o + 2P$. The remaining entries in Table 1.3 were calculated in a similar fashion. The values of CFSE for crystal fields of other symmetries can easily be obtained using the energies of the orbitals given in Table 1.1. As described above, calculation of S and CFSE assumes that there is no configuration interaction; that is, electrons in the pattern operate as independent

Table 1.3 *Crystal field effects for weak and strong octahedral fields*[a]

Weak Field				Strong Field			
d^n	Config.	S	CFSE	d^n	Config.	S	CFSE
d^1	t_{2g}^1	1/2	$-0.4\,\Delta_o$	d^1	t_{2g}^1	1/2	$-0.4\,\Delta_o$
d^2	t_{2g}^2	1	$-0.8\,\Delta_o$	d^2	t_{2g}^2	1	$-0.8\,\Delta_o$
d^3	t_{2g}^3	3/2	$-1.2\,\Delta_o$	d^3	t_{2g}^3	3/2	$-1.2\,\Delta_o$
d^4	$t_{2g}^3 e_g^1$	2	$-0.6\,\Delta_o$	d^4	t_{2g}^4	1	$-1.6\,\Delta_o + P$
d^5	$t_{2g}^3 e_g^2$	5/2	$0\,\Delta_o$	d^5	t_{2g}^5	1/2	$-2.0\,\Delta_o + 2P$
d^6	$t_{2g}^4 e_g^2$	2	$-0.4\,\Delta_o$	d^6	t_{2g}^6	0	$-2.4\,\Delta_o + 2P$
d^7	$t_{2g}^5 e_g^2$	3/2	$-0.8\,\Delta_o$	d^7	$t_{2g}^6 e_g^1$	1/2	$-1.8\,\Delta_o + P$
d^8	$t_{2g}^6 e_g^2$	1	$-1.2\,\Delta_o$	d^8	$t_{2g}^6 e_g^2$	1	$-1.2\,\Delta_o$
d^9	$t_{2g}^6 e_g^3$	1/2	$-0.6\,\Delta_o$	d^9	$t_{2g}^6 e_g^3$	1/2	$-0.6\,\Delta_o$
d^{10}	$t_{2g}^6 e_g^4$	0	$0\,\Delta_o$	d^{10}	$t_{2g}^6 e_g^4$	0	$0\,\Delta_o$

[a] The entries in the table are based on the energies of the levels as if only one electron were present in the system. Placing more than one electron in the levels produces electron–electron interactions, which change the energies of the levels, but this has been neglected in the values given. From Table 9.3, p. 374 of Huheey, J.E. (1983) *Inorganic Chemistry: Principles of Structure and Reactivity*, 3rd edn, Harper & Row Publisher, New York.

non-interacting entities, in which case the one-electron energies in Table 1.1 can be used to estimate the energies of multiple electron cases.

1.2 Molecular orbital theory

The most important theory for discussing bonding in chemistry is molecular orbital (MO) theory. This theory uses the atomic orbitals (AOs) of all of the atoms in a compound to construct molecular orbitals (MOs) the shapes of which depend on the size and shape of, and distance between, atomic orbitals. A critical element of the theory is the wave function, denoted by ψ, which is a mathematical expression that when properly manipulated provides information on a physical system. If the wave function is associated with an atom, it can be used to calculate important properties of the atom, such as the probability of finding an electron in regions of space about the nucleus. If the wave function is associated with a molecule, it can be manipulated to provide information on the energies of bonds, the distribution of electrons in the molecule and other important quantities.

The wave function for an atom contains information on the *principal quantum number* (n), for example the '3' in $3d$, the *orbital angular momentum quantum number* (l), for example the number associated with the letter 'd' in $3d$, and the *magnetic quantum number* (m_l), which distinguishes orbitals of a given n and l. For a d-'subshell', m_l has five values, which give rise to the five d-orbitals shown in Figure 1.2. For a p-subshell, m_l has three values, which produce the three p-orbitals, p_x, p_y and p_z, often discussed in organic chemistry. The fourth quantum number, called the *magnetic spin quantum number*, m_s, is associated with the spin angular momentum of the electron. As we have seen, it has values of $+1/2$ and $-1/2$ in units of $h/2\pi$, which refer to the spin moment of the electron relative to an electric or magnetic field. If two electrons occupy the same orbital, the values of n, l and m_l are the same but the spins of the electrons must be oppositely aligned; that is, m_s for one electron must be $+1/2$ and m_s for the second must be $-1/2$ $h/2\pi$.

1.2.1 MO diagram of molecular hydrogen

The best way to describe MO theory at an elementary level is to show what happens when two hydrogen atoms (H_A and H_B) come together to form a hydrogen molecule (H_A–H_B) (Figure 1.9). While molecular hydrogen

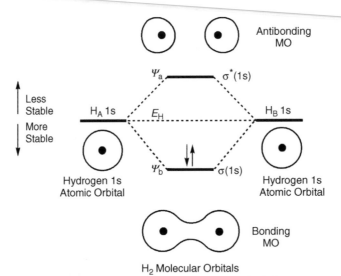

Figure 1.9   *Molecular orbital diagram for molecular hydrogen, H_2. The black dots represent the hydrogen nucleus, a proton*

has little relevance to metallo-drugs, qualitatively understanding how MO theory addresses the bonding in this simple molecule paves the way for understanding the bonding in more complicated metal complexes used in medicine.

MO theory starts with writing the complete (many-electron) wave function in terms of wave functions for individual electrons and then each one-electron wave function is written in terms of atomic orbitals. The specific approach for molecular hydrogen is to carry out *linear combinations of atomic orbitals* (LCAO) by taking the sum and difference of the AOs on both hydrogen atoms to find new, *molecular*, wave functions that are associated with the H_2 molecule. For molecular hydrogen, LCAO gives (1.1) and (1.2).

$$\Psi_b = [c_1\psi_A(1s) + c_2\psi_B(1s)] \tag{1.1}$$

$$\Psi_a = [c_1\psi_A(1s) - c_2\psi_B(1s)] \tag{1.2}$$

In these expressions, Ψ_b is called the *bonding* molecular wave function and Ψ_a the *antibonding* molecular wave function for the hydrogen molecule, while ψ_A and ψ_B are the atomic wave functions for the 1s orbitals on the two hydrogen atoms that were joined together to form the hydrogen molecule. The quantities, c_1 and c_2, which are positive, are simply weighting coefficients that determine the fraction (amount) of each atomic wave function to be used in making the molecular wave function. Conceptually, the expressions say that there are two ways to combine the 1s orbitals on the two hydrogen atoms, H_A and H_B, when the hydrogen molecule is formed. One gives the bonding MO associated with the bonding wave function, (1.1). This MO gives a spatial volume that allows the electron to be found a significant fraction of the time between the two nuclei (Figure 1.9). The second way to do the combination gives the antibonding MO associated with the antibonding wave function, (1.2). This MO produces a shape that *prohibits* the electron from being found midway between the two nuclei (Figure 1.9). Both of these MOs are referred to as being *sigma type*; that is, they are associated with σ *bonds*, meaning that electron density for these MOs is on the bond axis of the system. In MO language, the bonding MO is labeled σ(1s), while the antibonding MO is denoted σ*(1s). For comparison, the other type of MO commonly encountered in chemistry is the *Pi type*, or simply π-*bond*, for which electron density is *not* on the bond axis of the system.

As shown in Figure 1.9, the energies of the 1s AO for both H_A and H_B, E_H, must be identical because both atoms are identical. The energy of $\sigma(1s)$ MO is lowered from E_H and it becomes more stable than the $\sigma^*(1s)$ MO, which is raised in energy from E_H. For this situation in which the energies of the two AOs that formed the MOs are identical, the 'makeup' or composition of the $\sigma(1s)$ MO and the $\sigma^*(1s)$ MO must have exactly equal contributions from both $\psi_A(1s)$ and $\psi_B(1s)$; that is, the coefficients c_1 and c_2 in (1.1) and (1.2) are the same and positive. This implies that electrons in this MO spend equal fractions of their time on either hydrogen atom of H_2 Once the molecular orbitals are formed and their approximate energies determined, the last step is to place the appropriate number of electrons into the diagram, filling the levels from bottom up, so that each MO holds two electrons with their spins oppositely aligned (Figure 1.9). Since each hydrogen atom brings one electron to the bonding scheme, the two electrons are placed in the lowest-energy MO, giving the electronic configuration for H_2 of $\sigma(1s)^2$, which is called the *ground state* of the hydrogen molecule. This stabilizes the hydrogen molecule relative to the two atoms and produces the well-known two-electron single covalent bond for molecular hydrogen. As is shown in Figure 1.9, the antibonding MO is not occupied by electrons when H_2 is in the ground state. However, it can be occupied by an electron if one of the electrons in the $\sigma(1s)$ MO is moved to the $\sigma^*(1s)$ MO. This event, which would obviously require the addition of energy, is called an *electronic transition*, which could happen if H_2 were to absorb a photon with an energy equal to the energy spacing between $\sigma(1s)$ and $\sigma^*(1s)$. The resulting electronic configuration, which is written as $\sigma(1s)^1\sigma^*(1s)^1$, is referred to an *excited state* of the hydrogen molecule, which when produced effectively eliminates the bond between the two hydrogen atoms; that is, one electron is in a bonding MO and one is in an antibonding MO. However, since the lifetime of the excited state is short, $\sim 10^{-9}$ s, dissociation of the hydrogen molecule into hydrogen atoms does not appreciably occur under conditions for which the transition can be observed. Located at higher energy (less stable) than the 1s atomic orbital is the 2s atomic orbital of hydrogen. These atomic orbitals interact with one another to produce another set of MOs, $\sigma(2s)$ and $\sigma^*(2s)$, which, if H_2 is in the ground state, are at very high energy and are not occupied.

The basic features of the diagram shown in Figure 1.9 are the same as the features in the diagrams obtained from the application of CF theory. In an octahedral crystal field, a level, e_g, is raised in energy relative to some reference point, E_o, and a second level, t_{2g}, is lowered relative to E_o (Figure 1.4). Placement of electrons in the lower level *stabilizes* the system, but adding electrons to the upper level *destabilizes* the system. While there are some similarities between the two theories, they are quite different in the way that they handle the role of the ligands in the bonding model. In crystal field theory the ligands are assumed to be point charges and their presence splits the d-orbitals into some pattern. The electrons that were originally on the metal ion before it was involved in complex formation are still 100% on the metal ion and any electronic transitions that are possible can only take place within the split d-orbital set. Molecular orbital theory, on the other hand, considers that all atomic orbitals on all atoms can potentially interact to produce new orbitals called molecular orbitals. The extent to which these interactions occur is determined by the shapes of the atomic orbitals and their energies, and the amount of each AO that is 'mixed into' the MO, which is in turn controlled by the weighting coefficients in expressions similar to (1.1) and (1.2). In reality, if all of the orbitals on all of the atoms are addressed in this manner, the mathematics and the subsequent electronic structure become very complicated. Fortunately, inorganic chemists have learned to focus on only those parts of a complete MO diagram that are most relevant to the property of the system being addressed.

1.2.2 MO diagram for [Co(NH$_3$)$_6$]$^{3+}$

Consider the *partial* MO diagram for the octahedral complex, [Co(NH$_3$)$_6$]$^{3+}$, shown in Figure 1.10. While the complete diagram is more complicated than that given, Figure 1.10 contains enough detail to show how a diagram for a metal complex is created and what information it contains. The MO diagram is divided into three

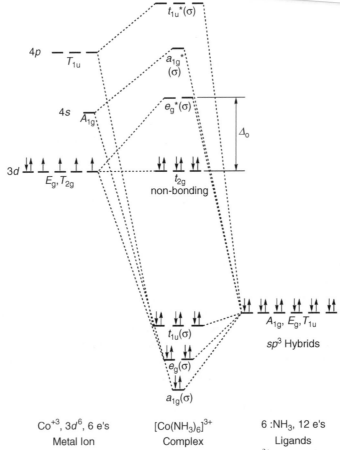

Figure 1.10   *Molecular orbital diagram for [Co(NH$_3$)$_6$]$^{3+}$, σ bonding only*

parts. The left side shows the relative energies of the $3d$, $4s$ and $4p$ AOs on Co^{+3}, the right side gives the orbitals on the six ammonia molecules that interact with the orbitals on the metal ion and the center shows the MOs that form from all of the AOs that are brought to the bonding scheme by the bonded partners. The orbitals on the right side of the diagram are actually the hybrid orbitals, sp^3, associated with the lone pair of electrons on the ammonia molecule. Since there are six ammonia molecules, there are six lone pairs, each having the same energy. While the fundamental 'units' used in MO theory are the atomic wave functions, 'mixing' some combination of AOs – for example, combining parts of the $2s$ orbital with the $2p_x$, $2p_y$, and $2p_z$ on nitrogen to form a 'hybrid' orbital – and entering the resulting hybrid orbital into the molecular orbital scheme produces the same mathematical result as using the unhybridized orbitals.

The dotted lines on the diagram show which metal and ligand AOs have been combined to form a particular MO. In order for atomic orbitals on M and L to combine and form a molecular orbital, the AOs on each must have shapes that allow the orbitals to overlap with one another. The labels used on the AOs (upper-case letters) and MOs (lower-case letters) are derived from *group theory*, which is branch of mathematics concerned with the symmetry properties of structures. In addition, the MOs indicate the type of orbital formed; in this case all are σ type. The MOs in the lower portion of the diagram are bonding MOs, while those in the upper part of the diagram, indicated with an asterisk, are antibonding MOs.

The composition of a molecular orbital is determined by the relative energies of the atomic orbitals. If the energies of two interacting AOs are the same, as would occur with a homonuclear diatomic molecule like H_2, the values of c_1 and c_2 in (1.1) and (1.2) are the same. If the energies of the two interacting orbitals are not the same, as would occur when two different atoms are bonded to each other – that is, a heteronuclear diatomic molecule – and the AO on atom 'B' is more stable (at a lower negative energy) than the AO of atom 'A', c_2 is greater than c_1. For the opposite situation, where the energy of A > B, then c_2 is less than c_1. For transition metal complexes which contain many atoms there are often more than two AOs contributing to a particular MO, which means that the wave functions are much more complicated than (1.1) and (1.2). While it is not easy to determine how much each wave function contributes to these MOs – that is, the coefficients, c_n, in expanded versions of (1.1) and (1.2) – it is straightforward to determine which centers, metal ions or ligands, make the *major* contribution to the MO – that is, the largest value of c_n. For example, consider the bonding MO, t_{1u} (σ), and its antibonding complement, t_{1u} (σ^*), shown in Figure 1.10. The fact that energy of t_{1u} (σ) lies closer to the energy of the ammonia atomic orbitals than any of the metal atomic orbitals in the diagram means that this MO contains *more* ligand character than metal character; that is, the weighting coefficients, c_n, for the wave functions from L are greater than the coefficients for the wave functions associated with M. Conversely, t_{1u} (σ^*), which is close in energy to AOs on the metal ion – for example $4p$ – must have contributions that are weighted heavily in favor of the AOs on the metal ion. All of this means that t_{1u} (σ) is *mostly ligand in character* and its complement, t_{1u} (σ^*), is *mostly metal in character*. With one exception, all of the MOs in the diagram are compositions of AOs from both L and M, with $a_{1g}(\sigma)$, $e_g(\sigma)$ and $t_{1u}(\sigma)$ being mostly ligand in character and $e_g(\sigma^*)$, $a_{1g}(\sigma^*)$, $t_{1u}(\sigma^*)$ being mostly metal in character. The t_{2g} set of d-orbitals, which are directed off the axes of the system, have no orbitals with which they can interact on the ammonia side of the diagram. The reason for this is not so easy to explain, but is related to the fact that none of the symmetry properties of the orbitals that are coming from the ligands on the right side of the diagram match the symmetry properties of the t_{2g} set from the metal ion on the left side of the diagram. One of the requirements for the formation of an MO is that the symmetries of the interacting AOs must match and there must be no T_{2g} term on the right side of the diagram. Without any interacting orbital on the ligand side of the diagram, members of the t_{2g} set appear in the diagram as 'pure' d-orbitals and as such are not really molecular orbitals at all. In the MO description of the bonding, they are referred to as *nonbonding* orbitals. For certain ligands, for example CN^- and CO, there are antibonding π-type MOs on the *ligand* that can interact with the t_{2g} set on the metal. In this situation, the t_{2g} set *can* form a π-bond between M and the carbon atoms on the ligands, which results in bonding and antibonding π-type MOs in the MO diagram.

It should be evident that determining even the approximate composition of MOs depends critically on the relative energies of the metal and ligand orbitals. The property of the orbital that approximates its relative energy in an MO diagram is its *electronegativity*, and the hybrid orbital of the ammonia molecule is more electronegative (has greater negative energy) than are the $3d$ atomic orbitals of Co^{+3}. If the reverse were true – that is, if the energies of the ligands were raised, less stable, relative to the metal orbitals – the positions of the MOs would be shifted upward in the diagram and compositions of the resulting MOs, in terms of their contributing AOs, would be the reverse of those described above and shown in Figure 1.10.

A final point to make in connection with the MO diagram is the appearance of the molecular orbitals. Consider $a_{1g}(\sigma)$, which is the most stable bonding MO in Figure 1.10. The wave function for this MO would have contributions from the six sp^3 hybrids on the ammonia molecules, with each hybrid being a combination of the $2s$ and $2p$ orbitals on nitrogen, and the $4s$ orbital on the metal ion. When this molecular wave function is used to calculate the boundary surface for the MO, a structure similar to that given in Figure 1.11 results, confirming that electrons in $a_{1g}(\sigma)$ would more likely be found on the ligands than on the metal ion. It should be evident that MOs for multi-atom systems are quite complex, and for the purposes of the brief overview of MO theory given here it is more important to understand the location, electron occupancy and approximate composition of an MO rather than its detailed shape.

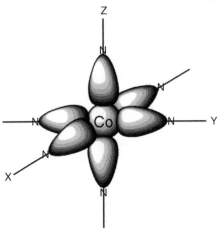

Figure 1.11   *Approximate shape of the $a_{1g}(\sigma)$ bonding molecular orbital for $[Co(NH_3)_6]^{3+}$*

The procedure for placing electrons into the diagram is to fill the diagram, from the bottom to the top, with the total number of electrons that were originally in the orbitals on M and L that created the diagram. In this case there are six electrons from the Co^{+3} ion and 12 electrons from the six lone pairs on the ammonia molecules. Since each individual MO can only hold two electrons with their spins oppositely aligned, the occupancy of the MOs for the *ground state* of the system is that shown in Figure 1.10. A critical feature of the diagram is that the central portion of the MO field is exactly the splitting pattern predicted by CF theory. There is a triply degenerate set, t_{2g}, which is *nonbonding* in the MO diagram, that is separated from a doubly degenerate set at higher energy, which is the antibonding MO, e_g^*. Calculations show that the distance between these levels is Δ_o, as indicated on the diagram. A photon with the correct energy entering the system could cause one of the electrons in the t_{2g} set to be promoted to a vacant level in the e_g^* set, which, aside form the slight change in the labels of the levels, is exactly what is predicted by CF theory. However, unlike CF theory, MO theory provides a much more comprehensive picture of bonding in that transitions outside the *d*-level of the metal ion are possible. For example, since the composition of an MO depends on the amount and types of AOs that created it, an electron in an excitation process could start out from an MO that is mainly ligand in character, for example $t_{1u}(\sigma)$, and end up in an MO which is mainly metal in character. This type of transition, which cannot be accommodated by simple CF theory, is called a *charge transfer transition*, in which charge is transferred from the ligand to the metal ion during the absorption process. Since the direction of the charge transfer is from L to M, the transition carries the acronym *LMCT*, for *ligand-to-metal charge transfer*. While the diagram in Figure 1.10 does not have this as a possibility, an electron could also start out in an MO that is mainly metal in composition and terminate in an MO that is mostly ligand in character. Such a transition carries the acronym *MLCT*, for *metal-to-ligand charge transfer*.

While there is no doubt that molecular orbital theory is the most comprehensive way to analyze bonding in chemistry, it requires many parameters and even in its qualitative form is quite challenging. Fortunately, crystal field theory along with a basic knowledge of molecular orbital theory will suffice to understand the physical and chemical properties of most of the metal complexes encountered in this text.

1.3   Absorption spectra of metal complexes

One of the most striking characteristics of transition metal complexes is that they are brightly colored. Once it was learned that the absorption bands giving rise to the color were associated with electronic transitions mainly

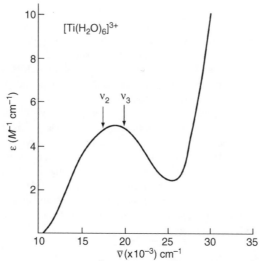

Figure 1.12 *Absorption spectrum of [Ti(H$_2$O)$_6$]$^{3+}$ in water. Adapted from Lever, A.B.P., Inorganic Electronic Spectroscopy, 1968, Elsevier*

within the *d*-level of the ion, efforts were made to understand the origin of the bands in terms of the electronic structure of the compounds using crystal field theory. Shown in Figure 1.12 is the absorption spectrum of [Ti(H$_2$O)$_6$]$^{3+}$, $3d^1$, a metal complex that is purple in color. This complex is an octahedral structure with the single *d*-electron on the Ti^{+3} ion in the lower triply degenerate set of *d*-orbitals; that is, the electronic configuration is, $t_{2g}^1 e_g^0$ (Figure 1.4). Since there is a higher-energy unoccupied level, e_g, it is possible to promote the electron in the lower level to the higher level if the wavelength of light falling on the sample is of the correct energy to bridge the energy gap, Δ_o, between t_{2g} and e_g. Thus, simple CFT predicts that [Ti(H$_2$O)$_6$]$^{3+}$ should have a single electronic transition, designated as $t_{2g}^0 e_g^1 \leftarrow t_{2g}^1 e_g^0$, which should be equal to the crystal field splitting parameter, Δ_o. Note that the spectroscopist writes this transition by listing the excited state first and having the transition arrow point to the excited state from the ground state, which is listed last.

Box 1.1 Absorption spectra of metal complexes [1]

A UV–visible spectrophotometer is used to measure the absorption spectrum of a complex in solution. The instrument consists of a light source which emits many different wavelengths of radiation, a grating which systematically selects wavelengths to be passed through the sample and a detector which measures the amount of radiation passing through the sample. If the solute in the solution absorbs light in a certain region of the spectrum, the photons which fall on the sample in this region will be reduced in number relative to the corresponding case in which only solvent is present in the light beam. While the absorbed radiation is quickly reemitted by the solute, the emitted photons are distributed in all possible directions in space, causing the detector to 'see' fewer photons than were directed at the sample.

Each chemical compound has a characteristic absorption spectrum that shows how the compound absorbs light as a function of wavelength. The absorption of radiation by the sample is given by the Beer–Lambert law, $A = \log I_o/I_t = \varepsilon c l$. In this law, the absorbance or optical density, A, is the logarithm of the ratio of the intensity of radiation falling on the sample, I_o, to the intensity of radiation exiting the sample, I_t. The absorbance is also equal to the product of ε, the molar extinction coefficient (molar absorbtivity), c, the concentration of the solute and l, the path length of the beam passing through the sample. If the unit of c is moles/liter (M) and the unit of l is centimeters (cm), the unit of ε is $M^{-1}\,cm^{-1}$, showing that the absorbance, A, is a *unitless* quantity. Each absorption band in a chemical compound has a characteristic value of ε, which is in effect a measurement of the 'photon-absorbing power' of that band. Since an absorption band is caused by a transition between states (levels) within the electronic structure of the compound, the magnitude of ε is a measure of the 'allowedness' of the transition between the states. The value of ε, which varies over many orders of magnitude, allows the spectroscopist to assign the transition in terms of a specified set of *selection rules*.

There are a number of ways to display the absorption spectrum of a compound. The x-axis of the plot can be in *wavelength*, λ (usually in nm, 10^{-9} m), but the spectroscopist prefers *wavenumber*, $\tilde{\nu}$ (usually in cm^{-1}), where $\tilde{\nu} = 1/\lambda$. The latter is used because the distance between states (levels) in the energy-level diagram of a compound is in units of energy and since $E = h\nu$ and $\nu = c/\lambda$, energy, E, is proportional to $1/\lambda$, not λ. The y-axis of the plot is usually given in ε (in units of M^{-1} cm^{-1}). Sometimes, if the values of ε span a large range, log ε will be given or the units will be in mM^{-1} cm^{-1}, where mM is M multiplied by 10^{-3} or millimoles. Yet another variation is to multiply the actual value of ε by some factor, for example 10^{-3}, in order to reduce the number of 'zeros' that need to be given on the y-axis of the plot. Thus, if the actual value of ε is $3000 \, M^{-1}$ cm^{-1}, the label on the y-axis may be $3.0 \, M^{-1}$ cm^{-1}, and indicate that the values shown should be multiplied by 10^3 in order to obtain the actual value of ε. If absorbance, A, is given on the y-axis, one cannot determine the values of ε unless c, the concentration, and l, the path length, are also given. In this case, the values of ε must be calculated using the Beer–Lambert law.

A careful look at the spectrum in Figure 1.12 shows that the spectrum of $[Ti(H_2O)_6]^{3+}$ is a bit more complicated than expected: the absorption band is slightly broadened to its low-energy side. In fact, $[Ti(H_2O)_6]^{3+}$ experiences what is called a *Jahn–Teller distortion*, in which two of the water molecules that are *trans* to each other move slightly closer to the metal ion than the remaining four. This of course slightly changes the symmetry of the crystal field away from 'pure' octahedral to a *tetragonal* distortion of a kind similar to that described in connection with the generation of the square planar crystal field (Figure 1.5). However, in this case the two ligands (charges) on the z-axis move closer to rather than farther away from the metal ion.

The Jahn–Teller effect applies to all chemical systems that have more than one way to indicate an electronic configuration. In this case the single electron could be placed in any one of the three orbitals of the t_{2g} set – that is, d_{xz}^1, d_{yz}, d_{xy} or d_{xz}, d_{yz}^1, d_{xy} or d_{xz}, d_{yz}, d_{xy}^1 – and all would yield the same CFSE and total energy. This situation is called a *triple orbital degeneracy* and the *Jahn–Teller theorem states that the system must structurally distort in some way to remove the degeneracy and lower the overall energy of the system.* If the two water molecules on the z-axis are slightly 'pushed in' toward the metal ion relative to the remaining four in the plane, the octahedral splitting pattern would distort to the one shown in Figure 1.13, which has the d_{xy} orbital at a lower energy (more stable) than d_{xz} and d_{yz} orbitals. This is called a '*z-in*' tetragonal Jahn–Teller distortion because the two ligands on the z-axis of the system have been moved closer to the metal ion than the four in the plane. Of course, any structural change of this type would affect the e_g set, causing the $d_{x^2-y^2}$ and d_{z^2} to shift to new positions in the manner shown in Figure 1.13. Since the single electron must occupy the lowest orbital when the system is in the ground state of, for example, d_{xy}^1, there are now *three* transitions possible – $d_{xz}, d_{yz} \leftarrow d_{xy}$, ν_1; $d_{x^2-y^2} \leftarrow d_{xy}$, ν_2; $d_{z^2} \leftarrow d_{xy}$, ν_3 – where there was only one – $t_{2g}^0 e_g^1 \leftarrow t_{2g}^1 e_g^0$ – in the pure octahedral crystal field. It should now be clear that the hidden band to the low-energy side of the main band at $\sim$17 400 cm^{-1} in the spectrum of $[Ti(H_2O)_6]^{3+}$ is actually ν_2, while the main band at $\sim$20 100 cm^{-1} is ν_3 [2]. The lowest energy transition, ν_1, which is quite small in energy, is in the infrared region of the spectrum and is not easily observed. Since the energy of the $d_{x^2-y^2} \leftarrow d_{xy}$, ν_2, transition is the crystal field splitting parameter, Δ for water bound to Ti^{+3} is $\sim$17 400 cm^{-1} or $\sim$345 kJ mol^{-1} (Figure 1.12). From the above presentation it is clear that CF theory works well in explaining the absorption spectrum of $[Ti(H_2O)_6]^{3+}$, even accounting for a distortion in the structure of the complex due to the Jahn–Teller effect.

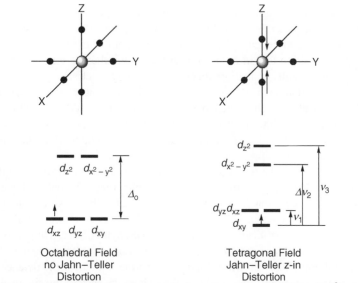

Figure 1.13   *Jahn–Teller (tetragonal) distortion, 'z-in,' for $[Ti(H_2O)_6]^{3+}$, $3d^1$*

1.3.1   Band intensity/selection rules

Quantum mechanically, all transitions between electronic states (levels) of chemical compounds are controlled by *selection rules*, which cause some transitions to be strong and others in the compound to be weak. For the kinds of compounds to be discussed in this text, two types of selection rule are important: one that pertains to the orbital motion of the electron, associated with the *orbital angular quantum number*, l, and a second associated with the spin of the electron specified by the *magnetic spin quantum number*, m_s. For an example of how the selection rule associated with orbital angular momentum works, consider the absorption band associated with the transition, $d_{x^2-y^2} \leftarrow d_{xy}$, v_2, for $[Ti(H_2O)_6]^{+3}$, which has the electron moving between two different d-orbitals. Since the electron starts out in a d-orbital, d_{xy}, which has $l = 2$, and it ends up in another d-orbital, $d_{x^2-y^2}$, which also has $l = 2$, the value of l does not change in the transition; that is, $\Delta l = 0$. In this case the transition is forbidden by the *Laporte selection rule*, or to put it another way, the transition is *Laporte-forbidden*. The word 'forbidden' does not imply that the transition is totally absent – it is clearly visible in the spectrum shown in Figure 1.12 – but rather, it simply means that the transition has some restrictions on it that make it weaker (less likely to occur) than other possible types of transition. If the quantum number l changes in a transition – that is, $\Delta l \neq 0$ – the transition is said to be *Laporte-allowed* and the absorption band in this case would be much stronger (more likely to occur) than if $\Delta l = 0$. In discussing the bonding in $[Co(NH_3)_6]^{3+}$ using molecular orbital theory (Figure 1.10), it was pointed out that an electron located in an MO that was mainly p in character (strong ligand contribution) could move to an MO that was mainly d in character (strong metal contribution). An example of such a transition is the LMCT transition, $e_g^* \leftarrow a_{1g}$ (Figure 1.10), in which the electron starts out in an MO that is mainly ligand in character, a_{1g}, and terminates in an MO, e_g^*, that is mainly metal in character. Since $\Delta l \neq 0$ – that is, $d \leftarrow p$ – this transition is Laporte-allowed and its intensity is much greater than a transition with $\Delta l = 0$. While the MO diagram shown in Figure 1.10 does not have MLCT as a possibility (for example, $p \leftarrow d$), such a transition would also have $\Delta l \neq 0$, would be Laporte-allowed and would be strong in intensity.

As will soon be evident, the net spin, S, can also change in an electronic transition. This happens when the net spin associated with the starting configuration (ground state) is different than the net spin of the ending configuration (excited state). The *spin selection rule* address what happens to the spin during a transition; that is, it indicates the difference between the spins of the excited and ground states of the system, referred to as ΔS. If $\Delta S = 0$, the transition is said to be *spin allowed* and the net spin of the system is unchanged during the transition. If $\Delta S \neq 0$, the transition is said to be *spin forbidden* and the net spin of the system must change during the transition. Experimentally, transitions that have $\Delta S = 0$ are much stronger (have a higher probability of occurring) than transitions that have $\Delta S \neq 0$. A simple explanation of why spin-forbidden transitions are weaker than their allowed counterparts is related to the fact that a spin-forbidden transition requires two events to take place: one is the promotion of the electron to a higher energy level and the other is a change in the spin of the electron during the promotion. This two-event situation is less probable than the one-event situation of simply promoting the electron to another level without changing its spin.

A final comment on the intensity of absorption bands observed for transition metal complexes pertains to the symmetry of the complex. Mathematicians recognize that all objects in nature have certain symmetry properties called *symmetry elements* that allow them to be systematically classified into a *point group*, which specifies a certain set of symmetry elements. Although the details of how this is done will not be given here, one symmetry element called the *center of inversion*, which is given the symbol i, is an important determinant of the intensity of a band in the absorption spectrum of a transition metal complex. Consider the structures of the tetrahedral and octahedral complexes shown in Figure 1.1, and note that for the octahedral arrangement of ligands the complex has a special point (the metal ion) through which each of the six ligands can be passed to reach an equivalent point: that is, another identical ligand. Carrying out this symmetry operation on the complex, passing all ligands through the center, will result in a structure which looks identical to the starting complex; that is, it will appear as if nothing was done to the complex. When this condition is met, the object is said to possess the center of inversion, i. However, the situation with the tetrahedron is different in that carrying out the same symmetry operation, in this case passing each of the four ligands through the metal ion to an equivalent point on the other side, will result in a structure which, although still a tetrahedron, looks as if the original has been rotated to a new position. In this case the object *does not* possess the center of inversion, i. Quantum mechanically, the presence or absence of i in a metal complex has a significant effect on the intensity of the d-d transitions for the complex. Consider a d-d transition with $\Delta l = \Delta S = 0$ for two complexes, one of which is an octahedral complex and the other of which has tetrahedral geometry. Experimentally it is found that all of the transitions for the tetrahedral complex, which does not possess the symmetry element i, are more intense (they are more allowed) than are the transitions for the octahedral complex, which possesses i.

Clearly, the intensities of bands observed in transition metal complexes provide important information on the selection rules operating in the transition, which ultimately paves the way for assigning the bands to specific transitions in the energy-level diagram of the compound. A summary of the type of transition, its expected intensity and an example of the type of complex exhibiting the transition is given in Table 1.4. While the total integrated area under an absorption band is the quantity that is proportional to the probability that the transition will occur, all absorption bands regardless of their origin have approximately the same ratio of the width of the band at half the maximum height of the band to the maximum height of the band; that is, the width at half height. This means that ε_{max}, the molar extinction coefficient at the maximum of the band in units of $M^{-1}\,cm^{-1}$, is a good measure of the intensity (probability) of the transition and thus is a useful metric for determining which selection rules apply to a transition.

1.3.2   Spectroscopic and crystal field terms

Analysis of the spectrum of a compound with a d^1 electronic configuration, after accounting for distortions due to the Jahn–Teller effect, proved to be relatively simple and straightforward. However, for complexes with more than one d-electron, indeed for most of the known metal complexes, analysis of spectra is more complicated. The problem with multi-electron systems is that once there is more than one electron in a given 'shell' or level of the atom, quantum mechanics dictates that not all spin and orbital angular momentum

Table 1.4 *Absorption properties of metal complexes*[a]

Type of Transition	Δl	ΔS	Center of Inversion, i	ε_{max} (M^{-1} cm^{-1})	Example Complexes[b]
Laporte-forbidden Spin-forbidden	0	$\neq 0$	yes	<1	$[Mn(H_2O)_6]^{2+}$ O_h, $S = 5/2$, $3d^5$
Laporte-forbidden Spin-allowed	0	0	yes	1–100	$[Ru(H_2O)_6]^{2+}$ O_h, $S = 0$, $4d^6$
Laporte-forbidden Spin-allowed	0	0	no	100–1000	$[NiCl_4]^{2-}$, T_d, $3d^8$
Laporte-allowed Spin-allowed	$\neq 0$ MLCT LMCT	0	no	1000–50 000	$[MnO_4]^{-}$, T_d, $3d^0$ $[AuCl(terpy)]^{2+}$, s.p., $S = 0$, $5d^8$

[a] The ranges of molar absorbtivity, ε, given are considered *approximate*. Values outside these ranges have been observed for the various types of transition.
[b] Abbreviations: O_h, octahedral; T_d, tetrahedral; *s.p.*, square planar; *terpy*, 2, 2′, 2″ terpyridine.

values for the system are possible. For example, suppose that the orbital motion of an electron about the nucleus of an atom is represented by a spinning bicycle tire with an axel or rod about which the tire is spinning. In a nonquantum mechanical world, the tire could spin with any angular velocity, setting up an angular momentum vector, l_i, with any magnitude, perpendicular to the angular velocity; that is, coincident with the axel of the spinning tire. Assume that for a multi-electron case there are many such tires, each with its axel randomly oriented in a different direction in space. The problem with this picture is that quantum mechanics simply does not allow all possible angular velocities or orientations of l_i and specifies that all of the vectors must be of a certain magnitude and must couple in such a way that only certain resultant vectors, L, where L is the vector sum of all l_i, are allowed. This is called *Russell–Saunders* (R-S) coupling and in simple terms it means that only certain angular velocities and orientations of the vectors are allowed. Not only are there restrictions on the orbital motion of the electron about the nucleus but if the electron is considered a particle spinning on its axis, the spinning velocity generates a spin moment, s_i, which is perpendicular to the spinning motion. Quantum mechanics specifies that not all s_i are allowed and that they must couple to give a resultant S which itself is quantized; that is, only certain values of S are allowed. To complicate matters further, L and S, which characterize the orbital and spin motion, can couple with each other to produce a resultant vector, J, which is characterized by the spin-orbit coupling constant, λ (not to be confused with wavelength, which is denoted by the same Greek letter). If the metal atom is relatively light, like the elements of the first-row transition series, coupling between L and S is generally weak, and $|\lambda|$ is in the range 50 to 800 cm^{-1}, which is relatively small. This means that the system can be described by L and S because their coupling together to define J is weak. However, for heavier elements like those in the third-row transition series and the lanthanides and actinides, $|\lambda|$ is greater than 10^3 cm^{-1}. With large $|\lambda|$, l and s, orbital and spin angular momenta associated with a single electron couple together to give j for a single electron. The coupling of the j vectors to produce the resultant J gives rise to a new coupling scheme, called *jj coupling*, for which only J has real physical meaning.

Fortunately, progress on the important issue of how to interpret absorption spectra of metal complexes can be made by embracing the language of the spectroscopist and discussing electronic states or levels of a metal ion in terms of spectroscopic or free-ion *terms* and crystal-field *terms*. The information present in these terms is summarized in Figure 1.14. As with the shapes of multicenter molecular orbitals, it is less important to know the makeup of a 'state' in terms of the different values of S and L that it may contain – called *microstates* – than the fact that the state exists and that it has a specific location in an energy-level diagram of the compound.

For the spectroscopist, analyzing the electronic spectra of transition metal complexes begins by finding the *spectroscopic terms* and their relative energies for the free ion with a specified number of *d*-electrons. These terms are found using a set of rules and while the precise order of states is determined by experiment, the *lowest state is always the state with the highest spin, S, and orbital, L, values*. Next, the spectroscopist applies a

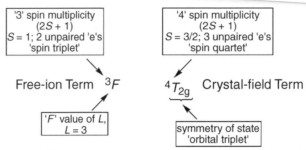

Figure 1.14 *Definitions of notations used for free-ion and crystal-field term symbols*

crystal field of some symmetry and determines how the *d*-orbitals split in the crystal field. In order to describe what happens to a free-ion term when a crystal field is applied, the spectroscopist uses *group theory* to determine how the free-ion term is *transformed* in the presence of the crystal field. Since the system has been changed from the free-ion case to a situation with a specified crystal field, the *language* used for the free-ion case – that is, free-ion terms – is abandoned and replaced with terminology derived from group theory, which specifies the symmetry properties of the states. These new terms are called crystal-field terms. A list of free-ion terms or spectroscopic terms and their counterpart octahedral crystal-field terms is given in Table 1.5. The crystal-field terms *A*, *E*, and *T* pertain to the number of ways equivalent electronic configurations in the presence of the octahedral field can be written. Since these configurations are related to each other by interchanging electrons in orbitals, *A*, *E*, and *T* indicate the *orbital degeneracy* of the state, while the subscripts on these states, for example 1g, 2g and so on, refer to the symmetry properties (g means that *i* is present) of the state. Like the splitting between the *d*-orbitals, the energies of the crystal-field terms derived from the split *d*-orbitals depend on the crystal field splitting parameter, Δ.

In 1954 two Japanese researchers, Y. Tanabe and S. Sugano, published a classic paper showing in diagram form the ground and excited states for all ions with an unfilled *d*-shell as a function of Δ for octahedral complexes (Figure 1.15). These diagrams, which are referred to as Tanabe–Sugano (T-S) diagrams, have become the mainstay in analyzing the electronic spectra of transition metal complexes. Not only do they allow one to determine the crystal-field terms for the ground and excited states of the ion, but by matching the positions of absorption bands obtained from spectral measurements to the diagram one can find the value of the crystal field splitting parameter, Δ, for the complex.

Table 1.5 *Conversion table for free-ion spectroscopic terms into octahedral crystal-field terms[a]*

Spectroscopic Term	Octahedral Crystal-field Term
S	A_{1g}
P	T_{1g}
D	$E_g + T_{2g}$
F	$A_{2g} + T_{2g} + T_{1g}$
G	$A_{1g} + E_g + T_{2g} + T_{1g}$
H	$E_g + 2T_{1g} + T_{2g}$
I	$A_{1g} + A_{2g} + E_g + T_{1g} + 2T_{2g}$

[a] Note that 'S', the spectroscopic term, is not in italics. This is different to the spin quantum number, '*S*', which *is* given in italics. In the case of the tetrahedron, the subscript 'g' in the crystal-field term is eliminated.
From Table 9.25, p. 443 of Huheey, J.E. (1983) *Inorganic Chemistry: Principles of Structure and Reactivity*, 3rd edn, Harper & Row Publisher, New York.

Box 1.2 Tanabe–Sugano diagrams

Tanabe–Sugano diagrams or T-S diagrams show the relative energies of crystal-field terms for an ion in an octahedral complex. Since these diagrams have been generated using a specific electronic configuration, for example d^3, they can be used for any metal ion with this configuration, for example $3d^3$, $4d^3$, $5d^3$, with any attached set of ligands, as long as the environment around the metal ion is octahedral (O_h) or nearly so.

An important feature of the T-S diagram is that the x-axis is the lowest energy state (ground state) of the ion and the energies of all other states are plotted relative to it. This axis, which is unitless, gives the value of Δ/B, where Δ is the crystal field splitting parameter in units of cm^{-1} and B is an energy parameter in units of cm^{-1} associated with how effectively electrons in an orbital repel one another. In a pure crystal field model, the value of B for the free ion and the ion in a complex must be the same since the size of the d-orbital is the same in both and thus e-e repulsions in the complex and the free ion should be identical. However, this is in fact rarely the case. This is because the bonding between the metal ion and the ligands cannot be described using pure electrostatic concepts, as required by the crystal field model, and some covalent interaction occurs which 'expands' the d-orbitals and reduces B from its free-ion value when the complex forms.

The fact that the splitting pattern for the d-orbitals in a tetrahedral (T_d) crystal field is the *inverse* of that in an O_h field, $t_{2g}e_g$ vs. et_2 (Table 1.1), means that the T-S diagrams are also applicable to complexes with T_d symmetry. While the order of the free-ion terms is of course independent of the type of crystal field present (when $\Delta = 0$ there is no crystal field), the relative energies of the crystal-field terms which emanate from the free-ion term are *inverted* in changing the geometry of the complex from O_h to T_d. The 'conversion code' for this symmetry change is $d^n (O_h) = d^{10-n} (T_d)$, where $n = $ the number of d-electrons. Thus, the T-S diagram for $d^4 (O_h)$ is also applicable to $d^6 (T_d)$. Although the T-S diagrams can be used for both octahedral and tetrahedral complexes, the symmetry properties of both types of complex are not the same in one important respect: while the octahedron possesses the symmetry element, i, *the center of inversion*, the tetrahedron does not. To indicate this, a crystal-field term associated with an octahedral complex carries the subscript 'g', which denotes the presence of this symmetry element in the complex. For example, the ground-state crystal-field term for the Cr^{+3} ion ($3d^3$) in the *octahedral* complex, $[CrF_6]^{3-}$, is '$^4A_{2g}$', although the subscript 'g' is not given on the T-S diagram. This same diagram can also be used for the *tetrahedral* complex, $[CoF_4]^{2-}$ ($3d^7$), in which case the ground-state crystal-field term is '$^4A_2$'. Because d^5 has a half-filled d-shell, octahedral and tetrahedral complexes use the same d^5 T-S diagram. Diagrams for the d^1 and d^9 cases are simple and are not presented in Figure 1.15. For d^1, the ground-state spectroscopic term is ^2D, which in an octahedral crystal field produces $^2T_{2g}$ as the ground state and ^2E_g as the only excited state. For d^9, the ground-state spectroscopic term is also ^2D, which in an O_h crystal field produces ^2E_g as the ground state and $^2T_{2g}$ as the only excited state. The tetrahedron–octahedron conversion given above applies to these diagrams as well.

While the origins of the symbols E_g, T_2 and so on which pertain to the symmetry properties of the state are not easy to describe, the superscript on the crystal-field term always denotes the net electron spin of the state, S. Thus, if in a given electronic configuration there are two unpaired electrons, for example V^{+3} (d^2) in an octahedral crystal field, the value of S is $1/2 + 1/2 = 1$. The spectroscopist indicates this as a superscript, with the value $2S + 1$, on the term, in this case '3', which denotes the *spin multiplicity* of the state. Moreover, this superscript is always the same for the crystal-field terms and the free-ion term from which they are derived; for example, a ^3F free-ion term will only produce $^3A_{2g}$, $^3T_{2g}$ and $^3T_{1g}$ crystal field terms (Table 1.5).

A striking feature of the T-S diagrams for systems with four to seven d-electrons is the presence of a vertical line in the diagram. Since the x-axis of the diagram is always the ground state (lowest energy state), a change in the ground state at some value of Δ means that a low-spin state will 'overtake' the high-spin ground state and replace the latter on the x-axis of the diagram. Thus, for example, the ground state for a *high-spin* octahedral complex of Co^{+3} ($3d^6$, $t_{2g}^4e_g^2$, $S = 2$) is $^5T_{2g}$, while the ground state for a *low-spin* complex ($t_{2g}^6e_g^0$, $S = 0$) with this ion is $^1A_{1g}$. Thus the left side of the T-S diagram shows the states that are possible for the high-spin case and the right side of the diagram shows the states possible for the low-spin situation.

As we will see, complexes with symmetries other than O_h and T_d have found their way into medicine. While a detailed description of the electronic structure of these complexes in terms of their crystal field states is beyond the scope of this chapter, the T-S diagrams shown in Figure 1.15 are starting points for analysis of the electronic structures of compounds with lower symmetries.

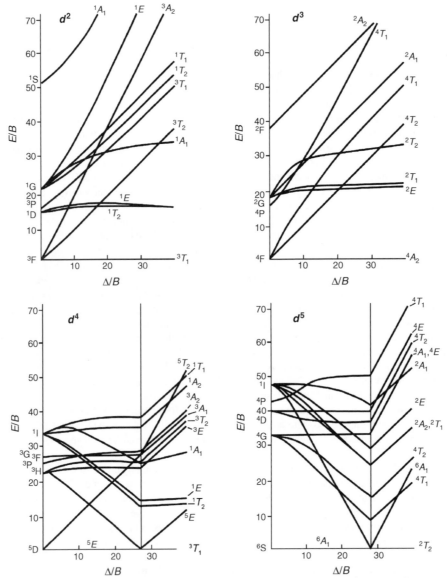

Figure 1.15   *Tanabe–Sugano diagrams. Adapted from Lever, A.B.P., Inorganic Electronic Spectroscopy, 1968, Elsevier*

1.3.3   Band assignments and Δ

The basic approach for obtaining the crystal field splitting parameter Δ from absorption spectra of transition metal complexes is to assign the absorption bands appearing in the spectrum of the compound to the crystal states given in the Tanabe–Sugano diagram. Once this is done it is possible to calculate the crystal field splitting parameter, Δ, and other parameters from the positions of the absorption bands of the complex using a fitting procedure. While the fitting process will not be discussed here, assigning the observed

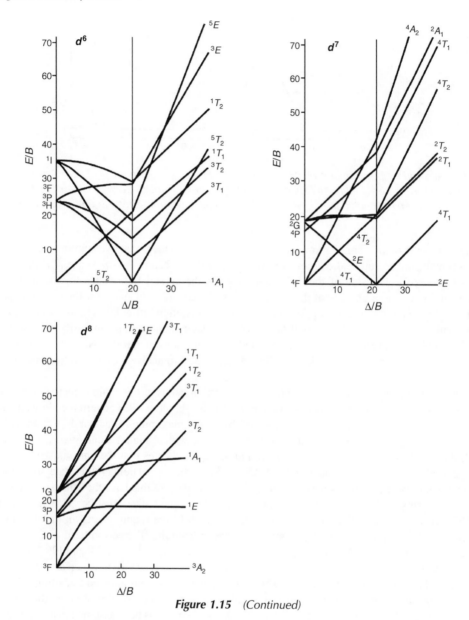

Figure 1.15   *(Continued)*

bands and estimating the value of Δ_o is a relatively simple procedure, which will be presented in this section.

Figure 1.16 shows the absorption spectrum of $[Co(NH_3)_6]^{3+}$, which contains six ammonia ligands bonded in an octahedral array to Co^{+3}, $3d^6$. This complex exhibits three absorption bands at 769 nm (0.2), 472 nm (56) and 338 nm (46), which are the *lowest-energy d-d* type absorptions for the complex [3]. In listing the bands, the wavelength of the band maximum in nanometers (10^{-9} m) is followed in parenthesis by the molar extinction

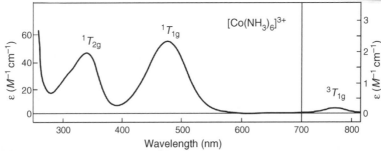

Figure 1.16   *Absorption spectrum of [Co(NH₃)₆]³⁺ in water. High-energy part adapted from Riordan, A.R. et al., Spectrochemical Series of Cobalt(III). An Experiment for High School Through College, Chem. Educator 2005, 10, 115–119; low-energy part adapted from Lever, A.B.P., Inorganic Electronic Spectroscopy, 1968, Elsevier*

coefficient, ε, in units of $M^{-1}\,cm^{-1}$. From the d^6 T-S diagram it is evident that two spin states are possible for the ion in an octahedral crystal field, a *high-spin* case, $t_{2g}^4 e_g^2$, with $S = 2$ and a *low-spin* case, $t_{2g}^6 e_g^0$, with $S = 0$. As pointed out above, the left side of the T-S diagram gives the crystal field states for the high-spin case (low values of Δ), and the right side of the diagram gives the crystal field states for the low-spin case (high values of Δ). On the basis of the absorption spectrum alone it is often difficult to choose which spin state is associated with the compound, but as will be shown in a later section, this can easily be determined by conducting magnetic measurements on the compound. Since it is known that $[Co(NH_3)_6]^{3+}$ has the low-spin ($S = 0$) ground state, the right side of the d^6 T-S diagram is used to analyze the absorption spectrum of the complex.

The next step in the process for assigning the *d-d* absorption bands of the complex is to determine which selection rules apply to the observed bands. As pointed out above, these rules govern the allowedness of the transition, which is reflected in the value of ε for the absorption maximum of the band. A band with a large value of ε means that the transition is more allowed than in one with a smaller value of ε. As is evident from Table 1.4, the value of ε for the bands at 474 nm and 338 nm indicates that they are most likely Laporte-forbidden spin-allowed transitions; that is, $\Delta l = 0$, $\Delta S = 0$. Since the crystal field ground state of the compound is $^1A_{1g}$, these bands must be associated with transitions to excited states that have the same spin as the ground state; that is, the superscript on the crystal-field term for the excited state must be '1'. Checking the T-S diagram (Figure 1.15) reveals that the only possibilities which fit the requirements, in order of increasing energy, are $^1T_{1g} \leftarrow {}^1A_{1g}$ and $^1T_{2g} \leftarrow {}^1A_{1g}$, which are derived from the 'I' free-ion term of Co⁺³. Thus, the band assignments are $^1T_{1g} \leftarrow {}^1A_{1g}$ at 474 nm (21 200 cm⁻¹) and $^1T_{2g} \leftarrow {}^1A_{1g}$ at 338 nm (29 550 cm⁻¹). The T-S diagram predicts that there should be *two* spin-forbidden transitions ($\Delta S = 0$), which are $^3T_{1g} \leftarrow {}^1T_{1g}$ and $^3T_{2g} \leftarrow {}^1A_{1g}$, and that these should occur at lower energy than the first spin-allowed transition, $^1T_{1g} \leftarrow {}^1A_{1g}$. Because these transitions are spin forbidden they should be much weaker in intensity than the spin-allowed transitions. The band at 769 nm (13 000 cm⁻¹) with $\varepsilon = 0.2\,M^{-1}\,cm^{-1}$ has been assigned to the spin-forbidden transition $^3T_{1g} \leftarrow {}^1A_{1g}$. Since the $^3T_{2g} \leftarrow {}^1A_{1g}$ spin-forbidden transition is at higher energy than $^3T_{1g} \leftarrow {}^1A_{1g}$ and overlaps the strong spin-allowed band at 474 nm, this transition appears as a shoulder on the strong band and for this reason is difficult to observe.

Once the band assignments have been made, the next step is to note the energy of the *lowest-energy spin-allowed* transition, which in this case is $^1T_{1g} \leftarrow {}^1A_{1g}$. While the theoretical energy of this transition is somewhat complicated, it is *approximately equal* to the splitting between the t_{2g} and e_g levels of the complex; that is, the crystal field splitting parameter, Δ. In practice, all of the observed bands in the complex are employed in a fitting procedure that uses Δ, B and other interelectronic repulsion parameters as variables. Although the location of the $^1T_{1g} \leftarrow {}^1A_{1g}$ transition suggests that Δ is 21 200 cm⁻¹ (474 nm), the actual value of this parameter obtained from fitting is $22\,870\,\text{cm}^{-1}$ and the value of B is $615\,\text{cm}^{-1}$. The latter value, which is significantly reduced from the free-ion value of $1100\,\text{cm}^{-1}$, shows that while the crystal field model works well for analyzing the spectrum of $[Co(NH_3)_6]^{3+}$, complex formation clearly affects repulsions between electrons in orbitals, an observation which cannot be accommodated by a simple electrostatic model.

The general approach given above for estimating the value of Δ can be used for all d^1-d^4 and d^6-d^9 complexes. This means that the location of the absorption band for the *lowest-energy spin-allowed Laporte-forbidden transition approximates the spacing (in energy) between the t_{2g} and e_g levels of the octahedral complex*. Since the splitting for the tetrahedral arrangement of ligands is the inverse of the octahedral case, the location of the lowest-energy spin-allowed transition for tetrahedral complexes is also a good measure of the splitting between the e and t_2 levels of the tetrahedral complex, which in this case is Δ_t.

As pointed out above, the T-S diagrams have broad application and can be used for all octahedral and tetrahedral transition metal complexes. While assigning bands for all of the possibilities will not be given here, a few interesting cases are worth noting. Experimentally, solutions of most high-spin complexes of Mn^{+2} ([Ar] $3d^5$) are very pale in color. The absorption spectrum of one of these complexes, $[MnF_6]^{4-}$, which has $S = 5/2$, is shown in Figure 1.17. Since this complex has the electronic configuration $t_{2g}^3 e_g^2$, transitions are not possible without changing the net spin of the system; that is, all transitions are spin-forbidden transitions. As is evident from the d^5 T-S diagram, the crystal field ground state for the complex is $^6A_{1g}$ and all excited states have spin multiplicities that are not spin hextets; that is, excited-state crystal-field terms do not carry the superscript '6'. This condition, where the spin multiplicity of the ground state is different from the spin multiplicities of *all* excited states, is not encountered in any other T-S diagram. Since all transitions must be spin forbidden, $[MnF_6]^{4-}$ is weakly colored; that is, the absorption bands in the visible region of the spectrum are of low intensity. In this case none of the positions the bands are good estimates of the value of Δ.

If the crystal field around the Mn^{+2} ion is increased, for example $[Mn(CN)_6]^{4-}$, the spin state of the metal ion changes from high spin to low spin, $S = \frac{1}{2}$, $t_{2g}^5 e_g^0$. In this case the crystal field ground state is $^2T_{2g}$, which allows transitions to higher-energy states with the same doublet spin multiplicity, making these complexes more intensely colored than their high-spin counterparts.

An important message from this analysis is that crystal field theory does a good job of explaining the optical properties of transition metal complexes. With absorption data for the complex, knowledge of basic selection rules and a Tanabe–Sugano diagram, the observed bands in the spectrum of the compound can be assigned and the approximate value of the crystal field splitting parameter Δ determined. Since Δ is a measure of the

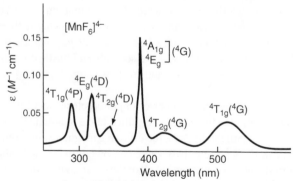

Figure 1.17 *Absorption spectrum and crystal field band assignments of $[MnF_6]^{4-}$. Adapted from Lever, A.B.P., Inorganic Electronic Spectroscopy, 1968, Elsevier*

thermodynamic stability of the complex, it influences the rate at which the compound undergoes a substitution reaction, which in a medical context is often the critical factor determining the efficacy of a compound as a useful drug or diagnostic agent.

A second important message is that the value of B for a complex is often much different to the value of B for the free ion; for example, $B = 615\ cm^{-1}$ for $[Co(NH_3)_6]^{3+}$ vs. $1100\ cm^{-1}$ for free Co^{+3}. Since this parameter is a measure of the interactions between electrons in the orbitals of the metal ion, this means that the interactions in the complex are very different from (they are much less than) the interactions in the free ion. Obviously, the fact that B is different in the complex versus the free ion is inconsistent with a crystal model which assumes that the ligands are simply present to 'perturb' the d-orbitals of the metal ion to a new splitting arrangement. This breakdown of the model prompted a 'patch' for crystal field theory, which ultimately became known as *ligand field theory*. In the latter, the simple electrostatic concepts that lead to the crystal field model are partially abandoned and some form of overlap between the orbitals on the ligands and metal ion – that is, covalent bonding – is allowed. While ligand field theory has proved useful for analyzing the bonding and spectra of complexes where covalent bonding is significant – for example, when the ligands are CN^-, CO, phosphine, thiolate and so on – it is ultimately only a patch to the model. The correct method for analyzing the bonding in these compounds is, of course, molecular orbital theory. But since MOT is quite rigorous, and is sometimes cumbersome to use, inorganic chemists have found ways to 'simplify' the bonding analysis of metal complexes. It is remarkable that CFT works as well as it does and it shows that many metal complexes are basically 'salt-like' substances and that simple electrostatic arguments can be used to explain the structure and spectral properties of these interesting materials.

1.4   Magnetic properties of metal complexes

As we have seen, the quantity S is a useful way to indicate the number of unpaired electrons on the metal ion in a complex. If a metal complex has no unpaired electrons on the metal ion, $S = 0$, and if it is placed near a magnetic field, the applied field will induce circulation of electronic currents, which cause the substance to be repelled by the field. In this case the substance exhibits *diamagnetism* and is said to be *diamagnetic*. However, if the metal ion in the complex has one or more unpaired electron, $S \neq 0$; placing the compound near a magnetic field will set up an electronic current causing it to be *drawn into* the field. This property of being drawn into the field is termed *paramagnetism* and the compound is referred to as being *paramagnetic*. In terms of the strengths of the two forces, the paramagnetic force, on a per atom basis, is the stronger of the two.

The degree to which the material exhibits either effect is called the *effective magnetic susceptibility* of the material, which is denoted by the quantity μ_{eff} and which is given in a unit called the *Bohr magneton* (BM), where $1\ BM = 9.27 \times 10^{-24}\ JT^{-1}$. While μ_{eff} can vary in a complicated way with temperature, paramagnetic complexes of the first-row transition metal elements have values of μ_{eff} that are relatively independent of T near room temperature. However, for complexes of the second- and third-row transition metal elements, λ, which is the spin-orbit coupling constant, is large and for these complexes μ_{eff} is usually quite sensitive to temperature over all temperature ranges.

The general approach for determining the number of unpaired electrons of a paramagnetic complex is to measure the force with which a sample of the material is pulled into a magnetic field. After making corrections for diamagnetic effects associated with the ligands, the effective magnetic susceptibility, μ_{eff}, associated with the paramagnetic metal ion can be obtained. This value can be compared with the magnetic susceptibility, calculated using a simple formula which assumes that all of the paramagnetism is due only to the spin moments of the unpaired electrons on the metal ion, with no orbital motion. This *spin-only magnetic susceptibility*, μ_{so}, is given in two forms: (1.3), which is in terms of S, and (1.4), which is in terms of the number of unpaired electron on the metal ion, n. For example, for a complex with three unpaired electrons, $n = 3$ and $S = 3/2$, which gives $\mu_{so} = 3.87$ BM.

$$\mu_{so} = 2\sqrt{S(S+1)} \text{ BM} \tag{1.3}$$

$$\mu_{so} = \sqrt{n(n+2)} \text{ BM} \tag{1.4}$$

Experimentally, it is found that metal ions in complexes with an A crystal field ground-state term, for example Co^{+2} in a tetrahedral field with the electronic configuration $e^4 t^3$ and the ground-state crystal $^4A_2$, have values of μ_{eff} that are nearly the same as μ_{so}. This means that the only property contributing to the observed paramagnetism of the compound is the spins associated with the unpaired electrons on the metal ion. However, for metal complexes with E or T ground-state crystal-field terms, for example Fe^{+2} in a tetrahedral field with the electronic configuration $e^3 t^3$ and the ground-state crystal-field term ^5E, and Fe^{+3} in a octahedral field with the electronic configuration $t_{2g}^5 e_g^0$ and the ground-state crystal-field term $^2T_{2g}$, the value of μ_{eff} is usually *larger* than μ_{so}. This is because there are really two important contributions to the paramagnetism for the sample; one is associated with the spin moment of the electron but the second is associated with the orbital motion of the electron about the nucleus. For example, for T_d, Fe^{+2}, $e^3 t^3$, which has the 'orbital doublet' ground state E, the electronic configuration which produced the crystal-field term can be written *two* ways: $d_{x^2-y^2}^2, d_{z^2}^1$, $d_{xz}^1, d_{yz}^1, d_{xy}^1$ (a) or $d_{x^2-y^2}^1, d_{z^2}^2, d_{xz}^1, d_{yz}^1, d_{xy}^1$ (b), both of which have identical energies. In this case the electronic configuration and the crystal-field term are said to be orbitally *doubly degenerate*. Inspection of these two electronic configurations reveals that the unpaired electron in the e level of (a) has moved from the d_{z^2} orbital to the $d_{x^2-y^2}$ orbital of (b). While this may seem trivial, it in effect constitutes a motion of the unpaired electron from one orbital to the other; that is, the electron appears to *rotate* about the nucleus. This rotation produces additional paramagnetism, which increases the magnetic susceptibly above the value calculated from the spin-only formula. A similar argument can also be made for a T ground-state crystal-field term, which is an *orbital triplet* and which was derived from an electronic configuration with *three* energy equivalent arrangements. These can be seen by simply rearranging the electrons in the t_{2g} (O_h) or t_2 (T_d) levels of the electronic configuration that produced the T crystal-field term. In the end, all complexes which have E or T crystal-field ground states accrue additional magnetic susceptibility from the orbital motion of electrons, so for these cases μ_{eff} is usually grater than μ_{so}. Fortunately, the contribution made by the orbital motion is not so large as to preclude using the simple spin-only formula to determine the value of S for the complex from μ_{eff}.

The above description of magnetic properties of complexes assumes that the ions with unpaired electron spin are well separated from each other and that there is no interaction between the spins; that is, the systems are said to be *magnetically dilute*. However, for certain substances, such as the iron oxides, in which iron ions are imbedded in an oxide crystal lattice and the separation between ions with unpaired electrons is not large, the individual magnetic moments of ions can couple to produce substances with large net magnetic moments. While the degree of coupling varies in a complicated way with the nature of the paramagnetic ions involved, their separation in the lattice and the physical size of the particle, one case – called *ferromagnetism* – has a large group of spin moments oriented in the same direction, resulting in a large net magnetic moment, which produces a permanent magnetic field for the material. These substances are referred to as *ferromagnetic* materials.

1.5 Reactions of metal complexes

1.5.1 Forward and reverse rates and equilibrium

Most of the metal complexes that have found their way into medicine exert their biological effects by reaction with nucleophiles present in the body. While metal complexes can react through a variety of different

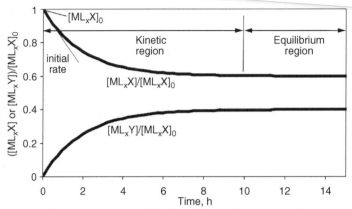

Figure 1.18   *Rate curves, concentration vs. time, for the reaction of ML_xX, starting concentration $[ML_xX]_0$, with nucleophile Y to form product ML_xY*

mechanisms, a substitution reaction, in which a ligand bonded to the metal ion is displaced by an attacking ligand, is the most common. In the hypothetical reaction shown in (1.5), one of the ligands originally attached to the metal ion is replaced by the attacking nucleophile to form a product with the same general structure as the starting material.

$$ML_xX + Y \underset{k_{-2}}{\overset{k_2}{\rightleftharpoons}} ML_xY + X \qquad (1.5)$$

Suppose that an attacking nucleophile, Y, is added to a solution containing ML_xX. At $t = 0$, the time of addition, there can be no product ML_xY and the concentration of ML_xX is its initial concentration, $[ML_xX]_0$, with the initial concentration of Y being $[Y]_0$. *While activities are required for rate and equilibrium expressions, this work will assume that activities of substances are equal to their concentrations, as would be the case when concentrations approach zero, and 'concentration' will be used throughout the text.* As time passes, the original concentration of starting material, ML_xX, decreases at some rate and the concentration of the product, ML_xY, increases at the same rate. This is depicted in Figure 1.18, which shows how the ratios $[ML_xX]/[ML_xX]_0$ and $[ML_xY]/[ML_xX]_0$ for the reaction change with time. The rate of decrease of ML_xX as a function of time is given by the *rate law*, (1.6):

$$\text{rate} = k_2[ML_xX][Y] \qquad (1.6)$$

In this expression, k_2 is the second-order *forward rate constant*, and $[ML_xX]$ and $[Y]$ are the concentrations of ML_xX and Y, respectively. If there is some mechanism by which starting material can be reformed from the product, the reaction will eventually reach a point where the concentrations of all of the components in the reaction medium will not change with time and the system is said to be at *equilibrium*. In the hypothetical reaction shown in Figure 1.18, equilibrium is reached after $\sim$10 hours. In a similar way to the process in the forward direction, the rate in the reverse direction is given by the *rate law*, (1.7):

$$\text{rate} = k_{-2}[ML_xY][X] \qquad (1.7)$$

In this expression k_{-2} is the second-order *reverse rate constant* and $[ML_xY]$ and $[X]$ are the concentrations of ML_xY and X, respectively.

At small intervals of time – that is, just after the nucleophile Y has been added to the medium – the observed rate of disappearance of ML_xX is said to be the *initial rate* of disappearance, which is the *initial slope* of the rate curve for ML_xX. In this region of the rate curve there is very little product and the possibility of making some starting material through the reaction of X with ML_xY (1.5) is negligible. Since the early part of the rate curve contains no significant 'back reaction', this part of the curve can be used to calculate the true forward *rate constant*, k_2, for the reaction from the *initial rate* of disappearance of ML_xX, and the initial concentrations, $[ML_xX]_0$ and $[Y]_0$, according to (1.8).

$$\text{Initial rate} = k_2[ML_xX]_0[Y]_0 \tag{1.8}$$

When the system reaches equilibrium, the rates in the forward and reverse directions must be the same to give (1.9) in which the concentrations are the values at equilibrium, indicated by $[ML_xX]_{eq}$, $[ML_xY]_{eq}$, $[X]_{eq}$ and $[Y]_{eq}$. Rearranging this expression gives (1.10), which shows the relationship between the equilibrium concentrations, the equilibrium constant, K, and the forward and reverse rate constant for the reaction. Since the free energy, ΔG, is related to the equilibrium constant through (1.11) or its exponential form (1.12), the free energy of the reaction can be calculated from K and the temperature. Experimentally, a common way to obtain k_{-2} is to determine k_2 by measuring the initial rate of the reaction, allowing the reaction to reach equilibrium, and determining K through the equilibrium concentrations.

$$k_2[ML_xX]_{eq}[Y]_{eq} = k_{-2}[ML_xY]_{eq}[X]_{eq} \tag{1.9}$$

$$\frac{k_2}{k_{-2}} = \frac{[ML_xY]_{eq}[X]_{eq}}{[ML_xX]_{eq}[Y]_{eq}} = K \tag{1.10}$$

$$\Delta G = -RT \ln K \tag{1.11}$$

$$K = e^{-\frac{\Delta G}{RT}} \tag{1.12}$$

1.5.2   Water exchange rates for metal ions

An important indicator or 'benchmark' of the rate at which a metal ion will be expected to undergo a substitution reaction is the rate at which it exchanges water that is bound to the ion with water that is free in solution. This rate, called the *water exchange rate*, can easily be measured using nuclear magnetic resonance, NMR. The general approach in studying water exchange kinetics using NMR is to employ water that has been labeled (enriched) with ^{17}O, which is a stable oxygen isotope, is nonradioactive and has high NMR sensitivity. Moreover, since ^{17}O has a nuclear spin, I, of $1/2$, which is the same as the nuclear spin of a proton, the isotope produces relatively simple, readily interpretable NMR spectra.

One experimental approach for determining water exchange rates using ^{17}O NMR is to form the *aqua* complex in medium which contains only $H_2{}^{16}O$, add a known amount of $H_2{}^{17}O$ and observe the displacement of the bound $H_2{}^{16}O$ by $H_2{}^{17}O$ using NMR (1.13).

$$[M(H_2{}^{16}O)_n]^{z+} + H_2{}^{17}O \underset{k_{-2}}{\overset{k_2}{\rightleftharpoons}} [M(H_2{}^{16}O_{n-1})(H_2{}^{17}O)]^{z+} + H_2{}^{16}O \tag{1.13}$$

At $t=0$, the time of addition of $H_2{}^{17}O$, all of the labeled water in solution will be in its free unbound form. While the observed NMR spectrum depends on the nature of the metal ion and the rate of the exchange process, if the metal ion is diamagnetic and if the water exchange rate is slow compared to the NMR measurement time, two peaks will be observed in the spectrum. If this is the case, the system is said to be in *slow exchange* on the NMR time scale; that is, rate constants for the chemical exchange are $k < {\sim}10\,s^{-1}$. One

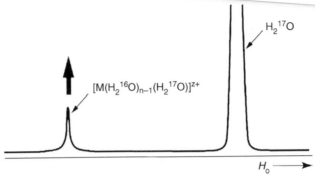

Figure 1.19 Simulated ^{17}O NMR spectrum of a slow exchanging diamagnetic aqua complex, $[M(H_2{}^{16}O)_n]^{z+}$, in $H_2{}^{16}O$, a short time after the addition of excess $H_2{}^{17}O$ to the medium. The intensity of the product peak, $[M(H_2{}^{16}O)_{n-1}(H_2{}^{17}O)]^{z+}$, increases with time

NMR peak is due to *free* $H_2{}^{17}O$ and a second peak, the intensity of which increases with time, is due to *bound* $H_2{}^{17}O$. A hypothetical ^{17}O NMR spectrum at some time after the addition of a large excess of $H_2{}^{17}O$ to a diamagnetic aqua complex is shown in Figure 1.19.

One way to measure the *rate constant* for the reaction to set up the experiment so that the concentration of labeled water, $[H_2{}^{17}O]$ (the brackets indicate concentration), is much larger than the concentration of the aquated metal ion, $[[M(H_2{}^{16}O)_n]^{z+}]$. While the exchange is a true second-order reaction and the rate law for the forward reaction is given by (1.14), if $[H_2{}^{17}O]$ is much larger than $[[M(H_2{}^{16}O)_n]^{z+}]$, the concentration of $H_2{}^{17}O$ can be considered constant during the course of the reaction.

$$\text{Rate} = k_2[[M(H_2{}^{16}O)_n]^{z+}][H_2{}^{17}O] \tag{1.14}$$

In this case the rate law can be rewritten by incorporating $[H_2{}^{17}O]$, which is a constant, into k_2, which is also a constant, to give a new rate expression, (1.15).

$$\text{Rate} = k_1[[M(H_2{}^{16}O)_n]^{z+}] \tag{1.15}$$

Since (1.15) shows that the rate of the reaction really only depends on the concentration of one component, $[M(H_2{}^{16}O)_n]^{z+}$, the equation is the rate expression for a *pseudo first-order* reaction, with k_1 as the *pseudo first-order rate constant* for the reaction. The word '*pseudo*' is used to denote the fact that $[H_2{}^{17}O] \gg [[M(H_2{}^{16}O)_n]^{z+}]$, and while the reaction is rigorously second-order (see (1.13)), the observed rate depends only on the concentration of *one* of the components ($[M(H_2{}^{16}O)_n]^{z+}$) and the reaction behaves as a true *first-order* reaction.

Since the NMR observation nucleus is ^{17}O, which is only present in the product complex and free bulk water, the way in which the concentration of starting material, $[M(H_2{}^{16}O)_n]^{z+}$, changes with time cannot be directly observed in the experiment. However, this concentration can be obtained using (1.16), which shows that for every molecule of $[M(H_2{}^{16}O)_{n-1}(H_2{}^{17}O)]^{z+}$ which forms and is detected by NMR, one molecule of the reactant, $[M(H_2{}^{16}O)_n]^{z+}$, must have reacted with labeled water. In (1.16), the subscript '0' indicates the concentration at $t = 0$, while the subscript 't' indicates concentration at some time t after addition of $H_2{}^{17}O$ to the reaction medium. Substituting (1.16) into (1.15) gives (1.17), which allows determination of k_1 from the starting concentration of $[M(H_2{}^{16}O)_n]^{z+}$ and the concentration of $[M(H_2{}^{16}O)_{n-1}(H_2{}^{17}O)]^{z+}$, measured by NMR, at various times.

$$[[M(H_2{}^{16}O)_n]^{z+}]_t = [[M(H_2{}^{16}O)_n]^{z+}]_0 - [[M(H_2{}^{16}O)_{n-1}(H_2{}^{17}O)]^{z+}]_t \tag{1.16}$$

$$\text{Rate} = k_1\{[[M(H_2{}^{16}O)_n]^{z+}]_0 - [[M(H_2{}^{16}O)_{n-1}(H_2{}^{17}O)]^{z+}]_t\} \tag{1.17}$$

The relationship between the concentration of unlabeled complex at $t = 0$, $[[M(H_2{}^{16}O)_n]^{z+}]_0$, or $[A]_0$, and at some time t, $[[M(H_2{}^{16}O)_n]^{z+}]_t$, or $[A]_t$, is given by (1.18) and the *half life*, $t_{1/2}$, the length of time required for $[[M(H_2{}^{16}O)_n]^{z+}]_0$ to decrease to half of its initial value, is given by (1.19).

$$\ln \frac{[[M(H_2{}^{16}O)_n]^{z+}]_t}{[[M(H_2{}^{16}O)_n]^{z+}]_0} = \ln \frac{[A]_t}{[A]_0} = -kt \tag{1.18}$$

$$t_{1/2} = \frac{0.6931}{k_1} \tag{1.19}$$

When using NMR to study kinetics it is necessary to convert the measured area under an NMR peak, which is proportional to the number of molecules in the solution that produced the peak, to the *concentration* of the substance in the medium. This can be done by having a known concentration of a ^{17}O-containing substance in solution – that is, an *internal standard* – and determining the relationship between integrated NMR peak area and concentration. Once this is done, $[[M(H_2{}^{16}O)_{n-1}(H_2{}^{17}O)]^{z+}]$ can be obtained from NMR peak areas and entered into (1.17) to calculate k_1.

Clearly, ^{17}O NMR is a useful way to measure the water exchange rate constant for a metal complex because it shows the build-up product as a function of time. While the hypothetical spectrum shown in Figure 1.19 is for a diamagnetic metal complex with a small water exchange rate constant, some metal ions have very fast water exchange rates and some ions are paramagnetic. For example, the pseudo first-order rate constant, k_1, for water exchange for K^+, which is diamagnetic, is $\sim 10^9\,s^{-1}$. This means that for this ion a given water molecule bound to the ion is exchanged with one from bulk solvent about 10^9 times per second! The net effect of the rapid chemical exchange is that the NMR instrument cannot make measurements fast enough to 'see' the bound and unbound water, so it reports the *average* of both. In this case the system is considered to be in *fast chemical exchange* on the NMR time scale, which means that the rate constant for water exchange is $k_1 > \sim 10\,s^{-1}$. If $[H_2{}^{17}O] \gg [[M(H_2{}^{16}O)_n]^{z+}]$, only a *broadened* ^{17}O NMR resonance (no separate peak) is observed. While the analysis is less straightforward than the previously described slow exchanging system, the water exchange rate constant, k_1, can be determined from the broadened ^{17}O NMR peak. If the ion is paramagnetic it is still possible to measure water exchange rates with NMR but in this case peaks are generally shifted and broadened and the analysis of the data is more complicated than that presented.

Figure 1.20 shows the water exchange rates for a number of metal ions, measured by ^{17}O NMR and other techniques [4]. The striking feature about the figure is that the pseudo first-order rate constant, k_1, for the exchange rate varies by 19 orders of magnitude, ranging from $10^{-10}\,s^{-1}$ for Ir^{+3} to $10^9\,s^{-1}$ for Cs^+. While the latter ion exchanges one of its bound water molecules with bulk water a *billion* times per second, the former exchanges its water molecules once every 300 years! Not surprisingly, a single bonding model cannot be used to explain all of the water exchange rates in Figure 1.20, but as will be evident in the following section, crystal field theory can be used to explain why some rates are very slow, and simple electrostatic arguments can be used to explain why others in Figure 1.20 are very fast.

1.5.3   Transition State Theory, the kinetic rate constant and equilibrium

Observations by Arrhenius and others showed that the kinetic rate constant for any reaction, k, is given by (1.20) and its exponential form, (1.21).

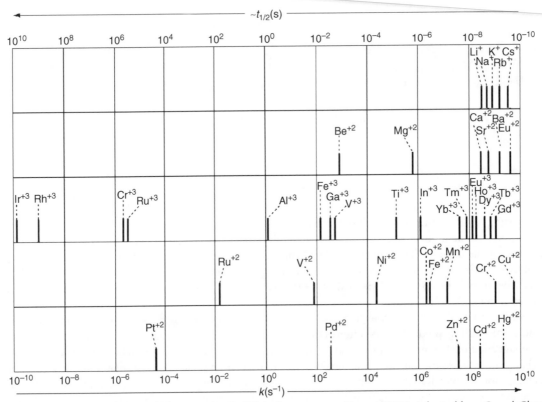

Figure 1.20   *Water exchange rate constants and half lives for some metal ions at 25°C. Adapted from Coord. Chem. Rev., 187, L. Helm & A.E. Merbach, Water Exchange on Metal Ions: Experiments and Simulations, 151–181. Copyright 1999, with permission from Elsevier*

$$\ln k = -\left(\frac{E_a}{RT}\right) + \ln A \qquad (1.20)$$

$$k = Ae^{-\frac{E_a}{RT}} \qquad (1.21)$$

In this empirical expression, called the *Arrhenius equation*, E_a is the *activation energy*, A is the preexponential factor, R is the gas constant and T is the absolute temperature. The Arrhenius equation shows that while an increase in temperature *increases* the rate constant for the reaction, and thus the rate of the reaction, an increase in E_a *decreases* the rate constant for the reaction.

In an attempt to more clearly define the factors that influence the rate of a reaction, Eyring and others analyzed the progress of chemical reactions using *transition state theory*. With this theory, a reaction is discussed using a plot with the *reaction coordinate* as the x-axis and the free energy as the y-axis. In order for a reactant to be converted into a product, the reaction must pass through a maximum in the energy, which is denoted as the *transition state* or the *activated complex* (Figure 1.21). Transition state theory considers the energy barrier between the reactants and the products as the *free energy of activation*, which if the reaction is moving in the 'forward' direction is given the designation $\Delta G^{\ddagger}_{f}$. Since in Figure 1.21 the reactants are less

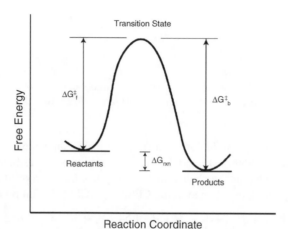

Figure 1.21   *Reaction coordinate based on transition state theory for a reaction with* $-\Delta G_{rxn}$

stable than the products, the free-energy change for the reaction, ΔG_{rxn}, is *negative*, which indicates that the reaction is thermodynamically favored in moving from left to right; that is, from reactants to products. If there is a route for products to return to reactants, the free energy of activation in the reverse, 'back', direction is $\Delta G^{\ddagger}_{b}$. Using (1.12), the equilibrium constant, K, for such a situation is greater than 1, and if enough time is allowed for the system to reach equilibrium, the concentration of products will be greater than the concentration of reactants.

The relationship between the kinetic rate constant, k, for a reaction and $\Delta G^{\ddagger}$ from transition state theory (1.22) is almost the same as with the Arrhenius equation, but the preexponential factor, A, in (1.21) is replaced with the product of a constant, b, and the temperature, T, in (1.22). Like the Arrhenius equation, (1.22) shows that an increase in temperature *increases* the rate constant of the reaction, while an increase in the activation free energy, $\Delta G^{\ddagger}$, *decreases* the rate constant of the reaction. It is also clear that the rate constant k in (1.21) has a slightly different temperature dependence than k in (1.21) but the main temperature dependence for both is in the exponential terms.

$$k = bTe^{-\frac{\Delta G^{\ddagger}}{RT}} \tag{1.22}$$

The water exchange reaction for a metal ion can take place through one of two possible mechanisms. One mechanism, called the *associative mechanism* (1.23), involves a water molecule from bulk solvent adding to the metal ion to form a transient high-energy *seven-coordinate* transition state (activated complex), which ultimately decays to the product by losing a water molecule. A second mechanism, called the *dissociative mechanism* (1.24), involves a *five-coordinate* complex in the transition state, which ultimately captures a water molecule to give the product.

$$[M(H_2O)_6]^{z+} + H_2O \rightarrow [M(H_2O)_7]^{z+} \rightarrow [M(H_2O)_6]^{+z} + H_2O \tag{1.23}$$

$$[M(H_2O)_6]^{z+} \rightarrow [M(H_2O)_5]^{z+} + H_2O \rightarrow [M(H_2O)_6]^{z+} \tag{1.24}$$

For transition metal ions with large crystal field stabilization energy (CFSE), changing the structure of the complex from a six- to either a seven- (associative mechanism) or a five-coordinate (dissociative mechanism) structure is generally energy expensive; that is, it costs energy to do this. Suppose for example the starting six-coordinate complex is d^6 ($S = 0$), which has CFSE $= -2.4\,\Delta_o + 3P$, and the geometry of the activated complex is a seven-coordinate structure; that is, the associative mechanism. While the exact geometry of the transitions

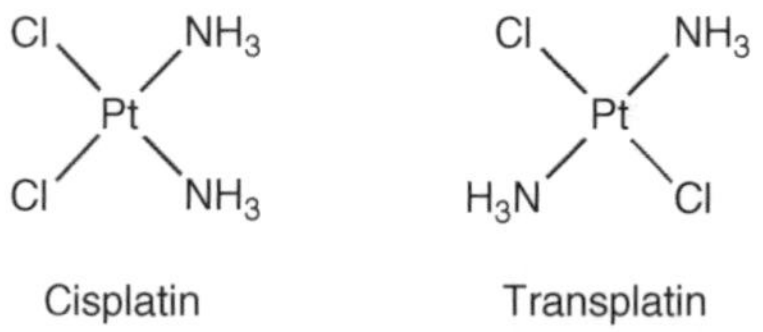

Figure 1.22 Structures of cisplatin and transplatin

state is not known, a pentagonal bipyramidal structure is a reasonable supposition (Figure 1.1). From the one-electron energy levels of the d-orbitals in Table 1.1, a pentagonal bipyramidal structure with $S = 0$ would have CFSE of $-1.548\,\Delta_o + 3P$. Thus, for the change in geometry from a six-coordinate octahedral to a seven-coordinate pentagonal bipyramid, the change in the CFSE, or ΔCFSE, given by (1.25), is $-0.852\,\Delta_o$.

$$\Delta\text{CFSE} = (-2.4\,\Delta_o + 3P) - (-1.548\,\Delta_o + 3P) = -0.852\,\Delta_o \tag{1.25}$$

The negative sign indicates that from the standpoint of CFSE, the change in the geometry in moving from the six-coordiante octahedral structure to the seven-coordinate pentagonal bipyramidal structure is *energetically unfavorable*; that is, discounting entropy, $\Delta G^{\ddagger}_f$ in Figure 1.22 is positive. Performing a similar calculation for the structural change from a six-coordinate octahedral to a five-coordinate square pyramidal, a reasonable structure for the transition state gives a ΔCFSE of $-0.4\,\Delta_o$, indicating that from the standpoint of crystal field effects, this *dissociative* process is also energetically unfavorable. Although factors other than simple crystal field effects are important in the rate of substitution of an ion, a useful guide is that when CFSE for the metal complex is large – for example, Ir^{+3} ($5d^6$, $S = 0$), Rh^{+3} ($4d^6$, $S = 0$), Cr^{+3} ($3d^3$, $S = 3/2$), Ru^{+3} ($4d^5$, $S = 1/2$) – $\Delta G^{\ddagger}_f$ is generally large and the rate constant for water exchange, indeed any substitution reaction, through (1.21) or (1.22), is small. Similar arguments also apply to complexes with a four-coordinate square planar geometry, for example Pt^{+2} ($5d^8$, $S = 0$). Square planar Pt^{+2} complexes are believed to substitute via an *associative* process (1.24) involving a five-coordinate transition state, which, if the structure of the intermediate is square pyramidal, has a value of ΔCFSE for reaching the transition state of $-0.628\,\Delta_o$. As is shown in Figure 1.20, these ions have relatively small water exchange rate constants, which is consistent with their large negative values of ΔCFSE. When Δ_o is small, which happens when the oxidation state of the metal ion is low and if the ion is in the first-row transition metal series, values of CFSE are small. In this case, CFSE of the starting complex and the ΔCFSE required to reach a transition state become less of a predictor of reaction rate and other factors such as the organization of solvent in the solvation sphere of the complex become important in determining the reaction rates of the compound.

For main group cations that have filled outer electronic shells, there is no CFSE. For these ions, simple electrostatic considerations involving the *electrostatic potential* on the surface of the cation that contacts the water molecule can be used to explain exchange rates. For example, if a cation has a large radius and a low net positive charge, the electrostatic potential on the surface of the cation (considered a sphere) that comes into contact with the dipole of the water molecule will be small, leading to a weak electrostatic bond between water and the cation. If the bond is weak, it will be easily broken, which will ultimately translate into rapid exchange of bound water molecules with those in solvent. Consider the series Al^{+3}, Ga^{+3}, In^{+3}, the series Be^{+2}, Mg^{+2}, Ca^{+2}, Sr^{+2}, Ba^{+2} and the series Li^+, Na^+, K^+, Rb^+, Cs^+, which are given in order of *increasing* ionic diameter or, since the charge on the cation in each series is the same, *decreasing* electrostatic potential on the surface of the cation. Since the smallest member of each series forms the strongest metal ion–water bond, it has the *largest* value of $\Delta G^{\ddagger}_f$ (Figure 1.21) and the smallest water exchange rate constant, k, through (1.21) and (1.22). The water exchange rate data given in Figure 1.20 show that k increases with atomic number for each series, which indicates that the simple electrostatic argument given above is probably correct.

Table 1.6 The trans effect series[a]

$CN^- \sim CO \sim C_2H_4 > PH_3 \sim SH_2 > NO_2^- > I^- > Br^- > Cl^- > NH_3 \sim py > OH^- > H_2O$

[a] In this series, CN^- is the strongest *trans* directing ligand and water is the weakest. This table is the series stated in text on p. 439 of Miessler, G.L., Tarr, D.A. (2004) *Inorganic Chemistry*, 3rd edn, Pearson Prentice Hall, Upper Saddle River, NJ.

1.5.4 Trans effect and substitution reactions

In even a brief presentation of the substitution properties of metal complexes, it is important to describe the *trans effect* observed for reactions of square planar compounds, especially those of Pt^{+2}, which are potent anticancer drugs. In simple terms, the *trans effect pertains to the ability of a ligand in a complex to direct substitution opposite, or trans, to itself*. This effect was discovered by Chernyaev [5], who after synthesizing many square planar complexes of Pt^{+2} found that the presence of certain ligands attached to the metal ion had the ability to cause the loss of the group that was in the position *trans* to the ligand in a substitution reaction. As might be expected, not all ligands have the same strength in '*trans* directing ability' and ultimately a series, called the *trans effect series* (Table 1.6), was created to rank common ligands by their ability to direct an incoming group (nucleophile) to the coordination site *trans* to themselves. In this series, the ligand with the greatest *trans* directing ability is CN^-, while the one with the *weakest* ability is water, H_2O.

A number of models have been proposed to explain the *trans* effect and why a given ligand is where it is in the series, but in the end none seem to provide an explanation of the ranking of all of the ligands in the series. Early models focused on the structure of the reactants and whether and to what extent groups that were *trans* to strong *trans* directing ligands have their bonds lengthened. These models, which used thermodynamic arguments, assumed that if a bond was slightly longer in the starting complex it would be the bond broken in the substitution reaction. Other models, which were based on transition state theory, and thus provided a kinetic focus, addressed the structure of the transition state or activated complex in the reaction. These models considered how the *electronegativity* of groups and/or their ability to form π-bonds with the metal ion in the transition state affected which groups would be lost when the product was formed. While these models addressed the *trans* effect from the kinetic perspective – that is, they considered the nature of the transition state and discussed what happened in terms of the reaction coordinate – difficulties in estimating the electronegativities of groups and a lack of knowledge of the structure of the transition state made them somewhat difficult to use.

In order to illustrate how the *trans* effect series works, consider the syntheses of the important anticancer drug *cisplatin*, *cis*-diamminedichloroplatinum (II), and its less active isomer, *transplatin*, *trans*-diamminedichloroplatinum (II), Figure 1.22 [6]. Note that when ammonia is incorporated into a metal complex as a ligand, scientific nomenclature requires that the term used in the name of the compound is 'ammine'. In the synthesis of cisplatin, which most often employs the *Dhara synthesis* [7] (Figure 1.23), a commercially available source of Pt^{+2}, potassium tetrachloroplatinate, $K_2[PtCl_4]$, is reacted with excess potassium iodide, KI, in water [4]. Most of the evidence for reactions of this type shows that the mechanism is *associative*; that is, an iodide ion attacks the platinum ion as one of the bound chloride ligands leaves. This substitution occurs sequentially until all four chloride ligands have been replaced by iodide to form $[PtI_4]^{-2}$. In this reaction, the attacking ligand approaches the platinum ion via the unobstructed z-axis of $[PtCl_4]^{-2}$; that is, it attacks the exposed d_{z^2} orbital of the complex (Figure 1.3) to produce a five-coordinate, most likely trigonal bipyramidal, transition state (Figure 1.1). After $[PtCl_4]^{2-}$ has been converted to $[PtI_4]^{2-}$, the next step in the Dhara synthesis of cisplatin is to add two equivalents of ammonia, NH_3, to the tetra-iodo complex. Note that when addressing a group bound to a metal ion which is an anion, the suffix 'o' is added to the name of the group. The interesting feature of the product of this reaction is that that while two geometric isomers are possible, *cis* and *trans* (Figure 1.23), and the *former* is less *thermodynamically stable* than the *latter*, only the *cis* isomer is formed in the reaction.

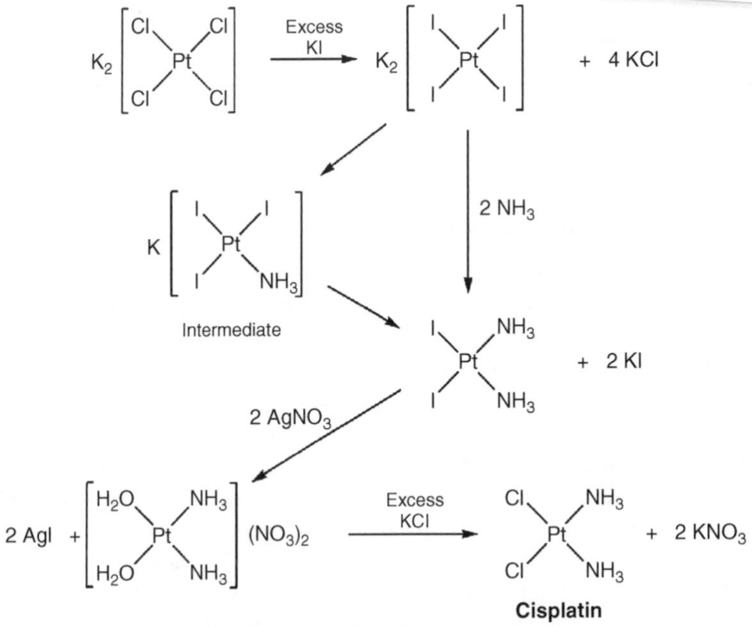

Figure 1.23 *Synthesis of cisplatin*

The key intermediate, which is normally not isolated in this reaction, is $[PtI_3(NH_3)]^-$ (Figure 1.23). The addition of the second equivalent of ammonia to this intermediate can proceed by displacing the iodide ion *trans* to the ammonia, or it can proceed by displacing one of the two iodides that are *cis* to the ammonia. If displacement were purely random, one might expect 67% *cis* (there are two ways to make this isomer) and 33% *trans* (there is one way to make this isomer), but the isolated product *is 100% cis*! This observation is a manifestation of the *trans* effect, which shows that the ability of iodide to direct substitution *trans* to itself is greater than that of ammonia to direct substitution *trans* to itself. Thus, as the reaction proceeds, the iodide ion that is *trans* to the other iodide is lengthening its bond, and it is eventually displaced from the platinum by ammonia to form the *cis* product, *cis*-$[PtI_2(NH_3)_2]$.

The next step in the Dhara synthesis is to replace the two iodides with two water molecules. This is done by adding two equivalents of silver nitrate, $AgNO_3$, which results in the formation of insoluble AgI. This reaction works because all metal complexes in aqueous solution exist in an equilibrium, which means that some fraction of the ligands bound to the metal ion must be in equilibrium with their unbound counterparts (Figure 1.24). If the complex is thermodynamically very stable, the equilibrium will be largely in favor of the

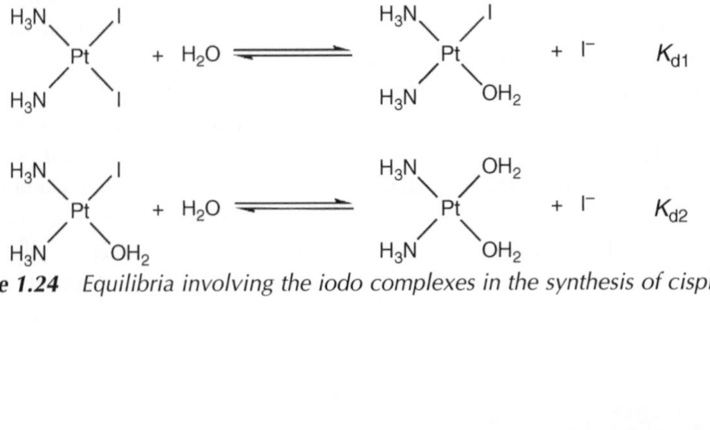

Figure 1.24 *Equilibria involving the iodo complexes in the synthesis of cisplatin*

intact complex, but the concentration of *unbound ligand* can never be zero. The equilibrium expressions (Figure 1.24) in the case of *cis*-[PtI$_2$(NH$_3$)$_2$] are written as dissociations (dissociation of I$^-$ from platinum) and are described by the dissociation constant, K_d, which is related to the normal (association) equilibrium constant, K, by (1.26).

$$K = 1/K_d \qquad (1.26)$$

While K_{d1} and K_{d2} (Figure 1.24) for this reaction are likely very small, addition of Ag$^+$ ion to the solution as AgNO$_3$, which is water soluble, causes insoluble AgI to precipitate from solution, which, from Le Chatelier's principle, drives the equilibrium to the right. This phenomenon is called a *phase change*, in which some of the material originally in the solution phase, Ag$^+$ and I$^-$, moves to a another phase, in this case insoluble AgI. Because nitrate ion, NO$_3^-$, is lower in the spectrochemical series (Table 1.2) than water, and the concentration of water is very high, $>50\,M$, nitrate ion is simply a 'spectator ion' and is not part of the equilibrium expressions given in Figure 1.24.

The last step in the synthesis is to simply displace the two coordinated water molecules of *cis*-[Pt-(H$_2$O)$_2$(NH$_3$)$_2$]$^{2+}$ by adding an excess of chloride ion as KCl, which forms yellow cisplatin, *cis*-[PtCl$_2$(NH$_3$)$_2$]. Even though Δ_o (H$_2$O) $> \Delta_o$ (Cl$^-$), addition of excess chloride ion to the medium drives the equilibrium to the right, and since cisplatin is only sparingly soluble in water, the complex precipitates from solution as its sparingly soluble dichloro form.

The synthesis of *transplatin* (Figure 1.25) also presents examples of the *trans* effect in coordination chemistry. As with cisplatin, the starting material for the synthesis of transplatin is K$_2$[PtCl$_4$], which, in this case, reacts with excess ammonia to produce the tetra-ammine complex, [Pt(NH$_3$)$_4$]$^{2+}$. The next step is to react this complex with excess chloride ion in hot hydrochloric acid, HCl. In this reaction, the first chloride ion displaces an ammonia molecule from platinum, which, because the solution is very acidic, is immediately protonated to form NH$_4^+$. Since the lone pair on NH$_4^+$ is bound by a proton, the released ammonia molecule is blocked from attacking the platinum ion. The second chloride can either occupy the position *trans* to the first chloride or one of the two positions *cis* to the first chloride. Since the *trans* effect series shows that chloride is a better *trans* directing ligand than is ammonia (Table 1.6), the second chloride displaces the ammonia molecule *trans* to the first chloride to produce transplatin. While the reaction conditions would almost certainly lead to

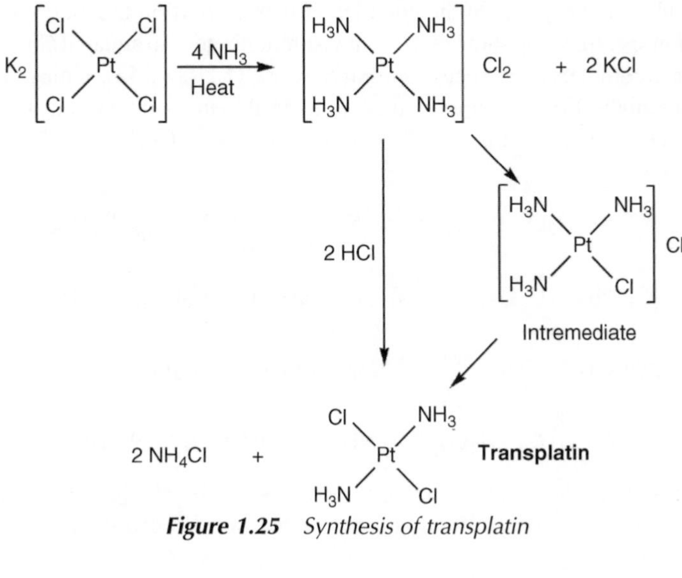

Figure 1.25   *Synthesis of transplatin*

further displacement of ammonia from platinum by chloride, transplatin is the least soluble component in the reaction medium and it precipitates from the medium before it has the chance to further react with chloride ion.

1.5.5   Stability of metal complexes

A property of metal complexes that is a measure of their stability in solution is called the *stability constant, K,* which is sometimes called the *formation constant, K_f.* In certain cases the stability of a complex may be sensitive to pH, especially in the physiological range, in which case the *conditional stability constant,* at some specified condition, can be indicated. While there are many different ways to determine the stability constant of a metal complex, one is to use absorption spectroscopy to measure spectral changes in solution as ligand is added to the system. For example, consider the equilibrium binding process in (1.27), in which one ligand reacts with a metal ion to produce a 1 : 1 complex. The stability constant (equilibrium constant) for this system is given by (1.28), where [ML] is the concentration of complex and [M] and [L] are the concentrations of *free metal ion* and *free ligand*, respectively.

$$M + L = ML \tag{1.27}$$

$$K = \frac{[ML]}{[M][L]} \tag{1.28}$$

The typical way to determine the stability constant, K, for this system is to first determine the molar extinction coefficient of the complex that forms, ML, by driving the system to the right by adding a large excess of the ligand which complexes all of the metal ion. Once ε_{max} for the complex is determined (Box 1.1), solutions with different concentrations of ligand and a constant concentration of metal ion are prepared and their absorption spectra measured to determine the concentration of complex, ML, in each. By knowing the concentration of ML present in each mixture and the total concentrations of metal ion and ligand in the mixtures, the value of K can be calculated from (1.28).

Most systems are much more complicated than the example given and more than one equilibrium expression is most often involved. For example, consider the reaction of ammonia, NH_3, with the aquated complex $[Cu(H_2O)_6]^{2+}$. Since Cu^{+2} has the electronic configuration $3d^9$, it is a Jahn–Teller distorted system with two of the *trans* water molecules on the z-axis of the complex farther away from the metal ion than the other four. Studies using absorption spectroscopy show that in water, ammonia adds to aquated Cu^{+2} in a stepwise manner to produce four distinct complexes, a process summarized by (1.29)–(1.32). Unless very unusual circumstances are present, the highest-order complex that forms in the ammonia system is $[Cu(NH_3)_4]^{2+}$, so for simplicity the two weakly-bound water molecules in the axial sites of $[Cu(H_2O)_6]^{2+}$ have been eliminated in the equilibrium reactions.

$$[Cu(H_2O)_4]^{2+} + NH_3 = [Cu(NH_3)(H_2O)_3]^{2+} + H_2O \tag{1.29}$$

$$[Cu(NH_3)(H_2O)_3]^{2+} + NH_3 = [Cu(NH_3)_2(H_2O)_2]^{2+} + H_2O \tag{1.30}$$

$$[Cu(NH_3)_2(H_2O)_2]^{2+} + NH_3 = [Cu(NH_3)_3(H_2O)]^{2+} + H_2O \tag{1.31}$$

$$[Cu(NH_3)_3(H_2O)]^{2+} + NH_3 = [Cu(NH_3)_4]^{2+} + H_2O \tag{1.32}$$

The equilibrium expression for the first step in this sequence is (1.33), which, if the concentration of water is eliminated in the reaction (it is a constant and is incorporated into K), gives (1.34).

$$\frac{[[Cu(NH_3)(H_2O)_3]^{2+}][H_2O]}{[[Cu(H_2O)_4]^{2+}][NH_3]} = K_1' \tag{1.33}$$

$$\frac{[[Cu(NH_3)(H_2O)_3]^{2+}]}{[[Cu(H_2O)_4]^{2+}][NH_3]} = K_1 \tag{1.34}$$

The equation which describes the formation of the final complex, called the *overall reaction*, from the free aqua complex and the free ligand is (1.35). By writing the individual equilibrium expressions (1.29)–(1.32) to obtain K_{1-4}, it would be easy to show that the equilibrium constant for the overall reaction (1.35) is the *product* of the individual stepwise equilibrium constants; that is, (1.36). This product is given the designation K_{1-n} or β_n, where n is the number of steps in the overall reaction. Since the values of K_{1-n} and β_n are often very large, the typical way to denote the overall stability constant is log K_{1-n} or log β_n.

$$[Cu(H_2O)_4]^{2+} + 4NH_3 = [Cu(NH_3)_4]^{2+} \tag{1.35}$$

$$\frac{[[Cu(NH_3)_4]^{2+}]}{[[Cu(H_2O)_4]^{2+}][NH_3]^4} = K_1 K_2 K_3 K_4 = K_{1-4} = \beta_4 \tag{1.36}$$

By systematically increasing the total concentration of ammonia in solution, collecting absorption spectra and fitting the data to a model, the stability constants for the copper–ammonia system at 30 °C were determined to be $K_1 = 1.78 \times 10^4\,M^{-1}$, $K_2 = 2.30 \times 10^3\,M^{-1}$, $K_3 = 9.77 \times 10^2\,M^{-1}$ and $K_4 = 63.1\,M^{-1}$, which in this case gives $K_1 \times K_2 \times K_3 \times K_4 = K_{1-4} = \beta_4 = 2.5 \times 10^{12}\,M^{-4}$, or log K_{1-4} or log $\beta_4 = 12.4$ [8]. Note that while the unit of concentration is indicated in the values of K and β, which is helpful for determining whether the equilibrium expression is written as an *association* or a *dissociation*, it is not possible to take the logarithm of the unit. As pointed out in Section 1.5.1, the concentration of a substance is assumed to be equal to the activity of the substance, which, since the latter is unitless, will not present a problem in obtaining the logarithm of the value.

Similarly to (1.12), the relationship between the stability constant of the overall reaction, K_{1-4} and β_n, and ΔG is given by (1.37), which is related to the change in enthalpy, ΔH, making and breaking bonds, and the entropy, ΔS, the amount of disorder in the reaction, through the well-known equation (1.38).

$$K_{1-4} = \beta_n = e^{-\frac{\Delta G}{RT}} \tag{1.37}$$

$$\Delta G = \Delta H - T\Delta S \tag{1.38}$$

1.5.6   Chelate effect

For many years scientists have recognized that ligands which form a ring structure that includes the metal ion, called a *chelate ring*, have exceptional thermodynamic stability. This 'extra' stability, beyond that which would be achieved with a similar system without a chelate ring, is called the *chelate effect* in inorganic chemistry. As with organic chemistry, chelate rings that have a total of five or six atoms in the ring (this includes the metal ion) are more stable than rings with four or seven members. In order to see how the chelate effect works, consider the reaction of two ammonia molecules with aquated Cu^{+2} (1.39) and, for comparison, the reaction of the bidentate ligand, ethylenediamine, $NH_2CH_2CH_2NH_2$, abbreviated 'en', with aquated Cu^{2+}, (1.40).

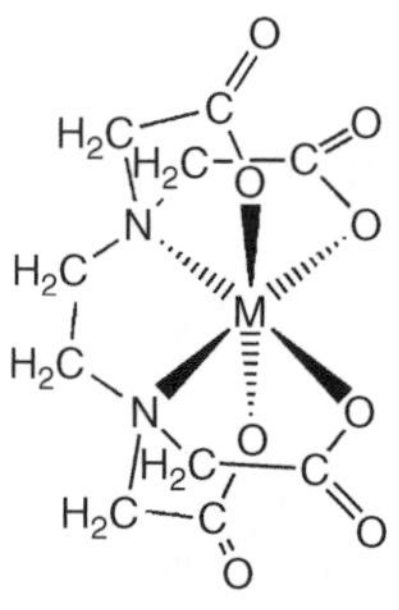

M-EDTA

Figure 1.26 *Structure of an octahedral complex with ethylenediaminetetraacetic acid, EDTA, which binds to the metal ion as a tetravalent anion*

$$[Cu(H_2O)_6]^{2+} \; + \; 2\,NH_3 \; \rightleftharpoons \; [Cu(H_2O)_4(NH_3)_2]^{2+} \; + \; 2\,H_2O \tag{1.39}$$

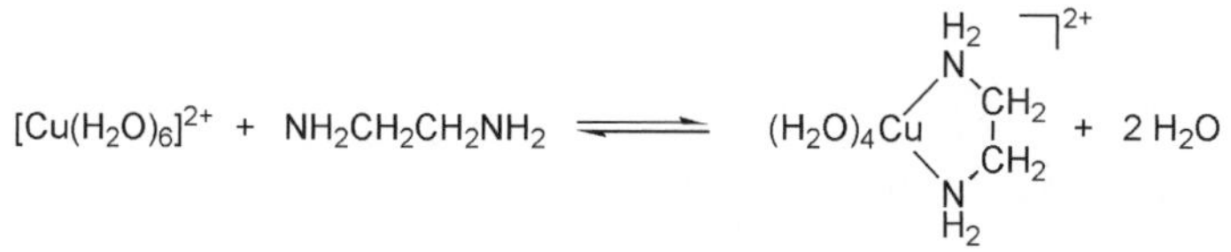

$$[Cu(H_2O)_6]^{2+} \; + \; NH_2CH_2CH_2NH_2 \; \rightleftharpoons \; (H_2O)_4Cu\!\!\left[\begin{array}{c} H_2 \\ N \\ \diagdown CH_2 \\ \mid \\ \diagup CH_2 \\ N \\ H_2 \end{array}\right]^{2+} \; + \; 2\,H_2O \tag{1.40}$$

Experimentally, the reaction with ammonia (1.39) has an overall stability constant for the formation of the *bis*-ammonia complex of log $\beta_2 = 7.61$, while the reaction with en (1.40), which results in the formation of a five-membered, en, chelate ring, has log $\beta_1 = 10.73$ at 25 °C [9]. This is an example of the chelate effect in that the presence of the ring increases the stability constant of the Cu^{2+}-en complex over that of the Cu^{+2} ammonia system by more than three orders of magnitude.

The nomenclature used in referring to the type of chelate ring formed by a ligand is to say the ligand is *bidentate* if it has two points of attachment (two donor atoms) to the metal ion, *tridentate* if it has three points of attachment, *tetradentate* if it four, and so on. While the arrangement of chelate rings, the nature of the donor atoms and their charge, and the size of the rings all affect the stability of the complex, in general ligands which have multiple points of attachment to the metal ion produce the most stable complexes. An example of a common *hexadentate* ligand which forms very stable complexes with a wide variety of metal ions is EDTA, ethylenediaminetetraacetic acid, which is shown in Figure 1.26.

The exceptional stability of metal chelate compounds implies that the binding constant of the ligand toward the metal ion, K, is large, and therefore the change in the free energy, ΔG, through (1.12) is large and negative. Since ΔG is related to ΔH and ΔS through (1.38), it is sometimes possible to determine which thermodynamic quantity makes the most important contribution to the free energy. However, since ΔH and ΔS reflect the *net changes* in these quantities in the reaction, and solvent molecules around the starting materials and products can play an important role, even a qualitative assessment of the relative importance of these quantities is challenging. The fact that there are two 'particles' on the left of reaction (1.40), the aquated complex and the bidentate ligand, and three particles on the right, the bidentate complex and two water molecules – the latter returning to bulk solvent – an increase in *entropy* in moving from left to right (ΔS is positive for the reaction) is often cited as the thermodynamic basis for the chelate effect.

1.5.7 Macrocyclic effect

As might be expected, totally surrounding the metal ion with a ligand that is cyclic and has no 'ends' further enhances the stability of the resulting complex. For example, consider the complexes $[Cu(232\ tet)]^{2+}$ and

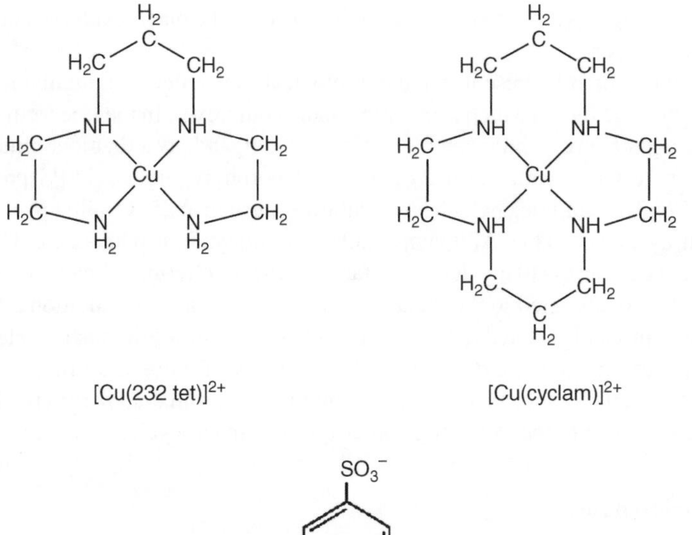

[Cu(232 tet)]$^{2+}$

[Cu(cyclam)]$^{2+}$

CuTPPS

Figure 1.27 *Structures of [Cu(232 tet)]$^{2+}$, [Cu(cyclam)]$^{2+}$ and [Cu(TPP)], where TPP is the di-deprotonated porphyrin core of meso-tetra-(p-sulfonatophenyl)porphyrin*

[Cu(cyclam)]$^{2+}$, shown in Figure 1.27, which are formed in the reaction of aquated Cu^{2+} with the *acyclic ligand, 232 tet*, or the *macrocyclic ligand, cyclam*. The stability constant for [Cu(232 tet)]$^{2+}$ at 25 °C is log $\beta_1 = 23.2$, while the stability constant for [Cu(cyclam)]$^{2+}$ is log $\beta_1 = 27.2$ [10]. The observation that the stability constant for a complex containing a macrocyclic ligand is greater than the stability constant for an analogous complex with an acyclic ligand is termed the *macrocyclic effect* in inorganic chemistry. The values of ΔH for [Cu(232 tet)]$^{2+}$ and [Cu(cyclam)]$^{2+}$ are -110.8 and -135.4 kJ mol^{-1} respectively, clearly showing that enthalpy contributes to the enhanced stability of the macrocyclic complex. The values of ΔS for the two compounds are 66.9 and 50.2 J mol^{-1} K^{-1}, respectively, showing that entropy actually favors the formation of the acyclic complex, [Cu(232 tet)]$^{2+}$. Although it is not easy to explain the observed values of ΔH and ΔS from the structures of the complexes alone, or what might be taking place in terms of bond breaking and formation and what the 'order' (entropy) is like on either side of the reaction, the macrocyclic effect appears to be largely governed by *enthalpic* factors.

The porphyrin ligand, which is present in many biological molecules, is a highly versatile tetraaza (four nitrogen atoms) macrocyclic ligand which forms very stable complexes. In the free form, the porphyrin ligand has two pyrrole hydrogen atoms, which can be lost when the ligand, as a dianion, binds to a metal ion. For example, the reaction of Cu^{2+} with the *tetra* sulfonated porphyrin ligand TPPS produces the complex, Cu-TPPS (Figure 1.27). This complex, which has a stability constant at $25\,°C$ of $\log \beta_1 = 38.1$, is considerably more stable than $[Cu(cyclam)]^{2+}$ [11]. Although detailed thermodynamic data for Cu-TPPS are not available, the high stability of the complex is likely due to the fact that the porphyrinato ligand has two negative charges and thus is electrostatically attracted to the metal ion, which is a cation. An additional factor underlying the stability of Cu-TPPS is probably related to the high rigidity of the porphyrin macrocycle. Once a metal ion is bound in the cavity of the ligand, removing the porphyrin requires folding the structure in some manner that would be energetically unfavorable. Thus, highly conjugated, aromatic, porphyrin ligands enhance the macrocyclic effect to an even higher level than other types of macrocyclic structure.

1.5.8 Hard–soft acids–bases

The concept of hard and soft acids and bases, given the acronym HSAB, was introduced by the renowned inorganic chemist Ralph G. Pearson [12]. This concept addresses the rate of formation and thermodynamic stability of metal complexes in terms of the nature of the metal ion and ligand that come together to form a bond. Pearson observed that a metal ion which is a so-called 'hard acid' forms a strong complex with a ligand that is a 'hard base', and a metal ion that is a 'soft acid' also forms a strong complex with a ligand that is a 'soft base'. If the bonding partners are interchanged – that is, soft acid with hard base or hard acid with soft base – complexes with lower stabilities result. This early remarkable observation was based on the measured rates and stabilities of complexes and on an assessment of the 'deformability or polarizibility' of the 'electron clouds' on both the metal ion and the ligand. If a metal ion or ligand has a large radius and low net charge, the nucleus has less control over the frontier (outer) electrons and the electron distribution of the ion can be more easily distorted (polarized). A metal ion or ligand donor atom with these properties is considered a *soft acid* (M) or *soft base* (L). If, on the other hand, an ion has a high charge and a small radius, the electronic shape of the ion cannot be easily polarized or distorted. A metal ion or ligand with these properties is considered a *hard acid* (M) or a *hard base* (L). While the observation of which combinations produce the highest stabilities and reaction rates is irrefutable, explaining the effect in terms of the physical properties of atoms and ions is not straightforward and is beyond the scope of the presentation given here. However, one point concerning the connection between the rate of a reaction (a kinetic property) and the stability of the product (a thermodynamic property) is worth noting. From transition state theory, the rate constant for a reaction depends in an inverse exponential way on the activation free energy of the reaction, $\Delta G^{\ddagger}$, through (1.22). On the other hand, the stability constant for the products formed in the reaction, K, is determined from the differences in the free energies of the products and reactants, ΔG_{rxn}, through (1.12). Using the principles of transition state theory, this means that if ΔG_{rxn} is large and negative, $\Delta G^{\ddagger}_f$ is small and positive and the rate constant, k, for reactants moving to products is large (1.22). In the context of the HSAB concept, this means that if all other factors are equal, a soft ligand will react *faster* with a soft metal ion to produce a complex with high stability, while a hard ligand will react with a soft metal ion more *slowly* to produce a complex with low stability. While the exact reasons why hard–hard and soft–soft combinations are better than hard–soft/soft–hard combinations is challenging to explain, simply knowing which combinations lead to products with high stabilities and rates of formation is very useful for assessing whether and to what extent a metal complex introduced into the body will react with nucleophiles found in the biological milieu. A list of the hard–soft acids–bases to be encountered in this text is given in Table 1.7.

Table 1.7 *Hard–soft acids–bases*

Metal Ions (Acids)			Ligands (Bases)		
Soft	Intermediate	Hard	Soft	Intermediate	Hard
Au^+	Au^{+3}	Tc^{+5}, Tc^{+7}	CN-R	Pyridine	H_2O
Tc^+	Ru^{+3}	Gd^{+3}	CO	Imidazole	HO^-
Pt^{+2}	Ti^{+2}	Ga^{+3}	$R\text{-}S^-$	Br^-	acac
Pt^{+4}	Fe^{+2}	Ti^{+4}	R_2S		NH_3
Cu^+	Zn^{+2}	Fe^{+3}	Cp		RNH_2
Ru^{+2}	Cu^{+2}	Re^{+5}, Re^{+7}	R_3P		R_2NH
Re^+	Tc^{+3}	V^{+4}, V^{+5}	dmso		Cl^-
	Re^{+3}	Rh^{+3}	CN^-		CO_3^{2-}, HCO_3^-
		Mn^{+2}			$H_2PO_4^-$, HPO_4^{2-}
					RCO_2^-
					O^{-2}

Ligands (Bases) on DNA/RNA			Ligands (Bases) on Proteins		
Soft	Intermediate	Hard	Soft	Intermediate	Hard
	Nitrogen sites on, A, G, T, C, U	$ROPO_2OR$	$R\text{-}S^-$, Cys	Imidazole	Lys, RNH_2
			$RSCH_3$, Met RS^-, GSH $R\text{-}Se^-$, Sec	RSSR, Disulfide	RCO_2^-, Asp, Glu $\phi\text{-}O^-$, Tyr

Cp, cyclopentadienyl; dmso, dimethylsulfoxide; acac, acetylacetone; GSH, glutathione.

Problems

1. Give the number of *unpaired electrons* in the complexes $[Ni\,(en)_3]^{2+}$, octahedral, $[NiCl_4]^{2-}$, tetrahedral and $[Ni(CN)_4]^{2-}$, square planar. The abbreviation 'en' is $H_2NCH_2CH_2NH_2$ or ethylenediamine, which acts as a bidentate ligand with two points of attachment to the Ni^{+2} ion.

2. Of the following pairs of complexes, which one has the *larger* crystal field spitting parameter, Δ? Which complex of the pair has the absorption maxima of its *d-d* electronic transitions at *higher* energy? Give reasons for your choice.

   a. $[CoF_6]^{4-}$ vs. $[CoF_6]^{3-}$
   b. $[Fe(CN)_6]^{4-}$ vs. $[Os(CN)_6]^{4-}$
   c. $[CoF_6]^{3-}$ vs. $[CoCl_6]^{3-}$
   d. $[Co(H_2O)_6]^{2+}$ vs. $[CoCl_4]^{2-}$
   e. $[TiF_6]^{3-}$ vs. $[VF_6]^{2-}$.

3. Which crystal-field terms (states) are produced from a free-ion 'F' term (state) in octahedral and tetrahedral crystal fields?

4. Using an appropriate diagram, give the ground-state crystal-field term, for example $^1A_{1g}$, for the ions given below. For each possibility, indicate the *two lowest-energy excited states* that have the same spin multiplicity as the ground state; that is, Laporte-forbidden–spin-allowed transitions.

   a. Co^{+2}, tetrahedral
   b. Fe^{+3}, octahedral, high spin

c. Fe^{+3}, octahedral, low spin

d. Cu^{+2}, tetrahedral.

5. Consider that $[Au\,Cl_4]^-$, $5d^8$, can exist in either of two possible geometries, square planar, $S=0$ or tetrahedral, $S=1$. Using the energies of the d-orbitals in Table 1.1, estimate the value of P/Δ_o at which both geometries, from crystal field considerations, would be *equally likely*.

6. Determine the crystal field stabilization energy, CFSE (in units of Δ_o), and the electronic configuration, $t_{2g}^3 e_g^1$ and so on, of the following. The quantity S, the total spin quantum number, is the number of unpaired electrons divided by 2.

    a. $[FeBr_6]^{4-}$, octahedral, $S=2$

    b. $[Fe(CN)_6]^{4-}$, octahedral, $S=0$

    c. $[FeCl_4]^{2-}$, tetrahedral, $S=2$.

7. The complex $[Co(en)_3]^{3+}$ (diamagnetic, $S=0$) exhibits two absorption bands at $21\,550\,cm^{-1}$ (ε, $88\,M^{-1}\,cm^{-1}$) and $29\,600\,cm^{-1}$ (ε, $78\,M^{-1}\,cm^{-1}$). Using a Tanabe–Sugano diagram and assuming that the symmetry about the cobalt ion is octahedral, assign the two transitions (for example $^3T_{2g} \leftarrow {}^4A_{1g}$) for this complex.

8. What is the ground-state *free-ion term symbol* for an ion with the $3d^1$ electronic configuration? What *crystal-field terms* are associated with d^1 in an octahedral crystal field?

9. Draw and label the approximate molecular orbital diagrams for an octahedral complex with only σ-bonding between the metal ion and ligands. Indicate which levels in the diagram can be derived using simple crystal field splitting arguments.

10. The magnetic moment for a metal complex is 3.92 BM. How many unpaired electrons are there on the metal ion in the complex?

11. A complex of a first-row transition metal complex has an octahedral geometry with $S=1$ and $\mu_{eff}=3.42$ BM. Answer the following:

    a. How many unpaired electrons does the complex have?

    b. What is the ground-state crystal-field term, including the spin and orbital multiplicity, of the state?

    c. Briefly explain why μ_{eff} is greater than μ_{so} for this complex.

12. The pseudo first-order rate constant for the reaction of the thiol-containing ligand glutathione with a metal complex is $1.2 \times 10^{-2}\,s^{-1}$ at $37\,°C$. If the concentration of thiol in the reaction medium is $1\,mM$ and the concentration of metal complex is $1\,\mu M$, calculate the *second-order rate constant* for the reaction from the data given.

13. Carbonic anhydrase (CA) is a zinc-containing enzyme located in the cytoplasm and mitochondria that catalyzes to conversion of carbon dioxide to carbonic acid by reaction with water. If the concentration of carbon dioxide in solution decreases from $220\,mM$ to $55.0\,mM$ in 1.22×10^4 seconds, and the reaction is far from reaching equilibrium, calculate the $t_{1/2}$ (half life) for the first-order reaction.

14. The DNA nucleotide adenosine triphosphate, ATP, reacts with the metal complex $[ML_3Cl]$ in aqueous media by loss of the chloride ligand according to the reaction below:

$$[ML_3Cl] + ATP = ML_3-ATP + Cl^-$$

    a. If the initial rate for this reaction is $3.6 \times 10^{-2}\,M\,s^{-1}$ at $37\,°C$ and the initial concentration of both the nucleotide and the metal complex in the reaction medium is $3.0 \times 10^{-3}\,M$, calculate the *second-order rate constant*, k_2, for the reaction.

    b. After 24 hours the reaction reaches equilibrium and the equilibrium constant, K, is determined to be, 10^2. Calculate the free energy, ΔG, for the reaction R $=8.314\,J\,K\,mol^{-1}$.

    c. Calculate the second-order *reverse rate constant*, k_{-2}, to reform starting materials for the reaction.

15. Using the *trans* effect series, predict the products of the following reactions. Assume a ratio of starting material to reactant of $1:1$.

    a. $[Pt(CO)Cl_3]^- + NH_3 \rightarrow$
    b. $[Pt(NH_3)Br_3]^- + NH_3 \rightarrow$.

16. The reaction of a nucleophile with a metal complex produces two products, A and B, which are in equilibrium with each other with $[A]_{eq} = 2[B]_{eq}$ at 37 °C. Calculate the free-energy difference between A and B, $R = 8.314\,J\,K\,mol^{-1}$.

17. The equilibrium constants for the addition of the first and second ligand chloride ligand to a metal complex are $K_1 = 3.6 \times 10^4$ and $K_2 = 4.1 \times 10^3$. Calculate $\log \beta_2$ for this complex.

18. The linear two-coordinate Au^+ complex $[AuCl(PEt_3)]$, where PEt_3 is triethylphosphine, is implicated in the reactions of the antiarthritic drug auranofin. When this complex reacts with nucleophiles by losing the chloride ligand, will it form a more stable complex with the ε-amino group (NH_2-R) of the amino acid lysine or the thiol group (HS-R) of the amino acid cysteine? Briefly explain your choice.

References

1. Tinoco, I., Sauer, K., Wang, J.C., and Puglisi, J.D. (2002) Physical chemistry, *Principles and Applications in Biological Chemistry*, 4th edn, Prentice Hall, Upper Saddle River, NJ.
2. Paul, A. (1975) Optical and esr spectra of titanium (III) in Na_2O-B_2O_3 and Na_2O-P_2O_5 glasses. *Journal of Materials Science*, **10**, 692–696.
3. Riordan, A.R., Jansma, A., Fleischman, S. *et al.* (2005) Spectrochemical series of Cobalt(III). An experiment for high school through college. *Chemical Educator*, **10**, 115–119.
4. Helm, L. and Merbach, A.E. (1999) Water exchange on metal ions: experiments and simulations. *Coordination Chemistry Reviews*, **187**, 151–181.
5. Chernyaev, I.I. (1926) The mononitrites of bivalent platinum. I. *Ann Inst Platine (USSR)*, **4**, 243–275.
6. Alderden, R.A., Hall, M.D., and Hambley, T.W. (2006) The discovery and development of cisplatin. *Journal of Chemical Education*, **83**, 728–734.
7. Dhara, S.C. (1970) A rapid method for the synthesis of cisplatin. *Indian Journal of Chemistry*, **8**, 193–194.
8. Trevani, L.N., Roberts, J.C., and Tremanine, P.R. (2001) Copper(II)-Ammonia complexation equilibria in aqueous solutions at temperatures from 30 to 250 °C by visible spectroscopy. *Journal of Solution Chemistry*, **30**, 585–622.
9. Spike, C.G. and Parry, R.W. (1953) Thermodynamics of chelation. II. Bond energy effects in chelate ring formation. *Journal of the American Chemical Society*, **75**, 3770–3772.
10. Liang, X. and Sadler, P.J. (2004) Cyclam complexes and their applications in medicine. *Chemical Society Reviews*, **33**, 246–266.
11. Jiménez, H.R., Julve, M., Moratal, J.M., and Favs, J. (1978) Stability constants of metalloporphyrins. A study of the protonation, deprotonation, and formation of Copper(II) and Zinc(II) complexes of meso-tetra-(p-sulfonatophenyl) porphyrin. *Journal of the Chemical Society. Chemical Communications*, 910–911.
12. Pearson, R.G. (1963) Hard and soft acids and bases. *Journal of the American Chemical Society*, **85**, 3533–3539.

Further reading

Huheey, J.E., Keiter, E.A., and Keiter, R.L. (1993) Inorganic chemistry, *Principles of Structure and Reactivity*, 4th edn, Benjamin-Cummings Publishing Co., San Francisco, CA.
Miessler, G.L. and Tarr, D.A. (2004) *Inorganic Chemistry*, 3rd edn, Pearson Prentice Hall, Upper Saddle River, NJ.

Housecroft, C.E. and Sharpe, A.G. (2007) *Inorganic Chemistry*, 3rd edn, Pearson Education Limited, Edinburgh Gate, UK.

Lever, A.B.P. (1968) *Inorganic Electronic Spectroscopy*, 1st edn, Elsevier Publishing Co., Amsterdam, Netherlands.

Lever, A.B.P. (1984) *Inorganic Electronic Spectroscopy*, 2nd edn, Elsevier Science Publishers B. V., Amsterdam, Netherlands.

Atkins, P. and De Paula, J. (2006) *Atkins' Physical Chemistry*, 8th edn, W. H. Freeman and Co., New York, NY.

Ebbing, D.D. and Gammon, S.D. (2009) *General Chemistry*, 9th edn, Houghton Mifflin Co., Boston, MA.

2

Metallo-Drugs and Their Action

2.1 Introduction

Introduction of metallo-drugs into the body exposes them to reaction with many substances in the biological system. While the main targets for these agents are proteins and DNA, passage through the blood and eventually into the cell allows the metal complex to come in contact which substances that can modify its composition. The main focus of this chapter is on giving a brief overview of the structures of proteins and DNA and summarizing the ways in which these important biomolecules react with metal ions. Since many metal-containing agents used in medicine have easily-displaced ligands, the manner in which complexes react with simple ions found in the blood and cells will also be presented and discussed. In addition to highlighting the biological chemistry of metallo-drugs, it is also important to outline how their pharmacological effects are measured and, since there is increasing demand for more effective agents, how a discovery made in the laboratory makes its way through the approval process to become a drug.

2.2 Proteins as targets for metallo-drugs

The human body has tens of thousands of different proteins, involved in a variety of different catalytic, transport and structural roles. If a metal ion introduced into the body in the form of a metallo-drug attaches itself to a part of the protein that is critical for function, the ability of the protein to perform its biological task could be impaired, which could be enough to kill the cell. For example, many catalytically-active proteins called *metallo-enzymes* have naturally-occurring metal ions such as Zn^{+2} or $Cu^{+1/+2}$ in their active sites. Since the same ligands on the protein that bind to the naturally-occurring metal ion could also serve as donor atoms to the metal ion in a metallo-drug, addition of the drug to the protein could result in displacement of the natural metal ion, causing the enzyme to be catalytically inactive. Many proteins in the body use metal ions as a means of organizing their structure. Since some of these *metallo-proteins* are involved in gene expression and interact directly with DNA, disruption of their structure by substituting the natural metal ion with one supplied by a metallo-drug could affect the specificity and affinity of the interaction, thus influencing the ability of the cell to make the proper proteins for survival.

Metals in Medicine   James C. Dabrowiak
© 2009 John Wiley & Sons, Ltd

2.2.1   Protein structure

Structurally, proteins are long polyamide polymers that are made up of monomer units called *amino acids*, the structures of which, along with their three- and one-letter amino acid codes, are shown in Figure 2.1. As is evident in the figure, all of the amino acids have an amino, a carboxylic acid and a hydrogen atom attached to the α-carbon atom of the compound, but the forth group on the α-carbon, called the *side chain*, characterizes each amino acid. Except for glycine, all of the common amino acids have four different groups attached to the α-carbon atom, which makes the atom, and thus the amino acid, chiral. With the exception of cysteine, which has the *R* absolute configuration, all of the remaining naturally-occurring amino acids have α-carbon atoms that have the *S* absolute configuration.

A protein is formed by connecting a large number of amino acids together in a *sequence* by reacting the α-carboxylic acid group of one amino acid with the α-amino group of another amino acid to form a *peptide* or *amide bond* with the elimination of a water molecule (Figure 2.2). If the chain is very short, up to ~20 amino acids for example, the molecule is called a *peptide*. These are named by taking the name of the amino acid at the amino terminal end, adding a 'yl' ending to the name, and proceeding toward the carboxyl terminal end in the same manner, except that the last amino acid *residue* in the chain retains its unmodified name. For example, the *tripeptide* with the sequence Gly-His-Ser is named gylcylhistidylserine, while the *tetrapeptide* with the sequence AGFT is named alaninylglycylphenylalaninylthreonine.

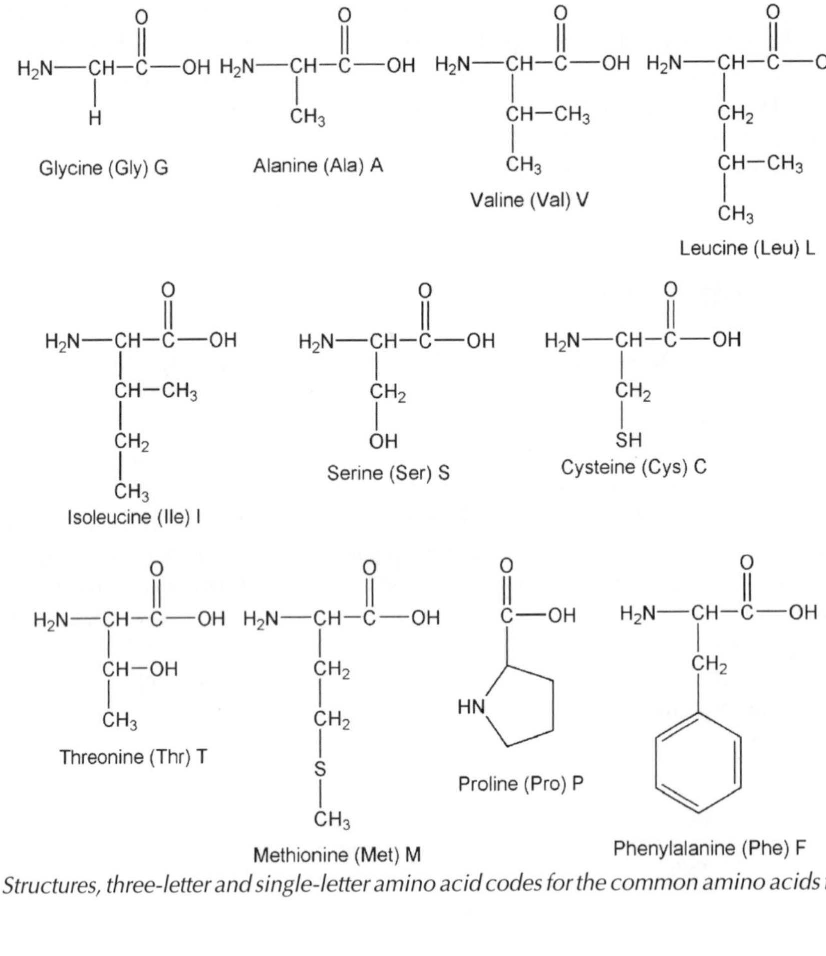

Figure 2.1   *Structures, three-letter and single-letter amino acid codes for the common amino acids found in proteins*

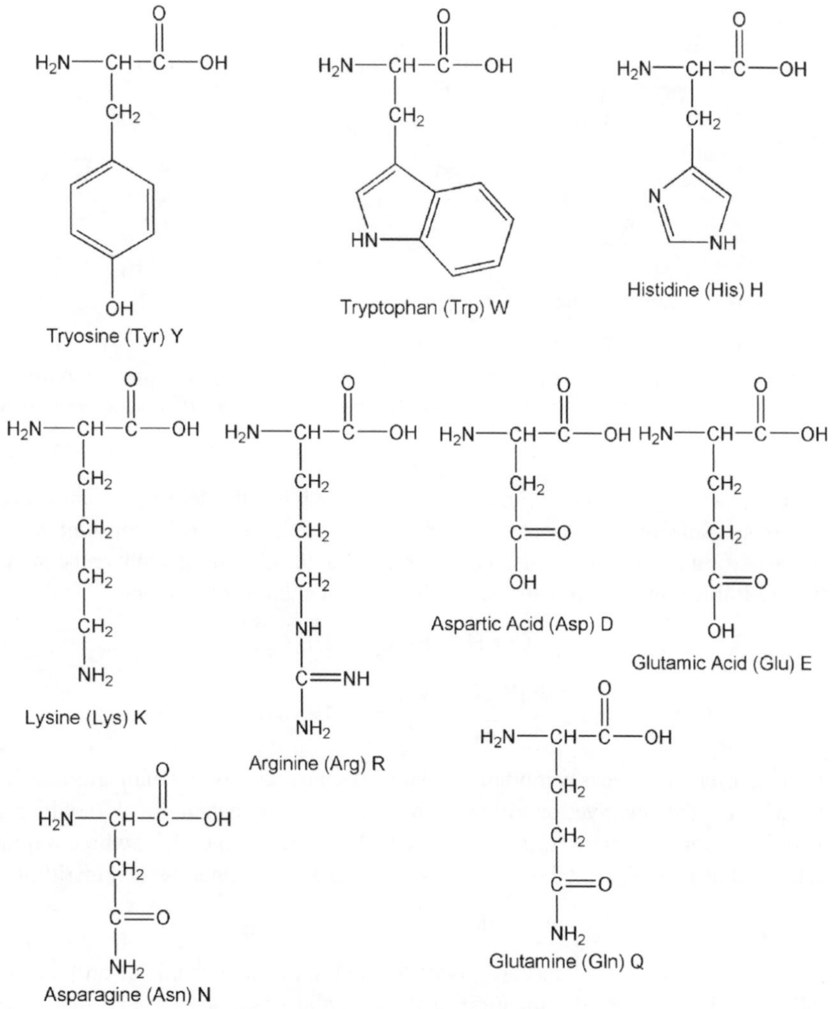

Figure 2.1   *(Continued)*

Proteins have two basic organizational features, both of which involve systematic *hydrogen bonding* between the atoms that are in the peptide bond in the *backbone* of the polymer. One of these features, called the *α-helix*, constrains the backbone of the polymer to form a helical, spring-like structure, while the other, called a *β-sheet*, connects sections of the protein backbone through hydrogen bonds in a flat sheet-like motif. Examples of each of these structural features will be apparent in the various metallo-drug protein interactions presented and discussed in subsequent sections of the text.

2.2.2   Metal binding sites on proteins

Of the 20 common amino acids found in proteins (Figure 2.1), only seven, aspartic and glutamic acids, histidine, lysine, methionine, cysteine and tyrosine, have donor atoms in their side chains that, from the standpoint of coordination chemistry, are potential targets for metallo-drugs. The first two, aspartic and glutamic acid, have carboxylic acid side chains, which can ionize through the loss of a proton to give the

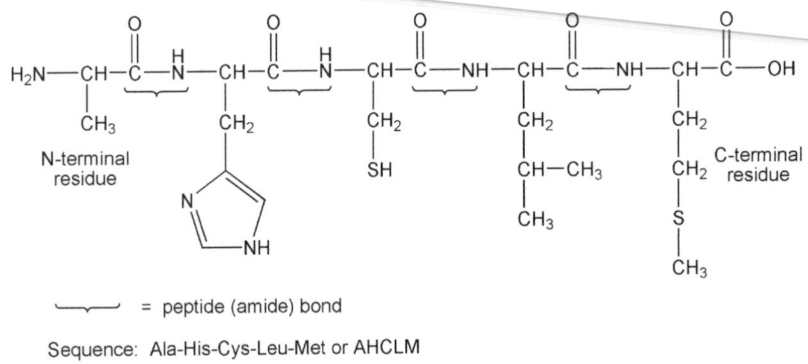

= peptide (amide) bond

Sequence: Ala-His-Cys-Leu-Met or AHCLM

Figure 2.2 *Connectivity in a peptide or protein, showing the peptide (amide) bond and the sequence, using three- and one-letter amino acid codes written from the N-terminal to the C-terminal residue of the peptide/polymer, which is convention*

corresponding carboxylate ion, shown in Figure 2.3a. The reaction for this deprotonation, which is important for allowing a metal ion to bind to the site, can be written as (2.1), with equilibrium constant, K_a, given by (2.2). In the case of both aspartic and glutamic acids, which are weak acids, the equilibrium constant K_a is $\sim 10^{-4}\,M$, showing that the equilibrium in (2.1) lies to the left in favor of the un-ionized acid.

$$R-CO_2H = R-CO_2^- + H^+ \tag{2.1}$$

$$K_a = \frac{[R-CO_2^-][H^+]}{[R-CO_2H]} = 10^{-4}\,M \tag{2.2}$$

Since equilibrium constants for weakly ionizing substances span many orders of magnitude, it is convention to report the pK_a, which is the negative logarithm of the equilibrium constant, or (2.3), which in this case is $pK_a = 4$. As was pointed out in Chapter 1, in working which equations like (2.3), which requires K_a to be a *unitless number*, the unit of M on K_a in (2.2) is simply eliminated when evaluating the logarithm of the number.

$$pK_a = -\log K_a \tag{2.3}$$

An important aspect of the ability of certain groups located in the side chains of proteins to act as donor atoms to a metallo-drug is the state of ionization of the group at physiological pH; that is, pH ~ 7. If the carboxylic acid is *deprotonated* at neutral pH, it would be easy to see that the carboxylate ion (an anion) would be a good binding ligand toward a metal ion (a cation) in a metallo-drug. The relationship that relates the state of ionization of an acid, or any site involved in a proton equilibrium, to the pH and the pK_a of the site is the *Henderson–Hasselbalch equation*, (2.4). In this equation, A^- is the ionized, deprotonated form of the acid, in this case R-CO$_2^-$, sometimes called the *conjugate base*, and HA is the protonated form of the acid, in this case, R-CO$_2$H, or the undissociated acid.

$$pH = pK_a + \log\frac{[A^-]}{[HA]} \tag{2.4}$$

From (2.4) it is clear that if $pK_a = pH = 4$, the logarithm part of (2.4) must be zero, showing that $[A^-] = [HA]$; that is, $log\ 1 = 0$. Simply stated, this means that at pH 4, half of the carboxylic acid functional groups in solution are protonated and half are deprotonated. Cleary, if the pH is raised to pH $= 7$, the approximate pH of biological fluids, the logarithm part of (2.4) is 3 and the *ratio* of deprotonated to protonated forms is 10^3, showing that most of the molecules in solution are in the *deprotonated*, carboxylate form and thus would be good donors toward a metal ion. As is evident from Figure 2.3, a metal ion interacting with the carboxylate group could be attached to one of the oxygen atoms, forming a *monodentate complex*, or it could be bound by *both* oxygens, producing a four-membered *chelate ring*. In the latter case the carboxylate ligand acts in a *bidentate* fashion to bind the metal ion. Because of resonance, both oxygens in the bidentate structure (Figure 2.3a) are identical and the negative charge is equally distributed between both atoms. Since carboxylate is a hard base, the side chains of aspartic and glutamic acids on a protein prefer to bind to metal ions that are hard acids; see Table 1.7.

Shown in Figure 2.3b is the five-membered heterocyclic compound *imidazole*, the functional group of the amino acid histidine. This group is a heterocyclic base that has two nitrogen atoms, one of which can be protonated to form the *imidazolium* ion. Since the β-methylene carbon atom of histidine is joined to the γ-carbon atom of imidazole, the two nitrogen atoms in the ring are in different locations relative to the γ-carbon. One is at the δ position of the ring, labeled N^{δ}, and the second is at the ε position, labeled N^{ε}. Since the ring is aromatic (counting two electrons on nitrogen and two carbon-carbon double bonds the ring has six π-electrons), the positive charge on the imidazolium ion can be placed on either of the two nitrogen atoms due to

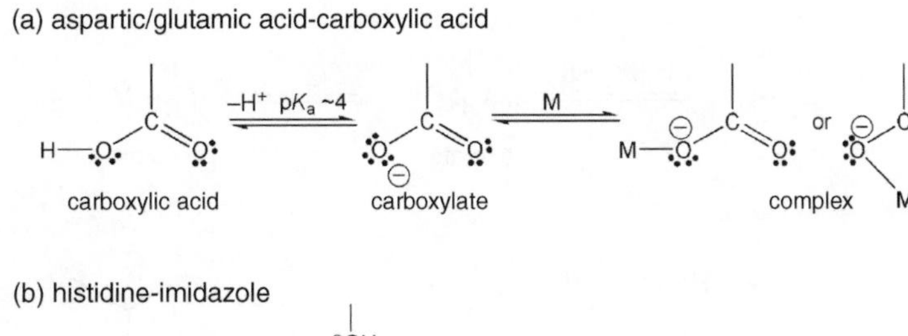

Figure 2.3   *Complexation of donor atoms located in the side chains of various amino acids to a metal ion (M) to form a 1:1 complex. The pK$_a$ values given are the approximate values for the free amino acid. (a) The carboxylic acid group of aspartic or glutamic acid, showing mono- and bidentate coordination to the metal ion. Due to resonance, the negative charge in the bidentate complex can be exchanged between the two oxygen atoms. (b) The imidazole group of histidine. The imidazolium ion of histidine has two resonance forms and the imidazole two tautomeric forms. A metal ion can bind to either of the tautomeric forms of imidazole, to form a M-N$^{\varepsilon}$ or M-N$^{\delta}$ complex. (c) The primary amine group of lysine. (d) The thioether group of methionine. (e) The thiol group of cysteine. (f) The phenol group of tyrosine*

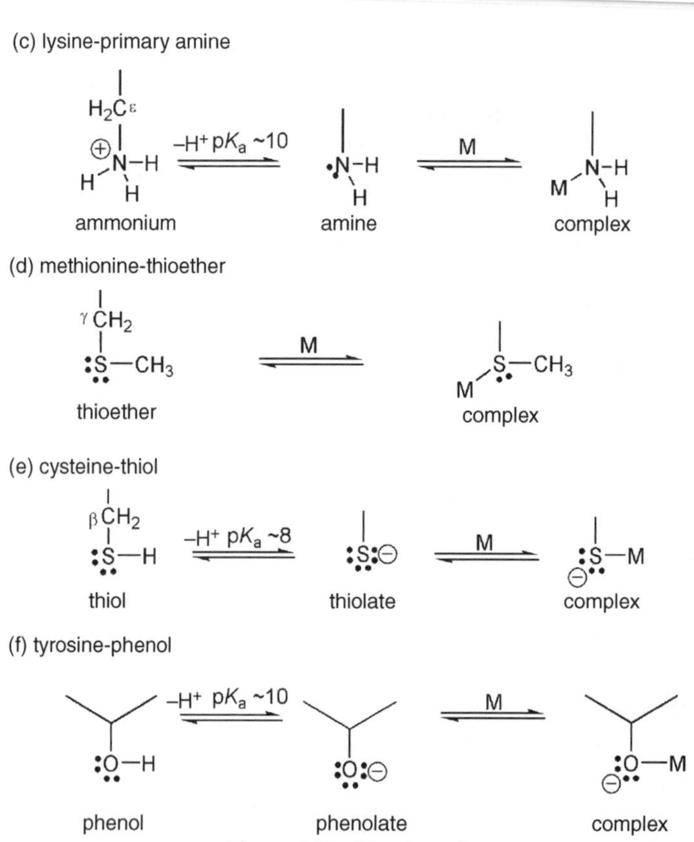

(c) lysine-primary amine

$$\text{H}_2\overset{|}{\text{C}}\varepsilon$$

ammonium $\xrightarrow{-\text{H}^+ \, pK_a \sim 10}$ amine $\xrightarrow{\text{M}}$ complex

(d) methionine-thioether

$$\gamma \overset{|}{\text{C}}\text{H}_2$$

:S—CH$_3$

thioether $\xrightarrow{\text{M}}$ complex

(e) cysteine-thiol

$$\beta \overset{|}{\text{C}}\text{H}_2$$

:S—H   thiol $\xrightarrow{-\text{H}^+ \, pK_a \sim 8}$:S:$^\ominus$ thiolate $\xrightarrow{\text{M}}$:S—M   complex

(f) tyrosine-phenol

:O—H   phenol $\xrightarrow{-\text{H}^+ \, pK_a \sim 10}$:O:$^\ominus$ phenolate $\xrightarrow{\text{M}}$:O—M   complex

Figure 2.3   *(Continued)*

resonance (Figure 2.3b). When imidazolium is *deprotonated* to form imidazole, a lone pair of electrons on nitrogen is exposed, which is a good binding site for metal ions, but as Figure 2.3b also shows, the deprotonated form has two *tautomers*, allowing placement of the lone pair on either nitrogen. As is evident from the figure, the pK_a for the deprotonation is $\sim$6, which, from the Henderson–Hasselbalch equation (2.4), means that at neutral pH most of the molecules in solution are in the *deprotonated* form and available for binding to a metal ion. While it may seem reasonable that only that fraction of the molecules that have a free lone pair would react with a metal ion, this is usually not the case. If the metal ion forms a very stable complex with imidazole, and if the metal ion is in excess, the system will be driven to the right in favor of completely-bound imidazole with the release of a proton to solution. Where the metal ion binds, N^δ or N^ε is mainly determined by the local environment around the imidazole in the protein, and binding to either site is known. Imidazole is considered an intermediate base and as such prefers to bind to intermediate-type metal ions (Table 1.7).

The functional group of the amino acid lysine is a primary amine group located on the ε-carbon atom in the side chain (Figure 2.3c). Since the pK_a for deprotonation of this group is $\sim$10, only $\sim$0.1% of the molecules in solution at pH 7 exist in the fully deprotonated form. However, if the binding constant of the metal ion for the primary amine is high, as with imidazole, reaction with excess metal ion will drive the equilibrium to the right in favor of the metal complex releasing a proton to solution. Since a primary amine is a hard base, it prefers to bind to metal ions that are hard acids.

The amino acid methionine, which is present in many proteins and enzymes, contains the thioether functional group (Figure 2.3d). Unlike the previously discussed cases, the lone pairs of electrons on the thioether group are not very basic toward a proton, so this group is not easily protonated and its metal complexation chemistry is *independent* of pH. Being a soft base, the thioether sulfur atom of methionine forms stable complexes with metal ions that are soft acids, and since there are two lone pairs, it is possible for the thioether group to simultaneously bind to *two* metal ions.

An important metal-binding site on proteins is the thiol functional group of cysteine (Figure 2.3e). Since the pK_a for the deprotonation of this group is slightly above neutral pH, $pK_a \sim 8$, the thiolate ion, an anion, can react with the metal ion of a metallo-drug to form a complex, which, if the stability is high, will drive the system in favor of the complex with the release of a proton to solution. However, it is also possible that the *protonated thiol* also serves as the nucleophile in the reaction, and that the intermediate with the metal ion *and* the proton bound to the sulfur atom will decompose to the thiolate complex with the loss of a proton in a *metal-assisted deprotonation* process. Because the thiolate ion has *three lone pairs*, it could bind up to three metal ions simultaneously, but only the mono-adduct, or the first-formed complex, is shown in (Figure 2.3e). Thiolate is a soft base which forms its most stable complexes with metal ions that are soft acids.

The phenol residue is in the side chain of the amino acid tyrosine (Figure 2.3f). Although the pK_a for deprotonation of this group is ~ 10, and thus very little *phenolate* is present at neutral pH, interaction with a metal ion through a lone pair of electrons of the protonated phenol, followed by a metal-assisted deprotonation, or via direct attack of the phenolate ion, would be expected to drive the equilibrium in favor of the complex. Phenolate, which is a hard base, would prefer to react with metal ions that are classed as hard acids. As with thiolate, phenolate, because it has three lone pairs of electrons, can simultaneously bind three metal ions.

While it is possible for some side chains other than those discussed to bind to metallo-drugs, all are less ideal metal-binding sites than those given in Figure 2.3. For example, the side chains of the amino acids glutamine and asparagine each contain a primary amide nitrogen atom with a lone pair of electrons that could potentially be used in binding to a metal ion (Figure 2.1). However, since the lone pair is extensively involved in resonance stabilization of the amide linkage (discussed below), it is not readily available for binding to a metal ion. At neutral pH the *guanidine* group of the amino acid arginine is protonated – that is, the positively-charged *guanidinum ion* – and on the basis of simple electrostatic arguments it would not be expected to bind to a metallo-drug containing a metal cation. Although serine and threonine have alcohol functional groups, the pK_as for proton removal from these groups are many pK_a units higher than the phenol of tyrosine, making them less likely targets for metallo-drugs. Tryptophan has the heterocyclic indole ring as a side chain, which contains a nitrogen atom with a lone pair of electrons. However, since indole is aromatic, and the lone pair on nitrogen is directly involved in the aromaticity of the indole ring, it is not available for binding to a metal ion.

A final point to make concerns the ability of an amide bond in the backbone of a protein to provide donor atoms toward a metal ion. As is shown in Figure 2.4, the amide nitrogen atom has a lone pair of electrons that could potentially be a metal-binding site but, as is also shown in the figure, it is involved in resonance stabilization of the amide linkage and as such not available for metal binding. This resonance, which constrains the amide linkage to a rigid planar arrangement (the *trans* arrangement of NH relative to CO is shown in Figure 2.4) is a critical structural feature of the backbone of a protein. However, as is indicated in Figure 2.4, it is possible to *deprotonate* the amide nitrogen atom to form an anion at the site that would be a good donor to a metal ion, but the pK_a for deprotonation of the amide nitrogen is very high, ~ 15, which is ~ 8 pK_a units more basic than neutral pH [1]. It has been found that if the amide bond is 'forced' near a metal ion, as would occur if there were good ligating groups on either side of the amide that bound to the metal ion, and if it were part of a chelate ring, the amide deprotonates with surprising ease [2]. This metal-assisted deprotonation depends on the nature of the flanking ligands and the metal ion but pK_a values for the deprotonation of the amide hydrogen can be in the physiological range, showing that under certain circumstances the deprotonated amide nitrogen atom in the protein backbone can be donor atom to a metal ion.

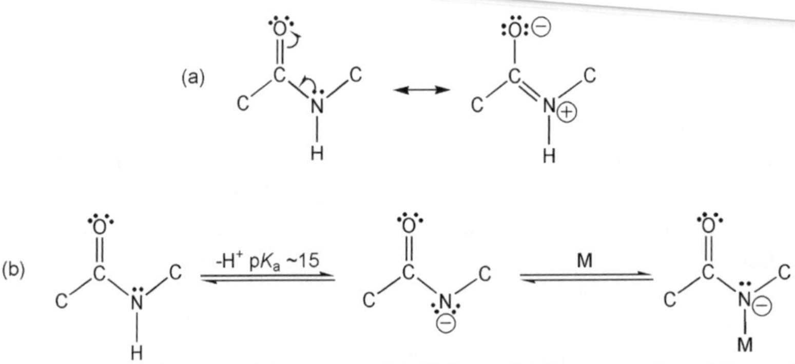

Figure 2.4   *(a) Resonance stabilization of the trans amide linkage. (b) Deprotonation of the amide nitrogen atom and binding of a metal ion to the deprotonated amide. All atoms shown lie in the same plane*

Box 2.1 Cellular guardians

Glutathione. Certain molecules in the cell are natural protecting agents against attack by metallo-drugs. For example, the simple tripeptide glutathione (γ-glutamylcysteinylglycine, GSH), which is found in the cell in relatively high concentrations, 1–5 mM, is mainly responsible for maintaining the reduction and oxidation, *redox*, state of the cell. As is shown in Figure 2.5a, GSH is an unusual peptide in that the carboxylic acid group on the γ-carbon atom of the side chain of glutamic acid is connected through an amide bond to the amino terminus of the dipeptide cysteinylglycine. Since the pK_a of proton removal from the thiol of GSH is 8.75, this group is largely protonated at neutral pH [3]. In the cell, glutathione exists in both its reduced and oxidized forms (Figure 2.5b) and in a healthy cell the ratio GSH/GSSG is $\sim$500, showing that the cytosol has a reducing environment; that is, the equilibrium lies to the left. When the cell is 'stressed', which can happen when it is attacked by toxic substances such as drugs, hydrogen peroxide (H_2O_2), hydroxyl radical (HO$^\bullet$) and superoxide ion ($O_2^\bullet$), referred to as *reactive oxygen species, ROS*, are produced, which can chemically damage biomolecules in the cell. Since GSH reacts with and inactivates ROS, and in the process is converted to the oxidized form, GSSG, glutathione serves as a protecting agent against foreign substances, generally called *xenobiotics*, that can harm the cell. Since reactions of this type change the normal GSH/GSSG ratio in the cell, the proper ratio of the forms can be restored by reducing GSSG to GSH, which is done by the enzyme *glutathione reductase*. The reducing agent in the reaction is the reduced form of *nicotinamide adenine dinucleotide, NADH*, which, with a proton source and glutathione reductase, supplies the two electrons and two protons required for the reduction of GSSG to two molecules of GSH (Figure 2.5b). Although GSH is mainly important for protecting the cell against oxidative stress, under certain conditions it can actually contribute to stress. For example, in the presence of Cu^{+2} and Fe^{+2}, GSH can be oxidized to the disulfide with the concomitant generation of ROS [4].

An interesting reaction carried out by GSH is a disulfide exchange reaction (Figure 2.5c). Often proteins have adjacent exposed cysteine thiols that need to be maintained in the thiol form for proper functioning of the protein [5]. If these were oxidized by reaction with cellular oxygen, ROS or other agents to disulfides, the function of the protein could be seriously impaired. As is shown in Figure 2.5c, two molecules of the reduced form, GSH, can convert the protein disulfide to an intermediate mixed disulfide of the type protein-SSG, which involves the loss of two electrons and two protons from two glutathione molecules. However, since there are two thiols and one disulfide on the reactant *and* product sides of the equation, there is no redox change for the reaction and the net effect of the reaction is to keep the thiol groups on the protein in their reduced state. Since the process changes the GSH/GSSG ratio, the proper ratio is again restored by the above-described enzymatic mechanism.

Many metallo-drugs contain soft acids, which can react with the thiol of GSH (Figure 2.3e) to produce adducts that are nontoxic to the cell and are often expelled from the cell by *efflux* mechanisms [6,7]. Since this cellular protection mechanism consumes GSH, the drug alters the balance between GSH and GSSG, which again needs to be restored by the cellular system. There is also evidence that continued exposure to the toxin (for example, platinum anticancer drugs) leads to an increase in the concentration of GSH, which makes the cell more *resistant* to the toxic effects of the agent [6,7].

Metallothionein. Metallothionein (MT), which is the collective name for a large family of low-molecular-weight proteins, is present in high concentration, $\sim$2 mM, in mammalian cells [8]. The protein consists of a single polypeptide chain 61–62 amino acids long, of which 20 residues, an unusually high fraction, are the amino acid cysteine. Although still being debated, the main functions of MT appear to be the storage of metal ions that are important for various metallo-enzymes and proteins in the cell, for example Zn^{+2} and $Cu^{+1/+2}$, binding of toxic metal ions that may enter the cell, for example Pt^{+2}, Cd^{+2}, Pb^{+2}, protection against ROS and maintenance of the redox balance in the cell.

Metallothionein has two metal-binding regions, which contain the 20 cysteine residues of the protein. The α-domain, which is farthest away from the amino terminus, contains 11 Cys, and the β-domain, which is closest to the amino terminus, has 9 Cys. When fully loaded with Zn^{+2} to form the Zn_7MT complex, the α-domain has four bound Zn^{+2} ions and the β-domain three Zn^{+2} ions. As is shown in Figure 2.6, the coordination geometries around all of the zinc ions are tetrahedral, with the 20-thiolate ions either *bridging* or *nonbridging* to the Zn^{+2} ions. Since not all of the MT molecules or domains on the molecules in the cell are fully loaded with metal ions, the protein can bind additional metal ions, which may enter the cell. If these ions are soft acids delivered by a metallo-drug, MT can serve as a cellular detoxifying agent by tightly binding the metal ions and not allowing them to bind to other molecules in the cell. As with GHS, cells that have been continually exposed to a toxic soft acid, Pt^{+2} or Cd^{+2}, elevate their levels of MT, which makes them more resistant to the toxin.

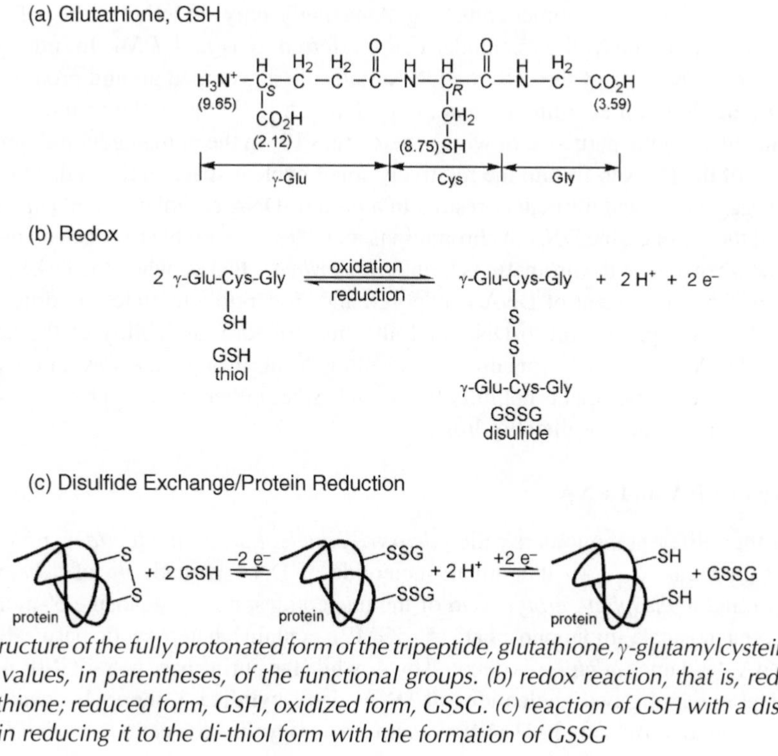

Figure 2.5   *(a) structure of the fully protonated form of the tripeptide, glutathione, γ-glutamylcysteinylglycine, GSH, showing the pK$_a$ values, in parentheses, of the functional groups. (b) redox reaction, that is, reduction/oxidation reaction, of glutathione; reduced form, GSH, oxidized form, GSSG. (c) reaction of GSH with a disulfide functional group on a protein reducing it to the di-thiol form with the formation of GSSG*

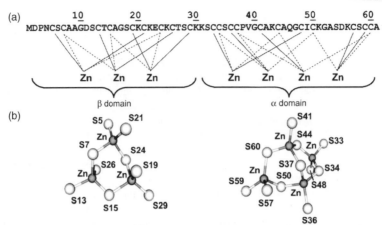

Figure 2.6 *(a) Amino acid sequence and Zn^{+2}-S-Cys connectivities of zinc human metallothionein-2, Zn$_7$MT-2. (b) Structures of the Zn$_4$(Cys)$_{11}$ cluster (α-domain) and the Zn$_3$(Cys)$_9$ cluster (β-domain) in Zn$_7$MT-2. Reprinted with permission from M. Knipp et al., Reaction of Zn7Metallothionein with cis- and trans-[Pt(N-donor)$_2$Cl$_2$] Anticancer Complexes: trans-PtII Complexes Retain their N-Donor Ligands, J. Med. Chem. 50, 4075–4086. Copyright 2007 American Chemical Society*

2.3 DNA as a target for metallo-drugs

An important biomolecule that can be targeted by metallo-drugs is *deoxyribonucleic acid* or *DNA*. This molecule, which contains all of the genetic information for all of the cells that make up the body, is a double-stranded polymer that is 3×10^9 monomer units long. Amazingly, only a fraction of the polymer, ∼5%, codes for the proteins found in the body, the remainder being referred to as *junk DNA*. In order to fit a very long polymer, ∼1 m, into the nucleus of the cell, the DNA molecule is wrapped around proteins called *histones*, and the resulting particle is called a *nucleosome*, one of which is shown in the chapter on platinum drugs (Figure 4.26). The nucleosome particles, of which there are ∼10^7 in the human genome, serve as organizing units that allow all of the DNA to fit into the relatively small nuclear space in the cell. This requires several higher levels of organization and ultimately results in a protein-DNA complex called *chromatin*. Due to the blocking effects of the histones, the DNA in chromatin is much less vulnerable to attack by metallo-drugs than purified DNA, but when a gene in chromatin is being *transcribed* – that is, when the DNA sequence is being read to make a protein – a segment of DNA is exposed and susceptible to attack by drugs and other small chemical agents. If these agents bind to DNA and alter its structure, the ability of the cellular system to accurately read the DNA code could be impaired, which often results in either the prevention of the synthesis of a protein or the production of one that is not fully functional. Since either result can have a disastrous effect on the cell, DNA is a prime target for metallo-drugs.

2.3.1 Structure of DNA and RNA

DNA consists of four different monomers called *deoxyribonucleotides* or *nucleotides* (nts), the structures of which are shown in Figure 2.7. An individual nucleotide of DNA is made up of a *heterocyclic base*, a *deoxyribose sugar* and a *phosphate group*. Two of the nucleotides, deoxyadenosine 5′-monophosphate (5′-dAMP) and deoxyguanosine 5′-monophosphate (5′-dGMP), contain a heterocyclic base called a *purine*. The purine base in 5′-dAMP, named *adenine*, is denoted by A, while the purine base in 5′-dGMP, named *guanine*, is denoted by G. The remaining two nucleotides of DNA, deoxycytidine 5′-monophosphate (5′-dCMP) and deoxythymidine 5′-monophosphate (5′-dTMP), contain a heterocyclic base called a *pyrimidine*, with the base

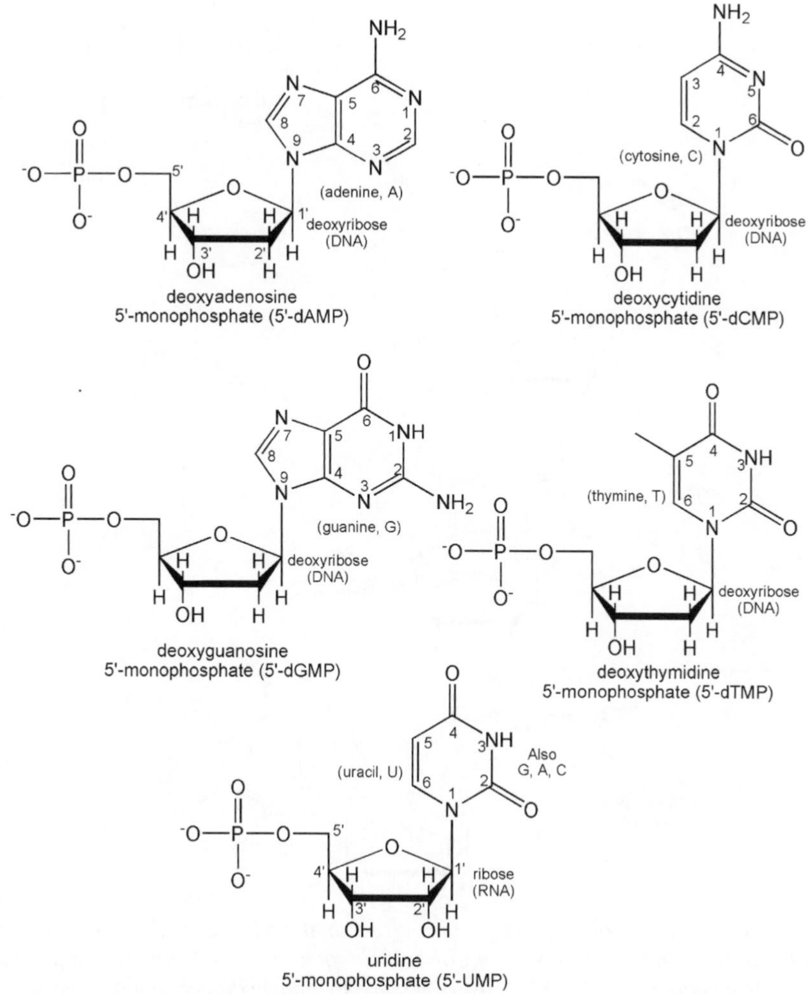

Figure 2.7 *The 5′-phosphate nucleotides of DNA and RNA. For simplicity, only 5′-UMP of RNA is shown*

in 5′-dCMP named *cytosine*, C, and the base in 5′-dTMP named *thymine*, T. As is shown in Figure 2.7, the C-1′ carbon atom of deoxyribose is connected to either N-9 of the purine or N-1 of the pyrimidine, with the phosphate group being attached to the C-5′ methylene carbon (CH_2) of deoxyribose.

The nucleotides of *ribonucleic acid, RNA*, are similar in structure to those of DNA, except the *ribose* sugar of a ribonucleotide has a hydroxyl group on the 2′ position of the ring, while deoxyribose has a hydrogen on this position (Figure 2.7). Three of the heterocyclic bases of RNA, G, C and A, are identical to those in DNA, while the fourth, *uracil* (U), which is a pyrimidine, is identical to the T of DNA but is missing the methyl group on the 5 position of the ring. The names of the 5′-phosphate ribonucleotides of RNA are adenosine 5′-monophosphate, guanosine 5′-monophosphate, cytidine 5′-monophosphate and uridine 5′-monophosphate.

In both DNA and RNA, the nucleotides are connected together by condensing the 5′-phosphate group of one nucleotide to the 3′-hydroxyl (alcohol) of another nucleotide, with the elimination of a water molecule, to form a *phosphodiester bond* (Figure 2.8a). Since the pK_a for proton removal from the oxygen on the phosphodiester linkage is ~1, at physiological pH each nucleotide in DNA and RNA has a negative charge localized on the

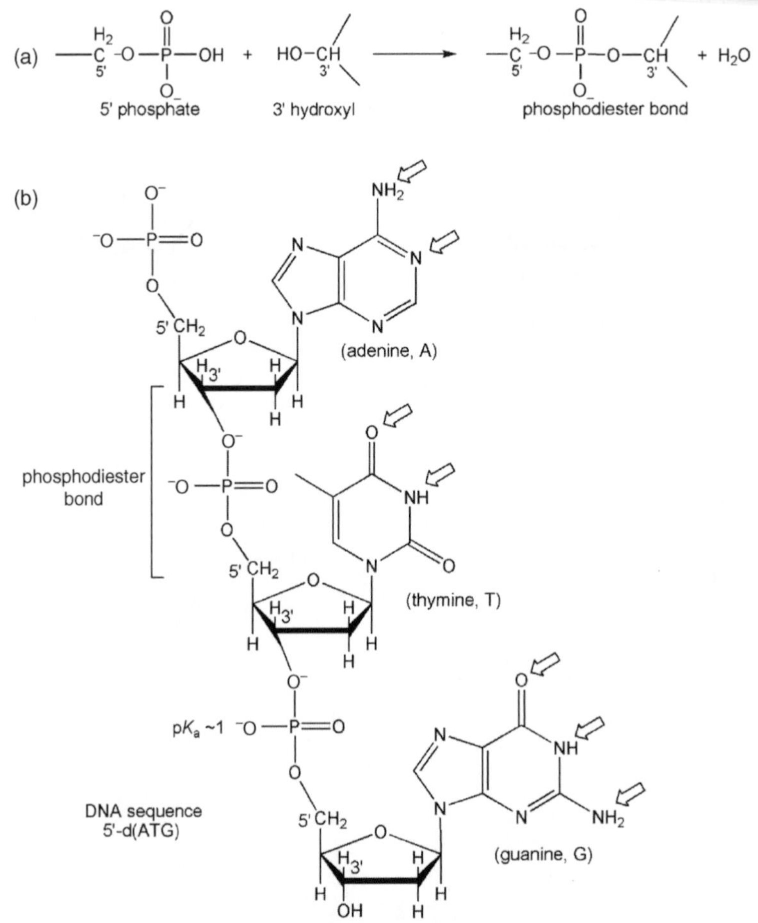

Figure 2.8 (a) Reaction of a 5'-monophosphate group with a 3'-hydroxyl group with the loss of a water molecule to form a phosphodiester bond in DNA or RNA. (b) Connectivity of a strand of DNA. The sequence, in the 5' to 3' direction, which is convention, is 5'-d(ATG). Also shown are the hydrogen bond sites (arrows) between a heterocyclic base and its compliment in a Watson-Crick double helix

phosphate linkage, making the polymers very water soluble (Figure 2.8b). The 'end' of DNA and RNA has a terminal phosphate group. Since the pK_{a1} and pK_{a2} values for successive deprotonation of the terminal phosphate are ∼1 and ∼6 respectively, this group in *dinegatively* charged at neutral pH. The presence of the negative charges on the phosphates requires that there be compensating positive charges for each nucleotide, which could be a simple ion like Na^+, Mg^{2+} (one per two nt) and so on, or in the case of genomic DNA in the nucleus of the cell, cationic groups, for example lysine, on the histone proteins that are bound to DNA.

The sequence of the nucleotides in DNA and RNA is written from the 5'-end toward the 3'-end of the polymer; an example of a trinucleotide DNA sequence, 5'-d(ATG), is shown in Figure 2.8b. In the case of DNA, the letter 'd' indicates that the sugar is deoxyribose not *ribose*, as occurs in RNA. However, if DNA is being exclusively addressed, the sequence is sometimes abbreviated using only the single-letter codes for the deoxyribonucleotides, writing from the 5'-end of the polymer, which for the sequence in Figure 2.8b is simply ATG.

Genomic DNA in the cell is *double-stranded*, where one polymer chain is connected to a second complimentary chain via specific hydrogen bonds on the heterocyclic bases. These hydrogen bond sites are

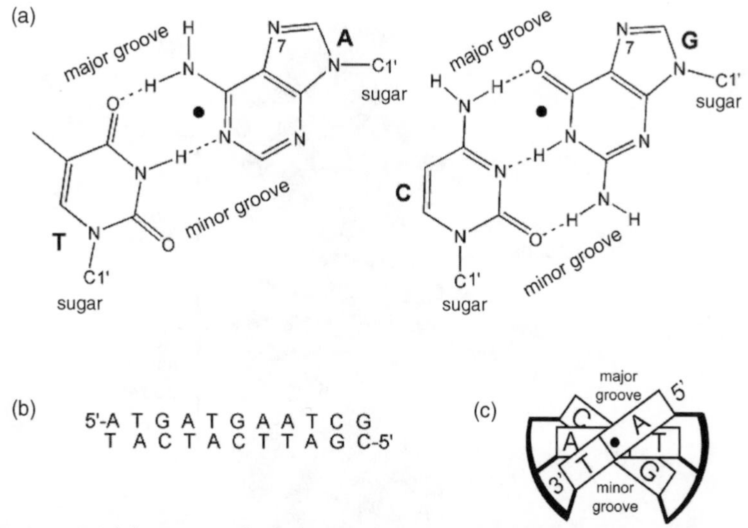

Figure 2.9   *(a) Watson–Crick base pairs (bp) of DNA showing adenine (A) paired with thymine (T) through two hydrogen bonds (dotted lines) and guanine (G) paired with cytosine (C) through three hydrogen bonds, with the approximate location of the helix axis indicated for B-DNA (black filled circle). (b) Sequence of a double-stranded DNA showing the antiparallel nature of the complimentary Watson–Crick strands. (c) View along the helix axis (filled black circle) of the double-stranded DNA in sequence (b) from its 5'-A end, showing the right-hand helical nature of the base pairs (elongated rectangles) and the locations of the major and minor grooves in B-DNA. The sugar-phosphate backbone of each strand is shown as a curved black line*

indicated by arrows in Figure 2.8b and in a more detailed fashion in Figure 2.9a. Adenine forms *two* hydrogen bonds with its complimentary base on the opposite strand, which is thymine, while guanine forms *three* hydrogen bonds with its complimentary base, cytosine. This pairing, referred to as Watson–Crick base pairing, gives rise to A-T and G-C *base pairs* (bp) in double-helical DNA. A typical way of denoting how the two strands are aligned in double-stranded, *duplex* DNA is shown in Figure 2.9b, which shows that the two hydrogen-bonded strands are *antiparallel* to each other, meaning that the 5' to 3' directions for both are oppositely aligned.

Double-stranded DNA has a helical structure, which is the origin of the term *double helix*. An observer viewing the DNA duplex in Figure 2.9b from the A-T base pair at the end of the sequence and looking directly down the *helix axis* would see that the base pairs of DNA have a 'spiral staircase' appearance and that the sense of the spiral is *right-hand* (Figure 2.9c). This characteristic feature of double-stranded DNA is due to conformations in the sugar-phosphate backbone and the fact that the base pairs are 'stacked' and in direct contact with one another; that is, the distance between the stacked base pairs is 0.34 nm, which is the *van der Waals* 'thickness' of the bases in the base pair. For the most stable form of DNA, called *B-DNA*, the average angle between the base pairs, called the *twist angle*, is ~36°, which means that there are ~10 base pairs for each turn of the helix (one turn is 360°).

Not shown in Figure 2.9c is the sugar-phosphate backbone, which has considerable bulk and, like the base pairs, also traces out a right-hand helix. Since the sugar-phosphate backbone extends far from the helix axis and the base pairs are approximately on the axis, the sugar-phosphate chain forms the edges of a *groove* of the DNA helix, with the base pairs (their exposed 'edges') forming the *floor* of the groove. Careful inspection of the geometries of both base pairs shows that the junction of the sugar to all four bases is such that if the base pairs are stacked in the manner shown in Figure 2.9c, the sugar-phosphate backbone actually forms two grooves, each with a different width. These grooves, called the *major* and the *minor groove*, are clearly evident in the structure of B-DNA shown in Figure 2.10.

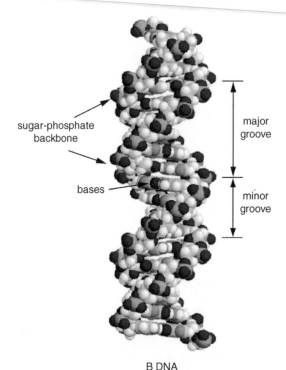

sugar-phosphate
backbone

major
groove

bases

minor
groove

B DNA

Figure 2.10 *Structure of B-DNA showing the major and minor grooves, the sugar-phosphate backbone and the edges of the stacked heterocyclic bases*

2.3.2   Metal binding sites on DNA

The atoms on the bases that can serve as nucleophiles toward metallo-drugs are located on the exposed edges of the bases pairs (Figure 2.9a), which form the floor of the grooves of DNA. The most important binding sites are N-7 of adenine and guanine, which are both located in the major groove of double-helical DNA [9]. Since the pK_as for proton removal from these sites are far below physiological pH – that is, the lone pairs on the nitrogen atoms are only weakly basic toward a proton – competition between a proton and a metal ion for these sites at neutral pH is not a factor and for this reason metallo-drugs are often bound to these donor atoms. Both of these atoms are considered intermediate bases on the HSAB scale (Table 1.7), and thus they form their most stable complexes with metal ions that are intermediate acids. Also available for binding metal ions are the negatively-charged oxygen atoms of the phosphodiester linkage, which are part of the backbone of DNA. Since this oxygen is a hard base, metal ions that are hard acids often bind to this site.

If the metal ion has more than one coordination site capable of interacting with DNA, the ion can bridge two bases on the same strand to produce an *intrastrand crosslink*. Alternatively, it could react with sites on both strands, effectively connecting the strands together, to form an *interstrand crosslink*, or it could link donor atoms in the side chains on a protein to DNA in a *protein–DNA crosslink*. These possibilities are shown in Figure 2.11a.

One of the interesting features of double-stranded DNA is that small molecules that are flat in shape (usually containing aromatic residues) and are hydrophobic can insert themselves between the base pairs of DNA through a process called *intercalation* (Figure 2.11b). However, in order to do so, the spacing between the base pairs of DNA must increase to accept the 'thickness' of the intercalating molecule, which increases the length

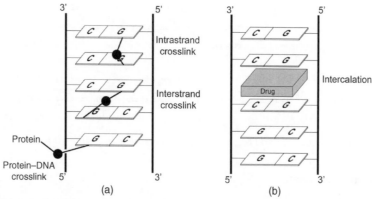

Figure 2.11   (a) Diagram showing some of the possible adducts that can form between a metal ion (black circle) and the bases of double-stranded DNA. (b) Intercalation of a drug molecule between the base pairs of DNA

of the DNA helix and usually leads to an increase in the *melting temperature* of the DNA; that is, the temperature at which the two Watson–Crick strands separate through disruption of the hydrogen bonding between the strands. The thermodynamic driving force for this interaction is the tendency for hydrophobic portions of drug molecules or other small structures to associate with the hydrophobic base pairs of DNA, which is a way to 'hide' the hydrophobic region of the drug molecule from water, which is very polar. Thus, aside from simply binding to a donor atom on DNA through a ligand displacement reaction involving the metal center, certain metal complexes that contain flat aromatic residues can bind to DNA by intercalation, which also affects the template function of DNA in gene expression.

2.4   Reaction of metal complexes in the biological milieu

Unlike purely organic-based drugs, metal complexes that are used for treating and diagnosing disease are exposed to different physiological environments in the body that could cause changes in their chemical composition as they travel from the site of administration to target molecules in the cell. Since some of the agents in blood and other fluids are common, well-known metal-binding ligands, a brief overview of the effects of the agents on the speciation of metal complexes in the biological milieu will be presented.

2.4.1   Reactions with chloride

The chloride concentration in blood is $\sim$104 mM. If a metal complex introduced into the body has exchangeable chloride ligands and the complex was preequilibrated in a medium that did not contain 104 mM chloride ion, passage of the compound through the blood will result in a shift of the chloro–aqua equilibrium. If the compound is stored in normal saline solution, $\sim$150 mM NaCl, which is common practice, introduction of the compound into the blood will shift the equilibrium in favor of aqua species. Since the concentration of chloride in the cytoplasm is 4–20 mM, even more extreme shifts in the equilibrium in favor of aqua complexes would be expected inside the cell. While the amount of equilibrium shift depends on the substitution kinetics of the metal complex and the length of time it remains in the different chloride environments, other substances in the body usually react with the compound, upsetting the simple chloro–aqua equilibrium. Since the drug concentration is usually in the low micromolar ($10^{-6} M$) to nanomolar ($10^{-9} M$) range, the reaction of the metal complex with water or chloride is pseudo first order.

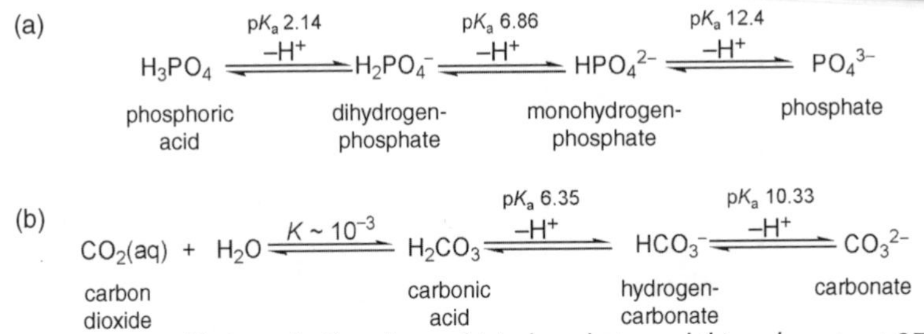

(a)

$$H_3PO_4 \xrightleftharpoons[]{\substack{pK_a\ 2.14 \\ -H^+}} H_2PO_4^- \xrightleftharpoons[]{\substack{pK_a\ 6.86 \\ -H^+}} HPO_4^{2-} \xrightleftharpoons[]{\substack{pK_a\ 12.4 \\ -H^+}} PO_4^{3-}$$

phosphoric   dihydrogen-   monohydrogen-   phosphate
acid   phosphate   phosphate

(b)

$$CO_2(aq) + H_2O \xrightleftharpoons[]{K \sim 10^{-3}} H_2CO_3 \xrightleftharpoons[]{\substack{pK_a\ 6.35 \\ -H^+}} HCO_3^- \xrightleftharpoons[]{\substack{pK_a\ 10.33 \\ -H^+}} CO_3^{2-}$$

carbon   carbonic   hydrogen-   carbonate
dioxide   acid   carbonate

Figure 2.12 *Equilibria and pK$_a$ values of (a) phosphate and (b) carbonate at 37°C*

2.4.2 Reactions with phosphate

In addition to chloride, fluids in the body contain phosphate, which acts as a buffer to stabilize pH. The term 'phosphate' or 'phosphate buffer' collectively refers to all of the chemical species shown in Figure 2.12a and should not be confused with one of the species, phosphate ion, which is PO_4^{3-}. The concentrations of phosphate in blood and the cytosol are ~5 and 80 mM respectively. At neutral pH, the main components present are dihydrogenphosphate, $H_2PO_4^-$, and monohydrogenphosphate, HPO_4^{2-}, with the latter, from (2.4), being slightly more abundant than H_2PO_4. Since the oxygen atoms on each species are considered hard bases (all of the oxygens are equivalent by tautomerism and resonance), the most stable phosphate complexes are formed with metal ions that are hard acids. Because of the high phosphate concentration in the cytosol, metal complexes inside the cell would be expected to be more susceptible to reaction with phosphate than complexes in the blood or the interstitial fluid that surrounds the cell. Both $H_2PO_4^-$ and HPO_4^{2-} react with metal ions to form monodentate or bidentate (chelate) complexes, which, depending on the pH, can be deprotonated to give coordinated bidentate (four-membered chelate ring) *phosphato* complexes with bound PO_4^{3-}.

2.4.3 Reactions with carbonate

Carbonate is a major component in blood, ~24 mM, and the cytosol, ~12 mM, where 'carbonate' or 'carbonate buffer' collectively refers to all of the species in Figure 2.12b. The interesting aspect of carbonate buffer is that one of the components that is in equilibrium with the other species is aquated carbon dioxide gas, $CO_2(aq)$. Since the equilibrium constant for the hydration of CO_2 to give carbonic acid, H_2CO_3, is small, $K \sim 10^{-3}$, there is very little of the latter in aqueous carbonate solutions [10]. The main component in carbonate buffer at neutral pH is hydrogen carbonate, HCO_3^-(~90.8%), with lesser amounts of CO_2 (~8.8%) and CO_3^{2-} (~0.3%), and only a trace amount of H_2CO_3. The kinetics of hydration of CO_2 in water to form HCO_3^- plus a proton are slow, with a half life of a few seconds. Since this would be far too slow for the many processes that require rapid interconversion of CO_2 and HCO_3^- in the body, the Zn^{+2}-containing enzyme *carbonic anhydrase* found in red blood cells and in tissue is used to catalyze the rapid hydration of CO_2 and dehydration of HCO_3^- at a very fast rate. Hydrogencarbonate and carbonate ion are hard bases and as such they would form stable complexes with metal ions that are hard acids.

Hydrogencarbonato and carbonato complexes can be made by two routes. As is shown in Figure 2.13a, one of the routes involves the addition of carbon dioxide gas, which is in solution, directly to the oxygen atom of a metal *hydroxo* species – that is, bound hydroxide ion – to give a hydrogencarbonato complex [10]. Since the reaction of CO_2 with a metal-hydroxo species does not involve a metal–ligand bond-breaking step, the kinetics of the reaction are relatively independent of exchange rates involving the metal center, which, as shown in Figure 1.20, vary over many orders of magnitude. For hydroxo complexes of Co^{+3}, Rh^{+3}, Ir^{+3} and other slow-exchanging complexes, the second-order rate constants for CO_2 addition to the complex to give the hydrogencarbonato complex are in the range 37–590 $M^{-1}\,s^{-1}$. The initially-formed complex can lose a

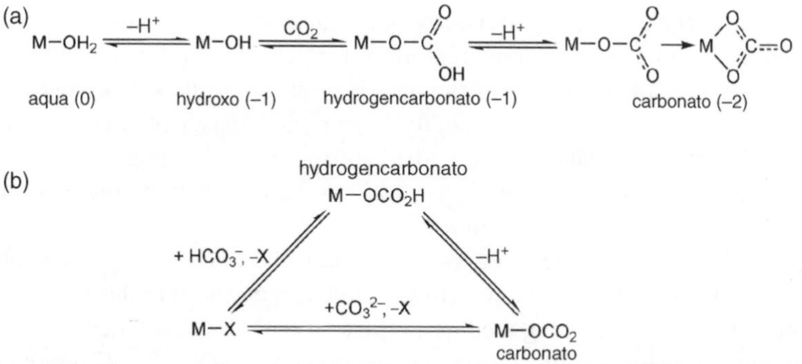

Figure 2.13 *Reaction of carbonate with metal complexes. (a) Reaction of a metal hydroxo complex with carbon dioxide to produce a hydrogencarbonato complex. The latter can deprotonate to form a monodentate carbonato complex, which, if there is a displaceable ligand in the position cis to the carbonate, can form a bidentate carbonato complex. The charge on the ligand bound to the metal ion is given in parenthesis. (b) Formation of hydrogencarbonato/carbonato complexes by a ligand substitution reaction. The group X is any displaceable monodentate ligand*

proton to give the monodentate carbonato complex. However, if there is a displaceable ligand in the coordination position *cis* to the carbonate, a second oxygen of the carbonato ligand can attack the metal center to give a bidentate carbonato complex with a four-membered chelate ring.

As is indicated in Figure 2.13a, which complex is present depends on the pH of the medium. If the basic solution containing the carbonato complex is acidified, the system moves to the left to form hydrogencarbonato complexes, some of which have been isolated [11], but these compounds are usually unstable and lose carbon dioxide to form hydroxo or, if the medium is very acidic, aqua species.

A second route for forming hydrogencarbonato/carbonato complexes is a ligand substitution reaction, in which hydrogencarbonate or carbonate acts as a nucleophile in a displacement reaction. Since this route involves metal–ligand bond breaking, substitution rates are highly dependent on the exchange kinetics of the metal ion involved.

2.5   Evaluating the pharmacological effects of agents

In addition to highlighting some of the chemical and biochemical aspects of metallo-drugs in the biological system, it is important to briefly discuss how the pharmacological effects of these agents are measured.

2.5.1   Measuring the cytotoxicity of a drug

The first approach used to evaluate a potential new drug or diagnostic agent is to measure its *cytotoxicity* in a *cell culture assay*. Human cells for studies with chemotherapeutic agents are usually grown in an incubator that provides a constant temperature, 37 °C, and a humidified atmosphere containing 5% carbon dioxide. Since the container with the cells is open and exposed to the atmosphere in the incubator, the humidified atmosphere suppresses the loss of water from the culture medium above the cells, while CO_2 in the atmosphere prevents the loss of dissolved carbon dioxide, which is present in the carbonate buffer, from escaping from the culture medium.

The composition of the culture medium, which approximates the composition of blood, varies somewhat with the cell type being studied, but media usually contain simple salts and ions, for example NaCl, Na_2HPO_4, $NaHCO_3$, the common amino acids, various vitamins, glucose, GSH and other additives. A small amount of a pH-sensitive dye, for example phenol red, is also usually present, which provides a quick visual check that the pH of the medium remains constant during the course of the incubation. Most studies also contain fetal bovine serum (FBS), which has a number of proteins important for cell growth. Since cells studies are inherently 'noisy' experiments and statistical analysis of the data is required, studies involving cytotoxic agents and adherent cells (cells that grow on a surface) are usually done in small plastic plates that contain many wells, for example 96, into which various substances to be studied are applied to the culture medium above the growing cells. For cells that do not adhere to a surface and grow while suspended in the medium, it is sometimes better to use small culture flasks for the cytotoxicity studies.

Although conditions for measuring the cytotoxicity of compounds vary widely, one approach is to make up stock solutions of the compound, expose the cells to various concentrations of the material in culture medium for 1–2 hours, remove the drug-containing medium, replace it with drug-free medium and allow the cells to grow for a period of time, sometimes called the *recovery period*, for example 24–48 hours. In a very crude way, this simulates what may be happening in chemotherapy in that a 'pulse' of agent enters the cellular environment, as would occur in intravenous administration, and is quickly eliminated by the body after the drug administration is terminated. After this period of growth, the number of live or dead cells is determined, and compared to the number of cells present in wells of the plate that have not been exposed to the toxic agent; that is, the *control* cells.

A convenient way to determine the number of live cells or the *viability* of the cell population is to expose the cells after the growth period to a solution containing a specially-designed organic tetrazolium salt, which is taken up by live cells. Once the compound is inside the cell, it is bioreduced by nicotinamide adenine dinucleotide (NADH) with the help of enzymes called hydrogenases to a water-soluble formazan dye, which has a strong absorption band in the visible region of the spectrum. Since the amount of dye produced is proportional to the number of live cells, the viability of the cell population can be obtained by measuring the absorbance of the solution in the well of the culture plate containing the cells, using a specially designed plate-reading device. When the absorbance from a drug-treated well is compared to the absorbance of a non-drug-treated well, the viability of the cell population that was exposed to drug can be easily determined.

Another method for determining the viability of a cell population is to stain the cells with the dye trypan blue, which emits blue fluorescence when it is bound to DNA. Since trypan blue cannot cross the cell membrane, it cannot enter a live cell. However, if a cell is dead and the outer membrane is disrupted, the dye can enter the cell and bind to DNA, which produces blue fluorescence. By counting the number of blue-stained cells using a specially equipped microscope called a *hemocytometer*, the number of *dead cells*, and thus the viability of the cell population, can be easily obtained.

The quantity that is typically reported in cytotoxicity studies is the IC_{50}, which is *the concentration of toxic agent needed to reduce the growth of a cell population relative to a control by 50%*. Since the concentration of the toxic agent in the culture medium and its exposure time to the cells influence the number of encounters between the agent and the cells, and thus how many cells will survive, reporting the details of the cell studies is important, especially if comparisons through IC_{50} with other studies are to be made.

2.5.2 Measuring drug uptake

In addition to measuring the toxic effects of compounds on cells, it is important to measure how much of an agent is taken up by or adheres to the cells, usually referred to as *uptake*. Such studies are done by exposing cells to an agent for a period of time and, after washing the cells, analyzing them for their drug content. The analytical techniques most often used in studies with metallo-drugs are *atomic absorption spectroscopy, AAS*, or a more sensitive technique, *inductively coupled plasma mass spectrometry, ICP-MS*. While it is not necessary to describe in detail how these techniques work, both destroy the material being analyzed, either by heating it to a very high temperature (AAS) or by bombarding it with a stream of atoms moving at a high velocity (ICP-MS). Since all chemical composition is destroyed in the sample, the *metrics* that are used in reporting the results of these studies include the *weight* of the metal found per *weight* of protein obtained from the cells (protein is assayed separately from the metal analysis), the *number or weight of metal atoms per cell*, the effective *concentration of metal* in the cell (a cell volume is assumed) and the *number of metal atoms (ions) per DNA nucleotide*, if DNA has been isolated from the cells and specifically assayed for its metal content [12–19]. In these studies, the exposure time to agent can be held constant and different compound concentrations applied, or the agent concentration can be kept constant and the exposure time varied. The latter approach has the advantage that the *rate* of the uptake is measured – that is, time is on one axis of an *uptake plot* – but often many cells are dead at the end of the experiment if the exposure time to agent is long.

Uptake of a drug by the cell requires that a drug molecule, which is in solution, collide with the surface of the cell, which is in the medium. This collision might result in the drug sticking to the surface or being internalized by the cell, or the collision could be unproductive in the sense that the drug molecule returns to solution. If the agent crosses the lipid bilayer of the cell by *passive diffusion*, a plot of bound drug vs. time is linear, so long as the system is far from equilibrium; equilibrium is when the drug concentration outside and inside are the same. However, since drug concentrations in the medium are typically in the micromolar range and the cell would likely die before the drug concentration inside reached this value, equilibrium involving a live cell is probably never reached in these studies. Since dead cells might bind drug (metal) as effectively as healthy cells, uptake plots could be linear with time regardless of the health (viability) of cell population during the experiment.

If drug enters the cell using a protein that spans the outer membrane, called a *transporter*, there is often an energy requirement to move the compound from the medium outside the cell, through the transporter, to inside the cell. If the energy source required for the transport is in excess, a plot of the amount of bound drug vs. time will be linear, but if the energy supply becomes depleted during the experiment, the curve will deviate from linearity. Since certain chemical agents are known to specifically block the pore or opening of a transporter, the presence of a blocking agent in the experiment could provide information on which transporter is important in uptake. It is also possible to create cell lines that synthesize unusually high amounts of a certain transporter. If these cells take up more of a given compound than cells with the normal number of transporter molecules, this is evidence that the selected transporter is important for the uptake of the agent into the cell.

2.5.3   Animal studies

While animals, mainly mice and rats, have been the mainstay of studies into the effects of drugs in the biological system, it is becoming increasingly apparent that the results obtained using animals often do not translate well to humans [20]. Thus, in recent years there has been increasing emphasis on the use of *in vitro* assays such as cytochrome P-450 enzymes, the *Ames test*, which measures the mutagenicity of compounds, and even computer modeling, referred to as *in silico*. The latter approach attempts to model the *pharmacokinetic* behavior of agents; that is, the rate of change of drug concentration in the body. However, in spite of these innovations, the toxicity, pharmacokinetics and efficacy of new agents are still extensively examined in animals and most regulatory agencies require that compounds be tested or *screened* in at least one rodent and one nonrodent animal model. Since animal studies raise ethical, safety and handling issues, there exists an extensive network of oversight for these studies that ranges from the institutional level, where the testing is usually carried out, to the federal level, which establishes the guidelines for research involving animals.

2.6   From discovery to the marketplace

2.6.1   Drug approval process

The road that leads from the laboratory bench where new drugs are discovered to the clinic where they are used for treating patients is marked with many pitfalls [20]. Only a tiny fraction of new compounds with therapeutic or diagnostic potential make it over the regulatory hurdles and become approved pharmaceutical agents. In the United States the regulatory agency is the *Food and Drug Administration, FDA*, which has similar counterparts in Europe: the *European Agency for the Evaluation of Medicinal Products, EMEA*; and in Japan: the *Ministry of Health, Labor, and Welfare, MHLF*. The United States *National Cancer Institute, NCI*, which identifies new agents for treating cancer, provides the following general summary of what is necessary for gaining FDA approval of an anticancer drug [21].

2.6.1.1 Pre-clinical phase

In this phase of the approval process, the *drug candidate*, a term applied to a material undergoing approval, is tested against three human cancer cell lines. If the candidate inhibits growth in one or more of the cell lines it is tested against a panel of 60 human tumor lines, which include a wide range of different types of cancer. A compound that is unique in some way, with high cytotoxicity for example, is more selective for some cell types or is thought to have a unique mechanism of action is tested using mice with human tumor *xenografts*; that is, human tumors that have been grafted under the skin of the animal. If the results of these studies show promise and the agent slows tumor growth with minimal toxicity to the *host*, the pharmacology, formulation and toxicology of the candidate are examined in animals. These studies determine where the agent is metabolized, the best formulation to use, the most effective dose to apply, and the frequency and method of administration.

2.6.1.2 Investigational new drug, IND, application

The next step in the approval process is for the *sponsor* of the drug candidate, which can be a pharmaceutical company, an academic institution or the NCI, to file an *Investigational New Drug, IND*, application with the FDA. This application provides the chemical structure of the agent, how it is believed to work in the body, its toxic effects in animals, and where and how human trials, called *clinical trials*, with the agent will be conducted. The plan for clinical trials, which describes the institutions involved, the number of people to take part in the study, various medical tests to be given and so on, must be approved by the *Institutional Review Boards, IRBs*, of each of the sites at which the trials will be conducted. The IRBs, which are generally composed of health professionals, clergy and consumers, are responsible for making sure that the people involved in the study are not exposed to undue risk. The clinical trials involving the drug candidate are carried out in three phases.

2.6.1.3 Phase I clinical trials

In the first phase of the trials, a small number of mostly healthy volunteers, 20–80, but sometimes also patients with cancer, as occurred with the anticancer drug cisplatin [22], are given the drug candidate, and the dosage, metabolism, side effects and other basic properties of the agent are determined.

2.6.1.4 Phase II clinical trials

In phase II, a limited number of patients, some of whom have a specific type of cancer, are treated with the agent. This phase usually involves a larger number of subjects than phase I.

2.6.1.5 Phase III clinical trials

In this phase, hundreds to thousands of people in nationally-located clinics, cancer centers, doctors' offices and so on are involved in the study. Most of the subjects in phase III have some form of cancer, but healthy volunteers are also included. At random, the subject receives either the drug candidate, a standard drug treatment or a placebo; that is, a solution like 154 mM NaCl, normal saline solution. Despite all of the efforts to select the best drug candidates, poor pharmacokinetics, lack of absorption, rapid metabolism and/or elimination, and even drug-induced death cause the failure rate for new drug candidates in phase III to be very high, ~50% [23,24].

2.6.1.6 New drug application, NDA

The second application to be filed with the FDA at the end of the clinical trials is the *New Drug Application or NDA*. In the NDA, the sponsor must provide all of the efficacy and safety information on the agent, show that the benefits of using it outweigh the risks, indicate that the labeling information for the agent is appropriate and show that the manufacturing methods used to obtain the material are adequate for ensuring the purity and integrity of the drug candidate. After the NDA is approved by the FDA, which typically takes 6–12 months, the drug candidate becomes a *drug* that is available for physicians to prescribe for patients. The length of time for the entire process, from discovery to final approval, can vary considerably, but 8–15 years can be expected.

2.6.1.7 Phase 0 clinical trials

It has been found that many promising new drug candidates often fail in phase I because the pharmacokinetics and *pharmacodynamics* – that is, the effects of the drug on the body – frequently do not correlate very well between animals and humans [25]. Since phase I trials are expensive and time consuming, it has been proposed to institute *phase 0 clinical trials*, in which human subjects are given very low, *therapeutically ineffective*, doses of the agent and the pharmacokinetics and pharmacodynamics of the drug candidate are evaluated. Since low dose lowers the risk of toxicity, these studies would be less expensive to carry out than those in phase I and might shorten preclinical testing, reduce the overall time period for drug development and provide a better way to quickly identify candidates that could successfully pass all of the trials in the approval process.

2.6.2 Profits

In the United States, the term of a patent is 20 years. If the discoverer of a new drug candidate files a US patent shortly after the discovery of the agent and proceeds through the steps outlined above, it is clear that long testing/approval times take up much of the time during which the sponsor has patent protection for the drug and the profits from its sale. Since development costs for a new drug are enormously high, estimated to be about one billion dollars [23], a successful drug needs to regain not only the money spent on its development but also make a profit for the sponsor, which is one of the reasons many new drugs are very expensive for the consumer.

Problems

1. The tripeptide glycylglycylglycinamide, L, reacts with aquated $Cu(ClO_4)_2$, where ClO_4^- is the very weakly-binding perchlorate ion, at pH 10.8 to produce a metal complex anion that has the formula $[CuL]^-$. The absorption data for the complex are: $\lambda_{max} = 517$ nm, $\varepsilon = 144\,M^{-1}\,cm^{-1}$.

   a. Propose the structure of this complex, clearly showing which donor atoms of the tripeptide are bound to the Cu^{2+} ion.

   b. Calculate the absorbance of a 10 mM solution of $[CuL]^-$ in a 1 cm path-length cell.

c. What is the origin of the electronic absorption band in the complex?

$$H_2N-\underset{H_2}{C}-\underset{\overset{\parallel}{O}}{C}-\underset{H}{N}-\underset{H_2}{C}-\underset{\overset{\parallel}{O}}{C}-\underset{H}{N}-\underset{H_2}{C}-\underset{\overset{\parallel}{O}}{C}-NH_2$$

glycylglycylglycinamide

2. A solution containing 50 μM of the anticancer drug cisplatin, cis-$[PtCl_2(NH_3)_2]$ (Figure 1.22) and 50 μM GSH in water at pH $= 7.0$ was allowed to stand at 37 °C.

   a. Propose a structure for the initially-formed product, indicating the total charge on the complex.
   b. After standing for 5 hours, free ammonia is observed in solution. Briefly explain why this occurs.

3. Compounds of gadolinium, Gd^{3+}, which are most often nine-coordinate, are used to enhance the contrast of the diagnostic image obtained in magnetic resonance imaging (MRI). Since these compounds usually have one coordinated but displaceable water molecule, suggest which functional groups on a protein could serve as potential donor atoms toward Gd^{3+}.

4. Complexes of Au^+ and Au^{+3} exhibit anticancer effects. Which ion would be more likely to be bound to N-7 of guanine of DNA?

5. A very sensitive analytical technique was used to measure the amount of Pt bound to genomic DNA in cells that were treated with a platinum-containing anticancer drug. The study found 200 platinum adducts per 10^7 nucleotides (nts) of DNA

   a. If the DNA is double-stranded, calculate the number of platinum adducts per 10^7 base pairs of DNA.
   b. If the length of the double-stranded genome is 3×10^9 base pairs, calculate the average number of platinum adducts per genome.

6. Answer the following pertaining to the ionization of functional groups:

   a. An imidazolium group on histidine in a protein has a pK_a for deprotonation of 6.3. What is the pH of the solution if the imidazolium is 20% deprotonated?
   b. Calculate the equilibrium constant, K, for the deprotonation of imidazolium group in (a) above.
   c. Measurements using NMR show that a thiol group of cysteine residue in a protein is 12% protonated at pH $= 6.9$. Calculate the pK_a for the group.

7. A metal hydroxo complex (metal-bound hydroxide ion) reacts with carbon dioxide gas in water to form a hydrogencarbonato complex with a second-order rate constant $k = 320\,M^{-1}\,s^{-1}$ at 37 °C. If the concentrations of the metal hydroxo complex and aquated CO_2 are 1 μM and 2 mM respectively, calculate the half life for the conversion of the hydroxo complex to the hydrogencarbonato complex under the conditions given.

References

1. Sigel, H., and Martin, R.B. (1982) Coordinating properties of the amide bond. Stability and structure of metal ion complexes of peptides and related ligands. *Chemical Reviews*, **82**, 285–426.
2. Neubecker, T.A., Kirksey, S.T., Chellappa, K.L., and Margerum, D.W. (1979) Amine deprotonation in copper(III)-peptide complexes. *Inorganic Chemistry*, **18**, 444–448.
3. Dawson, R.M.C. (1986) *Data for Biochemical Research*, 3rd edn, Clarendon Press, Oxford, UK.
4. Albro, P.W., Corbett, J.T., and Schroeder, J.L. (1986) Generation of hydrogen peroxide by incidental metal ion catalyzed autooxidation of glutathione. *Journal of Inorganic Biochemistry*, **27**, 191–203.

5. Shaked, Z., Szajewski, R.P., and Whitesides, G.M. (1980) Rates of thiol-disulfide interchange reactions involving proteins and kinetic measurements of thiol pK_a values. *Biochemistry*, **19**, 4156–4166.

6. Hall, M.D., Okabe, M., Shen, D.-W. *et al.* (2008) The role of cellular accumulation in determining sensitivity to platinum-based chemotherapy. *Annual Review of Pharmacology and Toxicology*, **48**, 495–535.

7. Balendiran, G.K., Dabur, R., and Fraser, D. (2004) The role of glutathione in cancer. *Cell Biochemistry and Function*, **22**, 343–352.

8. Knipp, M., Karotki, A.V., Chesnov, S. *et al.* (2007) Reaction of Zn_7Metallothionein with *cis-* and *trans-*[Pt(N-donor)$_2$Cl$_2$] anticancer complexes: *trans-*Pt^{II} complexes retain their N-donor ligands. *Journal of Medicinal Chemistry*, **50**, 4075–4086.

9. Richards, A.D. and Rodger, A. (2007) Synthetic metallomolecules as agents for the control of DNA structure. *Chemical Society Reviews*, **36**, 471–483.

10. Palmer, D.A. and van Eldik, R. (1983) The chemistry of metal carbonato and carbon dioxide complexes. *Chemical Reviews*, **83**, 651–731.

11. Mao, Z.-W., Liehr, G., and van Eldik, R. (2000) Isolation and characterization of the first stable bicarbonato complexes of Bis(1, 10-phenanthroline)copper(II). Identification of Lipscomb- and Lindskog-like intermediates. *Journal of the American Chemical Society*, **122**, 4839–4840.

12. Gateley, D.P. and Howell, S.B. (1993) Cellular accumulation of the anticancer agent cisplatin: a review. *British Journal of Cancer*, **67**, 1171–1176.

13. Di Pasqua, A.J., Goodisman, J., Kerwood, D.J. *et al.* (2007) Role of carbonate in the cytotoxicity of carboplatin. *Chemical Research in Toxicology*, **20**, 896–904.

14. Ghezzi, A., Aceto, M., Cassino, C. *et al.* (2004) Uptake of antitumor platinum(II)-complexes by cancer cells, assayed by inductively coupled plasma mass spectrometry (ICP-MS). *Journal of Inorganic Biochemistry*, **98**, 73–78.

15. Hah, S.S., Stivers, K.M., de Vere White, R.W., and Henderson, P.T. (2006) Kinetics of carboplatin-DNA binding in genomic DNA and bladder cancer cells as determined by accelerator mass spectrometry. *Chemical Research in Toxicology*, **19**, 622–626.

16. Hah, S.S., Sumbad, R.A., de Vere White, R. *et al.* (2007) Characterization of oxaliplatin-DNA adduct formation in DNA and differentiation of cancer cell drug sensitivity at microdose concentrations. *Chemical Research in Toxicology*, **20**, 1745–1751.

17. Yonezawa, A., Masuda, S., Yokoo, S. *et al.* (2006) Cisplatin and oxaliplatin, but not carboplatin and nedaplatin, are substrates for human organic cation transporters (SLC22A1-3 and multidrug and toxin extrusion family). *The Journal of Pharmacology and Experimental Therapeutics*, **319**, 879–886.

18. Zisowsky, J., Koegel, S., Leyers, S. *et al.* (2007) Relevance of drug uptake and efflux for cisplatin sensitivity of tumor cells. *Biochemical Pharmacology*, **73**, 298–307.

19. Tacka, K.A., Dabrowiak, J.C., Goodisman, J. *et al.* (2004) Effects of cisplatin on mitochondrial function in jurkat cells. *Chemical Research in Toxicology*, **17**, 1102–1111.

20. Ng, R. (2004) Drugs, *From Discovery to Approval*, John Wiley & Sons Inc., Hoboken, NJ.

21. http://www.cancer.gov/cancertopics/factsheet/NCIdrugdiscovery, accessed 11/13/08.

22. http://chemcases.com/cisplat/cisplat15, accessed 11/15/08.

23. DeNardo, G.L. (2007) The conundrum of personalized cancer management, drug development, and economics cancer biother. *Radiopharm*, **22**, 719–721.

24. Grass, G.M. and Sinko, P.J. (2002) Physicologically-based pharmacokinetic simulation modelling. *Advanced Drug Delivery Reviews*, **54**, 433–451.

25. Marchetti, S. and Schellens, J.H.M. (2007) The impact of FDA and EMEA guidelines on drug development in relation to phase 0 trials. *British Journal of Cancer*, **97**, 577–581.

Further reading

Housecroft, C.E. and Sharpe, A.G. (2007) *Inorganic Chemsitry*, 3rd edn, Pearson Education Limited, Edinburgh Gate, UK.

Huheey, J.E., Keiter, E.A., and Keiter, R.L. (1993) Inorganic chemistry, *Principles of Structure and Reactivity*, 4th edn, Benjamin-Cummings Publishing Co., San Francisco, CA.

Metzler, D.E. (2001) Biochemistry, *The Chemical Reactions of Living Cells*, 2nd edn, vol. **1**, Harcourt Academic Press, San Diego, CA.

Mathews, C.K., van Holde, K.E., and Ahern, K.G. (2000) *Biochemistry*, 3rd edn, Benjamin Cummings Publishing Co., San Francisco, CA.

Van Wynsberghe, D., Noback, C.R., and Carola, R. (1995) *Human Anatomy and Physiology*, 3rd edn, McGraw Hill, New York, NY.

Welling, P.G. (1986) Pharmacokinetics, *Processes and Mathematics*, ACS Monograph 185, American Chemical Society, Washington, DC.

3

Cisplatin

Cisplatin, *cis*-diamminedichloroplatinum(II), a square planar Pt^{2+} complex (Figure 3.1), was the first metal-based agent to enter into worldwide clinical use for the treatment of cancer. Currently, cisplatin is used either by itself or in combination with other drugs for treating lung, ovarian, bladder, testicular, head and neck, esophageal, colon, gastric, breast, melanoma and prostate cancer. Although sales of cisplatin are presently on the decline, with second- and third-generation analogs being more widely prescribed, cisplatin remains the 'gold standard' to which aspiring platinum and nonplatinum metal-based anticancer drugs are compared.

While it is not possible to cover the more than 40 years of research on cisplatin in one chapter, the important physical and chemical properties of the drug, aspects of its clinical applications and pharmacokinetics, as well as its biological mechanism of action will be presented and discussed here. More information on cisplatin regarding these and other topics can be found in the excellent books, review articles and publications of Rosenberg [1,2], Lippard [3–5], Reedijk [6,7], Farrell [8], Kelland and Farrell [9], Lippert [10], Hambley [11,12], Sadler [13], Schellens [14,15] and Boulikas and Vougiouka [16], and their coworkers.

Box 3.1 Discovery of cisplatin

The idea of investigating the antitumor properties of cisplatin was based on an accidental discovery made by Barnett Rosenberg, a professor of biophysics at Michigan State University, and his coworkers in the early 1960s [2,11,17,18]. Professor Rosenberg was studying the effects of electric fields on the growth of cells and he and his group constructed a special cell culture apparatus containing platinum mesh electrodes that allowed cells to grow and be harvested on a continuous basis. The goal of the study was to apply alternating currents of different frequencies to the electrodes and determine whether and to what extent electric current affected the ability of the cells to divide. While the investigators were really interested in mammalian cells, they tested the new apparatus using *E. coli* bacteria and found to their surprise that certain frequencies of current greatly reduced the number of cells growing in the culture apparatus. A check on the appearance of the bacteria that had been subjected to electric current revealed that bacterial cells were present but instead of having their normal 'sausage' shape, they were long spaghetti-like filaments, which indicated that the cells were growing but not dividing. Sensing that they were observing something new and very unusual, the investigators carried out a number of control experiments which showed that, while the electric current itself had no direct effect on cell division, the current was causing a chemical reaction to take place in the cell culture medium, which required oxygen, ammonium ion (NH_4^+) and chloride ion (Cl^-). After learning this, and after realizing that the current was causing a small amount of platinum metal on the surface of the electrodes to dissolve, they hypothesized that the

---

current was in some way producing platinum complexes in the culture medium and that these chemical species, not the electric current, were affecting the growth of *E. coli*. Suspecting that elemental platinum in the electrode was being oxidized to Pt^{+4}, they tested a sample of $[NH_4]_2[PtCl_6]$, which, in addition to having Pt^{+4}, also contained chloride and ammonium ion, but the compound had no effect on the growth of the bacteria. However, if solutions of $[NH_4]_2[PtCl_6]$ were exposed to ambient light for a long period of time on the laboratory bench top, or if they were deliberately irradiated using a light source, the resulting solutions were very effective in blocking division of *E. coli*. These observations led the investigators to irradiate solutions of $[NH_4]_2[PtCl_4]$, which contained Pt^{+2}, and after analyzing the solutions they concluded that *neutral chemical species* produced in the photoreactions were responsible for the lack of division of *E. coli* cells. By synthesizing authentic samples of many platinum complexes, they learned that six-coordinate octahedral *cis*-$[PtCl_4(NH_3)_2]$, which contains Pt^{+4}, and four-coordinate square planar *cis*-$[PtCl_2(NH_3)_2]$, which contains Pt^{+2}, were very effective in causing the bacterial cells to elongate. Thinking that a platinum complex capable of inhibiting cell division could also be useful for treating cancer, Rosenberg and his group studied the effects of the platinum compounds on tumor-bearing mice. They found that *cis*-$[PtCl_2(NH_3)_2]$, later to become known as cisplatin, was remarkably effective at arresting sarcoma 180 and leukemia L1210 in mice, while the *trans* isomer, *trans*-$[PtCl_2(NH_3)_2]$, exhibited little antitumor activity.

Turning a laboratory discovery into a useful commercial product, especially in the case of a drug, is not an easy task, and this was even more difficult in the case of cisplatin. At the time that Professor Rosenberg carried out his investigations, all of the anticancer drugs approved for use in the United States were either natural products or synthetic agents, but all were organic compounds. Anything which contained a heavy metal, like platinum, was viewed as a toxin that should be kept away from humans and certainly not be deliberately given to patients in the form of a drug. In order to turn his discovery into a useful pharmaceutical agent, Professor Rosenberg contacted the National Cancer Institute (NCI) and convinced the government organization to do more extensive animal tests on the platinum complex and to sponsor studies that would administer the compound to patients in clinical trials. Since the compound proved to be very effective for patients who failed other forms of chemotherapy, Rosenberg, with the help of Michigan State University and the Research Corporation, an organization that facilitates the transfer of technology from academia to industry, patented his new discovery. Since *cis*-$[PtCl_2(NH_3)_2]$ was synthesized and reported by Peyrone more than 100 years earlier [19], patenting on the basis of the compound being a new substance – that is, obtaining a so-called *composition of matter patent* – was not possible, so Rosenberg and coworkers filed a *methods of use patent* describing how the compound could be used to treat genitourinary cancer and possibly other types of tumors. The application was ultimately successful and in 1979 the United States Patent Office issued a patent for use of the compound in treating cancer. In this period, the pharmaceutical company Bristol-Myers (now Bristol-Myers Squibb), which has a long history of developing anticancer drugs, became interested in the new discovery and, after carrying out more tests, provided the necessary efficacy and safety information to the United States Food and Drug Administration (FDA) to have the compound, now named cisplatin, approved for use as a drug in 1978 [20].

Before the discovery of cisplatin, identifying new agents for treating cancer was the exclusive realm of organic and natural-product chemistry, and the finding that a metal complex was more effective than many organic compounds in the treatment of cancer energized inorganic chemistry. Professor Rosenberg seemed to be aware that cisplatin might have this effect on the field when, in the closing remarks in a paper that he published in 1971, he wrote, 'Without hesitancy I suggest it is now appropriate for inorganic chemists to join their organic brothers in submitting samples of their syntheses to appropriate Cancer Institutes for screening for anti-tumor activities' [1]. Almost overnight, cisplatin expanded the drug-discovery landscape to include inorganic chemistry, which has resulted in a number of new platinum-based antitumor agents that are in clinical use for the treatment of cancer, and many non-platinum-containing compounds that show promise in becoming useful anticancer agents.

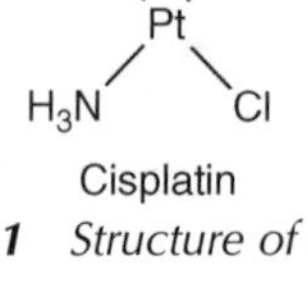

Cisplatin

Figure 3.1 *Structure of cisplatin*

3.1   Physical and chemical properties of cisplatin

Cis-diamminedichloroplatinum(II), *cis*-[PtCl$_2$(NH$_3$)$_2$], cisplatin, is a bright-yellow solid that was first synthesized by Peyrone in 1844 [19]. The modern synthesis of cisplatin usually employs the method of Dhara (Chapter 1), which uses the *trans* effect of iodide ion to efficiently direct two ammonia molecules into *cis* sites on the square planar platinum ion (Figure 1.23) [21]. Since Pt^{+2} has the electronic configuration 5d^8, and cisplatin is diamagnetic ($S = 0$), the electronic configuration involving the *d*-orbitals in the square planar complex, in order of increasing energy, is d_{xz}^2, d_{yz}^2 (degenerate), $d_{z^2}^2$, d_{xy}^2, $d_{x^2-y^2}^0$. The highest-energy orbital, $d_{x^2-y^2}$, is unoccupied because the energy difference between it and d_{xy}, which is Δ, is large, forcing the two electrons to pair in the more stable d_{xy} orbital.

The absorption spectrum of cisplatin in the polar organic solvent dimethylformamide, DMF, is shown in Figure 3.2 [22]. Since DMF is a very poor donor toward transition metal ions (DMF is low in the spectrochemical series), dissolved cisplatin in DMF retains the two bound chloride ligands that would otherwise be easily displaced in many other solvents, for example water. While the absorption spectrum of cisplatin is complicated and its transitions are overlapped, careful analysis using a number of optical techniques reveals that the compound has absorptions at 27 000 (25), 32 500 (170) and 35 500 (110), where the first quantity is the band maximum in wavenumbers, cm^{-1}, and the second, in parenthesis, is the molar extinction coefficient, ε, in units of M^{-1} cm^{-1}. From the general ranges given for the intensity of an absorption band and its probable origin (Table 1.4), all of the bands in Figure 3.2 are due to transitions within the *d*-level; that is, *d-d* transitions of the metal ion. Also, while it might be expected from the values of ε given that all three transitions are Laporte-forbidden ($\Delta l = 0$) and spin-allowed ($\Delta S = 0$) (Table 1.4), this is true for only the two highest-energy transitions. Detailed studies show that the lowest energy band at 27 000 cm^{-1} is in fact Laporte-*and* spin-forbidden, implying that while it is a transition associated with promotion of electrons within the *d*-level, the net spin of the system changes in the transition. The fact that this band is a bit stronger than expected for a spin-forbidden transition (Table 1.4) can be explained by how certain bands in the spectrum 'borrow' intensity from other bands, but this argument will not be given here. Since DMF strongly absorbs at energies greater than ~38 000 cm^{-1}, it cannot be used as a solvent to measure the highest energy bands of cisplatin. However, if the solvent is changed to water, which has an 'optical window' at high energy, and excess

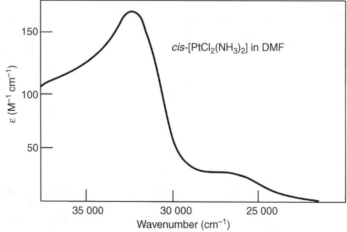

Figure 3.2   *Absorption spectrum of cisplatin, cis-[PtCl$_2$(NH$_3$)$_2$], in dimethylformamide (DMF). Reprinted with permission from H.H. Patterson et al., Luminescence, absorption, MCD, and NQR Study of the Cis and Trans isomers of dichlorodiammineplatinum(II), Inorg. Chem., 20, 2297–2301. Copyright 1981 American Chemical Society*

Table 3.1 *Rate, equilibrium and ionization constants for cisplatin in water*[a]

Parameter	Value	Temperature (°C)	Ionic Strength (I_c, M)
k_1	$5.18 \times 10^{-5}\,s^{-1}$	25	0.1
k_1	$1.84 \times 10^{-4}\,s^{-1}$	35.5	0.1
k_{-1}	$7.68 \times 10^{-3}\,M^{-1}\,s^{-1}$	25	0.1
pK_1	2.17	25	0.1
	2.07	37	0.1
k_2	$2.75 \times 10^{-5}\,s^{-1}$	25	1.0
k_{-2}	$9.27 \times 10^{-2}\,M^{-1}\,s^{-1}$	25	1.0
pK_2	3.53	25	1.0
pK_{a1}	6.41	27	
pK_{a2}	5.37	27	
pK_{a3}	7.21	27	

[a] Adapted from references [26–28]. I_c is the ionic strength of the medium, $I_c = \frac{1}{2}\sum M_i Z_i^2$, where M_i is the concentrations of the cation and the anion in solution and Z_i is their respective charges. See Figure 3.3 for the definitions of the rate and equilibrium parameters.

chloride ion is present in the medium to suppress the aquation of cisplatin through Le Chatelier's principle, two strong transitions at 45 500 (2000) and 49 500 (7500) can be observed. These have been tentatively assigned to MLCT transitions of the complex [22].

When dissolved in water, cisplatin, *cis*-[PtCl$_2$(NH$_3$)$_2$], **1** in Figure 3.3, slowly reacts with it to sequentially lose its two chloride ligands according to the scheme shown in Figure 3.3. The rate, equilibrium and ionization data for the complex in water are given in Table 3.1 [23–28]. As shown in Figure 3.3, the first step in the reaction is the attack of a water molecule on the Pt^{+2} ion of cisplatin with rate constant k_1, which results in the

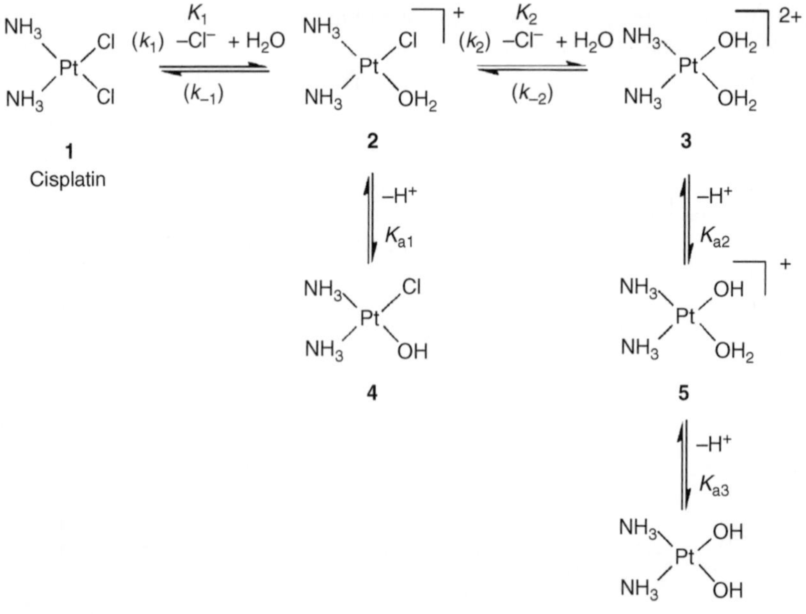

Figure 3.3 *Speciation of cisplatin in aqueous solution. Redrawn from Scheme 1.1 1.1p. 73 of Berners-Price, S.J., Ronconi, L., Sadler, P.J. (2006) Insights onto the mechanism of action of platinum anticancer drugs from multinuclear NMR spectroscopy. Prog. Nucl. Mag. Res. Sp., 49, 65–98*

displacement of one of the bound chloride ligands to produce the monoaqua species *cis*-[PtCl(NH$_3$)$_2$(H$_2$O)]$^+$, **2**. While this is really a bimolecular, second-order reaction for which the rate expression depends on the concentration of *both* cisplatin and water (see (1.6)), the fact that the concentration of water (55.5 *M*) is much larger than the concentration of the drug (a saturated solution of cisplatin in water at room temperature is ~4 m*M*) means that the concentration of water does not change much in the reaction and the conditions for the forward reaction are *pseudo first-order*, which is why k_1 carries units of s^{-1}. After the Cl$^-$ ion is displaced from cisplatin, it can recombine with the aquated form with rate constant $k_{-1} = 7.68 \times 10^{-3} M^{-1} s^{-1}$ to reform cisplatin. This is also a bimolecular reaction, but unlike the forward reaction it is a true second-order reaction in that its rate depends on the concentrations of *both* Cl$^-$ and the aquated form, which must be identical; that is, it cannot be argued that the concentration of one reactant is in large excess over the other. In this case the rate constant carries the *second-order* units of $M^{-1} s^{-1}$.

As is evident from Table 3.1, the pK_{a1} for the deprotonation of one of the protons on the bound water molecule of the monoaqua species, **2** in Figure 3.3, is 6.41, which from the Henderson–Hasselbalch equation (Chapter 2), means that at the pH of blood, pH = 7.4, most of the complex is in the monohydroxo form, **4**. Since the rate of protonation/deprotonation is fast compared to the rate of aquation, which involves metal–ligand bond-breaking/forming steps, as soon as the water molecule becomes bound to the platinum ion, the system immediately equilibrates between the two forms of the complex, **2** and **4**. The next step in the process is the loss of the chloride ion of **2** (**4**) by nucleophilic attack of water, which gives **3**, **5** and **6**. Clearly, the system is now quite complicated, but extensive studies by House [23–25], and by Berners-Price and Sadler [26–28] and their coworkers, have led to a detailed understanding of the system which provided the values given in Table 3.1.

Box 3.2 Heteronuclear single quantum coherence (HSQC) NMR

Nuclear magnetic resonance, NMR, spectroscopy is a powerful technique for characterizing the structure and properties of a molecule. One NMR technique that has proven to be especially useful in studying a wide range of platinum complexes in solution is *heteronuclear single (or multiple) quantum coherence*, HSQC or HMQC NMR. While it is beyond the scope of the presentation here to explain exactly how the NMR experiment itself is carried out, for example what pulse sequence is used, how signals are detected and so on, most modern NMR spectrometers are equipped with the proper hardware and software to make the measurement. The main challenge in using HSQC NMR (only one acronym will be used here) is the synthesis of a platinum complex with two different NMR active nuclei that are close together so that their nuclear spins strongly couple to each other. One of the benefits of HSQC NMR is that it is very sensitive and solutions in the 5 μ*M* range, which is near the physiological concentration of most drugs, can be studied. A second advantage is that it is a two-dimensional (2D) NMR technique, with the chemical shift values (δ) of both of the NMR active nuclei plotted on the x- and y-axes of a graph. This makes it easy to detect and correlate NMR changes with structural changes in the compound.

To show how the technique works, consider the structure of *cis*-[PtCl$_2$(NH$_3$)$_2$], cisplatin, which has bound ammonia and chloride ligands. If ^{15}N enriched ammonium chloride, ^{15}NH$_4$Cl, is used in the synthesis of cisplatin instead of normal ^{14}NH$_4$Cl (Figure 1.23), the two *cis*-ammonia molecules of the drug can be enriched with ^{15}N, an isotope of nitrogen that has the same nuclear spin as a proton – that is, $I = {}^1/_2$ – to give *labeled cis*-[PtCl$_2$(^{15}NH$_3$)$_2$]. Since the hydrogen atom bonded to the nitrogen atom also has $I = {}^1/_2$ and the hydrogen and nitrogen are directly connected to each other through a single bond, the nuclear spins of the two atoms are strongly coupled in a *heteronuclear* spin system. Note that since the three hydrogen atoms of ammonia are equivalent to each other by symmetry, it is only necessary to consider what happens with one N-H pair. Since platinum has an isotope, ^{195}Pt, that has a natural abundance of ~33% and the nuclear spin of this isotope is $I = {}^1/_2$, strong HSQC NMR peaks in the spectrum of a platinum complex often exhibit additional 'satellite' peaks caused by coupling to the nuclear spin of this isotope.

An interesting property of square planar complexes of Pt^{+2} is that they are examples of, and subject to, the *trans* effect. Since the two symmetry-equivalent ammonia molecules usually remain bound to the Pt^{+2} ion when cisplatin undergoes substitution reactions, they can be used as 'reporter groups' to identify the ligands that occupy the sites *trans* to the ammonias. For example, if one of the chloride ligands of cisplatin is replaced with a thiolate ligand, RS^-, the ^{15}N chemical shift of the ammonia molecule *trans* to the thiolate will be dramatically affected. Since bound thiolate ion removes electron density from Pt^{+2} through the same *d*-orbital that is directly involved in bonding to ammonia, the ammonia 'feels' the presence of the thiolate. The removal of electron density from the orbital by the thiolate (compared to chloride) causes a change in the *electronic shielding* of the ^{15}N nucleus of the bound ammonia – that is, the nucleus becomes *deshielded* – and the ^{15}N resonance shifts to a *lower field*, which is easily detected by HSQC NMR. Similar changes also take place in the electronic shielding of the ammonia protons, but since these nuclei are farther removed from the thiolate ion – that is, they are one bond more distant than nitrogen from the thiolate on the square planar platinum framework – changes in their chemical shift caused by *trans* effects are more modest.

Experimentally, it has been found that complexes of Pt^{+4} with bound ^{15}N ammonia molecules also exhibit HSQC NMR spectra but that the spectra are sensitive to pH. If the pH is in the neutral or basic range, no HSQC NMR spectrum is detected, but if the pH is below ~5, spectra can be observed. The reason for the pH sensitivity is associated with the ability of the highly-charged Pt^{+4} ion to act as a *Lewis acid* and make the protons of bound ammonia much more acidic than the protons of free, unbound ammonia. When the pH of the solution containing the Pt^{+4}- ammonia complex is made basic, one of the protons on a bound ammonia molecule is removed by the base; that is, bound NH_3 is converted into bound NH_2^-. Since this also increases the opportunity for the protons on water molecules in solvent to exchange with the protons on the bound ammonia ligands, the *residence time* of a specific proton on the nitrogen atom of the ammonia effectively *decreases*. This rapid chemical exchange process involving one of the nuclear spins that is required to make the measurement causes the HSQC NMR signal to be lost; that is, exchanges take place faster than the NMR instrument can 'see' the coupled nuclei. Thus, lowering the pH, which stops the proton–proton exchange process, 'activates' the heteronuclear spin system, which allows HSQC NMR spectra to be observed.

Berners-Price and Sadler and their coworkers have measured the ^1H-^{15}N HSQC NMR spectra of many complexes of Pt^{+2} and Pt^{+4} that contain ^{15}N-labeled ammonia and found that the ^1H and ^{15}N chemical shifts for these compounds are sensitive to the nature of the *trans* ligand [26,28]. A summary of expected chemical shift ranges for different ligands *trans* to ammonia for Pt^{+2} and Pt^{+4} complexes, which is useful for assigning spectra for new complexes and reaction products with biological molecules, is given in Figure 3.4.

A powerful technique for studying the kinetics and mechanism of the reaction of cisplatin and other metal complexes in solution is *heteronuclear single quantum coherence* (*HSQC*) NMR. For an example of how this technique can be used to study the hydrolysis of cisplatin, consider the HMQC NMR spectral data (HMQC, *heteronuclear multiple quantum coherence*, is related to HSQC) shown in Figure 3.5.

Figure 3.5a shows the HMQC NMR spectrum of *cis*-$[PtCl_2(^{15}NH_3)_2]$ at pH 4.72, 300 K, after the drug has been allowed to hydrolyze in water. As is evident from the figure, four HMQC NMR peaks, called *cross peaks*, are observed, corresponding to the dichloro, monochloro, monoaqua and diaqua forms of the drug (Figure 3.3). Cisplatin and *cis*-$[Pt(^{15}NH_3)_2(H_2O)_2]^+$ both have *symmetry-equivalent* ammonia molecules so, as expected, they each have only a single peak in the HMQC NMR spectrum (Figure 3.5a). As is evident from Figure 3.4, the peaks for both complexes are in the correct locations for NH_3 *trans* to Cl^-, cisplatin, and NH_3 *trans* to H_2O for *cis*-$[Pt(^{15}NH_3)_2(H_2O)_2]^{2+}$. The monochloro, monoaqua complex, *cis*-$[PtCl(^{15}NH_3)_2(H_2O)]^+$, which has *nonequivalent* ammonia molecules, exhibits *two cross peaks*; one, **2a**, is for NH_3 *trans* to Cl^-, and the other, **2b**, is for NH_3 *trans* to H_2O.

Figure 3.5b and c shows that the proton chemical shifts of *cis*-$[PtCl(^{15}NH_3)_2(H_2O)]^+$ and *cis*-$[Pt-(^{15}NH_3)_2(H_2O)_2]^{2+}$ are sensitive to pH, with the NH proton chemical shifts of both complexes moving from low field (high ppm) to high field (low ppm) as the pH changes from acidic to basic values. These changes in chemical shift occur because the pK_a values for the deprotonation of the bound water molecules in the complexes are in the range of the pH change; that is, adding base to the system results in the reaction of the base

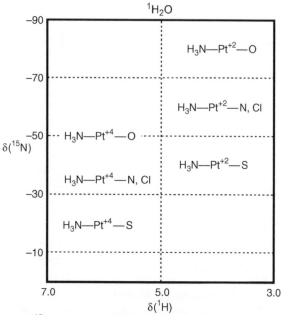

Figure 3.4   *Approximate ^1H and ^{15}N chemical shifts of NH_3 bound to Pt^{+2} and Pt^{+4} with different donor atoms trans to the bound NH_3 ligand. The ^{15}N chemical shift scale is relative to ammonium ion, NH_4^+, which is zero ppm. The vertical doted line is the approximate chemical shift of the protons of water. Reprinted from Prog. Nucl. Mag. Res. Sp., 49, S.J. Berners-Price et al., Insights onto the Mechanism of Action of Platinum Anticancer Drugs from Multinuclear NMR Spectroscopy, 65–98. Copyright 2006, with permission from Elsevier*

with the acidic proton on the bound water (Figure 3.3). Consistent with the *trans* effect, Figure 3.5b shows that the change in the chemical shift for the protons on the ammonia molecule that is *trans* to the water molecule, ammonia **2b**, for *cis*-[PtCl($^{15}NH_3$)$_2$(H$_2$O)]$^+$ is *greater* than the changes in the chemical shift for the protons on the ammonia molecule that is *cis* to the water molecule, ammonia **2a**. While not shown in Figure 3.5b, the ^{15}N chemical shift also changes when the bound water molecule in the complex *ionizes* and loses its proton.

Figure 3.5c shows that a similar *titration* effect is also observed for *cis*-[Pt($^{15}NH_3$)$_2$(H$_2$O)$_2$]$^{2+}$, which has *two* acidic water molecules. While it might be expected that deprotonation of one of the water molecules would immediately remove the symmetry equivalence of the two ammonia 'reporter ligands' that are opposite the water molecules, which the NMR measurement would detect, this is not the case. What actually happens is that when a proton is removed from one of the water molecules, there is an exchange of protons between both water molecules, and since this process is faster than the NMR instrument can detect (the chemical exchange is fast on the NMR time scale), both sites look identical. This means that rather than detecting separate peaks for the ammonia molecules, only a single peak is observed, which moves as the pH of the medium is changed; that is, loss of a proton occurs (Figure 3.5c).

The final point to make in connection with the reaction of cisplatin in water is that the midpoints of the titration-like curves shown in Figure 3.5b and c are the pK_as for the bound water molecules; that is, *the pH at which the concentrations of the fully protonated and fully deprotonated species are identical*. While this is strictly true for *cis*-[PtCl($^{15}NH_3$)$_2$(H$_2$O)]$^+$ (Figure 3.5b), which has only one water molecule that titrates to produce *cis*-[PtCl(HO)($^{15}NH_3$)$_2$], the diaqua complex, *cis*-[Pt($^{15}NH_3$)$_2$(H$_2$O)$_2$]$^{2+}$ (Figure 3.5c), has *two*

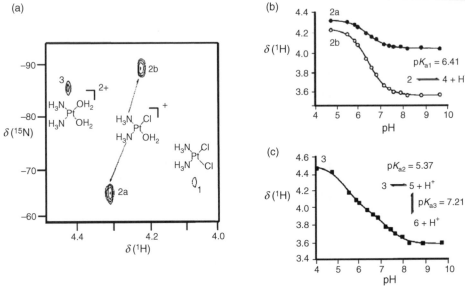

Figure 3.5 (a) [^{1}H, ^{15}N] HMQC NMR spectrum (^{1}H, 500 MHz) of a 5 mM solution containing cis-[PtCl$_2$(^{15}NH$_3$)$_2$] and its hydrolysis products in 95% H$_2$O/5% D$_2$O, pH 4.72, 300 K. Plots of the Pt-NH$_3$ ^{1}H NMR chemical shifts vs. pH for (b) cis-[PtCl(^{15}NH$_3$)$_2$(H$_2$O)]$^+$ and (c) cis-[Pt(^{15}NH$_3$)$_2$(H$_2$O)$_2$]$^{2+}$ allowed direct determination of the acid dissociation constants pK$_{a1}$, pK$_{a2}$, pK$_{a3}$. Reprinted from Prog. Nucl. Mag. Res. Sp., 49, S.J. Berners-Price et al., Insights onto the Mechanism of Action of Platinum Anticancer Drugs from Multinuclear NMR Spectroscopy, 65–98. Copyright 2006, with permission from Elsevier.

acidic protons, one on each water molecule. Since the pK$_a$ values for the two deprotonations are not the same but are not far apart, the titration curves for both overlap. In this case, the observed chemical shift change corresponds to the loss of a total of *two protons* and extracting pK$_a$ values for each from the data requires a fitting procedure.

3.2 Formulation, administration and pharmacokinetics

A common formulation of cisplatin, called Platinol-AQ, which is available in a ready-to-use form, is an aqueous solution of *cis*-[PtCl$_2$(NH$_3$)$_2$] containing 3.3 mM drug (1 mg ml^{-1}) and 154 mM sodium chloride, NaCl (normal saline solution) at a pH that has been adjusted to 3.5–4.5 by the addition of NaOH and/or HCl [29]. Some preparations also contain mannitol. Since additional chloride ion is added to the solution, the equilibria shown in Figure 3.3 lie to the left in favor of the dichloro form of the drug. Using the thermodynamic data given in Table 3.1, the ready-to-use Platinol-AQ solution at equilibrium contains ~95% *cis*-[PtCl$_2$(NH$_3$)$_2$], ~5% *cis*-[PtCl(NH$_3$)$_2$(H$_2$O)]$^+$ and a small amount of *cis*-[Pt(NH$_3$)$_2$(H$_2$O)$_2$]$^{2+}$. Although the presence of the aquated mononuclear forms can lead to the formation of multinuclear complexes that have [Pt(NH$_3$)$_2$]$^{2+}$ units bridged by hydroxide ligands, called *μ-hydroxo species*, the low pH suppresses the formation of these compounds [10].

Doses of cisplatin are based on the body surface area (BAS) of the patient (usually given in square meters, m^2; see Box 4.1), which can be estimated from the height and weight of the individual. Typical doses of cisplatin are in the range 20–140 mg m^{-2}; depending on the type of cancer and other factors, it can be administered as a single dose, a bolus injection or more commonly as an infusion into the blood over a period of hours. Patients are hydrated and then given an aqueous solution of the drug in reduced saline containing dextrose and mannitol in a intravenous drip. Depending on how long the infusion solution stands between its preparation and administration to the patient, and exactly what it contains, speciation would be expected to shift according to the equilibria and kinetics given in Figure 3.3 and Table 3.1.

The half life for the rapid phase of clearance of platinum from plasma after infusion of cisplatin is terminated is $\sim$0.5 hours, with most of the platinum being eliminated in the urine [29,30]. The term 'platinum' here is all chemical forms containing platinum that pass through a low-molecular-weight cutoff filter, sometimes called *ultrafilterable platinum*. Concentrations of platinum are highest in liver, prostate and kidney and lowest in bowel, heart and brain tissue, and detectable levels of platinum are present in tissue 180 days after administration [29].

A limitation of cisplatin therapy is nephrotoxicity (kidney damage), which becomes worse with continued use of the drug [29,31]. While the molecular basis for renal damage is not known, the fact that the drug accumulates in the kidney and may metabolized to other toxic platinum species that produce stress responses in renal tubular cells could be why it causes damage to the kidney.

3.3   Reaction of cisplatin in biological media

Introduction of the infusion solution of cisplatin into the blood causes the drug to undergo chemical changes. Since the pH of the infusion solution and of the blood are not likely to be the same (the pH of blood is $\sim$7.4) and proton transfer is rapid, the relative amounts of the aqua/hydroxo forms of cisplatin rapidly reequilibrate in blood according to the equilibrium constants (pK_a values) given in Table 3.1. In the case of **2** (**4**) of Figure 3.3, this proton equilibration is governed by pK_{a1}, which is 6.41 (27 °C), which means in blood the hydroxo form, **4**, predominates. While it is well known that hydroxide ion is a poorer leaving ligand than water in a substitution reaction and thus hydroxo complexes are less reactive than their aqua counterparts (negatively-charged OH$^-$ resists leaving positively-charged Pt^{+2}), this may not be much of a factor at pH 7.4 for the aqua/hydroxo forms of cisplatin. Since these two forms are in a proton equilibrium with each other, depletion of one form through a chemical reaction would result in the immediate replenishment of that form by proton transfer, so the total 'pool' of **2** (**4**) would be expected to react through the aqua species to form product.

Depending on the chloride concentration in the infusion solution, the distribution of chloro and aqua species in blood could change according to the scheme shown in Figure 3.3. If the infusion solution contains [Cl$^-$] > 105 mM (the chloride concentration in blood), Le Chatelier's principle states that the system will move to the *right* in favor of aquated species according to the rate parameters given in Table 3.1. If, in the infusion solution, [Cl$^-$] < 105 mM then the system will move to the *left* in favor of the dichloro form of the drug.

An interesting observation on what happens when cisplatin is introduced into cell culture media, which is similar in composition to blood, was published by Tacka *et al.* [32]. In this study, ^{15}N-labeled cisplatin was allowed to come to equilibrium in an aqueous solution containing 154 mM NaCl (normal saline), which displaced the equilibria in Figure 3.3 in favor of the dichloro form of the drug. Next, a portion of the solution was added to culture medium which contained RPMI (a complex mixture of amino acids, vitamins, various ions, buffers, with [Cl$^-$]$\sim$105 mM) and 10% fetal bovine serum (FBS), which contains a number of different proteins, and ^{1}H-^{15}N HSQC NMR data as a function of time were collected (Figure 3.6). As is evident in Figure 3.6a, the HSQC NMR spectrum shows the presence of two peaks, one at ^{1}H/^{15}N, $\delta = 4.09/-68.0$ ppm – due to the two equivalent ammonia molecules of cisplatin – which is overlapped with one of the peaks of *cis*-[PtCl(OH)(^{15}NH$_3$)$_2$] (the ammonia molecule *trans* to Cl$^-$), and a second, weaker one at ^{1}H/^{15}N, $\delta = 3.61/-80.5$ ppm – which is due to the ammonia molecule of *cis*-[PtCl(OH)(^{15}NH$_3$)$_2$] – which is *trans* to OH$^-$ (Figure 3.3). By measuring the HSQC NMR peak intensities with time (Figure 3.6b), and after correcting for

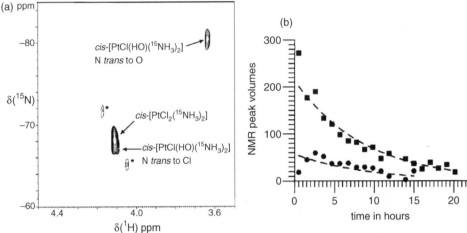

Figure 3.6 (a) ^1H-^{15}N HSQC NMR spectrum (500 MHz, ^1H) of 65 μM cis-[PtCl$_2$($^{15}NH_3$)$_2$] in RPMI-1640 cell culture medium plus 10% fetal bovine serum (FBS), pH 7.4, 3 hours following the addition of labeled cisplatin to the medium. (b) NMR peak intensities as a function of time in the culture medium with exponential fits to the data. Squares are intensities for the peak at $^1H/^{15}N$, δ = 4.09/−68.0 ppm, which is due to both cisplatin and the peak for N trans to Cl of cis-[PtCl(HO)($^{15}NH_3$)$_2$], while circles are for intensities of the peak at $^1H/^{15}N$, δ = 3.61/−80.5 ppm, due to N trans to O of cis-[PtCl(HO)($^{15}NH_3$)$_2$]. Reprinted with permission from K.A. Tacka et al., Experimental and Theoretical Studies on the Pharmacodynamics of Cisplatin in Jurkat Cells, Chem. Res. Toxicol, 17, 1434–44. Copyright 2004 American Chemical Society

the fact that two peaks for the two complexes are overlapped, the investigators determined that the pseudo first-order rate constant for the loss of the cisplatin in culture medium is $\sim 5.7 \times 10^{-5}\,s^{-1}$ (37 °C), which is nearly the same as the rate constant for the hydrolysis of cisplatin in water given in Table 3.1. Also, from the behavior of both NMR peaks with time, it appears that both complexes react with substances in the medium at about the same rate but, interestingly, no new HSQC NMR peaks for products appear. Since RPMI contains methionine ($\sim 115\,\mu M$), cysteine ($\sim 370\,\mu M$) and GSH ($\sim 3\,\mu M$), all of which contain nucleophilic sulfur, lack of observation of HSQC NMR peaks for Pt-S adducts is unexpected. While the agents in the medium causing the disappearance of the peaks are unknown, the study shows that both cisplatin and its monoaqua/hydroxo complex survive in a medium containing many potential platinum nucleophiles for a relatively long period of time.

If the infusion of cisplatin into the blood of the patient is over a short period of time, for example less than ~ 1 hour, a significant fraction of the drug remains in a low-molecular-weight, *ultrafilterable* form in blood [33,34]. However, if the infusion is over a longer period of time, for example 20 hours, all of the platinum in the blood is bound to high-molecular-weight protein molecules and is not ultrafilterable. Since the most abundant protein in blood is human serum albumin (HSA, $\sim 600\,\mu M$), Sadler and coworkers used ^{15}N-labeled cisplatin and HSQC NMR to study the interaction of cisplatin with HSA [35]. As shown in Figure 3.7, 13 hours after the reaction of cisplatin with HSA a number of product peaks with bound ammonia molecules and different ligands provided by HSA can be observed. Based on the location of the ^{15}N chemical shifts, three of the peaks are due to N *trans* to S (Figure 3.4). Human serum albumin has 35 cysteine residues, but 34 of these are involved in 17 disulfide bonds which are not very nucleophilic toward Pt^{+2}. The free cysteine (thiol), Cys$_{34}$, and the six Met (thioether) residues of HSA were considered as possible candidates for reaction with cisplatin. By chemically blocking the single free cysteine residue, Cys$_{34}$, on the protein, as well as methionine (Met) and histidine (His) residues, the researchers were able to show that the thioether sulfur atom in methionine (possibly Met$_{298}$) is most likely the ligand *trans* to N for the major peak, 'a', in the spectrum. Also, since the

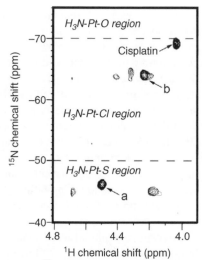

Figure 3.7 *^1H-^{15}N HSQC NMR of 1 mM^{15}N-labeled cisplatin plus 1 mM human serum albumin in 10 mM phosphate, 100 KCl, pH 6.4, 310 K, 13 hours after mixing. The chemical shift regions for NH$_3$ trans to X, where X = O, Cl or S are also indicated. Adapted from A.I. Ivanov et al., Cisplatin Binding Sites on Human Albumin, J. Biol. Chem., 1998, 273, 14721–30*

intensity of this peak increases with time at approximately the same rate as peak 'b' in the spectrum, both peaks, 'a' and 'b', are probably caused by the same platinum complex. Based on the observed chemical shifts, the investigators proposed that the cisplatin-HSA adduct has a thioether sulfur atom and a nitrogen atom from the protein opposite the two ^{15}N-labeled ammonia molecules in the complex. At later times, other cisplatin-HSA adducts form, with other Met residues and also Cys$_{34}$ being ligands. Also at later times, free ^{15}N-labeled ammonia is present in solution, indicating that HSA can eventually displace coordinated ammonia from the drug. Since sulfur is a strong *trans*-directing ligand, which would labilize the group *trans* to it in a substitution reaction, this behavior – that is, loss of $^{15}NH_3$ *trans* to S at later times – is expected.

3.4   Uptake, cytotoxicity and resistance

A major problem with cisplatin therapy is that repeated exposure to the drug causes the cancer being treated to become resistant to it, which ultimately forces termination of therapy. While it would seem reasonable to simply increase the dose of the cisplatin given to the patient, the drug has a relatively narrow *therapeutic window*, which means that the difference between the dose required for an anticancer effect and a dose that is toxic to the patient is small. When cells become resistant to the cisplatin by repeated exposure to the drug it is noticed that they bind much less platinum than normal or sensitive cells that have not been repeatedly exposed to it. From the molecular standpoint, resistance to cisplatin can be caused by cells becoming more efficient at blocking platinum from entering them, effectively moving it back across the membrane (sometimes in a chemically modified form), or if it has made its way to an important target such as DNA, removing it from the DNA via a DNA repair processes.

3.4.1   Influx and efflux of cisplatin

Cisplatin is believed to be internalized by the cell through a combination of passive diffusion and active transport mechanisms [36,37]. The former, because it relies only on the difference in the concentration of drug outside and inside of the cell, does not provide a basis for explaining why cells become resistant to the drug. In the case of passive diffusion, the platinum species being transported across the membrane are assumed to be the dichloro form and the monohydroxo form of the drug, both of which are neutral and carry no formal charge (Figure 3.3), and thus would be more lipophilic than the charged forms of the drug.

In 2000 Komatsu *et al.* reported that there may be a connection between the transporter proteins responsible for regulating the amount of copper that is internalized by the cell and the pharmacology of cisplatin [38]. These researchers found that cells that were resistant to cisplatin had higher amounts of the copper *efflux* proteins ATP7A and ATP7B, suggesting that in addition to removing copper from the cell, the transporters might also be able to efflux platinum. In subsequent work, Lin *et al.* [39] and Ishida *et al.* [40] showed that yeast cells which had fewer copies of the copper influx protein CTR1 were resistant to cisplatin, suggesting that cisplatin may be using the copper transporter protein to enter the cell. Since blocking platinum from entering the cell or removing it from the cell reduces the net amount of platinum available for disruption of cellular functions, recent attention has focused on the copper transport system as a means of explaining the resistance associated with cisplatin [41–43].

Jaehde and coworkers measured the intracellular platinum accumulation, DNA platination, DNA repair and cytotoxicity of cisplatin in two cisplatin-sensitive and -resistant cell line pairs (ovarian A2780/A2780cis and cervical HeLa/HeLaCK cells) [41]. The relative amounts of the copper influx protein CTR1 and two copper efflux proteins, ATP7A and APT7B, in each type of cell were measured using quantitative polymerase chain reaction (PCR) methods. The researchers found that the resistant cell lines, which were made resistant by continued prior exposure to cisplatin, had lower concentrations of the copper transporter CTR1 and, when compared to cisplatin-sensitive cells, also accumulated reduced amounts of platinum (Figure 3.8a and b). This showed that *reduced* amounts of CTR1 protein correlate with *increased* cellular resistance to cisplatin. In an effort to determine whether the copper efflux proteins ATP7A/B influence the amount of platinum that is removed from the cells by efflux mechanisms, the sensitive and resistant cells were treated with cisplatin-containing medium for a period of time, the medium was removed and replaced with drug-free medium, and, at different times, the cells were analyzed for their platinum content. As is shown in Figure 3.8c, the ovarian cancer line (A2780) released platinum within ~10 minutes, with both sensitive and resistant cells behaving the same. This contrasted with the cervical cancer line (HeLa), which did not release platinum, though, as in A2780, both its sensitive and its resistant cells did behave the same (Figure 3.8d). Since there are two copper efflux transporters, ATP7A and ATP7B, the relative amounts of which were different for the various cell types, the effect of each efflux transporter was difficult to determine. However, the fact that no difference in efflux between sensitive and resistant cells was observed led the researchers to conclude that ATP7A and ATP7B are as important as CTR1 in explaining the cisplatin resistance in the cell lines studied. The work also showed that the sensitive cells had more platinum bound to their genomic DNA than was the case for resistant cells, but that there was no difference in the rate at which platinum was removed from DNA due to repair for the sensitive and resistant cells. The study supported the earlier observations that the copper transporter CTR1 may be important in explaining the resistance developed by cells to cisplatin.

In a recent study, Lee and coworkers used fluorescence spectroscopy to study structural changes in the CTR1 protein imbedded in the membrane of yeast cells in the presence and absence of copper ions and cisplatin [44]. The approach used in the study was to attach two different florescent molecules – ECFP, a cyan dye, and EYFP, a yellow dye – to the carboxyl terminus of the CTR1 protein to produce two different conjugates, CTR1-ECFP and CTR1-EYFP. The optical properties of the dyes were such that the fluorescence spectrum of ECPT overlapped the absorption spectrum of EYFP. If the dyes in these conjugates are physically close to each other (10–100 Å), excitation of the donor dye, ECFP, results in the transfer of emitted fluorescence radiation to the acceptor dye, EYFP, which enhances the fluorescence of the latter. This distance-dependent energy transfer process, which is called *fluorescence resonance energy transfer* or *FRET*, is a useful technique for determining the proximity of the two dyes under different conditions. In this study, the

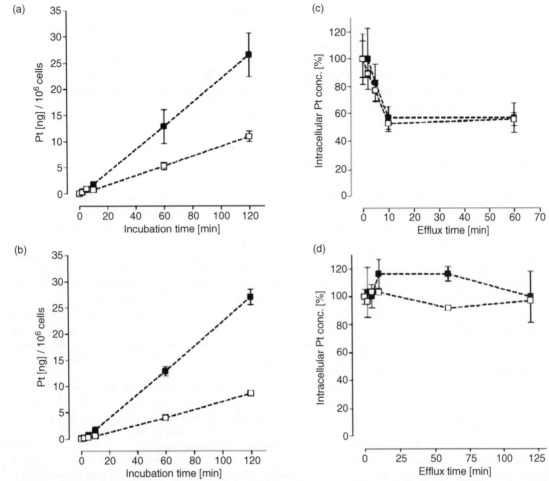

Figure 3.8 *Intracellular platinum concentration (amount of Pt absorbed due to influx), (a) sensitive ovarian A2780 cells, filled squares, cisplatin resistant A2780 cis cells, open squares; (b) cervical HeLa cells, filled squares, cisplatin resistant HeLa CK cells, open squares. Intracellular platinum concentration as a function of time after exposure to cisplatin (efflux of Pt from the cells), (c) sensitive ovarian A2780 cells, filled squares, cisplatin resistant A2780 cis cells, open squares; (d) cervical HeLa cells, filled squares, cisplatin resistant HeLa CK cells, open squares. Adapted from Fig. 4 p. 302, and Fig 5, 303, Zisowsky, J.; Koegel, S.; Leyers, S.; Devarakonda, K.; Kassack, M. U.; Osmak, M.; Jaehde, U. Relevance of Drug Uptake and Efflux for Cisplatin Sensitivity of Tumor Cells. Biochem. Pharmacol. 2007, 73, 298–307*

investigators incorporated CTR1-ECFP and CTR1-EYFP into the membrane of yeast cells and found that the FRET effect was increased, indicating that the labeled proteins are spatially close to one another when they are bound in the membrane. This observation is consistent with a structural analysis that indicates that CTR1 actually forms a homotrimer – that is, three proteins are spatially close to one another – when the active protein is imbedded in the membrane [45]. After establishing the presence of labeled proteins in the membrane, the investigators added Cu^{+2} and Cu^{+} to the medium above the cells and observed that the latter ion had a greater effect on FRET, suggesting that Cu^{+} may be the more important oxidation state of the ion for transferring copper into the cell. Interestingly, while CTR1 appears important for uptake of cisplatin into the cell, the presence of the platinum drug in the medium did not affect the FRET effect of copper-bound or copper-free CTR1-ECFP/EYFP in the membrane. The authors suggested that while CTR1 is important for the uptake of cisplatin into the cell, the drug may use the protein in a way that is different to that employed by copper. Since CTR1 forms a homotrimer with a pore or channel through the membrane, it could be that the drug and/or its aquated products pass through the pore to enter the cell, rather than interacting directly with the methionine thioether residues that are located in the copper-binding region of CTR1.

Box 3.3 Influx, efflux and drug transport

The cellular membrane contains many features that allow nutrients and essential elements to enter and waste products and toxins, called *xenobiotics*, to be eliminated. While there are many mechanisms for bringing materials into the cell – a process called *influx* – one transport system that brings copper into the cell may be important for the platinum drugs and other metal-containing agents that are used in medicine.

Since simple aquated Cu^+ and Cu^{+2} ions are toxic to most organisms, nature has developed a sophisticated transport mechanism, called copper *homeostasis*, for moving copper obtained from a food source to the interior of the cell where the ion is needed for the function of a number of metallo-proteins and enzymes [46–49]. Located in the cell membrane is the 190 amino acid *trans*-membrane copper transporter protein CTR1, which is responsible for moving copper, as Cu^+, from the extracellular space to the cytosol; that is, for *influx* of copper into the cell (Figure 3.9). The copper-binding ligands on CTR1 are two methionine residues (thioether donors) at positions 43 and 45 and two histidine regions (histidine has an imidazole side chain) near the amino terminus of the protein, which projects into the extracellular space outside the cell. An individual protein monomer has three *trans*-membrane domains and there is evidence that three monomers may bind next to each other to form a *homotrimer*, which is the active structural unit that is imbedded in the membrane [45]. While the details of the copper internalization mechanism are not known, the metal ion may pass through a pore-like structure formed by the trimeric protein in the membrane. Once inside the cell, the copper is sequestered by chaperone proteins, which transport the Cu^+ ions to two much larger, $\sim$1500 amino-acid proteins, ATP7A and ATP7B, which are *trans*-membrane proteins in the *trans*-Golgi network (TGN) in the cell (Figure 3.9). These two proteins have six cytosol-facing cysteine-rich copper-binding sites that capture the copper and move it inside the TGN by a process that requires ATP. When the copper is needed, it is transported from the TGN – a process called *efflux* – with the help of chaperone proteins. If efflux involving ATP7B is impaired, as in patients with *Wilson disease*, excessive copper is released to the cytosol of hepatocytes, which ultimately results in high copper levels in blood and accumulation of copper in the brain, kidney and cornea of the individual. If the gene for ATP7A is missing on the X-chromosome of humans, it causes *Menkes disease*, which is characterized by *lack* of available copper, especially in the brain. This fatal genetic disease results in progressive neurodegeneration associated with psychomotoric retardation and usually causes death at an early age.

While many of the chemical details concerning copper transport into the cell are unknown, there is evidence that changes in the oxidation state of the copper ion may be involved. Copper has two biologically-common oxidation states, Cu^+ and Cu^{+2}. The monovalent oxidation state mainly forms three-coordinate, trigonal complexes but two-coordinate, linear, and four-coordinate, tetrahedral, compounds of Cu^+ are also known (Figure 3.9). For the divalent oxidation state, four-coordinate, square planar, and its expanded variants, five-coordinate, square pyramidal, and six-coordinate, tetragonal (Jahn–Teller distorted), are often found. Since Cu^+ is a soft acid, it forms stable complexes with soft bases such as thioether sulfur (methionine) and thiolate (cysteine), while Cu^{+2} is of 'intermediate hardness', forming stable complexes with intermediate bases such as the nitrogen atoms of the imidazole group of histidine (Table 1.7). Thus, if the oxidation state of the copper ion is modulated by redox components in the cell, the affinities of the copper ion for different donor atoms can also be changed, making is possible to easily move the ion between sites, which is important in the transfer process.

The process of removing toxic substances from the cell is called *efflux*. In cancer chemotherapy continued exposure to drug usually causes cancer cells to find ways to rapidly remove the toxin from the cytoplasm and 'pump' it to the extracellular space. This phenomenon, called *multidrug resistance (MDR)*, which causes the cells to tolerate increasing concentrations of drug, severely limits the effectiveness of antitumor compounds and is the main reason that cancer chemotherapy is often discontinued. At the heart of MDR is the *trans*-membrane protein *p-glycoprotein, P-gp*, which is a member of the ABC transporters responsible for removing *xenobiotics* (toxins) from the cell [50]. The protein, which has a molecular weight of 170 kDa, consists of a single polypeptide chain with 12 *trans*-membrane α-helical segments that are grouped into two 'halves', each containing six helices. As is evident from the crystal structure of mouse P-gp, which has 87% sequence homology with human P-gp, the two halves of the protein form an arrowhead shape, with the point of the arrowhead protruding into the extracellular space (Figure 3.10). The drug-binding internal cavity of P-gp, which is located where the two halves meet on the cytosolic side of the protein, consists of a number of hydrophobic residues. These residues form a large cavity, ~6000 Å^3, capable of accepting a wide variety of different low-molecular-weight molecules in a nonspecific manner. Located at the ends of the two halves of the protein that project into the cytoplasm are nucleotide-binding domains (NBDs), which can bind ATP.

While the detailed mechanism of drug efflux from the cell by P-gp is not known, one proposal, shown in Figure 3.11, is that a drug molecule that is destined to be eliminated from the cell binds to the internal cavity of the protein, which causes binding of ATP to the NBDs [50]. It is known that drug efflux by P-gp is an energy-requiring process that involves ATP, which is hydrolyzed to ADP and inorganic phosphate. Binding of ATP to NBD1 and NBD2 causes the two 'arms' of the cytosolic portion of the protein to come together – dimerize – which induces a conformational change in the protein and exposes the captured drug molecule to the extracellular space, allowing the drug molecule to diffuse away from the cell. This mechanism explains the energy-driven unidirectional transport of a drug molecule from the cytosol to the extracellular space by P-gp.

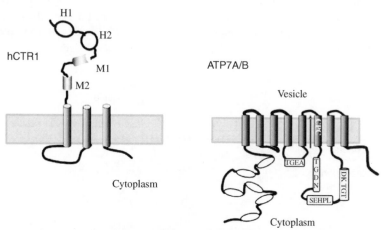

Figure 3.9 *Copper transporters. Left: schematic diagram of the monomeric form of the copper transporter protein hCTR1 in the plasma membrane (shaded rectangle). CRT1 is a 190-amino-acid protein with three trans-membrane domains and may form a homotrimer in the membrane. Transport of copper into the cell requires methionine residues at positions 43 and 45 (M1 and M2) and two histidine residues (H1 and H2). Right: schematic diagram of the domain structure of the copper transporter proteins ATP7A and ATP7B. The important regions of these proteins include: a phosphatase domain, TGEA; the CPC (cysteine and proline) and SEHPL sequences; a DKTGT motif that contains an invariant aspartate (D) residue; the TGDN that binds ATP; and a conserved `hinge' region that provides a flexible loop connecting the ATP-binding domain to the trans-membrane portion of the molecule. The N-terminal six metal-binding domains with CXXC motifs and the eight trans-membrane domains are also depicted. Reprinted from Cancer Lett., 234, R. Safaei, Role of Copper Transporters in the Uptake and Efflux of Platinum Containing Drugs, 34–39. Copyright 2006, with permission from Elsevier*

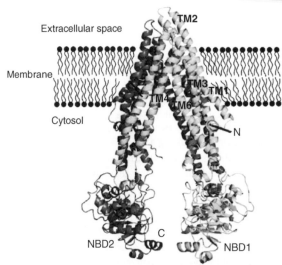

Figure 3.10 *Structure of the trans-membrane protein p-glycoprotein, P-gp. The protein has 12 trans-membrane α-helical portions, some of which are labeled TM, and two nucleotide-binding domains (NBD1,2). The amino and carboxyl termini of the protein are labeled N and C, respectively. Adapted from S.G. Aller et al., Structure of P-Glycoprotein Reveals a Molecular Basis for Polyspecific Drug Binding, Science, 2009, 323, 1718–22*

A recent study by Yonezawa *et al.* [51] measured the cytotoxicity and uptake of cisplatin and three other clinically-used platinum drugs, carboplatin, oxaliplatin and nedaplatin, using human embryotic kidney (HEK) 298 cells (Figure 3.12). By inserting a plasmid DNA *vector* carrying a gene for a specific human organic cation transporter (hOCT) protein into the cell, the researchers produced HEK 298 cells with an excess of a particular hOCT in the cell membrane. As is evident from Figure 3.12a, HEK 298 cells containing an abundance of the

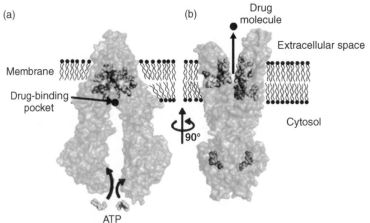

Figure 3.11 *Model of drug transport by P-gp protein. (a) The drug (black circle) gains entry into the cytoplasm and becomes bound to the internal cavity located between two 'halves' of the protein. (b) Rotation of the P-gp protein in the membrane by 90°, showing that ATP binding induces dimerization of the nucleotide-binding domains, NDB, causing a structural change in P-gp which exposes the drug molecule to the extracellular space. Adapted from S.G. Aller et al., Structure of P-Glycoprotein Reveals a Molecular Basis for Polyspecific Drug Binding, Science, 2009, 323, 1718–22*

human organic cation transporter hOCT2 are more susceptible to being killed by cisplatin, as measured by the amount of the enzyme lactate dehydrogenase (LDH) that is released to the medium from the dead cells. These cells also absorb more cisplatin (Figure 3.12b) than cells with an over-expression of the other transmitters, showing that hOCT2 is important in transferring cisplatin into and killing the cells.

3.4.2 Modification by glutathione and metallothionein

When cells develop resistance to cisplatin there is evidence that the concentrations of glutathione (GSH) and metallothionein (MT), which are normally in the low mM range, are increased to larger values Box 2.1 [37]. Since GSH and MT form very stable Pt^{+2}-S adducts, which effectively compromises the ability of the platinum ion to react with other molecules in the cell, the increase in the levels of these compounds in resistant cells is viewed as a molecular resistance mechanism for the platinum drugs. While the ultimate fate of cisplatin inactivated by GSH is not known with certainty, there is evidence that the modified platinum could be exported from the cell by *GS-X efflux pumps* [37].

Souid and coworkers used UV–visible spectroscopy and HPLC to study the kinetics of reaction of cisplatin with GSH and Zn_7MT under physiological conditions [52,53]. The second-order rate constants for the reactions were 0.013 and 0.65 M^{-1} s^{-1}, respectively, showing that MT is ~50 times more effective in binding cisplatin than GSH, suggesting that the protein could play an important role in resistance.

Vasák and coworkers also studied the reaction of cisplatin, as well as the *trans* isomer transplatin, with Zn_7MT using UV–visible absorption spectroscopy and mass spectrometry [54]. They found that cisplatin readily reacts with Zn_7MT, with the accompanying displacement of Zn^{+2} ions, and that due to the *trans* effect, cisplatin loses all four of its original ligands. The investigators suggested that thiolates of MT initially displace the chloride ligands of cisplatin, and because thiolates are strong *trans*-directing groups, the two ammonia molecules of the drug that are *trans* to the thiolates are rapidly lost by the attack of other MT thiolates on the platinum center. Interestingly, transplatin retains its two ammonia molecules upon reaction with Zn_7MT, which can also be explained by the *trans* effect. The investigators suggested that the reason that the two ammonia molecules remain in the transplatin-MT adduct is because thiolates of MT displace the two chloride ligands and thus are positioned *trans* to each other in the platinum adduct and not *trans* to the ammonia molecules as is the case for the MT adduct with cisplatin.

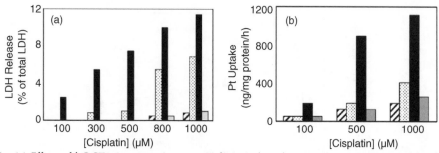

Figure 3.12 (a) Effect of hOCT1-3 expression on cisplatin-induced cytotoxicity as measured by the release of lactate dehydrogenase, LDH, to the culture medium. HEK298 cells were transfected with empty vector (stripe) hOCT1 (dots), hOCT2 (black), or hOCT3 (gray). (b) Uptake of cisplatin, ng Pt/mg protein/h, as measured by ICP-MS, by HEK298 cells expressing hOCT1, hOCT2, and hOCT3. Adapted from Fig. 4A and 5A p. 883 of; Yonezawa, A; Masuda, S; Yokoo, S; Katsura, T; Inui, K. Cisplatin and Oxaliplatin, but Not Carboplatin and Nedaplatin, are Substrates for Human Organic Cation Transporters (SLC22A1-3 and Multidrug and Toxin Extrusion Family). J. Pharmacol. Exp. Ther. 2006, 319, 879–886

3.4.3 DNA repair

Since the earliest reports of the anticancer effects of cisplatin by Rosenberg and coworkers, DNA has been suggested as the most likely target for the anticancer drug in the cell [1]. When DNA evolved as the molecule containing the genetic information for the organism, it became important to 'monitor' the polymer to make sure that both of its strands contained the correct sequence of nucleotides necessary for the organism to survive. Substances that chemically modify DNA, like UV light, reactive radicals and drugs, posed a threat to the survival of the organism, so nature developed a sophisticated system for checking and repairing DNA. One repair system, called *nucleotide excision repair* (*NER*), uses enzymes to remove chemical modifications that occur on DNA [55,56]. If one strand of double-stranded DNA is chemically modified in some way, NER enzymes remove a segment of DNA containing the damaged site and, by reading the opposing strand, replace the damaged segment with new DNA. Sometimes a chemical modification to DNA causes the nucleotide opposite the modification to be misread and when DNA is replicated the incorrect nucleotide can be inserted into the synthesized 'daughter' strand. This results in double-stranded DNA that has a mismatch of bases, or non-Watson–Crick base pairs, for example a guanine opposite a thymine, a guanine opposite a guanine and so on, which, if not corrected, could cause the cell to produce inactive or otherwise defective proteins. To correct this problem, the *mismatch repair* (*MMR*) system is activated and enzymes remove a section of DNA containing the incorrect nucleotide and replace it with the correct nucleotide.

The most common type of lesion on double DNA for cisplatin is the addition of the *cis*-$[Pt(NH_3)_2]^{2+}$ unit of the drug to the dinucleotide sequence, GG, which produces the adduct, Pt-GG, producing a *1,2 intrastrand crosslink* [57]. The Arabic numbers refer to two adjacent nucleotides in the sequence and '*intra*' means that the crosslink is on the same strand of DNA. While there are many studies addressing the repair of cisplatin-DNA adducts in the literature – see reviews by Jung and Lippard [57] and Kartalou and Essigmann [58] – a recently published investigation by Thomale and coworkers used an antibody specific for Pt-GG to determine the rate of platinum removal from DNA in single kidney cells derived from two types of mice after the mice were given an injection of cisplatin [59]. The investigators used an immunocytological assay (ICA) that combined the high specificity of a monoclonal antibody (*mab*, R-C18) that was able to recognize Pt-GG with a microscope imaging system that could quantitatively measure the amount of Pt-adduct present on DNA in the cell. As shown in Figure 3.13, the kidney cells of mice that have a normal DNA repair system were able to remove some

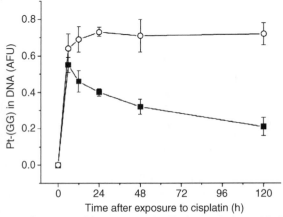

Figure 3.13 *Relative amounts of Pt-(GG) adducts on DNA of kidney tubular epithelium cells of mice to measure nucleotide excision repair, NER, as a function of time after the mice were exposed to cisplatin. XPC$^{+/+}$ (normal, wild-type mice, with intact NER, filled squares) and XPC$^{-/-}$ (mice that were deficient in NER, open circles). From B. Leidert et al., Adduct Specific Monoclonal Antibodies for the Measurement of Cisplatin-Induced DNA Lesions in Individual Cell Nuclei, Nucleic Acids Res., 2006, 34, e47. Reproduced by permission of Oxford University Press*

of the platinum from DNA over a 120 hour period after injection. However, those mice with a compromised NER system, referred to as XPC$^{-/-}$, were not able to remove Pt-GG from DNA in their kidney cells after injection of cisplatin (Figure 3.13), showing that the NER system is indeed important in removing platinum from DNA. The researchers further demonstrated the power of their analytical technique by measuring the platinum-adduct levels in tumor biopsies from patients that had been treated with cisplatin, finding that different cells within the same tissue section had different levels of DNA adduct formation. This type of analysis could be very important for determining which cells in a given population are more susceptible to damage by cisplatin.

3.4.4   Other resistance mechanisms

While the effort to understand why cells become resistant to cisplatin has focused on how the drug enters and is removed from the cell, and how the toxicity of cisplatin can be mitigated once the drug is inside the cell, an interesting report by Centerwall *et al.* [60] suggested that cisplatin may be modified in the medium *outside the cell*. The idea that cells may be able to release substances that threaten their survival to the extracellular space is not a new concept and Morin and coworkers have shown that tumors can recognize and reorganize their microenvironment and modify toxic substances before they cross the cell membrane [61,62]. Centerwall *et al.* showed that the intensities of the ^{1}H-^{15}N HSQC NMR signals for *cis*-[PtCl(OH)(^{15}NH$_3$)$_2$], *cis*-[Pt-(^{15}NH$_3$)$_2$(OH)$_2$] and to a certain extent *cis*-[PtCl$_2$(^{15}NH$_3$)$_2$] (compounds **2/4**, **3/5/6** and **1** of Figure 3.3, respectively) are rapidly diminished ($<\sim$15 minutes) when the compounds are in cell culture medium that has previously been exposed to Jurkat leukemia cells. While the nature of the products formed was not identified, and no ^{1}H-^{15}N HSQC NMR signals were observed for the products, the investigators showed that Jurkat cells that were made resistant to cisplatin modified more of the platinum compounds in the culture medium than sensitive (normal) Jurkat cells, suggesting that there may be a mechanism for modifying platinum compounds before they enter the cell [63].

3.5   Interaction of cisplatin with cellular targets

3.5.1   DNA as a target

In research reports shortly after the discovery of the anticancer properties of cisplatin, Rosenberg and coworkers [1] and Howle and Gale [64] suggested that the cellular target for cisplatin is DNA. In order to determine how the drug could interact with DNA, Rosenberg and coworkers first considered that cisplatin might bridge two strands of DNA, like certain organic bifunctional alkylating agents, to form an *interstrand crosslink*, but since the distance was greater than could be spanned by a single Pt^{+2} ion, the researchers proposed that cisplatin connected two adjacent purine bases on the same strand of DNA to form an *intrastrand purine dimer*. Subsequent studies by many investigators supported this initial idea and detailed studies by Fichtinger-Schepman, Reedijk and their coworkers showed that the major adduct is one with the *cis*-[Pt-(NH$_3$)$_2$]$^{2+}$ unit bound to the dinucleotide sequence, GG – that is, the adduct Pt-GG – with lesser amounts of the adduct Pt-AG, both of which are *1,2 intrastrand crosslinks* [65,66]. These investigators also showed that the drug reacts with the guanine bases in trinucleotide sequences of the type GXG to produce a *1,3 intrastrand crosslink* and that the platinum ion can also link both strands of DNA through guanines to form an *interstrand crosslink*. In all cases, the donor atom of the heterocyclic DNA base used for binding to Pt^{+2} is N-7 of either guanine or adenine.

While there are many studies in the literature addressing the kinetics and mechanism of cisplatin binding to DNA, an NMR investigation by Davies *et al.* [67] typifies the reaction of the drug with DNA, in this case the guanine sites of the oligonucleotide 5'-AATTGGTACCAATT-3'. This short segment of double-stranded

DNA is *palindromic*, meaning that the sequence on both strands, reading from the 5' to the 3' direction, is the same, and each strand is said to be *self-complimentary* in that one strand can *hybridize* with another strand like itself to form a Watson–Crick duplex. The investigators studied the reaction of *cis*-[PtCl$_2$(^{15}NH$_3$)$_2$] with the duplex DNA using ^{1}H-^{15}N HSQC NMR and by measuring the rates at which the NMR peak intensities of the platinum complex change in the presence of the duplex. By comparison with HSQC NMR spectra obtained from the reaction of ^{15}N-labeled cisplatin with other oligonucleotide duplexes, and by fitting the NMR kinetic data to various binding models, they proposed the scheme shown in Figure 3.14 for the reaction

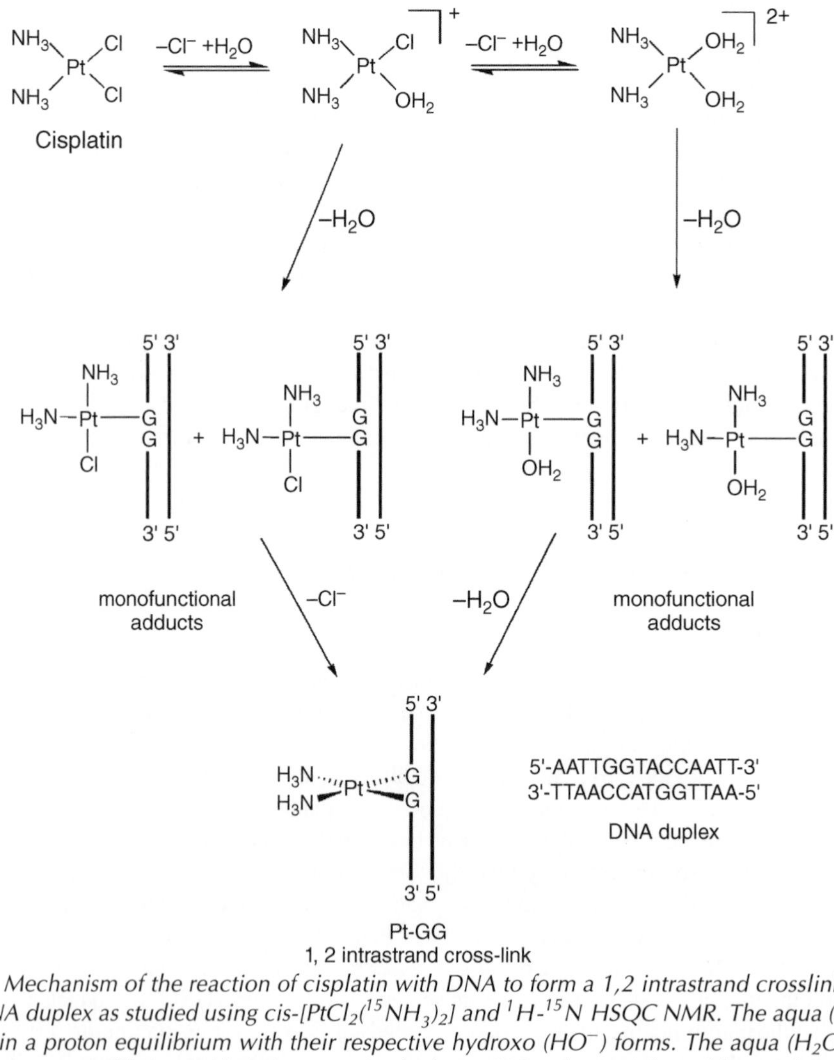

Figure 3.14 *Mechanism of the reaction of cisplatin with DNA to form a 1,2 intrastrand crosslink at GG of the indicated DNA duplex as studied using cis-[PtCl$_2$(^{15}NH$_3$)$_2$] and ^{1}H-^{15}N HSQC NMR. The aqua (H$_2$O) forms of the drug are in a proton equilibrium with their respective hydroxo (HO$^-$) forms. The aqua (H$_2$O) forms of the drug are in a proton equilibrium with their respective hydroxo (HO$^-$) forms. Adapted from Scheme 2.2, p. 5610 of Davies, M.S., Berners-Price, S. J., Hambley, T.W. (2000) Slowing of cisplatin aquation in the presence of DNA but not in the presence of phosphate: improved understanding of sequence selectivity and the roles of monoaquated and diaquated species in the binding of cisplatin to DNA. Inorg. Chem., 39, 5603–5613*

of cisplatin at a GG sequence of double-stranded DNA. The measured second-order rate constants for the reaction of the monoaqua complex to form a monofunctional adduct at the 3' and 5' guanine bases were 0.48 ± 0.19 and $0.16 \pm 0.06\,M^{-1}\,s^{-1}$, respectively, showing that reaction of the G that is toward the 3' end of the duplex is preferred. After the formation of the monofunctional adduct, N-7 on the adjacent guanine attacks the Pt^{+2} ion, which, with the loss of the bound Cl^- ion, produces the 1,2 intrastrand crosslink. The first-order rate constants for this closure are $(2.55 \pm 0.07) \times 10^{-5}$ and $(0.171 \pm 0.011) \times 10^{-5}\,s^{-1}$ for the 3' and 5' adducts, respectively, showing that closure toward the 5' end is much faster than closure toward the 3' end of the duplex. Chottard and coworkers used a kinetic analysis to suggest that the actual DNA binding species may be the diaqua complex, **3** in Figure 3.3 [68]. However, the NMR study by Davies *et al.* [67] showed that this species plays only a minor role (~1% of the platinum reacts with DNA by this route) in the platination of DNA. Interestingly, while cisplatin itself does not react with DNA, its hydrolysis is slowed by 30–40% when the dichloro complex is in the presence of DNA. The investigators suggested that there could be an association between the drug and DNA which limits access of water to the platinum ion and slows the hydrolysis rate of cisplatin.

A useful and easy way to determine whether a metal complex interacts with DNA is to study its effect on supercoiled DNA. If the complex binds to and changes the structure of the closed circular DNA, this effect can easily be detected by subjecting the metallated DNA to electrophoresis in a gel. Using this approach, Scovell and Collart studied the effects of both cisplatin and transplatin on the replicative form of the closed circular phage DNA, ϕX174 DNA [69]. They found that both platinum compounds unwound the closed circular DNA, with cisplatin producing the greater degree of unwinding. While transplatin does not have the proper geometry to form a 1,2 intrastrand crosslink, the compound forms 1,3 intrastrand and interstrand crosslinks, which also unwinds and change the writhe of closed circular DNA.

Keck and Lippard also studied the binding of cisplatin to the closed circular plasmid pUC 19 DNA using agarose gel electrophoresis [70]. As shown in Figure 3.15, pUC 19 DNA exhibits two bands in the gel. The faster-moving band is the closed circular form I DNA, while the slower minor component is the nicked circular form II DNA. The plasmid pUC 19 has 2686 base pairs or 5372 nucleotides. If any nucleotide in the DNA is attacked by some agent that severs the sugar-phosphate backbone, the energy stored in form I DNA will immediately be released, producing form II DNA, which is why, in practice, there is usually some form II DNA

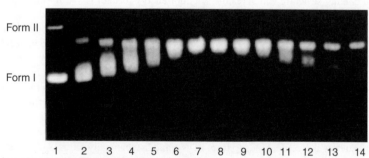

Figure 3.15 *Cisplatin binding to the closed circular DNA plasmid pUC19 as studied using agarose gel electrophoresis followed by staining the DNA with ethidium bromide. The lane number (in parenthesis) and the fraction of platinum bound to DNA per nucleotide are: (1) 0.0, (2) 0.019, (3) 0.032, (4) 0.041, (5) 0.046, (6) 0.061, (7) 0.074, (8) 0.078, (9) 0.087, (10) 0.093, (11) 0.106, (12) 0.123, (13) 0.135, (14) 0.159. The migration direction of the DNA is from top to bottom of the gel. Reprinted with permission from M.V. Keck and S.J. Lippard, Unwinding of Supercoiled DNA by Platinum-Ethidium and Related Complexes, J. Am. Chem. Soc., 114, 3386–90. Copyright 1992 American Chemical Society*

present in laboratory samples. As shown in Figure 3.15, binding cisplatin to form I DNA causes its mobility to decrease, pass through a minimum where it overlaps with form II and, at higher amounts of bound platinum, increase again. This effect is due to the platinum changing the writhe of the closed circular DNA molecule, causing the writhe to change from negative values (*left-hand super helix*), pass through zero (*no super helices present* in the molecule) and ultimately obtain positive values (*right-hand super helix*) (Figure 3.16). Interestingly, form II DNA increases its mobility as platinum is bound; the origin of this change in mobility for form II, which is devoid of super coiling, is unknown [70,71]. The unwinding angles for cisplatin and transplatin reported in the study were $13°$ and $9°$, respectively.

Box 3.4 Supercoiled DNA as a drug substrate

A sensitive and easy-to-use assay for studying drug–DNA interactions is the measurement of drug-induced changes in a *closed circular DNA* molecule using agarose gel electrophoresis. This technique uses a naturally-occurring closed circular DNA, usually a *plasmid*, as a drug binding substrate and measures how fast the DNA molecule with bound drug moves through the gel when current is applied and electrophoresis is carried out. Since the DNA in the gel can be stained with the dye ethidium bromide, EB, which brightly fluoresces under UV light when it intercalates between the base pairs of DNA, the location of DNA in the gel can easily be seen. However, in order for the assay to work with a DNA-binding drug, two important conditions must be met. Since the time required for electrophoresis is relatively long (hours) and the drug needs to be bound to the DNA molecule during migration in the gel, the 'off' rate constant of the drug from DNA must be small. A second important requirement for the technique is that the drug, when it is bound to DNA, must cause a structural change in the DNA that alters the mobility of the DNA in the gel.

When plasmids are created in the cell, the ends of a linear Watson–Crick double-stranded DNA molecule are covalently linked end-to-end to from a circular DNA which has no 'ends' [72]. Since these DNA molecules are quite long, 2–10 kb, the DNA double helix can be gently 'bent' so that the two ends of the strands can be joined together to form circular DNA. However, before covalently linking the ends together, the enzymatic 'machinery' in the cell uses energy to slightly alter the linear DNA molecule by taking out some of its turns; that is, the spiral that is characteristic of the double helix. This *reduces* the angle between individual base pairs of DNA, called the *twist angle*, from the optimal value of $\sim36°$. Since the cell needs to do work on the DNA to reduce the twist angle and seal the ends, the closed circular structure which results is a high-energy form of DNA. In order for the closed circular DNA to return the twist angle to the original value of $\sim36°$, the DNA distorts, introducing *super helical turns* wherein the Watson–Crick double helical stands, which remain intact, pass over one another in a *left-hand* sense to form a second higher-order helix called a *super helix*. This DNA, which is called *supercoiled DNA* or *form I DNA*, looks like a rubber band that you have twisted by rolling it between your fingers. From the standpoint of thermodynamics, the energy that went into making *supercoiled DNA* in the cell is now stored in the *supercoils* of the closed circular DNA. The parameter which indicates the number of *super-helical* turns in the *supercoiled DNA* is the *writhe* (w), with each super-helical turn corresponding to one unit of writhe. The writhe is negative for a left-hand super helix – all naturally-occurring supercoiled DNAs have $w = (-)$ – and is positive – $w = (+)$ – for a right-hand super helix.

If an agent, such as a drug molecule, binds to form I DNA and mechanically reduces the twist angle at the site of binding, the writhe (number of super-helical turns) will be reduced, making the DNA less like a twisted rubber band and more open or doughnut-like in shape. Since the shape of the DNA greatly influences the rate at which it moves through a gel, drug-induced changes in the writhe of form I DNA can easily be detected in a gel electrophoresis experiment.

Consider the binding of the platinum drug cisplatin to form I DNA shown in Figure 3.16. When cisplatin is incubated with DNA, the platinum ion bridges two guanine bases in the sequence GG to form a 1,2 intrastrand crosslink [70,71], which *reduces* the twist angle between the two GC base pairs at the adduct site. The amount by which the drug reduces the normal twist angle of the closed circular DNA is called the *unwinding angle*. This reduction in the twist angle at each platinum site causes the writhe of the DNA to become *less negative*, making the DNA more open or doughnut-like in shape, which slows the migration rate of the DNA in a gel relative to control DNA without bound platinum. Depending on the number of Pt^{+2} ions bound to form I DNA, the writhe of the platinated DNA can range from negative values (left-hand super helices), through zero (no super helices), to positive values (right-hand super helices). In addition to changing the twist angle of DNA, cisplatin also produces a 'kink' in the helix axis at the site of binding which also changes the shape of the closed circular DNA and affects the mobility of DNA in the gel.

While closed circular DNA is a convenient substrate for investigating the binding of drugs to DNA, it is also useful for studying drugs that can cleave the sugar-phosphate backbone of DNA. If an agent breaks the backbone at any point along either strand, either by hydrolyzing the phosphodiester linkage of the backbone or by chemically damaging the deoxyribose sugar, thus breaking the carbon chain of the backbone, all of the energy stored in supercoiling is immediately released and the DNA adopts an open circular structure with no supercoiling ($w = 0$). This form of closed circular DNA is called *nicked circular DNA*, *relaxed DNA* or *form II DNA* (Figure 3.17). If the cutting agent has low or no sequence specificity – that is, if it randomly cuts at all possible nucleotide positions of the DNA – and if it is allowed to cut for an extended period of time, a break in the backbone will eventually occur on one strand near an existing break on the opposing stand. When this occurs the short segment of duplex DNA between the two breaks will melt – that is, the Watson–Crick base pairs will separate – and the DNA will alter its form again to produce *linear DNA* or *form III DNA*. This DNA usually has many breaks in its sugar-phosphate backbone, but since it still has significant Watson–Crick regions it is basically a linear rod-like molecule which moves in the gel at a migration rate that is different from either form I or form II DNA.

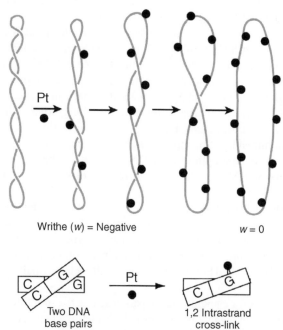

Figure 3.16   *Cisplatin (Pt, black circles) binding to supercoiled form I DNA, which decreases the writhe (w), supercoiling, of DNA. The broad solid line is a Watson–Crick DNA double helix. Also shown is a view down the DNA helix axis that indicates that the formation of the 1,2 intrastrand crosslink decreases the twist angle between the base pairs of DNA, which decreases the amount of supercoiling of the closed circular DNA*

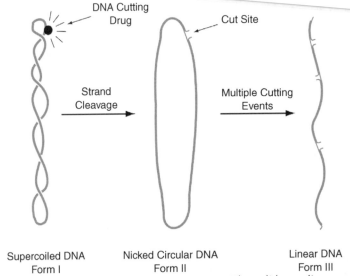

Figure 3.17 *Supercoiled DNA cleaved by a DNA cutting agent. The solid gray line represents Watson–Crick double-stranded DNA*

In 1995 Lippard and coworkers published the single crystal x-ray analysis of cisplatin bound to a 12-mer DNA duplex with the sequence CCTCT**GG**TCTCC/GGAGACCAGAGG, with the binding site of the $[Pt(NH_3)_2]^{2+}$ unit indicated in boldface type in the sequence. The platinated DNA duplex was made by reacting the diaqua form of cisplatin, **3** in Figure 3.3, with the C-rich oligonucleotide DNA strand, which, after purification, was hybridized to the G-rich complimentary strand. This ensured that the $[Pt(NH_3)_2]^{2+}$ unit was attached to only one site; that is, the central GG sequence of the duplex DNA. The structure shows that the platinum ion is bound to N-7 of both guanines located in the major groove and that it significantly distorts the axis of the double helix, bending it toward the major groove, which opens the 'face' of the minor groove of DNA on the opposite side, Figure 3.18. The investigators suggested that the exposed minor groove caused by platinum adduct formation might facilitate the binding of the high-mobility group (HMG) proteins, which have been implicated in DNA repair, to the site, and that there could be a connection between binding of HMG to the site and the tumor-suppressor protein p53, which is also involved in DNA repair [4,73,74].

As shown in Figure 3.18b, the platinum ion is bound to N-7 of both guanine bases, which, because the platinum ion requires square planar coordination with central angles of ~90°, forces the two guanines to tilt inward toward each other and away from their parallel, stacked-type arrangement in normal B-form DNA. While this creates strain in the hydrogen-bond network to their compliment cytosine bases, the Watson–Crick base pairing between G and C is strained but remains intact, with each base pair being 'propeller twisted' by a significant amount.

The idea that DNA is the target for cisplatin was in part based on the structures of cisplatin and transplatin (Figure 1.22) and their observed anticancer properties [1]. Cisplatin, which has *cis* leaving ligands – that is, ligands that are readily lost in a substitution reaction – exhibited potent anticancer effects, while transplatin, with *trans* chloride leaving ligands, exhibited no or only modest antitumor effects. If both compounds lose their chloride ligands in a reaction with a target molecule, and one compound is very active and the other

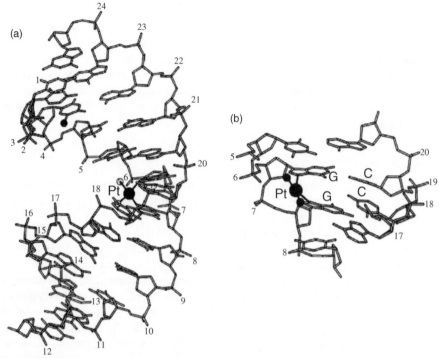

Figure 3.18 *Crystal structure of cisplatin bound to the DNA 12-mer duplex, CCTCT**GG**TCTCC/GGAGACCA-GAGG, with the site of binding, GG, of the [Pt(NH$_3$)$_2$]$^{2+}$ unit indicated in boldface type. (a) The binding of the [Pt(NH$_3$)$_2$]$^{2+}$ unit to the GG sequence on one strand bends the helix axis of the double-stranded DNA toward the major groove of DNA by ~40°. (b) View showing the [Pt(NH$_3$)$_2$]$^{2+}$ unit bound to N-7 of the guanine bases in the GG sequence. The two cis ammonia molecules are black circles above and below the platinum ion. Reprinted by permission from Macmillan Publishers Ltd: P.M. Takahara et al., Crystal Structure of Double Stranded DNA Containing the Major Adduct of the Anticancer Drug Cisplatin, Nature, 377, 649–52. Copyright 1995*

inactive, it makes sense to consider biological targets that could provide two properly spaced donor atoms that could coordinate to *cis* positions of the platinum ion. This thinking led to double-stranded DNA as the likely target because it could provide two donor atoms to the platinum ion by utilizing two adjacent bases on the same DNA strand; that is, the formation of a 1,2 intrastrand crosslink with the platinum ion. However, closer examination of the reason for transplatin only being weakly active showed that it hydrolyzed faster than cisplatin and that more of the compound was inactivated by substances in blood than was cisplatin [11,75,76]. This realization led to the synthesis and testing of many *trans* compounds – that is, complexes which have ligands that are easily displaced in *trans* positions – which were found to have potent cytotoxicities, *in vivo* antitumor properties and, most importantly, activity against cisplatin-resistant cells [77–83]. While these compounds do not have the geometry to form 1,2 intrastrand crosslinks, they can form 1,3 intrastrand and interstrand crosslinks with DNA, which, because the compounds are very active as anticancer agents, dramatically changed the paradigm on what type of DNA lesion was important for antitumor effects.

Another example of a paradigm shift is the recent report by Lippard and coworkers showing that the cationic complex *cis*-[PtCl(NH$_3$)$_2$(py)]$^+$, where py is the organic base pyridine, has significant anticancer activity in a mouse tumor model [84]. This compound has only one leaving ligand, the Cl$^-$ ion, which can be lost when the compound reacts with DNA, in this case forming a *monofunctional* adduct at N-7 of a guanosine residue in a B-form DNA dodecamer duplex. While N-7 of guanine has been implicated in nearly all of the various types of platinum adduct that form on DNA, including linking proteins to DNA through the platinum ion, Bierbach and coworkers designed platinum complexes that avoid guanine entirely and bind to N-3 of adenine located in the minor groove of DNA [85]. Since these compounds possess potent antitumor effects, they also show that the originally-described structure-activity relationships of the platinum compounds may need amending.

3.5.2 Non-DNA targets

Through the years studies have consistently shown that the amount of platinum bound to cellular DNA is small compared to the amount bound to other components in the cell; that is, only a small percentage of the total amount of bound platinum is bound to DNA [86–90]. While there is little doubt that modification of cellular DNA is a lethal event for the cell, DNA binding is probably not the only mechanism by which the platinum compounds express their anticancer effects. To underscore the point that mechanisms other than DNA binding can kill the cell, Bose and coworkers recently reported that square planar Pt^{+2} and six-coordinate octahedral Pt^{+4} complexes containing a pyrophosphate ligand, called *phosphaplatins*, are very effective in killing ovarian cancer cells that are resistant to cisplatin and carboplatin, and, surprisingly, that one of the Pt^{+2} complexes, which is an analog of oxaliplatin called pyrodach-2, shows no evidence of binding to DNA [91]. While the mechanism by which these compounds kill cells is not completely clear, they induce the production of fas and fas-related proteins in the cell, which are transcription factors involved in apoptosis, and also some genes such as Bak and Bax that initiate cytochrome c release from mitochondria, which causes apoptosis. Since this behavior is very different from that of cisplatin, the phosphaplatins appear to kill cells using a mechanism that is different from that of the former drug. The fact that pyrodach-2 reacts with cysteine and a nonapeptide containing two cysteine residues with a rate constant of $4.3 \times 10^{-3} M^{-1} s^{-1}$ (25 °C) suggests that reaction with sulfur residues on proteins may be important in the mechanism of action of the agent.

Research by Dimanche-Boitrel and coworkers recently showed that cisplatin may bind to and inhibit the Na$^+$/H$^+$ exchanger protein NHE1, which is located in the cell membrane [92,93]. Interaction with this protein causes a lowering of the intracellular pH, promotion of lipid rafts in the membrane (increased fluidity) and induction of apoptosis via the fas death pathway in human colon cancer cells. Since this inhibition occurs within ~5 minutes after addition of cisplatin to the culture medium (the membrane is the first barrier that the drug must cross), and platination of genomic DNA occurs at a much slower rate – ~1 hour for the appearance of Pt-GG adducts – interaction of cisplatin with membrane-bound NHE1 appears to be an important mechanism for killing cancer cells.

Another example of a potential protein target for cisplatin is zinc finger proteins, many of which bind to DNA and initiate gene expression in the cell. Bose and coworkers studied the reaction of cisplatin with a 31-amino-acid-long zinc finger peptide, zpp, which is the DNA binding domain of the enzyme DNA polymerase-α, one of the key enzymes responsible for accurate synthesis of genetic information in the cell (Figure 3.19) [94,95]. This enzyme, as well as *E. coli* DNA polymerase I, binds cisplatin, which blocks the function of the enzyme [96,97].

The zinc finger peptide zpp contains four cysteine residues, which are bound as thiolates to the zinc ion in a tetrahedral geometry. The arrangement of the cysteine residues is such that two residues are located near the amino (N) terminus of the peptide and two are positioned near the carboxyl (C) terminus of the peptide (Figure 3.19a). Using circular dichroism Box 4.2, fluorescence spectroscopies and

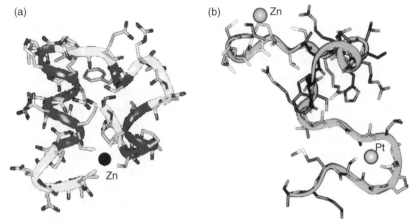

Figure 3.19 *(a) Average ribbon diagram of Zn^{+2} bound to the peptide sequence ICEEPTCRNRTRHLP-LQFSRTGPLCPACMKA, zpp. The Zn^{+2} ion is bound to four thiolate sulfur atoms of the four cysteine residues of the peptide in a tetrahedral coordination geometry. (b) Qualitative representation of the 1 : 1 cisplatin–zpp adduct with the Zn^{+2} bound to the amino terminus and the Pt^{+2} bound to the carboxyl terminus of the peptide. Adapted from Inorg. Chim. Acta, 358, R.N. Bose et al., Structural Perturbation of a C4 Zinc-Finger Module by cis-Diamminedichloroplatinum(II): Insights into the Inhibition of Transcription Process by the Antitumor Drug, 2844–2854. Copyright 2005, with permission from Elsevier*

two-dimensional NMR spectroscopy, it was shown that cisplatin reacts with zpp in a stepwise manner to displace the coordinated Zn^{+2} ion from the peptide. The first formed structure, zpp-Pt, has the Zn^{+2} ion coordinated to the N-terminus and the Pt^{+2} ion bound near the C-terminus of the peptide (Figure 3.19b). If additional cisplatin is added, the first formed complex further reacts with the loss of Zn^{+2} to produce a

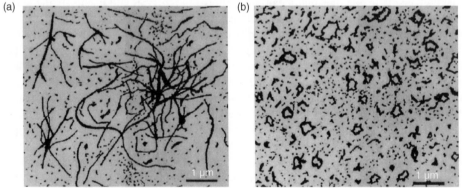

Figure 3.20 *Electron microscopy images of platinum-free (a) and platinum-added (b) microtubules. The microtubules in (a) were allowed to form and were stabilized by the addition of the anticancer drug taxol. In (b) the tubulin monomers (27 μM) were allowed to oligomerize in the presence of the diaqua form of cisplatin (27 μM) for 17 minutes prior to image capture. Adapted from Fig 2a and c, p. 194 of: Tulub, A. A.; Stefanov, V. E. Cisplatin Stops Tubulin Assembly into Microtubules. A New Insight into the mechanism of Antitumor Activity of Platinum Complexes. Int. J. Biol. Macro. 2001, 28, 191–198*

peptide product containing two bound platinum ions. In the study the investigators pointed out that the reaction of cisplatin with the zinc finger is faster than the reaction of cisplatin with DNA, suggesting that zinc fingers could also be targets for cisplatin in the cell. Since reaction with cisplatin changes the structure of the peptide, which will likely alter its DNA binding properties, the investigators suggested that this type of interaction could be the mechanism by which cisplatin inactivates human polymerase-α.

Tubulin, a protein that self-assembles to form microtubules, is directly involved in cell division when chromosomes separate during *mitosis*. Since microtubules require assembly and disassembly that is very rapid, agents that affect this process can cause the cell to enter into *cell-cycle arrest* and die. One compound that affects the process is the organic natural-product anticancer drug *taxol* (paclitaxel), which kills cells by blocking the ability of microtubules to disassemble [98]. Tulub and Stefanov observed that addition of the diaqua form of cisplatin, *cis*-$[Pt(NH_3)_2(H_2O)_2]^{2+}$ (Figure 3.3), to tubulin under conditions that would normally promote the formation of microtubules has a dramatic effect on the shapes of the structures that form [99]. As shown in Figure 3.20a, if tubulin is allowed to polymerize in the presence of guanosine triphosphate, GTP, the energy source required for tubulin assembly, and taxol is added to prevent disassembly, normal filamentous tube-like structures are present. However, if the assembly process is initiated in the presence of the diaqua form of cisplatin, nonfilamentous circular structures form, showing that the diaqua complex can dramatically affect microtubule assembly (Figure 3.20b). By studying microtubule formation in the presence of GTP and its N-7 methylated analog, Me-7-GTP, and by making NMR measurements, the investigators concluded that the platinum complex may be binding to N-7 of GTP, which could not occur with Me-7-GTP, and in this way *cis*-$[Pt(NH_3)_2(H_2O)_2]^{2+}$ could be blocking normal microtubule formation. The investigators suggested that interaction of cisplatin with tubulin assembly may be one of the mechanisms used by the drug to kill cancer cells.

The enzyme thioredoxin reductase (TrxR), a key enzyme in the disulfide/dithiol regulatory system (Figure 4.27), reduces the disulfides of thioredoxin (Trx), other proteins and a number of low-molecular-weight compounds in the cell. The enzyme has an unusual selenocysteine residue at its carboxyl terminus that is directly involved in the redox chemistry of the protein. Since the selenate ion, Se^-, is a soft base, it is an excellent target for the platinum drugs. Arnér and Holmgren and their coworkers showed that both cisplatin and transplatin can irreversibly bind to TrxR and inactivate the enzyme [100]. The second-order rate constants for the reaction of both compounds with TrxR are 21 ± 3 and $84 \pm 22\,M^{-1}\,s^{-1}$, respectively, showing that transplatin is about four times more reactive toward TrxR than is cisplatin. Since a large percentage of cisplatin that enters the cell is believed to be attacked and modified by GSH, the investigators synthesized and examined the inhibitory effects of the cisplatin-GSH adduct GS-Pt [101]. Surprisingly, they found that GS-Pt also inhibits the function of TrxR, suggesting that unreacted drug, as well as *cisplatin that has reacted with GSH*, can block the function of thioredoxin reductase in the cell [99]. Neither cisplatin, transplatin nor carboplatin had a significant inhibitory effect on *E. coli* Trx or human or bacterial glutaredoxin (Grx), which are part of the thioredoxin system.

Sheldrick and coworkers recently reported on the reaction of cisplatin with proteins in living *E. coli* bacterial cells [102]. In this interesting study, *E. coli* cells growing in culture medium were treated with a high concentration of cisplatin (1 mM) for 3 hours, which 'stresses' the cells and causes them to defend themselves against the stress by making proteins to combat the toxin, cisplatin, to which they are exposed. Next, the investigators lysed the cells, and after carrying out a number of separations they isolated fragments of proteins with bound platinum and identified the sequence of each using liquid chromatography–mass spectrometry. Since the sequences of all of the proteins in *E. coli* are known, the sequence of the peptide found could be matched to a specific sequence of a protein in *E. coli* to determine which proteins in the bacterium bound cisplatin. The investigators found that 31 proteins in *E. coli* bound platinum after the cells were treated with cisplatin. They discovered that low-abundance proteins like the DNA mismatch repair protein mutS, the DNA protein helicase II (uvrD) and topoisomerase I (top1), a DNA binding protein, were platinated proportionately more than other proteins in the cell, suggesting that they could be the main targets of the drug in the cell. They also found that two proteins, acrD and mdtA, which are involved in efflux of agents from the cell, thioredoxin 1 (thiO), which is in the disulfide/dithiol regulatory system, and fimA1, an external filament-like type-1fimbril protein, were also attacked by cisplatin. While all potential donor atoms on proteins, O, N and S, were bound to platinum, the thioether functional group of methionine was the most common metal-binding site.

In 40 years of research on the mechanism of action of cisplatin much has been learned but many questions, including how the drug kills cancer cells, have not been completely answered. Although most of the emphasis has been on DNA as a target, the high nucleophilic nature of sulfur suggests that platination of proteins could also be important in the mechanism by which cisplatin kills cancer cells. Investigating other possible modes of action of cisplatin would not only shed new light on the biochemical events associated with cell death but also provide important information for uncovering more effective platinum compounds for the treatment of cancer.

Problems

1. A patient is given a dose of cisplatin by intravenous infusion over a period of 30 minutes that results in a maximum concentration of ultrafilterable platinum (low-molecular-weight compounds), C_{max}, in blood plasma at the end of the infusion of $10 \, \mu M$.

    a. If the half life for elimination of ultrafilterable platinum from plasma is $\sim$30 minutes, estimate the concentration of ultrafilterable platinum in the plasma of the patient 1 hour after the infusion was terminated.

    b. If the hydrolysis rate constant for cisplatin is $1.84 \times 10^{-4} \, s^{-1}$ at 35 °C, estimate the *percentage* of cisplatin that is converted into the monoaquo form in water in 1 hour. Assume that the system never reaches equilibrium.

2. Treatment of cisplatin in water with two equivalents of silver nitrate, $AgNO_3$, followed by filtration to remove insoluble AgCl produces a solution containing the diaqua complex *cis*-$[Pt(H_2O)_2(NH_3)_2]^{2+}$.

    a. Would you expect the *d-d* bands in the absorption spectrum of the diaqua complex in a water solution at pH 3 to be shifted to higher energy, to lower energy or to remain unchanged compared to the locations of the *d-d* bands for cisplatin? Briefly explain.

    b. If the pH of the solution containing the diaqua complex is adjusted to pH 10, would the *d-d* absorption bands for the resulting complex be shifted to higher or lower energies or unchanged relative to the positions of the bands for cisplatin? Briefly explain your answer.

3. A solution of cisplatin at 27 °C at pH 5 contains $\sim$95% cisplatin, *cis*-$[PtCl_2(NH_3)_2]$, $\sim$5% *cis*-$[Pt(H_2O)Cl(NH_3)_2]^+$ and only a trace amount of *cis*-$[Pt(H_2O)_2(NH_3)_2]^{2+}$. Using the data in Table 3.1 and the Henderson–Hasselbalch equation, calculate the *relative concentrations* of the monoaqua and mono-hydroxo complexes, *cis*-$[Pt(H_2O)Cl(NH_3)_2]^+$ and *cis*-$[Pt_2Cl(HO)(NH_3)_2]$, immediately after the pH of the solution is changed to pH $= 7.4$.

4. The reaction of cis-$[Pt(H_2O)Cl(^{15}NH_3)_2]^+$ with the amino acid S-methionine in water was studied using ^1H-^{15}N HSQC NMR. The initially-formed platinum-containing product in the reaction has NMR peaks at $^1H/^{15}N$, $\delta = 4.41/-62.4$ and $4.19/-44.0$ ppm. Draw the structure of the product and assign the HSQC NMR peaks observed in the spectrum.

5. The transmembrane copper-transporting protein CTR1 is believed to be involved in the transport of cisplatin into the cell. The extracellular domain of CTR1 contains methionine and histidine (imidazole) residues that are believed to be involved in the transfer of Cu^+ and possibly also Pt^{+2} into the cell. Using HSAB concepts, explain which donors of the protein would mostly be involved in binding Cu^+ or Pt^{+2} in the transfer process. Which coordination geometries would most likely be found for Cu^+ and Pt^{+2} in the transfer process?

6. The reaction of cis-$[Pt(H_2O)Cl(NH_3)_2]^+$ with double-stranded DNA to form a monoadduct is a *second-order* reaction, while 'ring closure' to form a 1,2 intrastrand crosslink is a *first-order* reaction. Briefly explain why the orders of the reactions are as stated.

7. Incubation of closed circular DNA with cisplatin followed by electrophoresis in an agarose gel shows that the mobility of the platinated DNA in the gel (as detected by staining with ethidium bromide) is slower than that of control DNA without bound platinum. However, treatment of the platinated DNA with excess sodium cyanide, NaCN, for a period of time, followed by electrophoresis, shows that the mobility of the DNA is now identical to that of control DNA that did not have bound platinum. Using crystal field theory, briefly explain these observations.

8. It is known that cisplatin reacts with a zinc finger protein that has a Zn-S_4 site, where S is the thiolate ion of cysteine, to displace the bound Zn^{2+} ion. If the attacking platinum complex is cis-$[Pt(H_2O)Cl(NH_3)_2]^+$, propose a structure for the *first-formed* intermediate complex in the ejection of Zn^{+2} from the protein by the platinum complex.

References

1. Rosenberg, B. (1971) Some biological effects of platinum compounds. New agents for the control of tumors. *Platinum Metals Review*, **15**, 42–51.
2. Rosenberg, B. (1978) *Nucleic Acid–Metal Ion Interactions, vol.* **1** (ed. T.G. Spiro), John Wiley & Sons, Inc., New York, pp. 1–29.
3. Barnes, K.R. and Lippard, S.J. (2004) Cisplatin and related anticancer drugs: recent advances and insights. *Metal Ions in Biological Systems*, **42**, 143–177.
4. Wang, D. and Lippard, S.J. (2005) Cellular processing of platinum anticancer drugs. *Nature Reviews. Drug Discovery*, **4**, 307–320.
5. Jamieson, E.R. and Lippard, S.J. (1999) Structure, recognition, and processing of cisplatin-DNA adducts. *Chemical Reviews*, **99**, 2467–2498.
6. Bloemink, M.J. and Reedijk, J. (1996) Cisplatin and derived anticancer drugs: mechanism and current status of DNA binding. *Metal Ions in Biological Systems*, **32**, 641–685.
7. Reedijk, J. (1999) Why does cisplatin reach guanine-N7 with competing S-donor ligands available in the cell? *Chemical Reviews*, **99**, 2499–2510.
8. Farrell, N.P. (2004) Preclinical perspectives on the use of platinum compounds in cancer chemotherapy. *Seminars in Oncology*, **31** (Suppl 14), 1–9.
9. Kelland, L.R. and Farrell, N.P. (2000) *Platinum-Based Drugs in Cancer Therapy*, Humana Press Inc., Totowa, NJ.
10. Lippert, B. (1999) Cisplatin, *Chemistry and Biochemistry of a Leading Anticancer Drug*, VCHA & Wiley-VCH, Zurich.
11. Alderden, R.A., Hall, M.D., and Hambley, T.W. (2006) The discovery and development of cisplatin. *Journal of Chemical Education*, **83**, 728–734.

12. Hambley, T.W. (2001) Platinum binding to DNA: Structural controls and consequences. *Journal of The Chemical Society-Dalton Transactions*, 2711–2718.
13. Guo, Z. and Sadler, P.J. (2000) Medicinal inorganic chemistry. *Advances in Inorganic Chemistry*, **49**, 183–306.
14. Ma, J., Verweij, J., Kolker, H.J. *et al.* (1994) Pharmacokinetic-dynamic relationship of cisplatin *in vitro*: Simulation of an i.v. bolus and 3h and 20 h infusion. *British Journal of Cancer*, **69**, 858–862.
15. van den Bongard, H.J., Mathôt, R.A., Beijnen, J.H., and Schellens, J.H. (2000) Pharmacokinetically guided administration of chemotherapeutic agents. *Clinical Pharmacokinetics*, **39**, 345–367.
16. Boulikas, T. and Vougiouka, M. (2004) Recent clinical trials using cisplatin, carboplatin and their combination chemotherapy drugs (Review). *Oncology Reports*, **11**, 559–595.
17. Kotz, J.C. and Treichel, P.M. (1994) *Chemistry & Chemical Reactivity*, 2nd edn, Thomson Brooks, Cole, Ca.
18. Rosenberg, B., VanCamp, L., Trosko, J.E., and Mansour, V.H. (1969) Mansour, platinum compounds: A new class of potent antitumor agents. *Nature*, **222**, 385–386.
19. Peyrone, M. (1844) Über die einwirkung des ammoniaks auf platinchlorür. *Liebigs Annalen der Chemie*, **51**, 1–29.
20. Blumenstyk, G. (1999) How one university pursued profit from science–and won. *The Chronicle of Higher Education*, **12**, A39–A40.
21. Dhara, S.C. (1970) A rapid method for the synthesis of *cis*-[Pt(NH$_3$)$_2$Cl$_2$]. *Indian Journal of Chemistry*, **8**, 193–194.
22. Patterson, H.H., Tewksbury, J.C., Martin, M. *et al.* (1981) Luminescence, absorption, MCD, NQR study of the cis and trans isomers of Dichlorodiammineplatinum(II). *Inorganic Chemistry*, **20**, 2297–2301.
23. Miller, S.E. and House, D.A. (1989) The Hydrolysis products of cis-Diamminedichloroplatinum(II). I. The kinetics of formation and anation of the cis-Diammine(aquo)chloroplatinum(II) cation in acidic aqueous solution. *Inorganica Chimica Acta*, **161**, 131–137.
24. Miller, S.E. and House, D.A. (1989) The hydrolysis of cis-Dichlorodiammineplatinum(II) 2. The kinetics of formation and anation of the cis-Diamminedi(aquo)platinum(II) cation. *Inorganica Chimica Acta*, **166**, 189–197.
25. Miller, S.E., Gerard, K.J., and House, D.A. (1991) The hydrolysis products of cis-Diamminedichloroplatinum(II) 6. A kinetic comparison of the cis- and trans-isomers and other cis-di(amine)di(chloro)platinum(II) compounds. *Inorganica Chimica Acta*, **190**, 135–144.
26. Berners-Price, S.J. and Appleton, T.G. (2000) *Platinum-Based Drugs in Cancer Therapy* (eds L.R. Kelland and N. Farrell), Humana Press Inc., Totowa, NJ.
27. Berners-Price, S.J., Frenkiel, T.A., Frey, U. *et al.* (1992) Hydrolysis products of cisplatin: pK_a determinations via [^{1}H,^{15}N] NMR spectroscopy. *Journal of the Chemical Society. Chemical Communications*, 789–791.
28. Berners-Price, S.J., Ronconi, L., and Sadler, P.J. (2006) Insights into the mechanism of action of platinum anticancer drugs from multinuclear NMR spectroscopy. *Progress in Nuclear Magnetic Resonance Spectroscopy*, **49**, 65–98.
29. http://dailymed.nlm.nih.gov/dailymed/drugInfo.cfm?id=4915, accessed 5/09.
30. van der Vijgh, W.J.F. (1991) Clinical pharmacokinetics of carboplatin. *Clinical Pharmacokinetics*, **21**, 242–261.
31. Yao, X., Panichpisal, K., Kurtzman, N., and Nugent, K. (2007) Cisplatin nephrotoxicity: A review. *The American Journal of the Medical Sciences*, **334**, 115–124.
32. Tacka, K.A., Szalda, D., Souid, A.-K. *et al.* (2004) Experimental and theoretical studies on the pharmacodynamics of cisplatin in jurkat cells. *Chemical Research in Toxicology*, **17**, 1434–1444.
33. Gullo, J.J., Litterst, C.L., Maguire, P.J. *et al.* (1980) Pharmacokinetics and protein binding of cis-dichlorodiammine Platinum(II) administered as a one hour or a twenty hour infusion. *Cancer Chemotherapy and Pharmacology*, **5**, 21–26.
34. Vermorken, J.B., van der Vijgh, W.J., Klein, I. *et al.* (1984) Phramacokinetics of free and total platinum species after short term infusion of cisplatin. *Cancer Treatment Reports*, **68**, 505–513.
35. Ivanov, A.I., Christodulou, J., Parkinson, J.A. *et al.* (1998) Cisplatin binding sites on human albumin. *The Journal of Biological Chemistry*, **273**, 14721–14730.
36. Gateley, D.P. and Howell, S.B. (1993) Cellular accumulation of the anticancer agent cisplatin: A review. *British Journal of Cancer*, **67**, 1171–1176.

37. Hall, M.D., Okabe, M., Shen, D.-W. *et al.* (2008) The role of cellular accumulation in determining sensitivity to platinum-based chemotherapy. *Annual Review of Pharmacology and Toxicology*, **48**, 495–535.

38. Komatsu, M. *et al.* (2000) Copper transporting P-type adenosine triphosphatase (ATP7B) is associated with cisplatin resistance. *Cancer Research*, **60**, 1312–1316.

39. Lin, X., Okuda, T., Holzer, A., and Howell, S.B. (2002) The copper transporter CTR1 regulates cisplatin uptake in *Saccharomyces cerevisae*. *Molecular Pharmacology*, **62**, 1154–1159.

40. Ishida, S., Lee, J., Thiele, D.J., and Herskowitz, I. (2002) Uptake of the anticancer drug cisplatin mediated by the copper transporter Ctr1 in yeast and mammals. *Proceedings of the National Academy of Sciences of the United States of America*, **99**, 14298–14302.

41. Zisowsky, J., Koegel, S., Leyers, S. *et al.* (2007) Relevance of drug uptake and efflux for cisplatin sensitivity of tumor cells. *Biochemical Pharmacology*, **73**, 298–307.

42. Kuo, M.T., Chen, H.H.W., Song, I.-S. *et al.* (2007) The roles of copper transporters in cisplatin resistance. *Cancer Metastasis Reviews*, **26**, 71–83.

43. Safaei, R. and Howell, S.B. (2005) Copper transporters regulate the cellular pharmacology and sensitivity to Pt drugs. *Critical Reviews in Oncology/Hematology*, **53**, 13–23.

44. Sinani, D., Adle, D.J., and Lee, J. (2007) Distinct mechanisms for Ctr1-mediated copper and cisplatin transport. *The Journal of Biological Chemistry*, **282**, 26775–26785.

45. Aller, S.G. and Unger, V.M. (2006) Projection structure of the human copper transporter CTR1 at 6-Å resolution reveals a compact trimer with a novel channel-like architecture. *Proceedings of the National Academy of Sciences of the United States of America*, **103**, 3627–3632.

46. Prohaska, J.R. (2008) Role of copper transporters in copper homeostasis. *The American Journal of Clinical Nutrition*, **88**, 826S–829S.

47. Safaei, R. (2006) Role of copper transporters in the uptake and efflux of platinum containing drugs. *Cancer Letters*, **234**, 34–39.

48. Sarkar, B. (1999) Treatment of Wilson's and Menkes diseases. *Chemical Reviews*, **99**, 2535–2544.

49. Park, J.K., Jung, Y.-S., Kim, J.-S. *et al.* (2008) Structural insight into the distinct properties of copper transport by the *Helicobacter pylori* CopP protein. *Proteins*, **71**, 1007–1019.

50. Aller, S.G. *et al.* (2009) Structure of P-glycoprotein reveals a molecular basis for polyspecific drug binding. *Science*, **323**, 1718–1722.

51. Yonezawa, A., Masuda, S., Yokoo, S. *et al.* (2006) Cisplatin and oxaliplatin, but not carboplatin and nedaplatin, are substrates for human organic cation transporters (SLC22A1-3 and multidrug and toxin extrusion family). *The Journal of Pharmacology and Experimental Therapeutics*, **319**, 879–886.

52. Dabrowiak, J.C., Goodisman, J., and Souid, A.-K. (2002) Kinetic study of the reaction of cisplatin with thiols. *Drug Metabolism and Disposition: The Biological Fate of Chemicals*, **30**, 1378–1384.

53. Hagrman, D., Goodisman, J., Dabrowiak, J.C., and Souid, A.-K. (2003) Kinetic study of the reaction of cisplatin with metallothionine. *Drug Metabolism and Disposition: The Biological Fate of Chemicals*, **31**, 916–923.

54. Knipp, M., Karotki, A.V., Chesnov, S. *et al.* (2007) Reaction of ZN_7 metallothionein with *cis*- and *trans*-[Pt(N-donor)$_2$Cl$_2$] anticancer complexes: *trans*-Pt^{II} complexes retain their N-Donor ligands. *Journal of Medicinal Chemistry*, **50**, 4075–4086.

55. Martin, L.P., Hamilton, T.C., and Schilder, R.J. (2008) Platinum resistance: The role of DNA repair pathways. *Clinical Cancer Research*, **14**, 1291–1295.

56. Patrick, S.M., Tillison, K., and Horn, J.M. (2008) Recognition of cisplatin-DNA interstrand crosslinks by replication protein A. *Biochemistry*, **47**, 10188–10196.

57. Jung, Y. and Lippard, S.J. (2007) Direct cellular responses to platinum-induced DNA damage. *Chemical Reviews*, **107**, 1387–1403.

58. Kartalou, M. and Essigmann, J.M. (2001) Mechanisms of resistance to cisplatin. *Mutation Research*, **478**, 23–43.

59. Liedert, B., Pluim, D., Schellens, J., and Thomale, J. (2006) Adduct-specfic monoclonal antibodies for the measurement of cisplatin-induced DNA lesions in individual cell nuclei. *Nucleic Acids Research*, **34**, e47.

60. Centerwall, C.R., Kerwood, D.J., Goodisman, J. *et al.* (2008) New extracellular resistance mechanism for cisplatin. *Journal of Inorganic Biochemistry*, **102**, 1044–1049.

61. Sherman-Baust, C.A., Weeraratna, A.T., Rangel, L.B. *et al.* (2003) Remodeling of the extracellular matrix through overexpression of collagen VI contributes to cisplatin resistance in ovarian cancer cells. *Cancer Cell*, **3**, 377–386.

62. Morin, P.J. (2003) Drug resistance the microenvironment: Nature and nurture. *Drug Resist Update*, **6**, 169–172.

63. Centerwall, C.R., Tacka, K.A., Kerwood, D.J. *et al.* (2006) Modification and uptake of a cisplatin carbonato complex by jurkat cells. *Molecular Pharmacology*, **70**, 348–355.

64. Howle, J.A. and Gale, G.R. (1970) Cis-Dichlorodiammineplatinum(II). Persistent and selective inhibition of deoxyribonucleic acid synthesis *in vivo*. *Biochemical Pharmacology*, **19**, 2757–2762.

65. Fichtinger-Schepman, A.M., van der Veer, J.L., den Hartog, J.H. *et al.* (1985) Adducts of the antitumor drug cis-Diamminedichloroplatinum(II) with DNA: Formation, identification, and quantitation. *Biochemistry*, **24**, 707–713.

66. Blommaert, F.A., van Dijk-Knijnenburg, H.C.M., Dijt, F.J. *et al.* (1995) Formation of DNA adducts by the anticancer drug carboplatin: Different nucleotide sequence preferences *in vitro* and in cells. *Biochemistry*, **34**, 8474–8480.

67. Davies, M.S., Berners-Price, S.J., and Hambley, T.W. (2000) Slowing of cisplatin aquation in the presence of DNA but not in the presence of phosphate: Improved understanding of sequence selectivity and the roles of monoaquated and diaquated species in the binding of cisplatin to DNA. *Inorganic Chemistry*, **39**, 5603–5613.

68. Legendre, F., Bas, V., Kozelka, J., and Chottard, J.-C. (2000) A complete kinetic study of GG versus AG platination suggests that the doubly aquated derivatives of cisplatin are the actual DNA binding species. *Chemistry – A European Journal*, **6**, 2002–2010.

69. Scovell, W.M. and Collart, F. (1985) Unwinding of supercoiled DNA by cis- and trans-diamminedichloroplatinum-(II): Influence of the torsional strain on DNA unwinding. *Nucleic Acids Research*, **13**, 2881–2895.

70. Keck, M.V. and Lippard, S.J. (1992) Unwinding of supercoiled DNA by platinum-ethidium and related complexes. *Journal of the American Chemical Society*, **114**, 3386–3390.

71. Kobayashi, S., Nakamura, Y., Meahara, T. *et al.* (2001) DNA topology on an increase in positive writhing number of DNA: Conformation changes in the time course of *cis*-Diamminedichloroplatinum(II)-DNA adducts. *Chemical & Pharmaceutical Bulletin*, **49**, 1053–1060.

72. Mathews, C.K., van Holde, K.E., and Ahern, K.G. (2000) *Biochemistry*, 3rd edn, Benjamin Cummings, San Francisco, CA.

73. Takahara, P.M., Rosenzweig, A.C., Federick, C.A., and Lippard, S.J. (1995) Crystal structure of double stranded DNA containing the major adduct of the anticancer drug cisplatin. *Nature*, **377**, 649–652.

74. Imamura, T., Izumi, H., Nagatani, G. *et al.* (2001) Interaction with p53 enhances binding of cisplatin-modified DNA by high mobility group 1 protein. *The Journal of Biological Chemistry*, **276**, 7534–7540.

75. Hambley, T.W. (1997) The influence of structure on the activity and toxicity of Pt anti-cancer drugs. *Coordination Chemistry Reviews*, **99**, 2451–2466.

76. Wong, E. and Giandomenico, C.M. (1999) Current status of platinum-based antitumor drugs. *Chemical Reviews*, **99**, 2451–2466.

77. Radulovic, S., Tesic, Z., and Manic, S. (2002) Trans-platinum complexes as anticancer drugs: Recent development and future prospects. *Current Medicinal Chemistry*, **9**, 1611–1618.

78. Farrell, N. (1996) Current status of structure-activity relationships of platinum anticancer drugs: Activation of the trans geometry. *Metal Ions in Biological Systems*, **32**, 603–639.

79. Coluccia, M. and Natile, G. (2007) Trans-platinum complexes in cancer chemotherapy. *Anti-Cancer Agents in Medicinal Chemistry*, **7**, 111–123.

80. Kasparkova, J., Marini, V., Bursova, V., and Brabec, V. (2008) Biophysical studies on the stability of DNA intrastrand crosslinks of transplatin. *Biophysical Journal*, **95**, 4361–4371.

81. Najajreh, Y., Perez, J.M., Navarro-Ranninger, C., and Gibson, D. (2002) Novel soluble cationic trans-Diamminedi-chloroplatinum(II) complexes that are active against cisplatin resistant ovarian cancer cell lines. *Journal of Medicinal Chemistry*, **45**, 5189–5195.

82. Lippert, B. (1996) Trans-Diamineplatinum(II): What makes it different from cis-DDP? Coordination chemistry of a neglected relative of cisplatin and its interaction with nucleic acids. *Metal Ions in Biological Systems*, **33**, 105–141.

83. Zorbas-Seifried, S., Jakupec, M., Kukushkin, N. *et al.* (2007) Reversion of structure-activity relationships of antitumor platinum complexes by acetoxime but not hydroxylamine ligands. *Molecular Pharmacology*, **71**, 357–365.

84. Lovejoy, K.S., Todd, R.C., Zhang, S. *et al.* (2008) cis-Diammine(pyridine)chloroplatinum(II), a monofunctional platinum(II) antitumor agent: Uptake, structure, function, and prospects. *Proceedings of the National Academy of Sciences of the United States of America*, **105**, 8902–8907.

85. Guddneppanavar, R. and Bierbach, U. (2007) Adenine-N3 in the DNA minor groove-an emerging target for platinum containing anticancer pharmacophores. *Anti-Cancer Agents in Medicinal Chemistry*, **7**, 125–138.

86. Volckova, E., Dudones, L.P., and Bose, R.N. (2002) HPLC determination of binding of cisplatin to DNA in the presence of biological thiols: Implications of dominant platinum-thiol binding to its anticancer action. *Pharmaceutical Research*, **19**, 124–131.

87. Eastman, A. (1990) Activation of programmed cell death by anticancer agents: cisplatin as a model system. *Cancer Cells (Cold Spring Harbor, NY: 1989)*, **2**, 275–280.

88. Lindauer, E. and Holler, E. (1996) Cellular distribution and cellular reactivity of platinum(II) complexes. *Biochemical Pharmacology*, **52**, 7–14.

89. Hah, S.S., Stivers, K.M., de Vere White, R.W., and Henderson, P.T. (2006) Kinetics of carboplatin-DNA binding in genomic DNA and bladder cancer cells as determined by accelerator mass spectrometry. *Chemical Research in Toxicology*, **19**, 622–626.

90. Hah, S.S., Sumbad, R.A., de Vere White, R. *et al.* (2007) Characterization of oxaliplatin-DNA adduct formation in DNA and differentiation of cancer cell drug sensitivity at microdose concentrations. *Chemical Research in Toxicology*, **20**, 1745–1751.

91. Bose, R.N., Maurmann, L., Mishur, R.J. *et al.* (2008) Non-DNA binding platinum anticancer agents: Cytotoxic activities of platinum-phophato complexes toward human ovarian cancer cells. *Proceedings of the National Academy of Sciences of the United States of America*, **105**, 18314–18319.

92. Rebillard, A. *et al.* (2007) Cisplatin induced apoptosis involves membrane fluidation via inhibition of NHE1 in human colon cancer cells. *Cancer Research*, **67**, 7865–7874.

93. Rebillard, A., Lagadic-Gossmann, D., and Dimanche-Boitrel, M.T. (2008) Cisplatin cytotoxicity: DNA and plasma membrane targets. *Current Medicinal Chemistry*, **15**, 2656–2663.

94. Bose, R.N., Yang, W.W., and Evanics, F. (2005) Structural perturbation of a C4 Zinc-Finger module by cis-Diamminedichloroplatinum(II): Insights into the inhibition of transcription process by the antitumor drug. *Inorganica Chimica Acta*, **358**, 2844–2854.

95. Bose, R.N. (2002) Biomolecular targets for platinum anticancer drugs. *Mini Reviews in Medicinal Chemistry*, **2**, 103–111.

96. Kelly, T.J., Moghaddas, S., Bose, R.N., and Basu, S. (1993) Inhibition of immunopurified DNA polymerase-alpha from PA-3 prostate tumor cells by platinum (II) antitumor drugs. *Cancer Biochemistry Biophysics*, **13**, 135–146.

97. Duman, R.K., Heath, R.T., and Bose, R.N. (1999) Inhibition of *Escherichia coli* DNA Polymerase-I by the anti-cancer drug cis-Diamminedichloroplatinum(II): What roles do polymerases play in cis-platin-induced cytotoxicy? *FEBS Letters*, **455**, 49–54.

98. Singh, P., Rathinasamy, K., Mohan, R., and Panda, D. (2008) Microtubule assembly dynamics: an attractive target for anticancer drugs. *IUBMB Life*, **60**, 368–375.

99. Tulub, A.A. and Stefanov, V.E. (2001) Cisplatin stops tubulin assembly into microtubules. A new insight into the mechanism of antitumor activity of platinum complexes. *International Journal of Biological Macromolecules*, **28**, 191–198.

100. Arnér, E.S.J., Nakamura, H., Sasada, T. *et al.* (2001) Analysis of the inhibition of mammalian thioredoxin, thioredoxin reductase, and Glutarredoxin by cis-Diamminedichloroplatinum (II) and its major metabolite, the glutathione-platinum complex. *Free Radical Biology & Medicine*, **31**, 1170–1178.

101. Ishikawa, T. and Ali-Osman, F. (1993) Glutathione-associated *cis*-Diamminedichloroplatinum (II) metabolism and ATP-dependent efflux from leukemia cells. Molecular characterization of glutathione-platinum complex and its biological significance. *The Journal of Biological Chemistry*, **268**, 20116–20125.

102. Will, J., Sheldrick, W.S., and Wolters, D. (2008) Characterization of cisplatin coordination sites in cellular *Escherichia coli* DNA-binding proteins by combined biphasic liquid chromatography and ESI tandom mass spectrometry. *Journal of Biological Inorganic Chemistry: JBIC: a Publication of the Society of Biological Inorganic Chemistry*, **13**, 421–434.

Further reading

Metzler, D.E. (2001) Biochemistry, *The Chemical Reactions of Living Cells*, 2nd edn, vol. **1**, Harcourt Academic Press, San Diego, CA.

4

Platinum Anticancer Drugs

4.1 Carboplatin

Carboplatin, [Pt(cbdca-O,O')(NH$_3$)$_2$] (Figure 4.1), where cbdca is cyclobutane-1,1-dicarboxylate and O and O' are the ligand donor atoms, is a so-called second-generation platinum anticancer drug that is much less oto- and nephrotoxic than cisplatin [1]. The drug, which was reported by Cleare and Hoeschele in 1973 [2,3] and later patented [4], gained approval by the US Food and Drug Administration (FDA) under the brand name Paraplatin in 1989. The key difference between the structures of carboplatin and cisplatin is that the former possesses a six-membered dicarboxylate ring, which, because of the chelate effect, makes it much less chemically reactive than cisplatin. Carboplatin by itself or in combination with other drugs is in worldwide clinical use for the treatment of a variety of different cancers, including head and neck cancer, and ovarian, breast, small lung cell, testicular, bladder and brain tumors [1]. Although carboplatin is much less oto- and nephrotoxic than cisplatin, it is myelosuppressive, which leads to a reduction in the white cell count in the blood, thus exposing the patient to infection by various organisms.

4.1.1 Synthesis and properties of carboplatin

As shown in Figure 4.2, carboplatin can be synthesized by reacting potassium tetrachloroplatinate with excess KI in water, which produces the bright-orange complex anion [PtI$_4$]$^{2-}$ [5]. This complex precipitates from solution as its sparingly-soluble dipotassium salt, K$_2$[PtI$_4$]. Reaction of the tetraiodo complex with two equivalents of ammonia produces yellow cis-[PtI$_2$(NH$_3$)$_2$]. Due to the strong $trans$-directing ability of I$^-$, only the cis isomer is formed. Suspension of this complex in water containing one equivalent of Ag$_2$SO$_4$ in the dark with constant stirring results in the formation of insoluble silver iodide, AgI, and the formation of the very water-soluble cis diaquao complex, cis-[Pt(NH$_3$)$_2$(H$_2$O)$_2$]SO$_4$. Since sulfate anion (a hard base) is low in the spectrochemical series and Pt^{+2} is a soft acid, water (and not sulfate ion) binds to the platinum ion. The final step in the syntheses requires the addition of a solution containing one equivalent of Ba(cbdca), previously prepared by the reaction of Ba(OH)$_2$ with H$_2$-cbdca, to the solution containing the diaqua complex cis-[Pt-(NH$_3$)$_2$(H$_2$O)$_2$]SO$_4$. Since BaSO$_4$ is very insoluble in water it can be removed by filtration and carboplatin can be recovered from the filtrate by reduction of the volume of the solution. Carboplatin is a pale-white solid with a high aqueous solubility at ambient temperature of $\sim$18 mg ml^{-1} or $\sim$48 mM [6].

Metals in Medicine James C. Dabrowiak
© 2009 John Wiley & Sons, Ltd

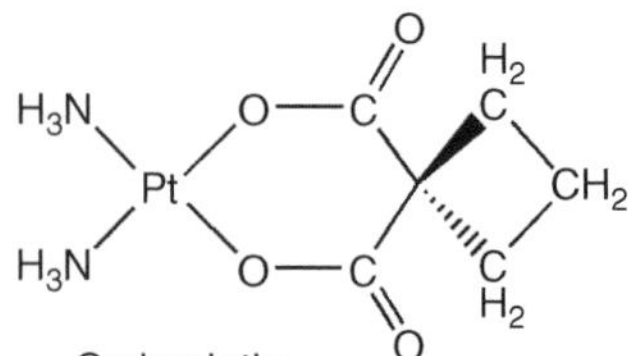

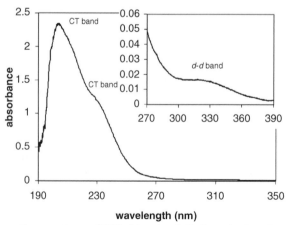

Carboplatin

Figure 4.1   *Structure of carboplatin*

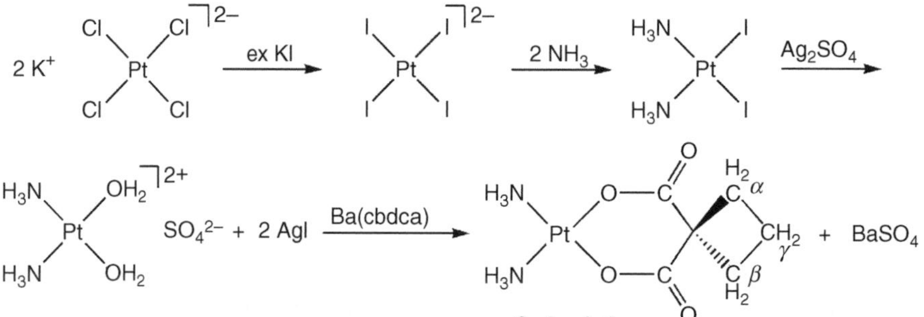

Figure 4.2   *Synthesis of carboplatin*

   The UV-visible absorption spectrum of carboplatin in water is shown in Figure 4.3 [7]. The drug has a complicated absorption envelope, having at least two strong absorptions at 203 ($\varepsilon = 4600\,M^{-1}\,cm^{-1}$) and 229 nm ($\varepsilon = 2300\,M^{-1}\,cm^{-1}$), and a weaker band at 330 nm ($\varepsilon = 40\,M^{-1}\,cm^{-1}$). Since the two strong bands at high energy have intensities consistent with a Laporte-allowed transition, $\Delta l \neq 0$, these bands are probably caused by charge transfer transitions, although the direction of the charge movement (MLCT or LMCT) is unknown. The lower-intensity band at 330 nm (30 300 cm^{-1}) is almost certainly due to a *d-d* transition. As was earlier pointed out, Pt^{+2} has the electronic configuration $5d^8$, which in the presence of a square planar crystal

Figure 4.3   *UV–visible absorption spectrum of 500 μM carboplatin in water in a 1 cm path-length cell. Reprinted with permission from A.J. Di Pasqua et al. Activation of Carboplatin by Carbonate, Chem. Res. Toxicol. 19, 139–149. Copyright 2006 American Chemical Society*

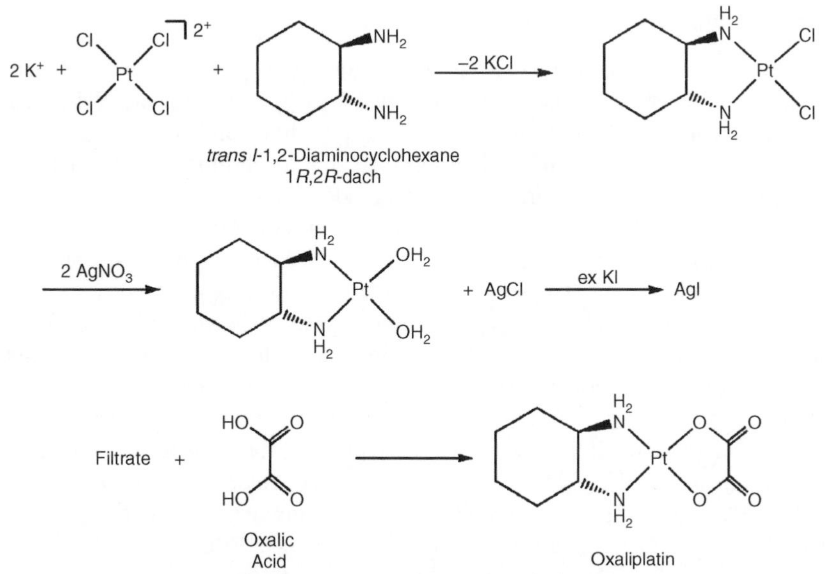

trans l-1,2-Diaminocyclohexane
1*R*,2*R*-dach

Oxalic
Acid

Oxaliplatin

Figure 4.18   *Synthesis of oxaliplatin*

4.2.3   Formulation, administration and pharmacokinetics of oxaliplatin

The clinically-used infusion solution of oxaliplatin, called Eloxatin, contains $5\,\text{mg ml}^{-1}$ (13.2 m*M*) oxaliplatin in water [49]. The drug is usually given by IV infusion over 2-6 hours and a dose of $\sim130\,\text{mg m}^{-2}$ ($\sim327\,\mu\text{mol m}^{-2}$), which is comparable to that for cisplatin [18], is employed. Typically, at the end of the infusion the concentration of platinum in *plasma ultra filtrate* (*PUF*), C_{max}, is $\sim3\,\mu\text{g ml}^{-1}$ ($\sim15\,\mu M$ Pt), and the amount of platinum in plasma ultrafiltrate decreases in a tri-exponential manner with half lives of $t_{1/2} = 0.28$, 16.3 and 273 hours. The shorter elimination phases may be caused by removal of intact oxaliplatin and/or its water or other low MW blood reaction products by filtration through the kidneys, while the long half-life elimination may be due to removal of low MW platinum-amino acid conjugates, which are slowly released after degradation of platinated proteins. In blood, platinum binds to albumin and γ-globulins and, while it accumulates in erythrocytes, binding to blood cells is not considered to be clinically significant.

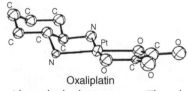

Oxaliplatin

Figure 4.19   *Structure of oxaliplatin without the hydrogen atoms. The spheres are thermo ellipsoids, which show the thermo movement (vibrations) of the atoms in the solid state. Reprinted from Inorg. Chim Acta, 92, M.A. Bruck et al., The Crystal Structures and Absolute Configurations of the Anti-Tumor Complexes Pt(oxalato)(1R,2R-cyclohexanediamine) and Pt(malonato)(1R,2R-cyclohexanediamine), 279–84. Copyright 1984, with permission from Elsevier*

4.2.4   Reaction of oxaliplatin in the biological milieu

Chaney and coworkers [50,51] characterized the biotransformation products formed when tritiated [³H]-oxaliplatin, which is radioactive, reacts in rat blood. This was done by adding the radiolabeled oxaliplatin to rat blood, allowing the platinum compound to react for various times and measuring the amounts of ³H-containing compounds which formed using high-pressure liquid chromatography, HPLC. The protocol involved centrifuging rat blood to separate the red blood cells (RBCs from the plasma and passing the plasma through a filter which separated molecules with values of MW greater than 30 kDa from smaller molecules to produce PUF. After 4 hours, 35% of the total Pt-dach was bound to plasma proteins, 12% in the PUF and 53% bound to RBC. The study showed that the disappearance of radiolabeled oxaliplatin in PUF was rapid, with a $t_{1/2}$ of 0.68 hours, and that the main products were [PtCl$_2$(dach)], [Pt(Cys)$_2$(dach)], [Pt(GSH)(dach)], [Pt(Met)(dach)], [Pt(GSH)$_2$(dach)] and free dach ligand. The compound [PtCl$_2$(dach)] formed transiently, while the sulfur-containing compounds were the major products 12 hours after the addition of drug to rat blood.

The reactions of radiolabeled oxaliplatin with RBCs and plasma proteins were also rapid, with $t_{1/2}$ values of 0.58 and 0.78 hours, respectively [50,51]. The researchers concluded that either the sulfur-containing nucleophiles in blood attack oxaliplatin to produce sulfur-containing products or else hydrogencarbonate, HCO_3^-, which is also present in blood, attacks oxaliplatin to produce [Pt(dach)(H$_2$O)$_2$]$^{2+}$, which then reacts with substances in the blood [52]. The proposed reactivity of oxaliplatin with carbonate may be similar to the reaction of carboplatin in carbonate buffer, recently studied by Di Pasqua *et al.* [7,25,26].

Jerremalm *et al.* [53] studied the kinetics of the reaction of oxaliplatin with biological sulfur-containing compounds and PUF using HPLC with detection at 254 nm in the UV region of the spectrum. Figure 4.20 shows the observed degradation rate constants, k_{obs}, of oxaliplatin for different concentrations of cysteine, methionine and glutathione in the reaction medium. While this reaction is second-order, the fact that the concentration of nucleophile is very much larger than the concentration of drug means that the reaction follows a pseudo first-order rate law, which explains why k_{obs} is linear with the concentration of nucleophile. The measured second-order rate constants for oxaliplatin reacting with GSH, Met and Cys, at 37 °C, pH 7.4, were 4.7, 5.5 and 15 M^{-1} min^{-1}, respectively. The researchers pointed out that all of these rate constants are higher than the corresponding rate constants for the reaction of cisplatin with the same sulfur-containing nucleophiles and that, in the case of oxaliplatin, direct nucleophilic attack on the intact drug is the likely mechanism.

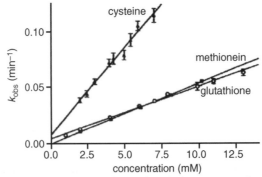

Figure 4.20   *Observed degradation rate constants (k_{obs}) of oxaliplatin (60 µM) for the reaction in the presence of glutathione (GSH), methionine (Met) and cysteine (Cys) at different concentrations (1–13 mM), 37°C, pH 7.4. Reprinted from Eur. J. Pharm. Sci., 28, E. Jerremalm et al., Oxaliplatin Degradation in the Presence of Important Biological Sulfur-Containing Compounds and Plasma Ultrafiltrate, 278–283. Copyright 2006, with permission from Elsevier*

Table 4.2 *Rate constants (k_2), mean ± SD (n) for cisplatin, oxaliplatin and carboplatin binding to Cd/Zn-thionein and GSH at 37°c, 10 mM HEPES, 4.62 mM NaCl, pH 7.4*

Pt drug	Cd/Zn-thionein ($M^{-1}s^{-1}$)	GSH ($M^{-1}s^{-1}$)
Cisplatin	0.75 ± 0.03 (3)	0.027 ± 0.0006 (2)
Oxaliplatin	0.44 ± 0.1 (2)	0.039
Carboplatin	0.0121 ± 0.0039 (6)	0.0017 ± 0.00037 (4)

From Table 2, p. 263 of reference [58].

In attempting to assess the importance of these nucleophiles in degrading oxaliplatin in PUF, a 'cocktail' was prepared which contained the sulfur nucleophiles as well as Cl⁻ at the concentrations estimated to be present in plasma ultrafiltrate [53]. By comparing the rate for the early part (0-60 minutes) of drug degradation in the cocktail with that for the early part of the degradation in PUF, it was concluded that the sulfur nucleophiles plus Cl⁻ are mainly responsible for the degradation of the drug in PUF. The fact that the long-term (1-24 hours) degradation rate in PUF is faster than in the cocktail suggests that substances other than the sulfur nucleophiles and Cl⁻ may also be responsible for degradation of oxaliplatin in plasma.

Both metallothionein and GSH are considered important detoxifying agents for the platinum drugs inside the cell. Souid and coworkers [55–58] measured the rates of binding of oxaliplatin as well as cisplatin and carboplatin to GSH and to a Cd^{+2}/Zn^{+2} form of metallothionein, Cd/Zn-thionein. As is evident from Table 4.2, oxaliplatin is more like cisplatin than carboplatin in its reactions with GSH and Cd/Zn-thionein, suggesting that the dach ligand is not a steric barrier to nucleophilic attack at the Pt^{+2} of oxaliplatin.

Sadler and coworkers [59] also studied the reaction of oxaliplatin with GSH and its oxidized form, glutathione disulfide (GSSG), using UV-visible absorption, HPLC, ESI-MS and NMR methods. The investigators found that the reaction of GSH with oxaliplatin produced two dinuclear complexes (complexes with two Pt^{+2} ions), the structures of which are shown in Figure 4.21. Product A has two thiolate anions from

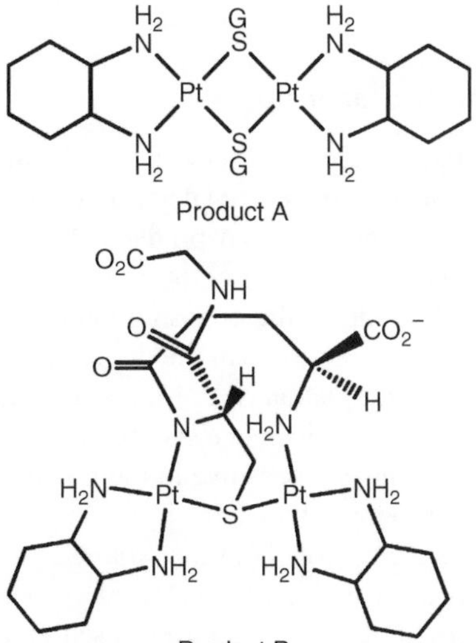

Product A

Product B

Figure 4.21 *Proposed structures for the products of the reactions of oxaliplatin with GSH and GSSG. Based on Figure 6b, p. 1210 of Fakih, S., Munk, V.P., Shipman, M.A., del Socorro Murdoch, P., Parkinson, J.A., Sadler, P.J. (2003) Novel adducts of the anticancer drug oxaliplatin with glutathione and redox reactions with glutathione disulfide. Eur. J. Inorg. Chem., 1206–1214*

two GSH molecules bridging two Pt-dach units, with two of the three lone pairs of electrons on each of the thiolates involved in the Pt-S-Pt bridge. Product B, which is also dinuclear, has only one GSH molecule but contains a Pt-S-Pt bridge, a coordinated deprotonated amide and a bound terminal amino group, forming an unusual nine-membered macrochelate involving the two platinum ions. Interestingly, the disulfide GSSG also reacts with oxaliplatin to give the same two products, which aside from the disulfide being a much poorer nucleophile toward Pt^{+2} than thiolate, implies that a redox process must be involved in the reaction. Although the mechanism of the reaction was not determined, the investigators speculated that water serves as the reducing agent for converting GSSG to GSH in the reaction.

4.2.5   Cytotoxicity and uptake of oxaliplatin

As shown in Figure 4.11, MCM-7 breast cancer cells absorb oxaliplatin from the culture medium above the cells in amounts comparable to those of cisplatin and much greater than those found in carboplatin [28]. Another similarity between oxaliplatin and cisplatin is that both drugs probably use organic cation transporters to enter the cell. By producing human embryonic kidney cells, HEK 293, which could overexpress certain cationic transporter proteins, Yonezawa *et al.* [60] showed that oxaliplatin is a good substrate for the transporters hOCT2, hOCT3, hMATE1 and hMATE2-K, and that when these transporters are present in the cell line, uptake and cytotoxicity of oxaliplatin are enhanced. Similar conclusions were reached by Zhang *et al.* [61], who showed that the cytotoxicity of oxaliplatin in six different colon cancer lines is reduced by the presence of the antiulcer drug cimetidine, which competes with oxaliplatin for transport via OCT. The study concluded that OCT1 and OCT2 are major determinants of the anticancer activity of oxaliplatin and these transporters may contribute to the antitumor specificity of the anticancer drug.

A recent study [62] used HPLC and mass spectrometry to show that the reaction of oxaliplatin with chloride ion produces [PtCl(ox)(dach)]$^-$, which contains monodentate oxalate and [PtCl$_2$(dach)]. Since [PtCl(ox)(dach)]$^-$ could be produced in the blood of patients receiving oxaliplatin, the researchers investigated the cytotoxicity of the compound toward human colon adenocarcinoma cells (HT-29), finding that the compound is as toxic as oxaliplatin toward the cells.

4.2.6   Interaction of oxaliplatin with cellular targets

As with other platinum drugs, the major cellular target for oxaliplatin is thought to be DNA. Using two-dimensional NMR methods, Chaney and coworkers [63] determined the solution structure of the adduct that forms when Pt(1*R*,2*R*-dach)Cl$_2$, the dichloro reaction product of oxaliplatin, is reacted with the single-stranded DNA oligomer 5'-CCTCAGGCCTCC. After purifying the platinated DNA using HPLC and gel electrophoresis, the investigators hybridized the opposing unplatinated DNA strand with the sequence 5'-GGAGGCCTGAGG to the platinated strand to form the duplex DNA shown in Figure 4.22a.

In order to solve the solution structure of the platinated DNA, Chaney and coworkers [63] collected NMR data on the duplex using a technique called *COSY*, which detects protons that are spin-spin coupled to each other through bonds. Since the coupling constants are sensitive to the torsion angle between coupled protons, the coupling constants provide information on torsion angles in the structure. Next, they used a second NMR technique called *NOESY*, which detects interactions between protons that are within ~0.5 nm of each other but are not necessarily connected to each other through a bond. Finally, when the information from both NMR techniques was combined and the distances and angles were used as *constraints* in a fitting procedure, a computer program was used to generate the most likely 3D structure of the molecule. As shown in Figure 4.22b, the [Pt(1*R*,2*R*-dach)]$^{2+}$ group of oxaliplatin forms an intrastrand crosslink between N7(G6) and N7(G7) of the cytosine-rich strand of the duplex. Since the solution structures of similar adducts with cisplatin had also been solved, the investigators compared the oxaliplatin-GG adduct on their duplex with a cisplatin-GG adduct on a

(a)   1 2 3 4 5 6 7 8 9 10 11 12

5'-d(CCTCA*GG*CCTCC)-3'
3'-d(GGAGTCCGGAGG)-5'

24 23 22 21 20 19 18 17 16 15 14 13

(b)

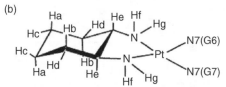

Figure 4.22 *Sequence and nomenclature used in the study. (a) Platinated DNA duplex characterized by NMR with the site of platination, GG, underlined. (b) Nomenclature and site environment of the oxaliplatin-DNA adduct. Reprinted from J. Mol. Biol., 341, Y. Wu et al., Solution Structure of an Oxaliplatin 1,2-d(GG) Intrastrand Crosslink in a DNA Dodecamer Duplex, 1251–1269. Copyright 2004, with permission from Elsevier*

similar duplex (Figure 4.23) [64]. While both types of Pt-adduct bend or kink the DNA molecule, there are significant differences between the two structures. For example, the oxaliplatin-GG structure has a narrow minor groove and the adduct forces a bend in the helix that is ∼31°, while the cisplatin-GG adduct produces a wide minor groove and a helix bend of 60-80° [64,65]. The investigators suggested that the differences in the structures of the DNA adducts may affect how mismatch repair and damage recognition proteins repair damaged DNA, which may ultimately explain the differences in the cytotoxicity and tumor range of the two drugs [66].

Lippard and coworkers [67] solved the solid-state (crystal) structure of the product of the reaction of [PtCl$_2$(1*R*,2*R*-dach)] with the single-stranded DNA oligomer CCTCT*GG*TCTCC, where the underlined dinucleotide sequence is the site of platinum binding. This study was carried out in a manner similar to that of Chaney and coworkers [63], except that the platinated duplex DNA duplex was crystallized and subjected to single crystal X-ray analysis. A novel feature of the structure shown in Figure 4.24 is the presence of a

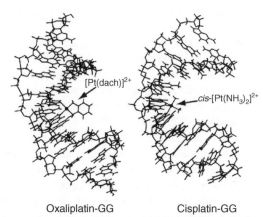

Oxaliplatin-GG   Cisplatin-GG

Figure 4.23 *Comparison of the average solution structure of the product of reaction of oxaliplatin with the DNA 12-mer CCTCA*GG*CCTCC and cisplatin with the DNA 12-mer CCTCT*GG*TCTCC, PDB 1A84. The underlined dinucleotide is the location of platinum binding. Reprinted from J. Mol. Biol., 341, Y. Wu et al., Solution Structure of an Oxaliplatin 1,2-d(GG) Intrastrand Crosslink in a DNA Dodecamer Duplex, 1251–1269. Copyright 2004, with permission from Elsevier*

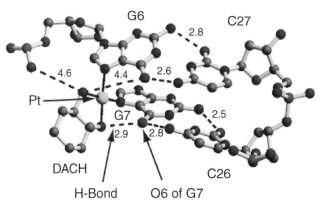

Figure 4.24 *View of the platination site in the DNA duplex CCTCTGGTCTCTCC/GGAGACCAGAGG. The pseudo-equatorial hydrogen atom on the nitrogen atom of 1R,2R dach, which is cis to N7 of G7, forms a hydrogen bond to O6 of G7 of the DNA duplex. Adapted with permission from B. Spingler et al., 2.4 A Crystal Structure of an Oxaliplatin 1,2-d(GpG) Intrastrand Crosslink in a DNA Dodecamer Duplex, Inorg. Chem, 40, 5596–5602. Copyright 2001 American Chemical Society*

hydrogen bond between the pseudo-equatorial NH hydrogen atom of the 1R,2R-dach ligand (the Hg hydrogen in Figure 4.22b) and the O6 atom of G7 at the site of platination. The investigators suggested that this feature of the structure shows the importance of chirality in the dach ligand of oxaliplatin in mediating the interaction of the drug with duplex DNA, which is also a chiral molecule.

By labeling one of the asymmetric carbon atoms of the dach ligand of oxaliplatin with radioactive ^{14}C, Hah et al. [68] used accelerator mass spectrometry (AMS) to measure the rate of adduct formation of oxaliplatin with purified salmon sperm DNA. They found that at 37 °C in a buffer containing 25 mM NaCl, 0.14 mMEDTA and 0.14 mM Tris.HCl, pH 7.4, the rate constant for binding of oxaliplatin to DNA was $3.36 \times 10^{-6}\,\mathrm{s}^{-1}$. They also identified the platinum-binding sites on DNA by enzymatically digesting and cutting the platinated DNA to deoxyribonucleotides and analyzing the resulting mixture using HPLC. With this approach they were able to show that while short incubation times (1 hour) produced monoadducts on DNA as expected, longer times (24 hours) yielded a higher percentage of bifunctional adducts, with the major adduct being a bifunctional adduct at the sequence GG.

These workers also studied drug uptake and DNA binding in the human cancer cell lines 833 K testicular cancer, MDA-MB-231 multidrug resistant breast cancer and T24 bladder cancer. In view of the very high sensitivity of AMS, they were able to use very low concentration of oxaliplatin (0.2 μM) in the culture medium, which likely induced little toxic 'stress' to the cells. The early parts of the uptake plots, 0-6 hours, were linear and the total amount of adduct on the isolated DNA from the cells after 6 hours was ∼200 oxaliplatin adducts (radiolabel) per 10^7 nucleotides of DNA. The fact that there were proportionally fewer 1,2-intrastrand crosslinks at GG from DNA extracted from the cells than was the case with purified salmon sperm DNA was attributed to more efficient repair of these lesions by repair enzymes inside the cells.

As shown in Figure 4.25, most of the radioactive ^{14}C label (which presumably remains bound to platinum as the dach ligand) that binds to cells treated with oxaliplatin is not bound to DNA. While the relative amount of label bound to proteins and other molecules varies with the cell type and the incubation time, the approximate amount bound to DNA from Figure 4.25 is only ∼1%.

The 'packaging' of DNA in the nucleus of a eukaryotic cells is very complex, but one repeat unit is a structure called a *nucleosome core particle (NCP)*, which consists of a segment of duplex DNA 146 base pairs in length wrapped around eight small histone proteins. Wu et al. [69] recently studied the binding of cisplatin and oxaliplatin to a NCP using x-ray crystallographic methods. They found that while oxaliplatin binds to the

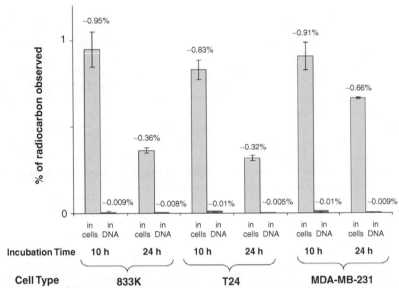

Figure 4.25 *Percentage of [^{14}C]-labeled oxaliplatin found bound in cells and in the DNA extracted from the cells determined by accelerator mass spectrometry, AMS. Cells of 833 K testicular cancer, MDA-MB-231 multidrug-resistant breast cancer and T24 bladder cancer were treated with 0.2 μM labeled oxaliplatin for 10 and 24 hours before analysis by AMS. Reprinted with permission from S.S. Hah, Characterization of Oxaliplatin-DNA Adduct Formation in DNA and Differentiation of Cancer Cell Drug Sensitivity at Microdose Concentrations, Chem. Res. Toxicol. 20, 1745–1751. Copyright 2007 American Chemical Society*

sequence GG, as well as GA and AG, both drugs also bind to specific methionine residues (Met) on the histone proteins (Figure 4.26). Since some of the binding sites are on the interior of the NCP, the study showed that either drug could easily penetrate the interior of the core particle, suggesting that these sites could also be attacked in genomic DNA. As was proposed by Reedijk [70], binding to methionine on the histone proteins could serve as a reservoir for platinum, which could ultimately facilitate the transfer of platinum ions to DNA.

Besides DNA, there exist many potential protein targets for the platinum drugs. Witte *et al.* [71] recently showed that oxaliplatin binds to and disrupts the function of the protein thioredoxin reductase (TrxR) [72,73]. While the identity of the specific binding site for platinum on the protein was not found, the authors suggested that selenocysteine is the residue being attacked by oxaliplatin and the aquated forms of cisplatin. Interestingly, carboplatin does not inhibit TrxR, which is probably a reflection of the low reactivity of this drug toward thiols, and by inference toward selenols as well.

Box 4.3 Thioredoxin reductase as a target for metallo-drugs

The enzyme thioredoxin reductase, TrxR, is important in cell proliferation, antioxidant defense and redox signaling [72,73]. A unique feature of TrxR is that it has the uncommon selenium-containing amino acid selenocysteine, Sec, at its active site. Except for substitution of selenium for sulfur, selenocysteine is identical in structure to the amino acid cysteine, Cys (Figure 4.27a). In the TrxR protein, these amino acids are adjacent to one another in the peptide backbone, and since they are on the surface of the protein they are readily accessible to various substrates, including small molecules such as drugs. One of the functions of TrxR is to help regulate the redox state of the cell; that is, it carries out *edox homeostasis* by keeping a proper balance between oxidized and reduced forms of thiol-containing molecules such as glutathione in the cell. The TrxR protein does this by changing its own oxidation state in the manner shown in Figure 4.27b. In the reduced state, the Cys and Sec residues of TrxR are in their reduced *thiol* and *selenol* forms – that is, RSH and RSeH – while in the oxidized state, the two residues are coupled via an S-Se *selenenyl sulfide* bond. After accounting for the loss of two acidic protons from the SH and SeH groups, the oxidation, as is indicated in Figure 4.27b, is a two-electron process. If a metallo-drug binds to the reduced form of TrxR and blocks the oxidation process, the ability of the enzyme to carry out cellular redox homeostasis will be compromised, which, if enough TrxR molecules are so modified, usually results in the death of the cell.

From the standpoint of hard–soft acid–base (HSAB) theory, the selenate ion, $R-Se^-$, is considered an even softer base than the thiolate ion, $R-S^-$. This is because Se^- has a larger diameter than S^- (Se is below S in the periodic chart) and thus the electron density of Se^- is more easily deformed, *polarized*, than the electron density of S^-. This suggests that Se^- is an excellent nucleophile for Pt^{+2} and other soft acids, and since metal binding to this amino acid would block the redox activity of TrxR, the enzyme is a prime target for many metallo-drugs.

4.3   New platinum agents

In addition to cisplatin, carboplatin and oxaliplatin, which are in wide clinical use throughout the world, some platinum agents have gained regional approval, and others are presently in clinical trails in a number of

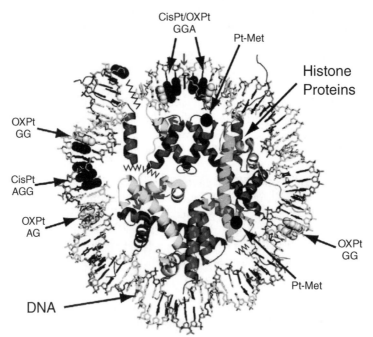

Figure 4.26   *A nucleosome core particle (NCP) which has been treated with cisplatin (CisPt) and in a separate experiment with oxaliplatin (OXPt) and their respective binding sequences under the abbreviation. Bases at which binding occurs are black or gray. The view shown is a 'slice' through the C_2 symmetry axis of the entire NCP, so that only half of the entire NCP is shown. The ribbon structures in the center are four of the eight histone proteins, around which are wrapped a DNA duplex 78 bp in length. Both drugs bind to methionine residues of the histone proteins, the locations of which are indicated by 'Pt-Met'. Reprinted by permission from Macmillan Publishers Ltd: Nat. Chem. Biol., 4, B. Wu et al., Site Selectivity of Platinum Anticancer Therapeutics, 110–2, Copyright 2008*

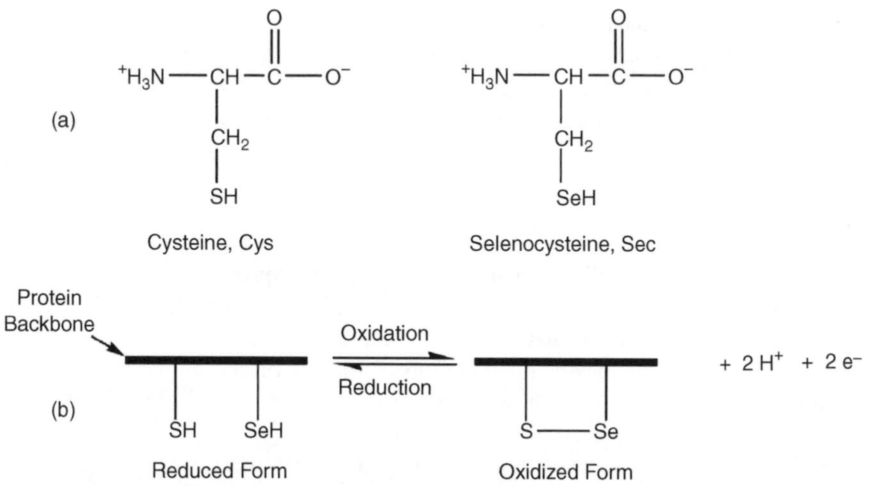

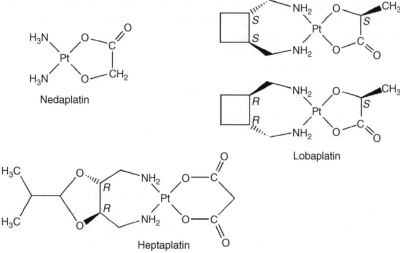

Figure 4.27 *(a) Structures of the zwitter ionic forms of the amino acids cysteine, Cys, and selenocysteine, Sec. (b) Reduced and oxidized forms of the active site region of thioredoxin reductase, TrxR*

countries. While a brief overview of some of these drugs and drug candidates will be given here, more details can be found in the excellent reviews by Wong and Giandomenico [74] and Galanski *et al.* [75].

4.3.1 Nedaplatin

The platinum-containing anticancer drug nedaplatin, *cis*-diammineglycolatoplatinum(II) (Figure 4.28) is in clinical use in Japan for the treatment of ovarian and cervix carcinomas, head and neck tumors, and esophageal and bladder cancer [75,76]. Nedaplatin is in some ways similar to carboplatin in that it does not appear to use organic cation transporters for entering the cell [60] and it is myelosuppresive, which limits the dose that can be administered to the patient [75].

Figure 4.28 *Structures of platinum antitumor agents that have gained regional approval for use as anticancer drugs*

Structurally, the nonleaving ligands of nedaplatin are two *cis*-ammonia molecules, which makes half of the drug identical to cisplatin and carboplatin. With nedaplatin, the leaving ligand is the dianionic form of glycolic acid, which forms a five-membered chelate ring with the Pt^{+2} ion. The two negative charges on the glycolic acid are due to the deprotonate carboxylic acid group and the deprotonated alcohol functional group; that is, an alkoxide.

4.3.2 Lobaplatin

Lobaplatin (*R,R/S,S*)-(1,2-cyclobutanedimethanamine)[(2*S*)-2-hydroxypropanoato, *O,O′*]platinum(II), which exhibits few of the side effects of cisplatin, is approved for use as a drug in China. As is evident from Figure 4.28, lobaplatin is a mixture of two compounds that are *diastereomeric* to each other; that is, stereoisomers, but not exact mirror images [75,77]. The nonleaving ligand of lobaplatin is a chiral cyclobutane diamine (the racemic form of the ligand is used in the synthesis), which forms a seven-membered chelate ring with the platinum ion. While a seven-membered ring is expected to be less thermodynamically stable than either a five- or a six-membered chelate ring, the strength of the metal-nitrogen bond suggests that the diamine remains bound to the metal ion when the compound is in the biological milieu.

The leaving ligand in the case of lobaplatin is the naturally-occurring optically-active form of lactic acid, which has the *S* absolute configuration. As with nedaplatin, the leaving ligand of lobaplatin is dianionic, with one negative charge on the carboxylate and the second on the alkoxide, which, since the diamine is uncharged, makes lobaplatin, like nedaplatin, electrically neutral. In the body, lobaplatin appears to remain largely intact until it is removed by glomerular filtration. The drug, which displays mild gastrointestinal and dose-limiting thrombocytopenia (reduced platelet levels in blood), is apparently not as nephro-, neuro- or oto-toxic as cisplatin. Lobaplatin is used to treat non-small-cell lung cancer, breast tumors and certain forms of leukemia.

4.3.3 Heptaplatin

Heptaplatin (eptaplatin), *cis*-malonato[(4*R*,5*R*)-4,5-bis(aminomethyl)-2-isopropyl-1,3-dioxolane]platinum-(II) (Figure 4.28) is approved for use as a drug in South Korea [75,78,79]. The nonleaving ligand in heptaplatin is a chiral form of a seven-membered diamine with a fused five-membered diether, while the leaving ligand of the complex is the dianion of malonic acid. Unlike carboplatin, the malonato ligand of heptaplatin has no bulky alkyl substituent on the center, C2, carbon atom. However, x-ray structural analyses [43] show that the malonato ligand bound to Pt^{+2} is significantly displaced from the metal-donor plane, which could affect substitution reactions and make the compound 'carboplatin-like' in its chemical reactivity. Heptaplatin appears to be effective against cisplatin-resistant cancer lines, which may be related to the levels of metallothionein in the cell. Heptaplatin is used to treat gastric cancer and, in combination with paclitaxel, head and neck squamous cancers. The drug has mild hepatotoxicity and myelosuppression, and the nephrotoxicity caused by the agent is dose-limiting.

4.3.4 Satraplatin (JM216)

Satraplatin, bis(acetato)amminedichloro(cyclohexylamine)platinum(IV) is a Pt^{+4} complex (Figure 4.29) that is currently in clinical trails in the United States, by itself and in combination with other agents, for treatment of non-small-cell lung cancer, various solid malignancies and refractory prostate cancer [74,75,80,81]. An attractive feature of satraplatin is that the agent can be taken orally in pill form. This feature, which eliminates health care personnel for administration, is viewed as an advance for platinum chemotherapy.

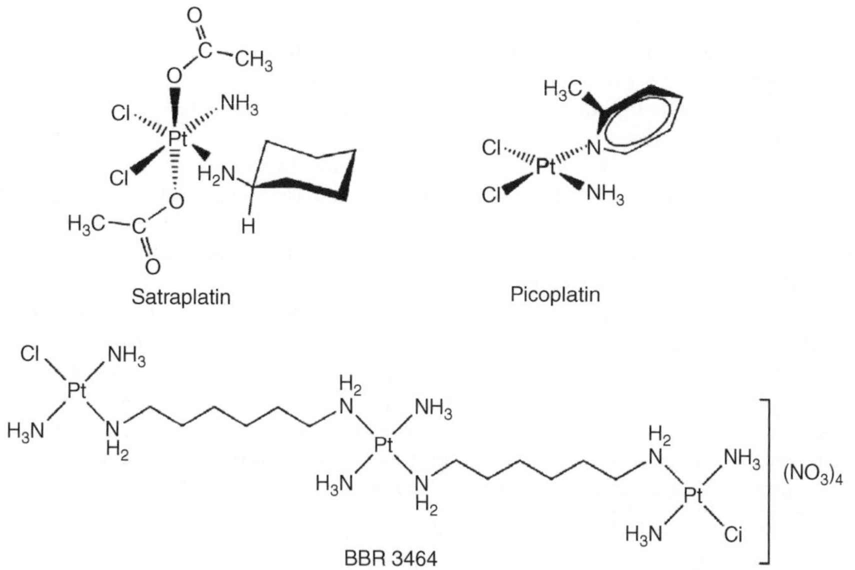

Satraplatin

Picoplatin

BBR 3464

Figure 4.29   *Some platinum-containing agents that are in clinical trials*

Satraplatin contains Pt^{+4}, which has an electronic configuration of $5d^6$. The ligands of the drug produce a crystal field about the metal ion which approximates an octahedron (in a true octahedral crystal field all six ligands are identical), and because the value of Δ is large, satraplatin is a low-spin diamagnetic compound with $S = 0$. The large magnitude of the crystal field causes the d-d transitions of the ion to lie in the near-UV region of the spectrum, 200-400 nm. For example, while the lowest-energy spin-allowed transition, $^1T_{1g} \leftarrow ^1A_{1g}$, for $[PtCl_6]^{2-}$ is at $28\,300\,cm^{-1}$ (353 nm), increasing the crystal field strength by substituting five of the chloro ligands with ammonia - for example $[PtCl(NH_3)_5]^{3+}$ - shifts the transition to $35\,400\,cm^{-1}$ (282 nm). Because the d-d transitions of Pt^{+4} complexes are in the UV region of the spectrum, the compounds of the ion are often colorless, or if a band 'tails' into the visible region, pale yellow in color. Another consequence of the large crystal field for Pt^{+4} is that ligand substitution reactions with the ion are generally slow. For example, the pseudo first-order water exchange rate constant for *trans*-$[Pt(ox)_2(H_2O)_2]$ at $25\,°C$ is 7.0 $(\pm 1.2) \times 10^{-6}\,s^{-1}$ [82].

Box 4.4 Synthesis of satraplatin

The synthesis of satraplatin [80], which demonstrates important aspects of the *trans* effect and the chemical reactivity of bound ligands, is shown in Figure 4.30. The synthesis begins with cisplatin, which is reacted with slightly more than one equivalent of tetraethylammonium chloride, Et_4NCl, which serves as the source of Cl^-, in hot dimethyl acetamide (DME). Since DME is a poor donor toward transition metal ions, it is unreactive to Pt^{+2} and thus does not interfere with the substitution chemistry at the platinum center. An additional important feature of the first step in the synthesis is that the ammonia that is liberated from the platinum ion escapes as a gas from the hot solvent, thus driving the equilibrium in favor of the trichloro product, which is orange in color. In the second step, an iodo ligand, I^-, is introduced into the coordination position adjacent to the ammonia molecule. This is done because the strong *trans*-directing ability of the iodide ligand will ensure that the next ligand into the coordination sphere, in this case cyclohexylamine, will occupy the position *trans* to the iodide; that is, *cis* to the ammonia molecule.

From the *trans* effect series, Cl⁻ is a slightly better *trans*-directing ligand than is NH₃, and thus it might be expected that the ammonia molecule, and not the Cl⁻ *cis* to the ammonia, would be displaced by the iodide ion. However, because the strength of a Pt-N bond is greater than that of a Pt-Cl bond – that is, Δ (NH₃) ≫ Δ (Cl⁻) – the iodide ion actually displaces one of the two chloride ligands that are *cis* to the ammonia molecule. Once the iodide is in place, one equivalent of cyclohexylamine is added, which ultimately occupies the coordination position *trans* to the iodide; that is, *cis* to the ammonia molecule. In step four of the synthesis, the iodide ligand, having done its '*trans*-directing', is removed by the addition of silver nitrate, AgNO₃. While both AgCl and AgI are insoluble in water, the latter is the less soluble of the two, so if slightly more than one equivalent of AgNO₃ is added to the reaction, AgI is the main precipitate that forms. By adding a source of chloride ion, in this case HCl, under mild conditions, Cl⁻ is inserted into the position vacated by the iodide ion. Excess aqueous hydrogen peroxide is used to oxidize Pt⁺² to Pt⁺⁴. This oxidation accomplishes two things: it expands the coordination sphere of the metal ion from four to six and, since Pt⁺⁴ is relatively exchange-inert, it 'locks' the ligands in the Pt⁺⁴ coordination sphere in place. An additional important feature of the product, which is a *trans* dihydroxo complex, is that the lone pairs of electrons on the oxygen atoms of the hydroxo ligands which are not used in bonding to the Pt⁺⁴ ion are nucleophilic. This allows reaction of the hydroxo ligands with acetic anhydride in a manner that is similar to the reaction of an organic alcohol with the anhydride to give an ester, only in this case the product is a bound acetate (acetato) ligand. Since acetic anhydride is in large excess – that is, it is the solvent in the reaction – both hydroxo ligands react to give satraplatin (Figure 4.30).

Satraplatin is the only orally-given platinum-containing drug in clinical trials. Because the stomach is a very aggressive chemical environment that is highly acidic and contains many digestive enzymes, metal complexes that undergo facile substitution reactions generally do not survive passage from the stomach into the upper gastrointestinal (GI) tract, where the pH is near neutral and adsorption of the drug into the blood occurs. However, satraplatin, which is relatively exchange-inert, very likely passes into the upper GI tract largely intact, where the drug is believed to enter the blood for distribution throughout the body.

The fate of satraplatin once it enters the blood is unknown, but one report indicates that it is rapidly reduced in fresh human whole blood *in vitro*, with a half life of only 6.3 minutes, and that the reduction may take place

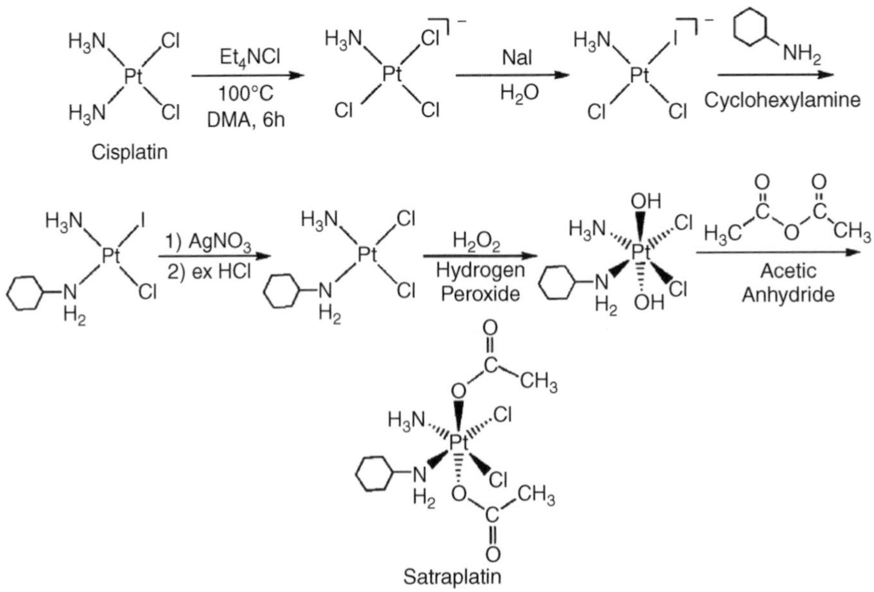

Figure 4.30 *Synthesis of satraplatin*

inside red blood cells [83]. Supporting this is a recent study which shows that satraplatin is reduced by hemoglobin, cytochrome c and liver microsomes, and that naturally-occurring reducing agent NADH is required for the reduction [84]. The product of the reaction is a square planar Pt^{+2} compound with ammonia, cyclohexylamine and two chloride ligands that were originally part of satraplatin [84].

The mechanism of action of satraplatin and other antitumor active compounds of Pt^{+4} has been reviewed by Hall *et al.* [85]. While DNA, like other platinum drugs, is believed to be an important target for satraplatin, the rate of binding of Pt^{+4} to DNA seems too slow to account for the antitumor effects of the drug, so bioreduction to Pt^{+2} is thought to be a critical step for the biological activity of the agent. Using x-ray absorption near-edge spectroscopy (XANES), a technique that can determine the oxidation states of metals present in low concentration, Hambley and coworkers [86] showed that Pt^{+4} antitumor agents can enter A2780 ovarian cancer cells and that the ratio of Pt^{+4} to Pt^{+2} inside the cell varies with the ease of reduction of the Pt^{+4} compounds to Pt^{+2} complexes, as measured using electrochemical techniques. The researchers found that the Pt^{+4} compounds that are the easiest to electrochemically reduce give the highest amount of Pt^{+2} inside the cell. While the cellular reducing agents are unknown, they could be sulfur-containing amino acids such as cysteine or methionine found on proteins or other small molecules in the cell. The second-order rate constants for the reduction of a Pt^{+4} complex similar in structure to satraplatin by cysteine and methionine at pH 7.0 at 32 °C are $(12.5 \pm 0.6) \times 10^{-3}$ and $(6.0 \pm 0.3) \times 10^{-6} M^{-1} s^{-1}$, respectively, showing that cysteine may be a bioreducing agent for Pt^{+4} in the cell [87]. In view of the kinetic inertness of Pt^{+4}, reduction probably takes place without a ligand substitution reaction by electron transfer from the reducing agent to one of the ligands directly bound to the platinum ion; that is, via an *outer sphere mechanism.*

4.3.5 Picoplatin (AMD473, ZD473)

The chemical innovation in picoplatin is the presence of a 2-methy pyridine ligand (Figure 4.29). Since this ligand must be positioned approximately perpendicular to the coordination plane of the metal ion, the methyl group on the 2-position of the heterocycle is over the z-axis of the complex. In view of the fact that substitution reactions on Pt^{+2} are bimolecular and that the nucleophile attacks via the z-axis to produce a five-coordinate transition state, substitution reactions with picoplatin would be expected to be slower than those with similar complexes not having this steric feature. As is evident from the crystal structure of picoplatin [88,89] (Figure 4.31), the pyridine ring is approximately perpendicular to the coordination plane of the platinum ion and the methyl group is over one of the axial sites of the metal ion. Measurements of the hydrolysis of picoplatin by Sadler and coworkers [89] showed that $t_{1/2}$ at 37 °C for displacement of the Cl^- *trans* to the

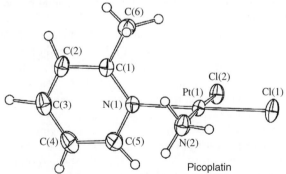

Picoplatin

Figure 4.31 *Structure of picoplatin showing thermo-ellipsoids of the atoms. Reprinted with permission from A.R. Battle et al., Platinum (IV) Analogs of AMD473 (cis-[PtCl₂(NH₃)(2-picoline)]): Preparative, Structural, and Electrochemical Studies, Inorg. Chem., 45, 6317–22. Copyright 2006 American Chemical Society*

pyridine ligand by water is 6.0 hours, while displacement of the Cl$^-$ *trans* to the ammonia molecule has a $t_{1/2}$ of 8.7 hour. Since both values are greater than $t_{1/2}$ for the first hydrolysis of cisplatin (1.05 hours [90]), the steric effects of the methyl group significantly affect the substitution kinetics of the complex.

Picoplatin is active against cisplatin-resistant cell lines and cell lines that have enhanced mechanisms for removing platinum from the cell, complexing platinum to cellular thiols or removing platinum from DNA [91]. While picoplatin is given by IV injection, the agent appears to have good bioavailability when given orally; that is, a significant amount of platinum reaches the blood. The side effects of picoplatin include nausea, vomiting, anorexia and a transient metallic taste, but there is no or very little nephro- or neurotoxicity with the agent. Picoplatin is currently in clinical trials by itself and in combination with other drugs in the United States for the treatment of solid tumors and lymphomas, and colorectal and prostate cancers.

4.3.6   BBR3464

Shown in Figure 4.29 is the unusual trinuclear platinum antitumor agent BBR3464, *triplatin tetranitrate*, developed by Farrell and coworkers [92]. The two platinum chloro-tri-ammine 'ends' of the molecule have a charge of 1 +, while the central platinum tetra-ammine 'core' has a charge of 2 +, making the net charge on the molecule 4 +. The molecule is assembled by first synthesizing the core part of the structure and then adding the two end groups. In terms of structure, BBR3464 is a radical departure from the paradigm that has dominated platinum-based chemotherapy. Early in the study of the structure activity relationships (SARs of the platinum anticancer drugs, it was determined that the most active structures, and hence those worth pursuing, were mononuclear: they contained *cis*, *nonleaving*, amine-type ligands and two *cis*, *leaving* ligands that could be relatively easily displaced. However, in recent years this central paradigm of platinum antitumor chemistry has been under attack and many platinum complexes not meeting these requirements, including BBR3464, are now known to exhibit potent antitumor effects [93–95].

In tissue culture, BBR3464 is cytotoxic at ten- to one hundred-fold lower molar concentrations than cisplatin and it displays activity in cisplatin-resistant cell lines [92]. The agent also exhibits antitumor activity against human tumors that have been grafted on to other living mammals (*xenografts*). Since these tumors also have an impaired *p53 gene*, a protein which is in part responsible for suppressing the expression of other genes that cause cancer, antitumor activity against these xenografts is considered significant. In clinical trails in the United States, BBR3464 is showing activity against melanoma, and against metastatic pancreatic and lung cancers, which are not normally treatable with cisplatin [81]. The dose-limiting toxicities with BBR3464 are bone depression - a common problem with most cytotoxic agents - and diarrhea.

The kinetic and equilibrium parameters of BBR3464 were measured using [^{1}H,^{15}N] HSQC 2D NMR by Berners-Price and Farrell and their coworkers [96]. While the tetra-ammine, core, portion of BBR3464 is considered exchange-inert and unreactive (two ammonia molecules and two primary amines), the terminal platinum complexes, which are independent of each other, are able to lose a chloride ligand in a substitution reaction. In water, the pseudo first-order rate constant at 25 °C for hydrolysis of either terminal platinum complex is $(7.1 \pm 0.2) \times 10^{-5}$ s^{-1}, while the equilibrium constant for the reaction is $K = 4.47 \times 10^{-4}$. Once a water molecule is bound at each of the terminal Pt^{+2} centers, the pK_a for deprotionation of water to form bound hydroxide from either center is the same, 5.62. In blood, if equilibrium with respect to chloride ($\sim$104 mM) is reached, the concentration of the aquated form of BBR3464 would be very small, $\sim$0.03%, and at pH 7.4 most of this would be in the less reactive hydroxo form (OH$^-$ is a poor leaving ligand).

The trinuclear compound, which is given by IV injection, binds to human serum albumin (HSA) in the blood. Since the molecule has a high charge of 4 +, there is immediate electrostatic binding to HAS, which is followed by a slower covalent-type interaction. The slower interaction is presumably due to displacement of ligands on BBR3464 by sulfur nucleophiles on the protein. The compound is readily taken up by L1210 leukemia cells in culture and it binds to cellular DNA, forming interstrand crosslinks (between two strands) [92,97]. Studies with purified DNA show that the two terminal platinum complexes of BBR3464 react with N7 of guanine by losing the chloro ligand, while the central cationic portion of the agent probably interacts electrostatically with DNA, which is an anion. While the structures of the DNA adducts have not been characterized, molecular modeling studies by Brabec and Farrell and their coworkers [98] suggest that BBR3464 binds to a duplex DNA 17-mer, forming 1,4-interstrand crosslinks, which can have the linker either inside or outside the major groove, or else it can form a 1,5-intrastrand crosslink.

Problems

1. The second-order rate constant for the reaction of carboplatin, $[Pt(cbdca)(NH_3)_2]$, with the amino acid methionine (Met) in water at $37\,°C$ is $k_2 = 2.7 \times 10^{-3}\,M^{-1}\,s^{-1}$. If RPMI culture medium, which contains $115\,\mu M$ Met, also contains $20\,\mu M$ carboplatin, calculate the following:

   a. The pseudo first-order rate constant, k_1, for the reaction of the drug with Met under these conditions.
   b. The concentration of carboplatin remaining after 1 hour in RPMI at $37\,°C$ due to the reaction of the drug with Met.

2. The nucleotide 5′-guanosine monophosphate (5′-GMP) reacts with ^{15}N-labeled carboplatin at $37\,°C$ at neutral pH to give *cis*-$[Pt(cbdca\text{-}O)(5'\text{-}GMP)(NH_3)_2]$, which contains monodentate cbdca, and 5′-GMP, which is bonded through N7 of G to the platinum ion.

   a. Draw the structure of the product complex.
   b. Carboplatin exhibits a Laporte-forbidden spin-allowed transition at 330 nm. Briefly explain whether this absorption band for the product is shifted to higher energy, lower energy, or is unshifted relative to 330 nm.
   c. The product complex exhibits two $[^1H,\ ^{15}N]$ HSQC NMR peaks at $\delta\ ^1H/^{15}N$, 4.37/−79.9 and 4.21/−68.1 ppm. Assign the peaks - for example, $^{15}NH_3$ *trans* to N-7 of G - of the complex.

3. The first-order rate constant for the decrease in the concentration of a drug in the blood of a patient after termination of IV administration of the agent is monoexponential with $k_1 = 0.17\,h^{-1}$.

   a. Calculate the time required for the drug concentration in the blood to reach half of its initial value.
   b. Calculate the time required to reach 10% of its original value.
   c. If $C_{max} = 11.2\,\mu M$, calculate the AUC for the drug.

4. Carbonate, CO_3^{2-}, is a versatile ligand that can bind to a metal ion to form mono-, di- and tri-nuclear complexes, containing one, two and three metal ions respectively. It can also form mono- and bi-dentate complexes with a single metal ion. There are also cases in which of one of the carbonate oxygen atoms acts as a bridge between two metal ions.

   a. Draw the electron-dot formula for carbonate CO_3^{2-} and its resonance forms.
   b. Using a generic 'M' for the metal ion, sketch the types of coordination mode indicated above.

5. A study of binding of carboplatin to cells reported that the rate of the drug binding to 3.5×10^6 cells in a total volume of $920\,\mu l$ culture medium which contains $130\,\mu M$ carboplatin is $17\,amol\,Pt\,h^{-1}\,cell^{-1}$ (amol: 10^{-18} mol).

   a. If the volume of a single cell is 2 pl (pl $= 10^{-12}$ l), calculate the effective *concentration* of platinum ions in a single cell after the cell population has been exposed to the drug solution for 1 hour.

b. If 2% of the total amount of platinum that is associated with the cell is bound to genomic DNA in the nucleus of the cell, calculate the binding density in terms of the number of platinum ions per 10^6 nucleotides (nt) of DNA after 30 minutes. Assume that there is one copy of a double-stranded DNA molecule in the nucleus that is 3×10^{-9} base pairs long.

6. Oxaliplatin is the most recent platinum drug to enter into worldwide clinical use for the treatment of cancer.

   a. While oxaliplatin is much less oto- and nephro-toxic than cisplatin, it has some side effects. Describe a side effect (reversible) that is associated with oxaliplatin.
   b. Most of the evidence suggests that oxaliplatin targets DNA in the cell. What is the preferred binding DNA sequence of oxaliplatin?

7. Agarose gel electrophoresis is a convenient and sensitive method for studying the binding of platinum drugs to DNA. Shown below is an agarose gel of the binding of BBR3464 to the supercoiled plasmid DNA pSP73. The plasmid was incubated with BBR3464, a trinuclear platinum complex with potent anticancer activity, for 48 hours, electrophoresed, and the DNA stained with ethidium bromide to give the gel shown. The lane number and value of r_b, the number of molecules of BBR3464 per DNA nucleotide, are: lanes 1, 9 (controls no BBR3464); 2, 0.001; 3, 0.005; 4, 0.01; 5, 0.02; 6, 0.03; 7, 0.04; 8, 0.05. The left side of the gel shows the charges on the electrodes in the electrophoresis experiment.

   a. Assign the bands in each lane to the correct forms of pSP73 DNA.
   b. Briefly explain why the DNA forms change their mobilities in the manner observed.
   c. Assuming that BBR 3464 is equally distributed between the two forms of DNA and the ratio of forms is 1 : 1, calculate the *number of BBR3464 molecules* bound per pSP73 DNA molecule for the forms for $r_b = 0.04$, lane 7, when both forms co-migrate in the gel. The length of the double-stranded pSP73 genome is 2464 base pairs.

8. The carrier (nonleaving) ligand in oxaliplatin is R,R-dach. In order to determine the effects of a chiral Pt-dach-type complex on the melting temperature of DNA, Brabec and coworkers (2008, *Chem. Eur. J.*, 14, 1330-1341) attached $[Pt(R,R\text{-dach})]^{2+}$, $[Pt(S,S\text{-dach})]^{2+}$ and cis-$[Pt(NH_3)_2]^{2+}$ (from cisplatin) to the two guanine bases (in bold) in the DNA duplex 15-mer, shown below. While this formed an *interstrand* crosslink, effectively 'stapling' the two strands of the duplex together at the bound guanines, heating the platinated DNA separated, except for the stapled part, the Watson-Crick base pairs of the duplex. The temperature at which at which half of the DNA molecules are in the duplex form and half are in the melted (separated) form is called the melting temperature, T_m, of the DNA. The values of ΔT_m found in the study were 6.0 °C (cis-$[Pt(NH_3)_2]^{2+}$), 3.4 °C ($[Pt(R,R\text{-dach})]^{2+}$) and 1.8 °C ($[Pt(S,S\text{-dach})]^{2+}$), where $\Delta T_m = T_m$ (platinated DNA) - T_m (unplatinated DNA).

   a. Briefly explain why $[Pt(R,R\text{-dach})]^{2+}$ and its mirror image, $[Pt(S,S\text{-dach})]^{2+}$, when bound to the same two G bases on the 15-mer produce DNA melting temperatures that are not identical.
   b. The values of ΔT_m for $[Pt(R,R\text{-dach})]^{2+}$ and $[Pt(S,S\text{-dach})]^{2+}$ are both smaller than the value of ΔT_m for cis-$[Pt(NH_3)_2]^{2+}$ bound to the 15-mer. Based on the structures of the adducts, suggest a reason why this result might be expected.

$$5'\text{-C} \quad \text{T} \quad \text{C} \quad \text{T} \quad \text{C} \quad \text{T} \quad \text{T} \quad \text{G} \quad \text{C} \quad \text{T} \quad \text{C} \quad \text{T} \quad \text{C} \quad \text{T} \quad \text{C-3}'$$
$$3\text{-G} \quad \text{A} \quad \text{G} \quad \text{A} \quad \text{G} \quad \text{A} \quad \text{A} \quad \text{C} \quad \text{G} \quad \text{A} \quad \text{G} \quad \text{A} \quad \text{G} \quad \text{A} \quad \text{G-5}'$$

15−mer duplex

9. The absorption bands for the transition $^1T_{1g} \leftarrow {}^1A_{1g}$ for the complexes $[PtF_6]^{2-}$, $[PtCl_6]^{2-}$ and $[PtBr_6]^{2-}$ are at 31 500, 28 300 and 23 000 cm^{-1}, respectively, while for $[PtCl(NH_3)_5]^{3+}$ and $[PtBr(NH_3)_5]^{3+}$ the transition is at 35 400, and 31 100 cm^{-1}, respectively. From this information, rank order the crystal field splitting parameters for the ligands F^-, Cl^-, Br^- and NH_3 attached to Pt^{+4}.

10. A 30 mM aqueous solution of an optically-active compound with a molecular weight of 429 g mol^{-1} was prepared and the optical properties of the solution were measured.

   a. At 25 °C the observed rotation, α, at 589 nm of the solution in a 10 cm cell is $+0.83°$ Calculate the specific rotation $[\alpha]_D^{25}$.

   b. This solution exhibits a positive CD band that has a value of the molar circular dichroism of $\Delta\varepsilon = +2.1$ in a 1 cm path-length cell. Calculate the difference in absorbance, $A_L - A_R$, for left and right circularly-polarized light for this CD maximum.

References

1. Boulikas, T. and Vougiouka, M. (2003) Cisplatin and platinum drugs at the molecular level. *Oncology Reports*, **10**, 1663–1682.
2. Cleare, M.J. and Hoeschele, J.D. (1973) Antitumor platinum compounds. Relationship between structure and activity. *Platinum Metal Review*, **17**, 2–13.
3. Cleare, M.J. and Hoeschele, J.D. (1973) Antitumor activity of group VIII transition metal complexes. I. platinum (II) complexes. *Bioinorganic Chemistry*, **2**, 187–210.
4. Cleare, M.J., Hoeschele, J.D., Rosenberg, B., and Van Camp, L.L.(Feb. 20, 1979) Malonato platinum anti-tumor compounds. United State Patent 4,140,707.
5. Harrison, R.C. and McAuliffe, C.A. (1980) An efficient route for the preparation of highly soluble platinum(II) antitumor agents. *Inorganica Chimica Acta*, **46**, L15–L16.
6. Sewell, G.J., Riley, C.M., and Rowland, C.G. (1987) The stability of carboplatin in ambulatory continuous infusion regimes. *Journal of Clinical Pharmacy and Therapeutics*, **12**, 427–432.
7. Di Pasqua, A.J., Goodisman, J., Kerwood, D.J. *et al.* (2006) Activation of carboplatin by carbonate. *Chemical Research in Toxicology*, **19**, 139–149.
8. Patterson, H.H., Tewksbury, J.C., Martin, M. *et al.* (1981) Luminescence, absorption, MCD, and NQR Study of the Cis and trans isomers of dichlorodiammineplatinum(II). *Inorganic Chemistry*, **20**, 2297–2301.
9. Di Pasqua, A.J. (2008) *Doctorial Dissertation*, Syracuse University.
10. Neidle, S., Ismail, I.M., and Sadler, P.J. (1980) The structure of the antitumor complex Cis-(diammino) (1, 1-cyclobutanedicarboxylato)-Pt(II): X Ray and NMR studies. *Journal of Inorganic Biochemistry*, **13**, 205–212.
11. Canovese, L., Cattalini, L., Chessa, G., and Tobe, M.L. (1988) Kinetics of the displacement of cyclobutane-1,1-dicarboxylate from diammine(Cyclobutane-1,1-dicarboxylato)platinum(II) in aqueous solution. *Journal of the Chemical Society Dalton Transactions*, 2135–2140.
12. Brandšteterová, E., Kiss, F., Chovancová, V., and Reichelová, V. (1991) HPLC analysis of platinum cytostatics. *Neoplasma*, **38**, 415–424.
13. Frey, U., Ranford, J.D., and Sadler, P.J. (1993) Ring-opening reactions of the anticancer drug carboplatin: NMR characterization of *cis*-[Pt(NH$_3$)$_2$(CBDCA-O)(5'-GMP-N7)] in solution. *Inorganic Chemistry*, **32**, 1333–1340.
14. Miller, S.E., Gerard, K.J., and House, D.A. (1991) The Hydrolysis of *cis*-diamminedichloroplatinum(II) 6. A Kinetic Comparison of the cis- and trans-isomers and other cis-di(amine)di(chloro)platinum(II) compounds. *Inorganica Chimica Acta*, **190**, 135–144.

15. Hay, R.W. and Miller, S. (1998) Reactions of platinum(II) anticancer drugs. Kinetics of acid hydrolysis of cis-diammine(cyclobutane-1,1-dicarboxylato)platinum(II) "Carboplatin". *Polyhedron*, **17**, 2337–2343.

16. Schnurr, B., Heinrich, H., and Gust, R. (2002) Investigations on the decomposition of carboplatin in infusion solutions II. Effect of 1,1-Cyclobutanedicarboxylic acid admixture. *Microchimica Acta*, **140**, 141–148.

17. Disirée van den Bongard, H.J.G., Mathôt, R.A.A., Beijnen, J.H., and Schellens, J.H.M. (2000) Pharmacokinetically guided administration of chemotherapeutic agents. *Clinical Pharmacokinetics*, **39**, 345–367.

18. van der Vijgh, W.J.F. (1991) Clinical pharmacokinetics of carboplatin. *Clinical Pharmacokinetics*, **21**, 242–261.

19. Ranford, J.D., Sadler, P.J., Balmanno, K., and Newell, D.R. (1991) ^{1}H NMR studies of human urine: urinary elimination of the anticancer drug carboplatin. *Magnetic Resonance in Chemistry*, **29**, S125–S129.

20. Welling, P.G. (1986) Pharmacokinetics, *Processes and Mathematics*, ACS Monograph 185, American Chemical Society, Washington, D.C.

21. Barnham, K.J., Djuran, M.I., del Socorro Murdoch, P. *et al.* (1996) Ring-opened adducts of the anticancer drug carboplatin with sulfur amino acids. *Inorganic Chemistry*, **35**, 1065–1072.

22. Barnham, K.J., Frey, U., del Socorro Murdoch, P. *et al.* (1994) [Pt(CBDCA-O)(NH$_3$)$_2$(L-Methionine-S)]: Ring-opened adduct of the anticancer drug carboplatin ("Paraplatin"). detection of a similar complex in urine by NMR spectroscopy. *Journal of the American Chemical Society*, **116**, 11175–11176.

23. Dedon, P.C. and Borch, R.F. (1987) Characterization of the reactions of platinum antitumor agents with biologic and nonbiologic sulfur-containing nucleophiles. *Biochemical Pharmacology*, **36**, 1955–1964.

24. Xie, R., Johnson, W., Rodriguez, L. *et al.* (2007) A study of the interactions between carboplatin and blood plasma proteins using size exclusion chromatography coupled to inductively coupled plasma mass spectrometry. *Analytical and Bioanalytical Chemistry*, **387**, 2815–2822.

25. Di Pasqua, A.J., Goodisman, J., Kerwood, D.J. *et al.* (2007) Role of carbonate in the cytotoxicity of carboplatin. *Chemical Research in Toxicology*, **20**, 896–904.

26. Di Pasqua, A.J., Centerwall, C.R., Kerwood, D.J., and Dabrowiak, J.C. (2009) Formation of carbonato and hydroxo complexes in the reaction of platinum anticancer drugs with carbonate. *Inorganic Chemistry*, **48**, 1192–1197.

27. Overbeck, T.L., Knight, J.M., and Beck, D.J. (1996) A comparison of the genotoxic effects of carboplatin and cisplatin in *Escherichia coli*. *Mutation Research*, **362**, 249–259.

28. Ghezzi, A., Aceto, M., Cassino, C. *et al.* (2004) Uptake of antitumor platinum(II)-complexes by cancer cells, assayed by inductively coupled plasma mass spectrometry (ICP-MS). *Journal of Inorganic Biochemistry*, **98**, 73–78.

29. Terheggen, P.M., Begg, A.C., Emondt, J.Y. *et al.* (1991) Formation of interaction products of carboplatin with DNA *in vitro* and in cancer patients. *British Journal of Cancer*, **63**, 195–200.

30. Parker, R.J., Gill, I., Tarone, R. *et al.* (1991) Platinum DNA damage in leukocyte DNA of patinets receiving carboplatin and cisplatin chemotherapy, measured by atomic absorption spectrometry. *Carcinogenesis*, **12**, 1253–1258.

31. Knox, R.J., Friedlos, F., Lydall, D.A., and Roberts, J.J. (1986) Mechanism of cytotoxicity of anticancer platinum drugs: evidence that *cis*-Diamminedichloroplatinum(II) and *cis*-Diammine-(1,1-cyclobutanedicarboxylato)platinum(II) Differ only in the kinetics of their interaction with DNA. *Cancer Research*, **46**, 1972–1979.

32. Blommaert, F.A., van Dijk-Knijnenburg, H.C.M., Dijt, F.J. *et al.* (1995) Formation of DNA adducts by the anticancer drug carboplatin: different nucleotide sequence preferences in vitro and in cells. *Biochemistry*, **34**, 8474–8480.

33. Natarajan, G., Malathi, R., and Holler, E. (1999) Increased DNA-Binding Activity of *cis*-1,1-Cyclobutanedicarboxy latodiammineplatinum(II) (Carboplatin) in the presence of nucleophiles and human breast cancer MCM-7 cell cytoplasmic extracts: activation theory revisited. *Biochemical Pharmacology*, **58**, 1625–1629.

34. Hah, S.S., Stivers, K.M., de Vere White, R.W., and Henderson, P.T. (2006) Kinetics of carboplatin-DNA binding in genomic DNA and bladder cancer cells as determined by accelerator mass spectrometry. *Chemical Research in Toxicology*, **19**, 622–626.

35. Sorokanich, R.S., Di Pasqua, A.J., Geier, M., and Dabrowiak, J.C. (2008) Influence of carbonate on the binding of carboplatin to DNA. *Chemistry & Biodiversity*, **5**, 1540–1544.

36. Kelland, L. (2007) The resurgence of platinum-based cancer chemotherapy. *Nature Reviews. Cancer*, **7**, 573–584.

37. Chaney, S.G., Campbell, S.L., Bassett, E., and Wu, Y. (2005) Recognition and processing of cisplatin- and oxaliplatin-DNA adducts. *Critical Reviews in Oncology/Hematology*, **53**, 3–11.

38. Graham, J., Mushin, M., and Kirkpatrick, P. (2004) Oxaliplatin. *Nature Reviews. Drug Discovery*, **3**, 11–12.

39. Wang, D. and Lippard, S.J. (2005) Cellular processing of platinum anticancer drugs. *Nature Reviews. Drug Discovery*, **4**, 307–320.

40. Argyriou, A.A., Polychronopoulos, P., Iconomou, G. *et al.* (2008) A review on oxaliplatin-induced peripheral nerve damage. *Cancer Treatment Reviews*, **34**, 368–377.

41. Cersosimo, R.J. (2005) Oxaliplatin-associated neuropathy: a review. *Annals of Pharmacotherapy*, **39**, 128–135.

42. Grolleau, F., Gamelin, L., Boisdron-Celle, M. *et al.* (2001) A possible explanation for a neurotoxic effect of the anticancer agent oxaliplatin on neuronal voltage-gated sodium channels. *Journal of Neurophysiology*, **85**, 2293–2297.

43. Bruck, M.A., Bau, R., Noji, M. *et al.* (1984) The crystal structures and absolute configurations of the anti-tumor complexes Pt(oxalato)(1R,2R-cyclohexanediamine) and Pt(malonato)(1R,2R-cyclohexanediamine). *Inorganica Chimica Acta*, **92**, 279–284.

44. Dabrowiak, J.C. and Bradner, W.T. (1987) Platinum antitumor agents. *Progress in Medicinal Chemistry*, **24**, 129–158.

45. Nakamoto, K. and McCarthy, P.J. (1968) *Spectroscopy and Structure of Metal Chelate Compounds*, John Wiley & Sons Inc., New York.

46. Tinoco, I., Sauer, K., Wang, J.C., and Puglisi, J.D. (2002) Physical chemistry, *Principles and Applications in Biological Chemistry*, 4th edn, Prentice Hall, Upper Saddle River, NJ.

47. Bennani, Yl. and Hanessian, S. (1997) trans-1,2-Diaminocyclohexane derivatives as chiral reagents, scaffolds, and ligands for catalysis: applications in asymmetric synthesis and molecular recognition. *Chemical Reviews*, **97**, 3161–3195.

48. Nakanishi, C., Ohnishi, Y., Ohnishi, J. *et al.* (Aug. 16, 1994) United State Patent 5,338,874.

49. Graham, M.A., Lockwood, G.F., Greenslade, D. *et al.* (2000) Clinical pharmacokinetics of oxaliplatin: a critical review. *Clinical Cancer Research*, **6**, 1205–1218.

50. Luo, F.R., Wyrick, S.D., and Chaney, S.G. (1999) Biotransformations of oxaliplatin in rat blood *in vitro*. *Journal of Biochemical and Molecular Toxicology*, **13**, 159–169.

51. Luo, F.R., Yen, T.-Y., Wyrick, S.D., and Chaney, S.G. (1999) High-performance liquid chromatograpic separation of the biotransformation products of oxaliplatin. *Journal of Chromatography B*, **724**, 345–356.

52. Mauldin, S.K., Gibbons, G., Wyrick, S.D., and Chaney, S.G. (1988) Intracellular biotransformation of platinum compounds with the 1,2-diaminocyclohexane carrier ligand in the L1210 cell line. *Cancer Research*, **48**, 5136–5144.

53. Jerremalm, E., Wallin, I., Yachnin, J., and Ehrsson, H. (2006) Oxaliplatin degradation in the presence of important biological sulfur-containing compounds and plasma ultrafiltrate. *European Journal of Pharmaceutical Sciences*, **28**, 278–283.

54. Jerremalm, E., Videhult, P., Alvelius, G. *et al.* (2002) Alkaline hydrolysis of oxaliplatin,-isolation and identification of the oxalato monodentate intermediate. *Journal of Pharmaceutical Sciences*, **91**, 2116–2121.

55. Hagrman, D., Goodisman, J., and Souid, A.-K. (2004) Kinetic study on the reaction of platinum drugs with glutathione. *The Journal of Pharmacology and Experimental Therapeutics*, **308**, 658–666.

56. Dabrowiak, J.C., Goodisman, J., and Souid, A.-K. (2002) Kinetic study of the reaction of cisplatin with thiols. *Drug Metabolism and Disposition: The Biological Fate of Chemicals*, **30**, 1378–1384.

57. Hargman, D., Goodisman, J., Dabrowiak, J.C., and Souid, A.-K. (2003) Kinetic study on the reaction of cisplatin with metallothionein. *Drug Metabolism and Disposition: The Biological Fate of Chemicals*, **31**, 916–923.

58. Goodisman, J., Hagrman, D., Tacka, K.A., and Souid, A.-K. (2006) Analysis of cytotoxicities of platinum compounds. *Cancer Chemotherapy and Pharmacology*, **57**, 257–267.

59. Fakih, S., Munk, V.P., Shipman, M.A. *et al.* (2003) Novel adducts of the anticancer drug oxaliplatin with glutathione and redox reactions with glutathione disulfide. *European Journal of Inorganic Chemistry*, 1206–1214.

60. Yonezawa, A., Masuda, S., Yokoo, S. *et al.* (2006) Cisplatin and oxaliplatin, but not carboplatin and nedaplatin, are substrates for human organic cation transporters (SLC22A1-3 and multidrug and toxin extrusion family). *The Journal of Pharmacology and Experimental Therapeutics*, **319**, 879–886.

61. Zhang, S., Lovejoy, K.S., Shima, J.E. *et al.* (2006) Organic cation transporters are determinants of oxiplatin cytotoxicity. *Cancer Research*, **66**, 8847–8857.

62. Jerremalm, E., Hedeland, M., Wallin, I. *et al.* (2004) Oxaliplatin degradation in the presence of chloride: identification of the cytotoxicity of the monochloro monooxalato complex. *Pharmaceutical Research*, **21**, 891–894.

63. Wu, Y., Pradhan, P., Havener, J. *et al.* (2004) NMR solution structure of an oxaliplatin 1,2-d(GG) intrastarnd cross-link in a DNA dodecamer duplex. *Journal of Molecular Biology*, **341**, 1251–1269.

64. Gelasco, A. and Lippard, S.J. (1998) NMR solution structure of a DNA dodecamer duplex containing a cis-Diammineplatinum(II) dGpG intrastrand crosslink, the major adduct of the anticancer drug cisplatin. *Biochemistry*, **37**, 9230–9239.

65. Marzilli, L.G., Saad, J.S., Kuklenyik, Z. *et al.* (2001) Relationship of solution and protein-bound structures of DNA duplexes with the major intrastrand cross-link lesions formed on cisplatin binding to DNA. *Journal of the American Chemical Society*, **123**, 2764–2770.

66. Chaney, S.G., Campbell, S.L., Bassett, E., and Wu, Y. (2005) Recognition and processing of cisplatin- and oxaliplatin-DNA adducts. *Critical Reviews in Oncology/Hematology*, **53**, 3–11.

67. Spingler, B., Whittington, D.A., and Lippard, S.J. (2001) 2.4 A Crystal Structure of an Oxaliplatin 1,2-d(GpG) intrastrand crosslink in a DNA dodecamer duplex. *Inorganic Chemistry*, **40**, 5596–5602.

68. Hah, S.S., Sumbad, R.A., de Vere White, R. *et al.* (2007) Characterization of oxaliplatin-DNA adduct formation in DNA and differentiation of cancer cell drug sensitivity at microdose concentrations. *Chemical Research in Toxicology*, **20**, 1745–1751.

69. Wu, B., Dröge, P., and Davey, C.A. (2008) Site selectivity of platinum anticancer therapeutics. *Nature Chemical Biology*, **4**, 110–112.

70. Reedijk, J. (2003) New clues for platinum antitumor chemistry: kinetically controlled metal binding to DNA. *Proceedings of the National Academy of Sciences of the United States of America*, **100**, 3611–3616.

71. Witte, A.-B., Anestal, K., Jerremalm, E. *et al.* (2005) Inhibition of thioredoxin reductase but not of glutathione reductase by the major classes of alkylating and platinum-containing anticancer compounds. *Free Radical Biology & Medicine*, **39**, 696–703.

72. Urig, S. and Becker, K. (2006) On the potential of thioredoxin reductase inhibitors for cancer therapy. *Seminars in Cancer Biology*, **16**, 452–465.

73. Becker, K., Gromer, S., Schirmer, R.H., and Muller, S. (2000) Thioredoxin reductase as a pathophysiological factor and drug target. *European Journal of Biochemistry*, **267**, 6118–6125.

74. Wong, E. and Giandomenico, C.M. (1999) Current status of platinum-based antitumor drugs. *Chemical Reviews*, **99**, 2451–2466.

75. Galanski, M., Jakupec, M.A., and Keppler, B.K. (2005) Update of the preclinical situtaion of anticancer platinum complexes: novel design strategies and innovative analytical approaches. *Current Medicinal Chemistry*, **12**, 2075–2094.

76. Totani, T., Aono, K., Komura, M., and Adachi, Y. (1986) Synthesis of (Glycolato-*O*, *O′*) Diammineplatinum(II) and its related complexes. *Chemistry Letters*, 429–432.

77. McKeage, M.J. (2001) Lobaplatin: A new antitumor palatinum drug. *Expert Opinion on Investigational Drugs*, **10**, 119–128.

78. Lew, J.W., Park, J.K., Lee, S.H. *et al.* (2006) Anti-tumor activity of heptaplatin in combination with 5-fluorouracil or pactitaxel against head and neck cancer cells *in vitro. Anti-Cancer Drugs*, **17**, 377–384.

79. Choi, C.-H., Cha, Y.-J., An, C.-S. *et al.* (2004) Molecular mechanisms of heptaplatin effective against cisplatin-resistant cancer cell lines: less involvement of metallothionein. *Cancer Cell International*, **4**, 6.

80. Giandomenico, C.M., Abrams, M.J., Murrer, B.A. *et al.* (1995) Carboxylation of kinetically inert platinum (IV) hydroxy complexes. An entrée into orally active platinum (IV) antitumor agents. *Inorganic Chemistry*, **34**, 1015–1021.

81. www.ClinicalTrials.gov (04/19/09).

82. Dunham, S.U. and Abbott, E.H. (2000) A nuclear magnetic resonance investigation of water exchange reactions and trans to cis isomerization for diaqua and Dihydroxobis(oxalate) and Bis(malonato)platinum(IV) complexes. *Inorganica Chimica Acta*, **297**, 72–78.

83. Carr, J.L., Tingle, M.D., and McKeage, M.J. (2002) Rapid biotransformation of satraplatin by human red blood cells *in vitro. Cancer Chemotherapy and Pharmacology*, **50**, 9–15.

84. Carr, J.L., Tingle, M.D., and McKeage, M.J. (2006) Satraplatin activation by hemoglobin, cytochrome C and liver microsomes *in vivo. Cancer Chemotherapy and Pharmacology*, **57**, 483–490.

85. Hall, M.D., Dolman, R.C., and Hambley, T.W. (2004) Platinum (IV) anticancer complexes. *Metal Ions in Biological Systems*, **42**, 297–322.

86. Hall, M.D., Foran, G., Zhang, M. *et al.* (2003) XANES determination of the platinum oxidation state distribution in cancer cells treated with platinum (IV) anticancer agents. *Journal of the American Chemical Society*, **125**, 7524–7525.

87. Chen, L., Lee, P.F., Ranfrod, J.D. *et al.* (1999) Reduction of the anti-cancer drug analog, *cis, trans, cis*-[PtCl$_2$(O-COCH$_3$)$_2$(NH$_3$)$_2$] by L-Cysteine and L-methionine and its crystal structure. *Journal of the Chemical Society Dalton Transactions*, 1209–1212.
88. Battle, A.R., Choi, R., and Hibbs, D.E. (2006) Hambley. Platinum (IV) analogs of AMD473 (cis-[PtCl$_2$(NH$_3$)(2-picoline)]): preparative, structural, and electrochemical studies. *Inorganic Chemistry*, **45**, 6317–6322.
89. Chen, Y., Guo, Z., Parsons, S., and Sadler, P.J. (1998) Stereospecific and kinetic control over the hydrolysis of a sterically hindered platinum picoline anticancer complex. *Chemistry–A European Journal*, **4**, 672–676.
90. McGowan, G., Parsons, S., and Sadler, P.J. (2005) Contrasting chemistry of cis- and trans-Platinum (II) diamine anticancer compounds: hydrolysis studies of the picoline complexes. *Inorganic Chemistry*, **44**, 7459–7467.
91. Kelland, L. (2007) Broadening the clinical use of platinum drug-based chemotherapy with new analogues: satraplatin and picoplatin. *Expert Opinion on Investigational Drugs*, **16**, 1009–1021.
92. Farrell, N. (2004) Polynuclear platinum drugs. *Metal Ions in Biological Systems*, **42**, 251–296.
93. Heringova, P., Woods, J., Mackay, F.S. *et al.* (2006) Transplatin is cytotoxic when photoactivated: enhanced formation of DNA crosslinks. *Journal of Medicinal Chemistry*, **49**, 7792–7798.
94. van Zutphen, S., Pantoja, E., Soriano, R. *et al.* (2006) New antitumor active compounds containing carboxylate ligands in *trans* geometry: synthesis, crystal structure and biological activity. *Dalton Transactions*, 1020–1023.
95. Wheate, N.J. and Collins, J.G. (2005) Multi-nuclear platinum drugs: a new paradigm in chemotherapy. *Current Medicinal Chemistry–Anti-Cancer Agents*, **5**, 267–279.
96. Davies, M.S., Thomas, D.S., Hegmans, A. *et al.* (2002) Kinetic and equilibria studies of the aquation of the trinuclear platinum phase II anticancer agent, [{trans-PtCl(NH$_3$)$_2$}$_2${μ-trans-Pt(NH$_3$)$_2$(NH$_2$(CH$_2$)$_6$NH$_2$)$_2$}]$^{4+}$ (BBR3464). *Inorganic Chemistry*, **41**, 1101–1109.
97. Di Blasi, P., Bernareggi, A., Beggiolin, G. *et al.* (1998) Cytotoxicity, cellular uptake and DNA binding of the novel trinuclear platinum complex BBR3464 in sensitive and cisplatin resistant murine leukemia cells. *Anticancer Research*, **18**, 3113–3117.
98. Brabec, V., Kašpárková, J., Vrána, O. *et al.* (1999) DNA modifications by a novel bifunctional trinuclear platinum phase I anticancer agent. *Biochemistry*, **38**, 6781–6790.

Further reading

Huheey, J.E., Keiter, E.A., and Keiter, R.L. (1993) *Inorganic Chemistry. Principles of Structure and Reactivity*, 4th edn, Benjamin-Cummings Publishing Co., San Francisco, CA.
Lever, A.B.P. (1968) *Inorganic Electronic Spectroscopy*, 1st edn, Elsevier Publishing Co., Amsterdam, The Netherlands.
Lever, A.B.P. (1984) *Inorganic Electronic Spectroscopy*, 2nd edn, Elsevier Science Publishers B. V., Amsterdam, The Netherlands.
Mathews, C.K., van Holde, K.E., and Ahern, K.G. (2000) *Biochemistry*, 3rd edn, Benjamin Cummings Publishing Co., San Francisco, CA.

5

Ruthenium, Titanium and Gallium for Treating Cancer

5.1 Ruthenium compounds for treating cancer

Complexes of ruthenium, Ru, which have been extensively used as metal catalysts for chemical transformations, are also showing considerable promise for the treatment of cancer [1–6]. While some of these agents directly attack and kill the primary tumor, others have the interesting property of preventing the tumor from releasing cells that spread to other parts of the body. The structures and properties of these compounds range from traditional coordination compounds that lose ligands by interacting directly with biological targets to organometallic compounds which, in addition to binding to molecular targets, can facilitate the generation of reactive oxygen species, ROSs, which chemically modify biological molecules in the cell. In addition to killing cells through chemical reactivity, some ruthenium compounds express their biological effects by acting as shape-specific inhibitors of enzymes that are important for the survival of the cell. A brief overview of the chemistry and mechanistic aspects of these compounds will be given in this chapter.

5.1.1 Chemistry of ruthenium in the biological milieu

Ruthenium, atomic number 44, has a number of known oxidation states, but only two, Ru^{+2} [Kr]$4d^6$, a soft acid, and Ru^{+3} [Kr]$4d^5$, an intermediate acid, have been extensively incorporated into new agents for the treatment of cancer. Both oxidation states prefer six-coordinate octahedral geometry, with the complexes of Ru^{+3}, due to its greater charge, having the higher value of CFSE, Δ. Since ruthenium is in the second-row transition metal series, below iron in the periodic table, most Ru^{+3}, t_{2g}^5, $S = \frac{1}{2}$, as well as Ru^{+2}, t_{2g}^6, $S = 0$, complexes are low-spin. The first-order water exchange rate constants, k, for aquated Ru^{+2} and Ru^{+3} are $\sim 10^{-2}$ ($t_{1/2} \sim 1$ minute) and $\sim 10^{-6}$ s^{-1} ($t_{1/2} \sim 19$ hours), respectively, at 25 °C (Figure 1.20), with the exchange mechanism most likely being associative (1.23) [7]. As expected, the ion with the higher charge, which has the higher value of Δ, and thus the greater free energy of activation, $\Delta G^{\ddagger}$, has the *smaller* exchange rate constant, k, through (1.22). While the exchange rate for aquated Ru^{+3} is too slow for reaction of the ion with biological targets, the exchange rate constant of Ru^{+2} is about an order of magnitude larger than that of Pt^{+2}, suggesting that complexes of Ru^{+2} may be better suited for reaction with biological targets in the body. However, as will be evident below, exchange rates for either ion can

Metals in Medicine James C. Dabrowiak
© 2009 John Wiley & Sons, Ltd

vary considerably depending on the nature of the attached ligands. An additional interesting aspect of ruthenium chemistry is that it appears that the two ruthenium oxidation states can be interconverted by substances in the body, suggesting that *redox* may play an important role in the pharmacology of ruthenium [5]. Although Ru^{+4} has been proposed as an intermediate in the reaction of certain Ru^{+3} complexes, the tetravalent oxidation state of the metal has been relatively little studied in connection with ruthenium metallo-pharmaceuticals.

5.1.2   Structure, synthesis and properties of ruthenium antitumor agents

A large number of ruthenium compounds have been synthesized and tested for their anticancer properties; some are presented in Figure 5.1. Two of the compounds, NAMI-A and KP1019, which contain Ru^{+3}, are in clinical trials [8,9], while two others, $[(\eta^6\text{-biphenyl})Ru(en)Cl]PF_6$ [10–13] and $[(\eta^6\text{-benzene})Ru(pta) Cl_2]$ [14,15], which contain Ru^{+2}, represent an interesting and very broad new class of *arene* compounds with antitumor activity. Like the platinum anticancer drugs, these compounds appear to exhibit their biological effects by losing one or more of their bound ligands and reacting with target molecules in the cell. However, some ruthenium compounds which exhibit antitumor effects are relatively exchange-inert (Figure 5.2) and are believed to exhibit their biological effects by 'nontraditional' mechanisms. One compound, $[(\eta^6\text{-}p\text{-cymene})Ru(azpy)I]PF_6$, appears to kill cells by generating ROS through a catalytic mechanism involving its attached ligands and GSH [16]. Another compound, DW12 (Figure 5.2), which contains exchange-inert Ru^{+2}, expresses its anticancer properties by mimicking the shape of a natural product, staurosporine, that stimulates the production of the cancer-protecting protein p53 in skin cancer cells (melanoma) [17,18].

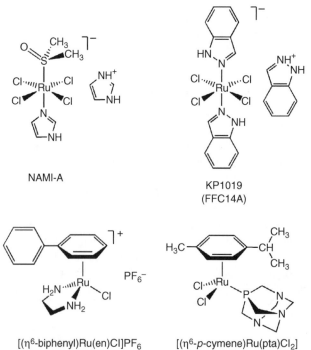

Figure 5.1   *Structures of some antitumor active ruthenium complexes*

Box 5.1 Organometallic chemistry

Organometallic chemistry is a broad area of science that studies the structures and properties of compounds with metal–carbon bonds. A characteristic feature of organometallic compounds, which distinguishes them from coordination complexes, is that their bonding is largely covalent in nature. Consider the MO diagrams for the organometallic compound $Cr(CO)_6$, *hexacarbonylchromium* (Figure 5.3), and the coordination compound $[Co(NH_3)_6]^{3+}$ (Figure 1.10). Inspection of these diagrams shows that while the atomic orbitals, AOs, on the metal sides for Co^{+3} and Cr^0 are the same – the $3d$, $4s$ and $4p$ are used in the MO bonding scheme – the right, ligand, sides are very different between the two compounds. For $[Co(NH_3)_6]^{3+}$ there are six hybrid sp^3 orbitals from the six ammonia molecules on the right side, while for $Cr(CO)_6$ there are six σ-orbitals and twelve π^*-orbitals from the six CO molecules on the ligand side of the diagram. The boundary surfaces – that is, where the electrons are found – for the σ ($2p$) and one of the π^* ($2p$) MOs on carbon monoxide, and the way in which these MOs interact with two of the AOs on the metal to form a metal–carbonyl bond, are shown in Figure 5.4. As is evident from Figure 5.4, the σ MO of CO interacts with the $d_{x^2-y^2}$ AO of Cr (also with the d_{z^2}, which is not shown) to form a σ-type bond between the carbon atom of CO and Cr, which produces the bonding, $e_g(\sigma)$, and antibonding, $e_g{}^*(\sigma)$, molecular orbitals for $Cr(CO)_6$, shown in Figure 5.3. While this Cr-CO σ-bond is similar to the σ-bond between the Co^{+3} ion and the lone pair of electrons in the sp^3 hybrid of NH_3 of $[Co(NH_3)_6]^{3+}$ (Figure 1.10), the fact that the energy of the σ MOs of CO is near the energy of the $3d$ orbitals on the metal means that the wavefunctions for the $e_g(\sigma)$ bonding and $e_g{}^*(\sigma)$ antibonding MOs for $Cr(CO)_6$ have significant contributions from the atomic wavefunctions of *both* M and L through expressions like (1.1) and (1.2). In simple terms, this means that electrons in these MOs for $Cr(CO)_6$ (there are four electrons in $e_g(\sigma)$) do not belong to either Cr or CO but rather are shared by both bonding partners; that is, the bond is covalent. This situation is different from that of $[Co(NH_3)_6]^{3+}$, for which the energies of the ammonia hybrids are much lower (more stable) relative to the $3d$ orbital of the metal, which make the wavefunction for the $e_g(\sigma)$ bonding MO heavily weighted in favor of the AOs of the ammonia hybrids. This shows that the four electrons in the $e_g(\sigma)$ bonding MO of $[Co(NH_3)_6]^{3+}$ spend more time on the ammonia molecules than on the cobalt ion, and the bond between ammonia and the metal ion is *ionic or salt-like* in nature. This change in the nature of the bonding from ionic to covalent means that in the latter case it is not possible to clearly assign oxidation state; that is, electrons in the bond spend significant fractions of their time on *both* M and L, and oxidation state is often *assigned* using a simple 'bookkeeping' procedure, with the value obtained referred to as the *formal oxidation state*.

Another important difference between the MO diagrams for $[Co(NH_3)_6]^{3+}$ and $Cr(CO)_6$ is that the t_{2g} set of orbitals in Figure 1.10, which had no orbitals on ammonia with which to interact and were labeled 'nonbonding', now interact with the π^* MOs on carbon monoxide in the manner shown in Figure 5.3. This interaction produces a bonding MO, $t_{2g}(\pi)$, which is lowered in energy relative to its contributing AOs, and an antibonding MO, $t_{2g}{}^*(\pi)$, which is raised in energy relative to its contributing AOs. As shown in Figure 5.4, this interaction results in a π-*bond* between the metal and the carbon atom of CO. Since the orbitals of the bonding partners bring a total of 18 electrons to the bonding scheme, and the electrons are placed into the MOs of $Cr(CO)_6$, filling from the bottom up, the highest occupied molecular orbital (*HOMO*) is $t_{2g}(\pi)$ and the lowest unoccupied molecular orbital (*LUMO*) is $t_{2g}{}^*(\pi)$. A transition metal can only contribute nine atomic orbitals to the bonding scheme shown in Figure 5.3 – that is, nd, $(n+1)s$ and $(n+1)p$ – which will, if there are σ and π MOs, interact with nine orbitals on the ligands to produce 18 MOs. Nine of these MOs are bonding MOs, which when occupied by electrons *stabilize* the system, and nine are antibonding, which when occupied by electrons *destabilize* the system. If the ligand is a good σ-donor and π-acceptor, as is the case with CO and many arene-type ligands, the energy gap between HOMO and LUMO in Figure 5.3 is large, which has important consequences for the overall stability of the compound. For compounds contributing *more* than 18 electrons to the bonding diagram, as will occur if the nature of the transition metal or the ligands is changed, one or more electron will be forced to occupy *antibonding* MOs in the diagram, which will lower the stability of the compound. For compounds contributing *fewer* than 18 electrons, not all of the possible bonding MOs will be occupied, which will also lead to a decrease in the stability of the system. Thus, the 18 electrons in the diagram produce a compound with the maximum stability, which is the basis for the *effective atomic number* rule or *18-electron* rule in organometallic chemistry. If a metal has an odd number of electrons in its valence shell, for example manganese, Mn, $3d^5 4s^2$, it will not be possible to achieve the even number of 18 electrons. In this case the carbonyl compound with maximum stability is *bis(pentacarbonymanganese)*, $Mn_2(CO)_{10}$, which has an Mn-Mn bond. Since the two electrons in the metal–metal bond are shared between both Mn atoms, at least some of the time an additional electron from the adjacent Mn atom can be added to the 17 electrons of each $Mn(CO)_5$ unit, allowing the compound to obey the 18-electron rule.

Aside from carbon monoxide, nitric oxide (NO) and other small molecules, there are a wide variety of unsaturated arene-type organic compounds that are good σ-donors and π-acceptors and that can react with transition metals to form organometallic compounds. One famous example is ferrocene, the synthesis of which is shown in Figure 5.5. Ferrocene can be made by treating cyclopentadiene with sodium hydroxide or another suitable base, which removes the acidic hydrogen of the diene to form the aromatic cyclopentadienyl ion, Cp. Reaction of two Cps with ferrous chloride produces the 'sandwich' compound ferrocene, $(\eta^5\text{-}C_5H_5)_2Fe$, which has two Cp ligands face-on-bonded to the iron ion. The nomenclature used to indicate the number of points of attachment of the arene ligand to the metal is the Greek letter η (eta), which carries as a superscript the number of ligand attachment points, which in this case is 5, for the five carbon atoms of Cp.

Ferrocene, which is very stable, obeys the 18-electron rule. Brought to the bonding scheme for ferrocene are 12 electrons from the two Cp ligands (each Cp is a six- electron donor; four π-electrons plus the two electrons in the negative charge) and six electrons from the valence shell of the Fe^{+2} ion, $3d^6$. Since the 18 electrons are in MOs that have significant contributions from both M and L, the concept of the *formal oxidation state* is applied when determining the oxidation of the iron in ferrocene. If the Cps are negatively charged and six-electron donors, the iron has a formal oxidation state of Fe^{+2}, but if the two Cp's are uncharged and five-electron donors, the formal oxidation state of the iron is Fe^0. As will be apparent, benzene, C_6H_6, and its analogs can also form sandwich-type organometallic compounds. In this case the benzene ring is considered as contributing six-electrons (the electrons in three π-bonds) to the bonding scheme.

For references, see the 'Further Reading' section at the end of this chapter.

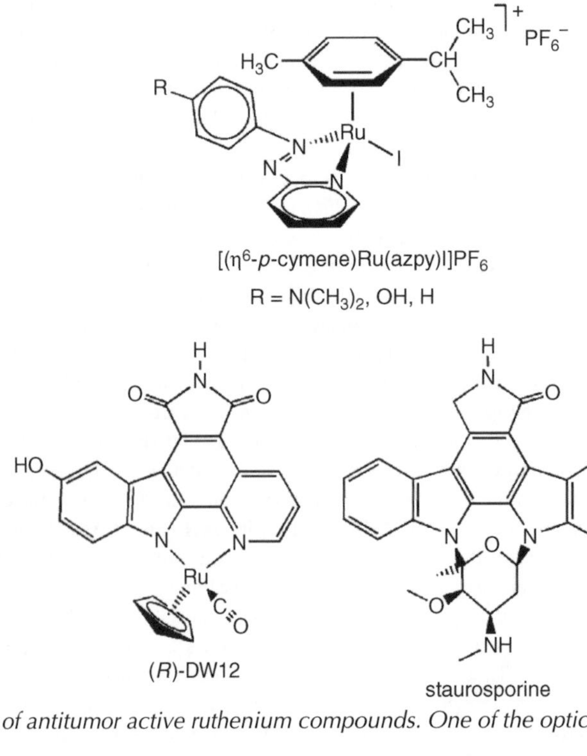

$[(\eta^6\text{-}p\text{-cymene})Ru(azpy)I]PF_6$

$R = N(CH_3)_2, OH, H$

(*R*)-DW12

staurosporine

Figure 5.2 *Structures of antitumor active ruthenium compounds. One of the optically-active forms, R, of DW12 is shown*

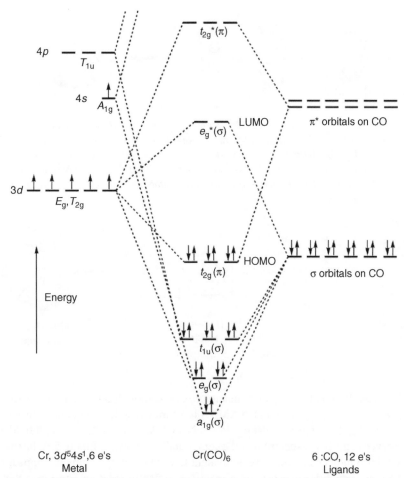

Figure 5.3 *Partial molecular orbital diagram for Cr(CO)$_6$. The energies of levels are only approximate. Antibonding MOs at high energy (less stable) are not shown in the diagram. While there are a total of 12 π* MOs on the six carbon monoxide molecules, for reasons pertaining to symmetry, only three π* MOs on the COs are used in interacting with the t$_{2g}$ orbitals (d$_{xz}$, d$_{yz}$, d$_{xy}$) on the metal to produce bonding, t$_{2g}$, and antibonding, t$_{2g}$*, MOs. The lowest unoccupied molecular orbital, LUMO, and highest occupied, HOMO, are also shown on the diagram*

5.1.3   Synthesis and biological properties of NAMI-A

The compound imidazolium *trans*-tetrachloro(dimethylsulfoxide)imidazoleruthenate(III), [ImH][*trans*-RuCl$_4$(dmso-*S*)(Im)], NAMI-A (Figure 5.1) has completed phase I clinical trials in the Netherlands for the treatment of metastatic cancer. As shown in Figure 5.6, NAMI-A is made by reacting the octahedral Ru^{+3} complex, RuCl$_3$·3H$_2$O, which contains bound chloride and water, with HCl and dimethylsulfoxide (dmso) in ethanol to give a six-coordinate complex with *trans*, dmso ligands [19]. Since *S*-bonded dmso can use unfilled 3d orbitals on the sulfur atom to overlap with filled 4d orbitals (t_{2g}^5) on the Ru^{+3} ion, like carbon monoxide, it is a good π-acceptor ligand and thus a good *trans*-directing ligand. Addition of excess imidazole to the *trans*-dmso complex suspended in acetone results in the displacement of one of the dmso molecules by imidazole and the formation of brick-red NAMI-A, which contains imidazolium as the counter cation for the anionic coordination compound.

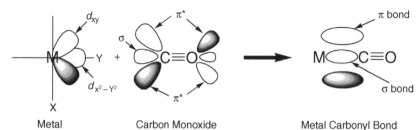

Metal   Carbon Monoxide   Metal Carbonyl Bond

Figure 5.4 *Bonding between carbon monoxide and a metal to form a metal carbonyl bond. For clarity, only portions of the $d_{x^2-y^2}$ and d_{xy} orbitals on the metal are shown. The metal carbonyl bond is formed using the σ (2p) MO on CO and the $d_{x^2-y^2}$ atomic orbital on M, while the π-bond between M and CO is formed using a π^* (2p) MO on CO and the d_{xy} atomic orbital on M. The shaded and unshaded orbitals indicate that the sign of the wavefunction is positive and negative, respectively*

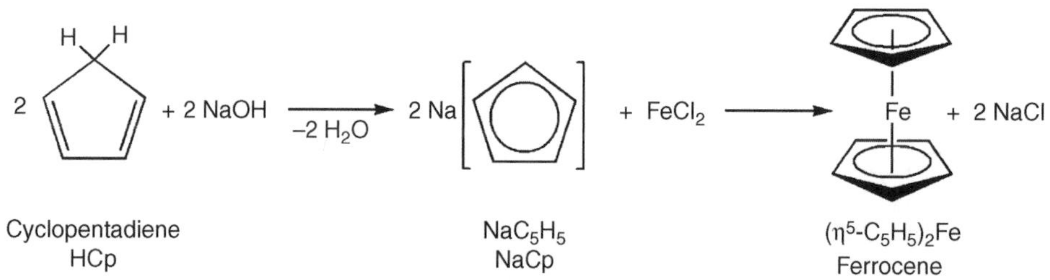

Cyclopentadiene NaC_5H_5 $(\eta^5\text{-}C_5H_5)_2Fe$
HCp   NaCp   Ferrocene

Figure 5.5   *Synthesis of ferrocene*

The hydrolysis of NAMI-A under conditions that approximate those found in blood was studied by Reedijk and coworkers using ^{1}H NMR [20]. Although NAMI-A is a paramagnetic compound with $S = \frac{1}{2}$, the protons on the coordinated imidazole and dimethylsulfoxide of NAMI-A, while shifted to high field and broadened, can be readily observed in the NMR spectrum of the compound. As shown in Figure 5.7, the two methyl groups of the coordinated dmso molecule, which are equivalent to each other by symmetry, appear as a broadened singlet at ~ -14.8 ppm, which is shifted *upfield* from free dmso in solution and dmso that is bound to a diamagnetic metal ion. The single proton attached to C5 of the coordinated imidazole, H_5, because it is farthest away on the imidazole ring from the paramagnetic Ru^{+3} ion, gives a relatively narrow, easily-observed resonance at ~ -3.2 ppm. Since the ruthenium ion is attached to N3 of imidazole (Figure 5.6a), H_2 and H_4, which are non-equivalent and are spatially closer to the paramagnetic metal ion, are severely broadened, but they are observed at ~ -5.6 and ~ -7.8 ppm. The cation of NAMI-A is the imidazolium ion. Since the pK_a for deprotonation of imidazole is 6.95, and the pH of the study was 7.4, the imidazolium is mostly in the unprotonated form. Due to resonance and tautomerization (Figure 2.3b), both of which are fast on the NMR time scale, the protons on C4 and C5 are in identical chemical environments and resonate at 7.49 ppm, while the single proton on C2 is at 8.96 ppm.

As shown in Figure 5.7, the spectrum of NAMI-A changes within minutes after the dissolution of the compound in buffer and the new peaks that appear are broadened, showing that the product is a paramagnetic complex; that is, Ru^{+3}. By making plots of the intensities (areas) of signals as a function of time, the researchers showed that one of the chloride ligands of NAMI-A is displaced by water and that the initially-formed complex is *mer*-$[RuCl_3(dmso\text{-}S)(Im)(H_2O)]$, indicated as NAMI-$A_{H2O}$ in Figure 5.7. However, this initially-formed complex quickly loses dmso, and since polynuclear species soon appear, further analysis of the reaction mechanism was not possible. The researchers reported that $t_{1/2}$ for the hydrolysis of NAMI-A is ~ 20 minutes at pH 7.4 but that the compound is much more stable at acidic values of pH.

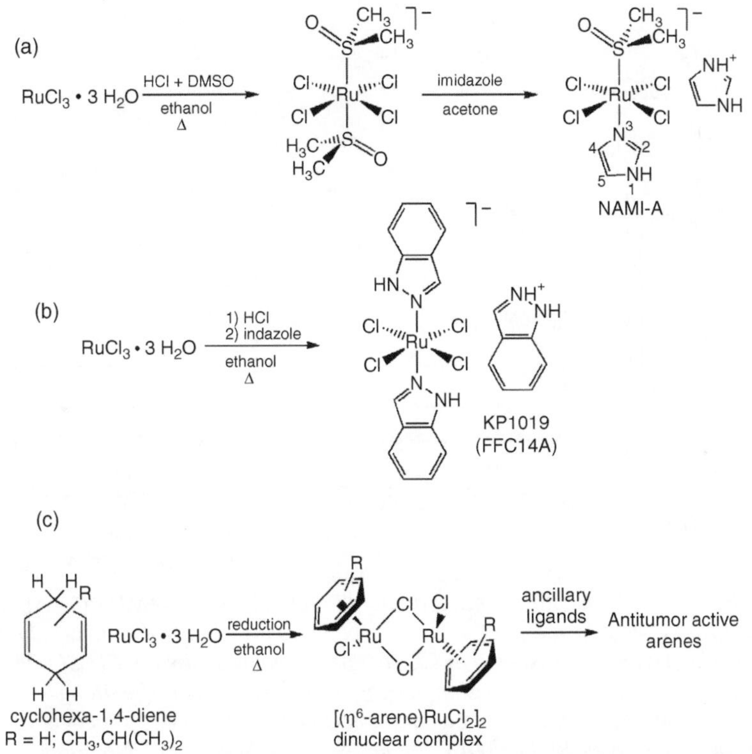

Figure 5.6   *Synthesis of some antitumor active ruthenium compounds*

The fact that NAMI-A hydrolyzes relatively quickly under physiological conditions suggests that the compound could react with substances in blood and/or targets inside the cell. However, in what has become known as the *activation by reduction* hypothesis, Clarke and coworkers suggested that the biological activity of Ru^{+3} compounds is related to their ability to be bioreduced to Ru^{+2}, which can happen in the hypoxic environment of tumors, thus producing products with rapid exchange kinetics more suitable for reaction with biological targets in the cell [5,6]. In this sense, NAMI-A could be considered a prodrug, like complexes of Pt^{+4}, for example satraplatin, that is activated by substances in the body to produce more-reactive species that attack targets in the cell.

In exploring the activation by reduction mechanism, Brindell *et al.* [21] employed stopped-flow kinetics measurements (see Section 6.2.3) to investigate the reduction of NAMI-A by *ascorbic acid*, vitamin C, a reducing agent that is found in relatively high concentration, 11–79 μM, in blood serum. Shown in Figure 5.8 are the UV–visible absorption spectra of NAMI-A before reduction and at various times after reduction with ascorbic acid. The absorption spectrum of NAMI-A shows a strong band at ~390 nm ($\sim 3240\, M^{-1}\, cm^{-1}$), which is probably a charge transfer band, and a weaker absorption at ~450 nm ($\sim 500\, M^{-1}\, cm^{-1}$), which is probably a *d-d* transition of the Ru^{+3} ion. Immediately after the addition of ascorbic acid, the intensity of the CT band of NAMI-A drops considerably and new weaker bands, consistent with *d-d* absorptions, appear at 430 and 350 nm. With time, these bands shift to higher energy and lower wavelength, indicating that the crystal field around the Ru^{+2} ion is slowly *increasing*.

The study reported that the second-order rate constant for the reduction of NAMI-A to its corresponding Ru^{+2} complex by ascorbic acid at pH 5 at 35 °C is $47 \pm 1.8\, M^{-1}\, s^{-1}$ and that the reduction takes place via an outer-sphere mechanism [21]. In this type of mechanism, the ascorbic molecule contacts one of the ligands of

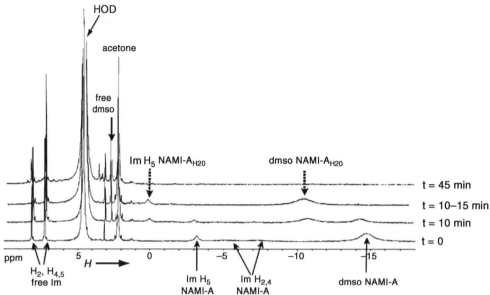

Figure 5.7   *300 MHz ^{1}H NMR spectra recorded during the hydrolysis of NAMI-A (5 mM) in phosphate buffer (0.05 M phosphate, 0.15 M NaCl, D_2O), pH 7.4, 37 °C. The times for various spectra are shown at the right. The numbering system for imidazole (Im) is shown in Figure 5.6a. Dimethylsulfoxide, CH_3SOCH_3, is dmso. Acetone, CH_3COCH_3, was used as an internal chemical shift standard in the experiment. The strong resonance labeled HOD is due to water having one proton and one deuterium. Adapted from J. Inorg. Biochem., 98, M. Bacac et al., The Hydrolysis of the Anti-Cancer Ruthenium Complex NAMI-A Affects its DNA Binding and Antimetastatic Activity: An NMR Evaluation, 402–412. Copyright 2004, with permission from Elsevier*

NAMI-A and transfers an electron directly to the ligand, which ultimately transfers the electron to the ruthenium ion to produce Ru^{+2}. The investigators also pointed out that the kinetics of the reaction are affected by the presence of excess chloride ion in the medium (Figure 5.8 inset) and suggested that the Ru^{+2} form of NAMI-A may have a catalytic role in the reaction. Due to Le Chatelier's principle, excess chloride ion may suppress the aquation of the Ru^{+2} form, thus blocking it from forming a dinuclear species with the unreduced Ru^{+3} complex that is critical for the reduction by ascorbic acid. Since perchlorate ion, ClO_4^-, is very low in the spectrochemical series, it is considered a 'spectator ion' in the reaction and would not be expected to bind to either Ru^{+3} or Ru^{+2}.

5.1.4   Control of tumor growth by NAMI-A

Unlike most of the platinum anticancer drugs, NAMI-A is not very toxic toward cells and it was recognized early that the main effect of the compound is to stop tumors from spreading to other parts of the body, a process called *metastasis* [1,2]. When tumors grow they release cells, which are carried by the blood and deposited at other sites in the body, causing new tumors to form. This process, which involves adhesion and movement of cells and invasion of healthy tissue environments, is impaired by NAMI-A, which has been found to be particularly effective in inhibiting metastasis in Lewis lung carcinoma, MCa mammary carcinoma and B16 melanoma [1,2,22]. Even though NAMI-A does not kill the primary tumor, finding agents like it that can slow metastasis so that the tumor can be surgically removed would expand the options available for treating the cancer patient.

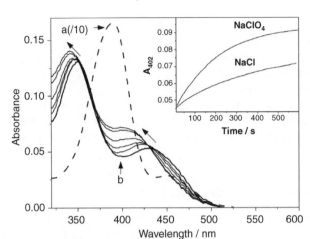

Figure 5.8   *UV–visible absorption spectra of NAMI-A [ImH][trans-RuCl₄(dmso-S)(Im)], recorded before (a), where the absorbance scale for the spectrum of NAMI-A has been divided by 10, and after reduction with ascorbic acid (H₂A) after 12 seconds (b), and during the hydrolysis process at various time intervals of 1, 2, 4, 8 and 10 minutes in the presence of NaCl. Inset: Absorbance at 402 nm as a function of time after reduction of NAMI-A by H₂A in the presence of NaCl or NaClO₄. Experimental conditions 0.5 × 10⁻³ M NAMI-A, 5 × 10⁻² M total H₂A, pH 5.0, 25 °C, I (ionic strength) = 0.6 M (0.24 M NaCl or NaClO₄). With kind permission from Springer Science + Business Media: J. Biol. Inorg. Chem., Kinetics and Mechanism of the Reduction of (ImH)[trans-RuCl4(dmso)(Im)] by Ascorbic Acid in Acidic Aqueous Solution, 12, 2007, 809–818, M. Brindell et al., Fig 1*

5.1.5   Clinical trials with NAMI-A

The results of phase I clinical trials with NAMI-A show that the compound is relatively nontoxic with doses of up to $500 \, \mathrm{mg \, m^{-2} \, day^{-1}}$, but that higher doses cause blisters on the hands, fingers and toes of patients with solid tumors [8]. This dose range is much higher than that of cisplatin, for which doses are in the 20–$140 \, \mathrm{mg \, m^{-2}}$ range. In the phase I clinical trials with NAMI-A, the agent was typically given by intravenous infusion over a 3 hour period in a solution that contained 0.9% NaCl, normal saline solution, (∼150 mM) at pH ∼4, a pH at which the compound is most stable [23]. A pharmacokinetic analysis showed that there is a linear relationship between the dose and the area under the concentration-versus-time curve (AUC), that the half life for the clearance of Ru from plasma was $t_{1/2} = 50 \pm 19$ hours, and that most of the Ru was bound to proteins in the blood (<95%). The clinical trials isolated the genomic DNA from the white blood cells (WBCs, leucocytes, of patients, finding that Ru binding to the DNA sequences GG and AG was below detection level at all doses studied. The investigators also measured the total amount of Ru bound to WBCs of the patients, finding no correlation between the amount bound and the dose of NAMI-A given.

5.1.6   Interaction of NAMI-A with potential biological targets

While little is known about the molecular mechanism by which NAMI-A exerts its metastatic effects, a number of studies have focused on the binding of the agent to human and bovine serum albumin, apo-transferrin and DNA [24–29]. The general consensus is that NAMI-A binds to exposed imidazole sites (histidine) on these proteins, but that since it only weakly interacts with DNA and its bases, DNA may not be the target for the agent in the cell [20,25,27,30].

In attempting to understand what happens to NAMI-A when it is introduced into blood, Brindell *et al.* [31] focused on the reduction of the compound by ascorbic acid, which is found in blood, and the interaction of the reduction products with human serum albumin, HSA, the most abundant protein ($\sim$600 µM) in blood. The investigators found that since NAMI-A is reduced within seconds by ascorbic acid at concentrations and conditions similar to those found in blood, the reduced complex, with its more rapid exchange kinetics, is the form of the agent that mainly interacts with HSA. However, the investigators pointed out that since NAMI-A is unstable at pH 7.4, some of the compound may undergo hydrolysis, producing Ru^{+3} aqua species that also bind to HSA. The reaction of reduced NAMI-A with HSA leads to the formation of stable adducts, with a binding saturation very similar to that of NAMI-A, 3.2 ± 0.3 (Ru^{+2}) and 4.0 ± 0.4 (Ru^{+3}) mol per mole of HSA, respectively, but the rate of reaction of the reduced compound with the protein is much greater than that with NAMI-A.

5.1.7 Synthesis and biological properties of KP1019

The compound *trans*-[tetrachlorobis(1H-indazole)ruthenate(III)], KP1019 or FFC14A (Figure 5.1), which was synthesized by Keppler and coworkers, is currently in phase I clinical trials for possible treatment of colon carcinoma and other types of cancer [9,32]. As shown in Figure 5.6b, the compound is made by dissolving commercially-available $RuCl_3 \cdot 3H_2O$ in refluxing ethanol-HCl and treating the solution with an excess of the heterocyclic base indazole, which yields the red-brown *trans* complex as the indazolium salt.

The rate of disappearance of KP1019 in water under various buffer conditions has been measured using HPCE (high-performance capillary electrophoresis), a sensitive analytical technique that separates molecules on the basis of their charge and size [32,33]. The disappearance rate of KP1019, assumed to be due to its hydrolysis, increases with the pH of the medium in a 5 mM phosphate buffer, with pseudo first-order rate constants at 37 °C at pH 6.0 and 7.0 of $3.4 \pm 0.2 \times 10^{-5}$ and $51.5 \pm 1.6 \times 10^{-5} \, s^{-1}$, respectively [33]. The compound is reduced by glutathione, GSH, which is mainly found inside the cell, and even more rapidly by ascorbic acid, which is found in blood [34]. As is the case with NAMI-A [21,31], the reduction of KP1019 by ascorbic acid suggests that the compound is very likely reduced to Ru^{+2} shortly after it is injected and that the reduction products circulate in blood and react with proteins.

5.1.8 Biological activity of KP1019

The compound KP1019 induces apoptosis in the colorectal tumor cell lines SW480 and HT29 mainly by blocking mitochondrial function and is cytotoxic against a number of different parental cell lines and their drug-resistant counterparts in the 56–179 µM range [32,35]. Cell lines with higher amounts of the drug-effluxing protein P-glycoprotein, Pgp, that are treated with KP1019 accumulate less ruthenium and are more resistant to the agent than control cells with the normal amount of Pgp. When HL60 cells are continually exposed to increasing concentrations of KP1019 over long periods of time, for example 1 year, they develop *acquired resistance* to the agent, but after this period they are only twice as resistant to KP1019 as cells that have not been previously exposed to the ruthenium compound. By comparison, this increase in resistance is very small, $\sim$10%, compared to the acquired resistance developed by most cell lines that are exposed to cisplatin. Interestingly, in the cells that have acquired resistance to KP1019 there seems to be no increase in the amount of Pgp or other efflux proteins that remove toxins from the cell, suggesting that the acquired resistance is not simply due to enhanced efflux of Ru from the cell. Since ruthenium is thought to be transported into the cell via the transferrin system, cells with a large number of transferrin receptors that were treated with KP1019 were found, as expected, to be sensitive to KP1019 [36]. Cell fractionation experiments showed that after a 2 hour treatment of human colon cancer cells with 10 µM KP1019, on average 55% of the intracellular Ru was found located in the nucleus, with the remainder of the cell-bound Ru in the cytosol and other cellular components.

5.1.9   Clinical trials with KP1019

In the phase I clinical trials involving KP1019, patients that were given doses of the compound in the range 25–600 mg twice weekly over three weeks exhibited no noticeable side effects from the agent. Most of the Ru that was administered was bound to proteins in the blood and the observed terminal half life was quite long, $t_{1/2} = 69$–284 hours.

5.1.10   Interaction of KP1019 with potential biological targets

The main focus on the interaction of KP1019 with potential targets in the body has been on its reaction with proteins in the blood, especially with the iron transport protein transferrin (Tf) Box 5.2. Since tumor cells have a greater number of transferrin receptors (TfRs) on their surface than normal cells, loading Tf with a metal ion that cannot perform the same biological functions as iron may be a means of delivering 'foreign' metal ions to a cancer cell and selectively killing it by interrupting critical metal-based cellular functions. Kratz *et al.* [37] studied the binding of KP1019 to apo-transferrin (apo-Tf) using UV–visible absorption, circular dichroism, CD, and NMR. These studies were done in a buffer containing 4 mM NaH$_2$PO$_4$, 100 mM NaCl and 25 mM NaHCO$_3$ (pH 7.4) that simulated some of the conditions found in blood. The investigators determined that KP1019 binds to apo-Tf, as evidenced by the change in the UV–visible absorption spectrum of KP1019 and the appearance of an induced CD spectrum for the coordination compound in the presence of the protein (Figure 5.9). The binding occurs within minutes; the presence of carbonate is required and apo-Tf binds two KP1019 complexes.

The absorption spectrum of the protein-bound KP1019 (Figure 5.9a) exhibits a band at $\sim$585 nm, which although the molar extinction is high ($\varepsilon \sim 1000\,M^{-1}\,cm^{-1}$) was assigned to a *d-d* transition of the bound Ru^{+3} ion. The bound metal complex also exhibits other poorly-defined absorptions in the region 340–550 nm, which appear to be *d-d* transitions and/or charge transfer transitions (LMCT) associated with indazole ligand which remains bound to the ruthenium ion in the protein complex. The absorptions due to apo-Tf itself, which are mainly from the side chains of the amino acids tryptophan, tyrosine and phenyl alanine are farther into the ultraviolet region at $\sim$280 nm and are not overlapped with absorptions due to the metal complex.

While KP1019 itself is not optically active, when the complex is bound to apo-Tf, the electronic transitions associated with the metal complex exhibit *induced optical activity* (Boxes, 4.2 and 5.3). This is because the

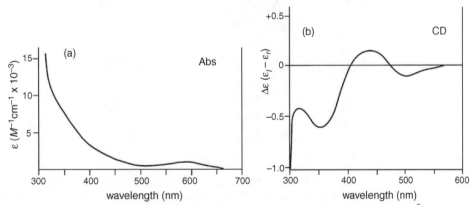

Figure 5.9   *(a) UV–visible absorption spectrum of KP1019 plus apo-transferrin (both 5×10^{-5} M) after 20 minutes and (b) circular dichroism (CD) spectrum of KP1019 plus apo-transferrin (both 1×10^{-4} M) in buffer containing 4 mM NaH$_2$PO$_4$, 100 mM NaCl and 25 mM NaHCO$_3$ (pH 7.4). Adapted from F. Kratz et al., The Binding Properties of Two Antitumor Ruthenium (III) Complexes to Apotransferrin, J. Biol. Chem. 1994, 269, 2581–2588*

complex is bound in a localized environment of protein that is very *chiral* and the electronic transitions on the complex 'sense' the chiral nature of the environment and exhibit CD effects. As shown in Figure 5.9b, the UV–visible absorptions associated with the bound metal complex exhibit *negative* CD bands at ~330 and ~505 nm and a *positive* CD band at ~435 nm. By systematically increasing the concentration of KP1019 in a solution containing apo-Tf and measuring CD spectra, the investigators were able to determine that apo-Tf binds *two* KP1019 molecules and that the CD signatures for each of the bound complexes are the same, strongly suggesting that KP1019 occupies the iron-binding sites of apo-Tf.

In order to show that the iron binding sites of transferrin were involved in binding KP1019, the investigators loaded apo-Tf with Al^{+3}, which because it has the same charge and nearly the same ionic radius as high-spin six-coordinate Fe^{+3} occupies the iron site of transferrin. When Al_2-Tf was treated with two equivalents of KP1019, no binding of ruthenium occurred, strongly suggesting that ruthenium complex binds to the iron site of transferrin.

Lactoferrin is a protein that has high sequence homology with human transferrin. Kratz *et al.* exposed crystals of apo-lactoferrin to a solution of KP1019 and the ruthenium compound slowly diffused into the crystal and bound to the protein. Solution of the resulting structure using x-ray crystallography revealed that KP1019 is bound to histidine-235 located in the iron-binding pocket of the N-lobe of apo-lactoferrin and that both indazole ligands remain bound to the ruthenium ion (Figure 5.10) [38]. Due to disorder in the crystal, the ligands in the plane of the ruthenium complex could not be located, but most likely they are a combination of the original chloro ligands and water.

The binding of KP1019 to human serum albumin, HSA, the most abundant protein in blood, has also been studied [39]. Circular dichroism studies show that binding of the ruthenium compound to HSA causes the loss of some of the α-helical structure of the protein, and since the fluorescence of a tryptophan residue at position 214 of the protein is reduced when KP1019 binds, the imidazole groups of the histidine residues in the region of

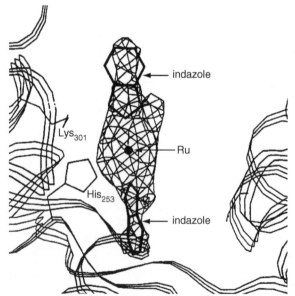

Figure 5.10 *Difference electron density map of KP1019 in the N-terminal site of human apo-lactoferrin (ribbon structure). The crystal was obtained by soaking a freshly prepared 5×10^{-4} M aqueous solution of KP1019 for 12 hours in the presence of crystallized apo-lactoferrin. Reproduced from F. Kratz et al., Protein-Binding Properties of Two Antitumor Ru(III) Complexes to Human Apotransferrin and Apolactoferrin, Metal-Based Drugs 1994, 1, 169–173*

the tryptophan – that is, histidine 242 or 246 – are suspected to be the ligands interacting with the ruthenium ion [9,40]. Although the rate of reaction of KP1019 with HSA is much lower than the rate of reaction with transferrin, the large amount of the former protein in blood and the fact that it has the higher binding constant for the ruthenium compound ($K_{HSA} = 9910\,M^{-1}$, $K_{Tf} = 6460\,M^{-1}$) suggests that HSA may be a reservoir and carrier of administered KP1019 and may thus mediate its accumulation in tumors [41].

Box 5.2 Transferrin

A group of proteins that are abundant in blood ($\sim35\,\mu M$) and are responsible for transporting iron to the cell are the *transferrins* (Tfs). These proteins, which have molecular weights of ~80 kDa and are structurally similar to each other, include *serum transferrin*, *lactoferrin* and *ovotransferrin* [42,43]. The transferrins consist of a single polypeptide chain 680–690 amino acids long that is folded into two lobes, the *N-lobe* in the N-terminal half of the protein and the *C-lobe* in the C-terminal half. Each lobe can bind one Fe^{+3} ion – a total of two Fe^{+3} ions per protein (Fe_2-Tf) – and each iron site has the same octahedral coordination environment (Figure 5.11). The Fe^{+3} ion in the site, which is high-spin six-coordinate, is bound by a nitrogen atom from the imidazole residue of histidine (His), two deprotonated phenol groups of tyrosine (Tyr), a carboxylate of aspartic acid (Asp) and a carbonato ligand, which form a four-membered chelate ring to the metal ion. Both sites on Tf are independent of each other – that is, far enough apart that there is no interaction between them – and at pH 7.4 in carbonate media (blood) the binding constant of Fe^{+3} for each site is $\sim10^{20}\,M^{-1}$. Since only $\sim30\%$ of the sites on Tf are normally occupied by Fe^{+3} in blood, the protein has considerable capacity to bind any additional metal ions that may be introduced into blood and are capable of binding to the site.

   Located on the surface of the cell are *transferrin receptors* (TfRs), which are responsible for recognizing and binding an iron-loaded transferrin and taking it into the cell. Since tumor cells have a high requirement for iron, they have more TfRs on their surfaces than noncancerous cells [44]. Each receptor can bind two transferrin molecules, but the binding constant depends on the number of Fe^{+3} ions bound to transferrin and K toward the receptor decreases in the order Fe_2-Tf > Fe-Tf > apo-Tf, where apo-Tf is the apo-protein that has no bound iron ions. This shows that the receptor prefers to bind and transport a completely loaded transferrin molecule with two bound Fe^{+3} ions into the cell. The adsorption process, which is *receptor-mediated endocytosis* and requires ATP, results in the iron–transferrin being deposited in an *endosome*, a small vesicle inside the cell. Since the pH in the endosome is in the acidic range, ~5.5, the release of the Fe^{+3} ion from transferrin likely involves protonation of the bound bidentate *carbonato* ligand to form a bidentate *hydrogencarbonato* ligand, which are often unstable and rapidly lose carbon dioxide gas, CO_2 [45]. In this way the iron can be liberated, making it available to other molecules in the cell.

   After the iron is released from transferrin, it is eventually transferred to the oligomeric protein *ferritin*, which is composed of 24 identical monomers assembled into a sphere-like structure. The interior of the sphere is large enough to accommodate $\sim4500\,Fe^{+3}$ ions in a complicated phospho/hydroxo matrix, which act as the storage reservoir for iron in the cell. When iron is required by the cell for various metallo-enzymes and -proteins and other functions, the iron reservoir in ferritin is accessed and a select number of iron ions are transferred by *chaperone proteins* to the parts of the cell where they are required. The underling theme of this pathway is that while Fe^{+3} is essential for many cellular functions, it is quite toxic and the ion is carefully sequestered by cellular systems from the point of its introduction into the blood to its incorporation into biological molecules in the cell.

5.1.11   Synthesis and properties of ruthenium arene compounds

In addition to NAMI-A and KP1019, a large number of ruthenium compounds containing bound hydrocarbons, referred to as arene compounds (some of which are shown in Figures 5.1 and 5.2), are being investigated for their anticancer properties [10–18]. These complexes, sometimes referred to as 'half-sandwich' compounds, can be synthesized by reaction of a nonaromatic precursor molecule such as cyclohexa-1,4-diene or one of its substituted derivatives with the $RuCl_3 \cdot 3H_2O$ in ethanol, which gives a dinuclear complex (a compound containing two metals) with $[(\eta^6\text{-arene})RuCl_2]$-type units connected by two

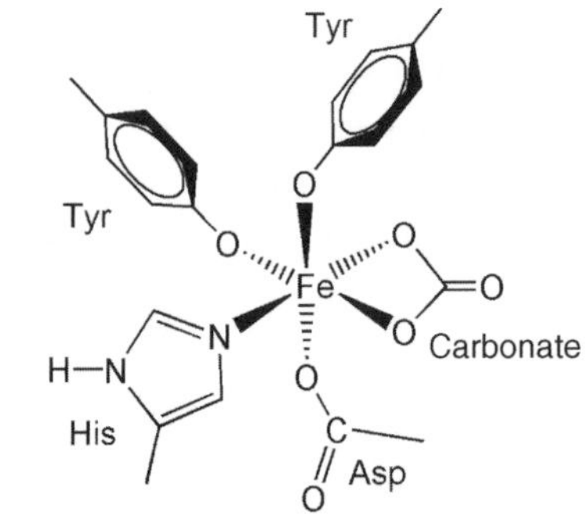

Figure 5.11 *The iron-binding site of transferrin, Tf. Transferrin has two identical sites each capable of binding an Fe^{+3} ion. From Baker, H.M., He, Q.-Y., Briggs, S.K., Mason, A.B., Baker, E.N. (2003) Structural and functional consequences of binding site mutations in transferrin: Crystal structures of the Asp_{63}Glu and Arg_{124}Ala mutants of the N-lobe of human transferrin. Biochemistry, 42, 7084–7089, and Kurokawa, H., Mikami, B., Hirose, M. (1995) Crystal structure of diferric hen ovotransferrin at 2.4 Å resolution. J. Mol. Biol., 254, 196–207*

chloride bridges (Figure 5.6c). A significant amount of the driving force for this reaction is the formation of an aromatic hydrocarbon from a nonaromatic precursor, which for the case shown in Figure 5.6c involves the conversion of cyclohexa-1,4-diene-type molecules into aromatic substituted benzenes. In this conversion, hydrogen (H_2) is formed from the diene, which is the reducing agent that converts Ru^{+3} to Ru^{+2}. The useful synthetic feature of the dinuclear compound is that it readily reacts with other (ancillary) ligands to produce mononuclear products that retain the arene and incorporate the added ligand. For example, reaction of the dinuclear compound with the bidentate ligand ethylenediamine (en) gives the mononuclear compound $[(\eta^6\text{-biphenyl})Ru(en)Cl]PF_6$, or if it is reacted with the bicyclic phosphine ligand pta, the compound $[(\eta^6\text{-}p\text{-}cymene)Ru(pta)Cl_2]$ – where pta is 1,3,5-triaza-7-phospaadamantane and *p*-cymene is 1-methyl-4-(1-methylethyl)benzene – can be obtained. The compound p-cymene, which occurs naturally, is found in cumin and thyme. If the dinuclear compound is treated with NaI, the chloride ligands can be replaced with iodide, which after reaction with a substituted bidentate azo-pyridine ligand gives the ruthenium arene $[(\eta^6\text{-}p\text{-}cymene)Ru(azpy)I]PF_6$ (Figure 5.2). Although the complete synthesis of DW12 (Figure 5.2) is multistep [17], the last step involves the reaction of the pyridocarbazole ligand with a ruthenium-Cp precursor to give the product.

From the drug-development standpoint, the ruthenium arenes are a very attractive class of compound. Since they contain difficult-to-displace aromatic ligands which can have a wide range of functional groups, many different types of hydrocarbon can be incorporated into the materials, which greatly expands the number of structures available for study. Also, since the compounds are often charged and hydrophobic groups can easily be attached, they can have both hydrophilic and hydrophobic properties, the latter often being important for transfer across the cell membrane.

The bonding in the ruthenium arenes is in many cases a mixture of the type found in organometallic compounds and that in coordination complexes. From MO theory, the bonds between ruthenium and the arene in these compounds, which occupy one of the *faces* of the octahedron (three coordination sites), are largely *covalent*, for the same reasons that the bonds between M and L in $Cr(CO)_6$ are covalent (Figure 5.3). However, the bonds between the ruthenium and the en and Cl^- ligands of $[(\eta^6\text{-biphenyl})Ru(en)Cl]PF_6$ and the two Cl^- ligands of $[(\eta^6\text{-}p\text{-}cymene)Ru(pta)Cl_2]$ are more *ionic* and similar to those between M and L of $[Co(NH_3)_6]^{+3}$ (Figure 1.10). Since these bonds are more salt-like in nature, the monodentate chloro ligands can be readily displaced in substitution reactions, but en, because it bidentate, is more difficult to displace. The pta ligand provides a strong σ-bond through the lone pair of electrons on phosphorous and since P has available empty $3d$ orbitals that accept electron density from the metal ion, the ligand also forms a π-bond. This stabilizes the Ru-P bond in $[(\eta^6\text{-}p\text{-}cymene)Ru(pta)Cl_2]$, making the pta ligand relatively resistant to displacement in a substitution reaction. The compound $[(\eta^6\text{-}p\text{-cymene})Ru(azpy)I]PF_6$ (Figure 5.2) contains a substituted bidentate azo-pyridine ligand, which can form a π-type interaction with the metal ion. Both the azo linkage (N = N) and the nitrogen atom of the pyridine have π^* molecular orbitals that interact with the t_{2g} set on the metal, forming a more covalent M-L bond, which makes the ligand more resistant to displacement. While iodide, I^-, in $[(\eta^6\text{-}p\text{-cymene})Ru(azpy)I]PF_6$ is a weak monodentate ligand, interestingly it resists displacement in a substitution reaction with, for example, the anion of GSH; that is, GS^- [16]. This may be due to the fact that I^- is large in size, making the formation of a seven-coordinate intermediate in an associative substitution process unfavorable. Since none of the ligands attached to $[(\eta^6\text{-}p\text{-cymene})Ru(azpy)I]PF_6$ are easily displaced, the antitumor effects of the compound are due to processes other than substitution at the metal center.

All of the groups of DW12 are difficult to displace by nucleophiles for the reasons given above. As with an azo-pyridine, the nitrogen donor atoms in the pyridocarbazole bidentate ligand form π-bonds with the ruthenium ion, which strengthens the M-L bond. An additional contributing factor to the stability of the bond is the fact that the pyridocarbazole ligand loses a proton from the indole-type nitrogen atom when it binds to the ruthenium ion, making the ligand negatively charged. As will be evident, DW12 is exchange-inert, and it exerts its biological effect by mimicking the structure (shape) of the natural-product staurosporine, which interacts with protein kinases in the cell [17,18].

Clearly, the bonding in the arene compounds is more complicated than the simple high-symmetry examples, $Cr(CO)_6$ and $[Co(NH_3)_6]^{3+}$, given earlier. However, the arene compounds are more organometallic in nature, in that they obey the 18-electron rule, which indicates that the spacing between HOMO and LUMO in their electronic structure is large, and based on the 18-electron rule the *formal oxidation state* of the ruthenium ion in compounds is +2.

The hydrolysis rate for the loss of Cl^- from $[(\eta^6\text{-biphenyl})Ru(en)Cl]PF_6$ to form $[(\eta^6\text{-biphenyl})Ru(en)(H_2O)]PF_6$ is $(1.23–2.59) \times 10^{-3}\,s^{-1}$ at 25 °C ($I = 0.1\,M$), which is about 20 times faster than that of cisplatin, and the equilibrium constant, K, for the reaction is $\sim 10^{-2}\,M$ [46]. By measuring the change in the UV–visible absorption spectrum of the aqua complex as a function of pH, the pK_a for deprotonation of the ruthenium-bound water molecule was found to be 7.71 ± 0.01, which is approximately one pK_a unit higher than that of the bound water of cis-$[PtCl(NH_3)_2(H_2O)]^+$ [46]. The rate constant for the hydrolysis of an arene compound similar in structure to $[(\eta^6\text{-}p\text{-cymene})Ru(pta)Cl_2]$ (Figure 5.1) was reported to be $\sim 1.7 \times 10^{-4}\,s^{-1}$ at 25 °C [47], and $[(\eta^6\text{-}p\text{-cymene})Ru(azpy)I]PF_6$ shows no detectable hydrolysis in one day at 37 °C [16].

5.1.12   Biological activity of the arene complexes

Sadler and Brabec and their coworkers reported that $[(\eta^6\text{-biphenyl})Ru(en)Cl]PF_6$ and similar compounds with different arene and bidentate ligands are cytotoxic toward A2780 human ovarian cancer cells, but uptake of ruthenium by the cell or the amount of Ru bound to DNA does not seem to correlate with cytotoxicity of the compounds [10,11,48,49]. One compound with bound p-terphenyl (three phenyl groups linked together through their *para* positions in a linear array), $[(\eta^6\text{-}p\text{-terp})Ru(en)Cl]PF_6$, proved to have cytotoxicity comparable to that of cisplatin in A2780 cells, human ovarian carcinoma, CH1 cells and human mammary carcinoma, SKBR3 cells.

The cytotoxicity of ruthenium compounds containing the pta ligand have been studied by Dyson, Sava and their coworkers [50]. The compounds are active against the TS/A mouse adenocarcinoma cancer cell line with $IC_{50} \sim 70\,\mu M$, but they are inactive against HBL-100 human, nontumor mammary cell line ($IC_{50} > 300\,\mu M$) in a 72 hour exposure to the cells. By comparison, the ruthenium compounds are much less cytotoxic than cisplatin, for which, under the same conditions, $IC_{50} = 0.53\,\mu M$. The compounds can also reduce the growth of lung metastases in CBA mice with MCa mammary carcinoma and do not affect the growth of the primary tumor; that is, they appear to have antimetastatic effects. The half life for clearance of the compounds from the blood of the mice is ~ 11 hours.

The ruthenium-azo-pyridine compounds are highly cytotoxic (more than cisplatin under the same conditions) toward human ovarian A2780 and human lung A549 cancer cells, with IC_{50} values generally in the low micromolar range after a 24 hour exposure to the compound and a 96 hour recovery period [16]. Studies with A549 cells show the compounds, which are relatively exchange-inert, probably kill the cells by generating ROS.

The compound DW12 inhibits various kinase enzymes, in the nanomolar range, that are important for phosphorylating serine and threonine residues on proteins [51]. The organometallic compound, which is similar in structure to the natural-product staurosporine, is very cytotoxic toward melanoma, an aggressive form of skin cancer that is generally resistant to chemotherapy [13,18,51].

5.1.13   Targets of the ruthenium arene compounds

While the biochemical basis for the antitumor effects of the ruthenium arene compounds is unknown, research efforts on the mechanism of action of the agents have focused on the interaction of the compounds with DNA and proteins.

For example, compounds like $[(\eta^6\text{-}p\text{-cymene})Ru(pta)Cl_2]$, in which the pta ligand has been replaced with a phosphite-carbohydrate, binds to human serum albumin and transferrin [47]. Shown in Figure 5.12 is a portion of the mass spectrum of the products that form when the ruthenium compound reacts with apo-Tf. The quantity m/z is the mass of the molecule in Daltons (Da) divided by the net charge on the molecule, z, which in this case is $+1$. Since the mass of ruthenium complex without one of its chloride ligands is $\sim$540 Da, the observed peak at 80 092 Da is due to the mono-adduct and the peak at 80 650 Da corresponds to the bis-adduct. The investigators suggested that the ruthenium arene compound is interacting with the sites on the protein that normally bind iron. The rate constant for the reaction of the Ru compound with apo-Tf is approximately $10^{-4}\,s^{-1}$ at 25 °C, which is about a factor of two less than that for the reaction of the compound with HSA under the same conditions [47].

A number of researchers have examined the binding of Ru-arene compounds to DNA, finding that the compounds bind to and unwind closed circular DNA and that N7 of guanine is the adduct site of the ruthenium

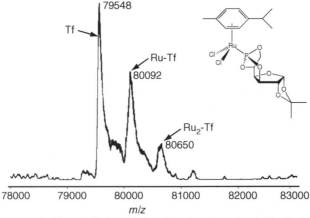

Figure 5.12 *Mass spectrum of apo-transferrin (Tf) incubated with a fourfold excess of the ruthenium arene compound shown in the figure in 20 mM ammonium carbonate buffer at 37 °C and pH 7.4, recorded after 30 minutes of incubation. The peak at m/z 79 548 is apo-Tf, the peak at m/z 80 092 corresponds to the mono-adduct and the peak at m/z 80 650 corresponds to the bis-adduct between the ruthenium compound and the protein. From I. Berger et al., In Vitro Anticancer Activity and Biologically Relevant Metabolism of Organometallic Ruthenium Complexes with Carbohydrate-Based Ligands, Chem. Eur. J. 2008, 14, 9046–9057. Copyright Wiley-VCH Verlag GmbH & Co. KGaA. Adapted with permission*

ion [10–15,47,52,53]. Since some of these compounds have an arene that can intercalate (insert) between the base pairs of DNA Figure 2.11, simultaneous intercalation *and* direct Ru-DNA adduct formation with the compounds is possible.

Besides DNA, proteins are targets for the ruthenium arenes in the cell. Casini *et al.* studied the ability of ruthenium arene compounds to affect the catalytic activity of the enzyme thioredoxin reductase, TrxR [54] (Box 4.3). This enzyme, which is important for maintaining the redox balance in the cell, contains a selenocysteine residue in its active site. The investigators found that the ruthenium compounds inhibit the function of the enzyme, with the most effective compound being [(η^6-*p*-cymene)Ru(pta)(cbdca)], which has the same di-carboxylate ligand found in carboplatin. Since the compound appears to bind to the protein without the loss of groups from the metal ion, the investigators suggested that binding is stabilized by specific interactions between the cbdca ligand and the protein.

Cathepsin B, cat B, is a cysteine protease that is capable of degrading the extracellular matrix in rheumatoid arthritis and other diseases and is also important in the invasion of cancer cells into normal tissue [54]. Casini *et al.* found that ruthenium arene compounds are even more effective in inhibiting the function of cat B than TrxR and the study suggested that in this case the ruthenium compounds lose ligands in binding to the protein [54].

In addition to directly binding to target molecules in the cell, Sadler and coworkers reported that certain cytotoxic ruthenium arenes containing phenylazopyridine ligands, for example [(η^6-*p*-cymene)Ru(azpy)I]PF$_6$ (Figure 5.2), can enter into a catalytic redox cycle that produces reactive radicals which kill the cell [16]. The mechanism involves GSH, which is converted to its disulfide GSSG through reaction with the coordinated phenylazopyridine ligand, and the subsequent generation of ROSs, which modify cellular targets, causing cell death (Figure 5.13). To probe the mechanism, the investigators used 2′,7′-dichlorodihydrofluorescein-diacetate (DCFH-DA), which is taken up by live cells and if ROS are present oxidizes to the highly-florescent product 2′,7′-dichlorofluorescein (DFC), which can be detected optically. By carrying out a number of control experiments, the investigators proposed that after the ruthenium arene enters the cell, the iodide ligand is partially displaced by GS$^-$, the anion of GSH, and that another molecule of GSH adds to the double bond of the azo linkage. While there is no reaction between GSH and the free phenylazopyridine ligand, when the azo-pyridine is bound to the metal ion one of the nitrogens of the coordinated azo linkage is attacked by GS$^-$, which forms a glutathione azo-pyridine adduct (Figure 5.13). This reaction takes place because the ruthenium ion removes electron density from the azo linkage, making one of the nitrogen atoms electropositive, which is then attacked by anionic GS$^-$. The investigators suggested that the initially-formed adduct is converted to a hydrazo group with the release of GSSG, and since the hydrazo is susceptible to oxidation by molecular oxygen, which is present inside the cell, the original ruthenium arene complex is regenerated. Because they were unable to detect the hydrogen peroxide formed in the reaction of O$_2$ with the hydrazo intermediate, the investigators suggested that the H$_2$O$_2$ which is formed is rapidly decomposed by the released iodide ion, which generates ROS. The latter chemically modifies biological molecules, which ultimately causes cell death.

The unique feature of the mechanism is that it does not involve direct binding of the metal ion to a macromolecule in the cell, but rather the ruthenium ion in the complex serves as an activator for chemistry that takes place on a coordinated ligand. Since *micromolar* concentrations of the ruthenium compound can convert *millimolar* concentrations of GSH to GSSG, the process is catalytic with respect to the ruthenium complex, raising the possibility that small amounts of compound might not only change the GSH/GSSG ratio in the cell but also produce radical damage to biological targets within it.

The ruthenium compound DW12 (Figure 5.2), synthesized and studied by Meggers and coworkers, is an interesting example of an exchange-inert ruthenium arene compound that exerts its biological effect by having the correct shape to block the function of kinase enzymes, which are important in the complex signaling system in cancer cells [17,18,51]. Shown in Figure 5.14 is the crystal structure of (*R*)-DW12, one

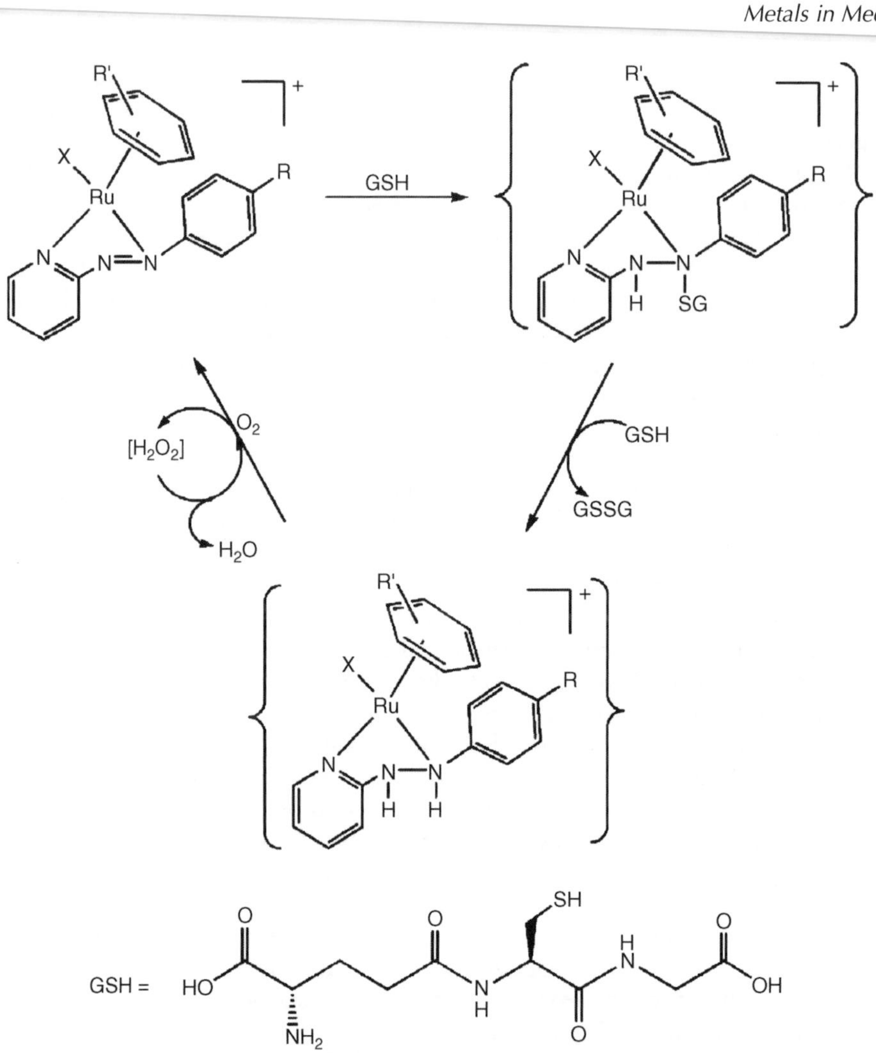

Figure 5.13 *Proposed cycle for the catalytic oxidation of GSH to its corresponding disulfide GSSG by ruthenium(II) arene phenylazopyridine complexes. Initially X is iodide, which is displaced by GS⁻ during the early stages. Released iodide can catalyze the decomposition of H_2O_2. Bound GSH at pH 7 has deprotonated thiol and carboxyl groups and a protonated amino group. Adapted from Figure 4, p. 11 631 of Dougan, S.J., Habtemariam, A., McHale, S.E., Parsons, S., Sadler, P.J. (2008) Catalytic organometallic anticancer complexes. Proc. Natl. Acad. Sci. USA, 105, 11628–11633*

of the optically-active forms of DW12, bound to a 297 amino acid fragment of the serine/threonine kinase PAK1, p21-activated kinase 1, which is important in regulating cell morphology. As is evident from the crystal structure, the maleimide moiety and the indole-OH group of the pyridocarbazole ligand are involved in four hydrogen bonds to either the amide carbonyl, C=O, or the amide hydrogen, N-H, in the peptide backbone of the protein. In addition, the bound carbon monoxide ligand has interactions with a glycine-rich loop of the protein. Not so evident from Figure 5.14 is the existence of unused (vacant) space below (*R*)-DW12 in the binding pocket, normally occupied by ATP, which that is required for the catalytic function of the enzyme. By retaining the pyridocarbazole and replacing the cyclopentadienyl ligand of DW12 with other, more bulky ligands, Meggers and coworkers showed that it is possible to fill this space with ligands

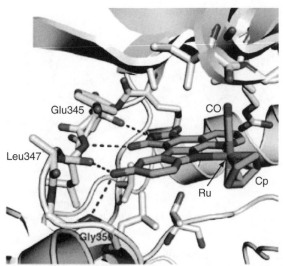

Figure 5.14   *Cocrystal structure at 2.35 Å of (R)-DW12 with PAK1, p21-activated kinase-1 (residues 249–545, with a mutation, Lys replaced by Arg at position 299). The hydrogen bonding between pyridocarbazole ligands of DW12 and specific residue on the protein are shown. Reprinted with permission from J. Maksimoska et al., Targeting Large Kinase Active Site with Rigid, Bulkey Octahedral Ruthenium Complexes, J. Am. Chem. Soc. 130, 15764–15765. Copyright 2008 American Chemical Society*

attached to Ru and to produce analogs that have a much better fit to the binding pocket, which produces a greater inhibitory effect on the PAK1 enzyme. This study nicely demonstrates the use of metal coordination properties in producing molecular shapes capable of targeting specific binding surfaces of proteins.

In addition to the compounds highlighted in this section, many other ruthenium complexes show promise for treating cancer. These complexes contain a variety of different ligands, including polypyridyl [55,56], bipyridyl [57,58], o-phenanthroline [59] and edta [60,61]; some have metal–metal bonds [62]; and some may express their biological effects outside the simple binding to biological targets such as proteins and DNA [63]. An important signaling molecule in the cell is *nitric oxide*, NO. Since disruption of the NO signaling pathway has been implicated in the pathophysiology of many disease states, including cancer, compounds that can potentially bind this small molecule and change its concentration in the cell may be useful for the treatment of cancer. For this reason, complexes of Ru^{+3}, which are known to readily bind NO, are being investigated as potential NO scavengers, and some may emerge as new agents for the treatment of cancer and other diseases [63].

In addition to ruthenium, osmium, Os, which is below Ru in the periodic chart, forms arene compounds that are isostructural and isoelectronic with Ru and are also anticancer-active [4]. Since Ru is in the second-row and Os the third-row transition metal series, as expected the complexes of the latter are much more resistant to ligand exchange than are the compounds of the former. Osmium compounds interact with polymeric DNA mainly at G sites and some are active against cisplatin-resistant cells, showing that the cellular detoxification mechanism used for the compounds may be different from that used by cisplatin.

5.2   Titanium compounds for treating cancer

The huge success of cisplatin as an anticancer drug stimulated research on many different types of metal-containing compound and in 1979, one year after cisplatin was approved for use as a drug, Köpf

Figures for the Titanium Section

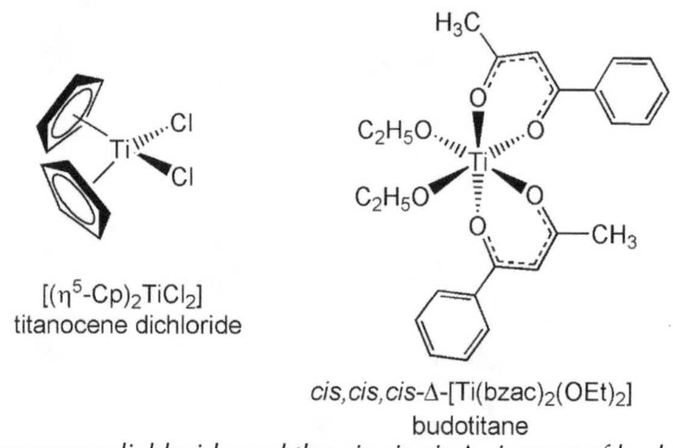

$[(\eta^5\text{-Cp})_2\text{TiCl}_2]$
titanocene dichloride

cis,cis,cis-Δ-$[\text{Ti(bzac)}_2(\text{OEt})_2]$
budotitane

Figure 5.15   *Structures of titanocene dichloride and the cis,cis,cis-Δ- isomer of budotitane are shown*

and Köpf-Maier reported that the simple metallocene dichlorobis(η^5-cyclopentadienyl)titanium(IV), titanocene dichloride (Figure 5.15), exhibited potent antitumor properties against Ehrlich ascites tumor cells and other cancers implanted in mice [64]. The motivation for examining the anticancer effects of the compound was that, like cisplatin, titanocene dichloride was neutral in charge and it contained two chloride-leaving ligands in *cis* positions. In subsequent years, Köpf-Maier and coworkers investigated the anticancer properties of related metallocenes containing Mo, V and Nb, and based on these studies titanocene dichloride, [Cp$_2$TiCl$_2$], which showed no nephro- or myelotoxicity, was ultimately selected as the first nonplatinum metal complex to enter clinical trials [65–67]. The compound is active against B16 melanoma and colon 38 carcinoma, and it is less toxic than cisplatin. The discovery of the anticancer properties of titanocene dichloride prompted Keppler and coworkers to investigate the anticancer properties of β-diketonate complexes of Ti^{+4}, finding that diacidobis(β-diketonato)metal(IV) complexes with M = Ti, Zr and Hf exhibited antitumor activity [67]. One of these compounds, diethoxybis(1-phenylbutane-1,3-dionato)titanium(IV), which was later named budotitane (Figure 5.15), was ultimately selected for clinical trials. While the clinical trials on both of these compounds have been discontinued, mainly because the compounds lack stability, the search for new antitumor active titanium compounds remains a promising area of research.

5.2.1   Structure, synthesis and properties of titanocene dichloride

Titanium, atomic number 22, has a number of known oxidation states, but compounds of the tetravalent oxidation state, Ti^{+4}, have been studied as potential drugs for treating cancer. Tetravalent titanium has the electronic configuration [Ar]3d^0, its complexes are diamagnetic ($S = 0$), and the most common coordination geometry for the ion is six-coordinate octahedral. However, since the ion has zero CFSE, the actual coordination geometry of Ti^{+4} can vary depending on the stereochemical demands applied by the bound ligands. Since the ion is highly charged it is considered a hard acid, which prefers to bind to hard bases. The high charge on the cation also causes the protons on coordinated water molecules to be very acidic and there is evidence that aquated Ti^{+4}, even in 1 M perchloric acid, has coordinated hydroxo (OH$^-$) ligands. The aquated ion is only sparingly soluble (micromolar concentrations) at neutral values of pH.

   The antitumor agent titanocene dichloride can be synthesized by the reaction of the sodium salt of the cyclopentadienyl anion with TiCl$_4$ in tetrahydrofuran, THF, which yields the bright red complex [(Cp)$_2$TiCl$_2$] (Figure 5.16) [68]. In unbuffered water containing 0.32 M KNO$_3$ at 37 °C, the compound loses the first chloride ligand to form the monoaquo complex with a large undermined rate, but the rate constant for the

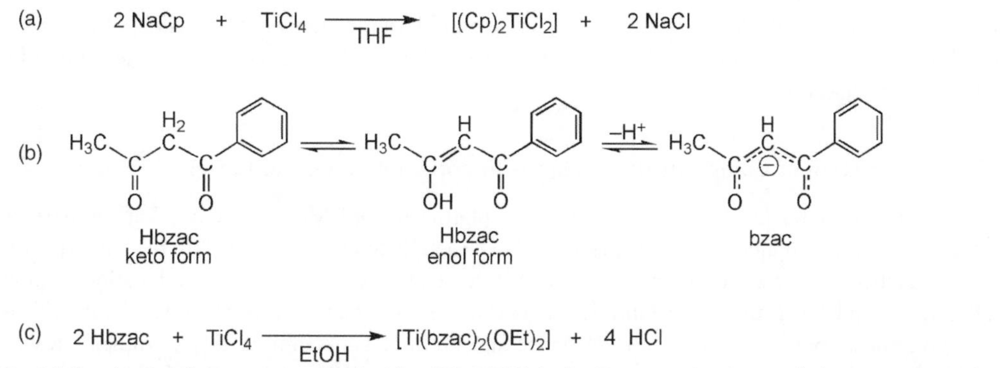

(a) $2\ NaCp\ +\ TiCl_4\ \xrightarrow{\ THF\ }\ [(Cp)_2TiCl_2]\ +\ 2\ NaCl$

(b)

Hbzac
keto form

Hbzac
enol form

bzac

(c) $2\ Hbzac\ +\ TiCl_4\ \xrightarrow{\ EtOH\ }\ [Ti(bzac)_2(OEt)_2]\ +\ 4\ HCl$

Figure 5.16   *(a) Synthesis of titanocene dichloride, $[(Cp)_2TiCl_2]$. (b) Tautomerization and deprotonation of 1-phenylbutane-1,3-dione, Hbzac. (c) Synthesis of budotitane. Since the β-diketone Hbzac is unsymmetrically substituted, it has two enol forms, but only one form is shown*

loss of the second chloride ligand to form the diaqua compound is 0.84 ($\pm$0.14) h^{-1}, $t_{1/2} \approx 50$ minutes [69]. The pK_as for the deprotonation of the bound water molecules of the diaqua complex are 3.51 ($\pm$0.05) and 4.35 ($\pm$0.09), both of which are lower than the corresponding protons of the bound water molecules of cis-$[Pt(NH_3)_2(H_2O)_2]^{2+}$ (Chapter 3). In unbuffered water, titanocene dichloride loses both Cp ligands, presumably due to the protonation of the Cp ligand, with the rate constant for the loss of the first Cp being $k = 6.4$ ($\pm$0.1) $\times 10^{-3}$ hour^{-1}, $t_{1/2} \approx 108$ hours [70].

Since titanocene dichloride has low stability in water, making stable preparations of the agent that would be suitable for administering to patients proved to be a major problem [70,71]. In some clinical trials, the compound was placed in a solution containing NaCl and mannitol and lyophilized to dryness, and the resulting powder was reconstituted in water prior to administration to the patient. While this procedure helped stabilize the compound prior to its use in therapy, the inability to find a suitable formulation was one of the factors that led to the abandonment of clinical trials with titanocene dichloride. In phase II clinical trials, involving patients with advanced renal carcinoma and breast metastatic carcinoma, titanocene dichloride did not perform well against other treatment regimes and it was eliminated from further testing. [70,72]

Since $[(Cp)_2TiCl_2]$ is administered by IV injection into the blood, it is important to consider whether and to what extent the compound interacts with transferrin and HSA. Sadler and coworkers used ^{1}H and ^{13}C NMR and other physical techniques to show that titanocene dichloride reacts with apo-transferrin under physiological conditions of 100 mM NaCl, 25 mM hydrogencarbonate and 4 mM phosphate, pH 7.4, 25 °C, and that the compound rapidly loses its original ligands in binding to the protein [73]. The first site on Tf to load with Ti^{+4} is the C-lobe, followed by the N-lobe, and NMR showed that the metal-induced conformational changes in the protein were similar to those produced by Fe^{+3}. The complex, Ti$_2$-Tf, is stable at the pH of the endosome, pH 5.0–5.5, but if ATP is present, which is normally required for iron transport into the cell, both Ti^{+4} ions are readily released from the protein in the endosomal pH range. The study also showed that since Ti$_2$-Tf blocked the uptake of radioactive ^{59}Fe-Tf into BeWo cells, the titanium-loaded transferrin probably binds to the transferrin receptor on the surface of the cell. Collectively, these results suggested that, while titanocene dichloride would quickly degrade in blood, titanium ions could be bound by apo-Tf or half-loaded Tf – that is, Fe-Tf – and thus Tf may be a route by which Ti^{+4} enters the cell.

Valentine and coworkers used ^{1}H NMR and other techniques to study the interaction of $[(Cp)_2TiCl_2]$ with HSA, finding that 20 molecules of the titanium complex bind to the protein [74]. While the binding of such a large number of molecules to the protein is not surprising (HSA has numerous specific and nonspecific sites that can bind a wide variety of different molecules), binding made the Cp ligands more resistant to loss due to hydrolysis. Similar observations of binding to and stabilization by HSA have also been reported for molybdocene dichloride, $[(Cp)_2MoCl_2]$, which also exhibits anticancer properties [75]. While HSA may serve as a binding/protecting agent for these metallocenes in blood, the ultimate fate of the protein-bound metal complexes is unknown.

5.2.2 Interaction of titanocene dichloride with potential biological targets

The early proposal that DNA is the target site for cisplatin led Köpf-Maier and coworkers to also consider DNA as the target site for titanocene dichloride and related metallocenes in the cell [76]. In studies using a modified form of transmission electron microscopy (TEM) that can specifically detect the location of titanium in the cell, it was found that Ti initially accumulated in the nuclei of human tumors grafted under the skin of mice that were given doses of titanocene dichloride. Since the compound also inhibits DNA synthesis, it was suggested that titanocene dichloride and/or its biotransformation products target genomic DNA.

The interaction of titanocene dichloride with DNA and proteins has been summarized by Meléndez [65]. The compound binds to double-stranded calf thymus DNA with a binding constant of 2.61 (± 0.08) $\times 10^5 M$ and there is one Ti complex per four base pairs of DNA when the DNA molecule is saturated with the complex. The binding interaction requires ~ 1 hour to reach equilibrium, which is much faster than cisplatin, which requires ~ 20 hours. In another study with calf thymus DNA and titanocene dichloride, using radioactive ^{3}H-labeled Cp ligands, the adduct formed at pH 5.3 was DNA-$(Cp)_2$Ti, but at pH 7.0 it was DNA-CpTi [77]. This indicates that at neutral pH the group bound to DNA is missing one of the Cp ligands. Although the site of binding on DNA was not determined, the fact that Ti^{+4} is a hard base suggests that the negatively-charged oxygen on the phosphate group of polymeric DNA may be an interaction site for the compound [65,73]. A number of binding studies involving titanocene dichloride and nucleotides have been carried out, but since the latter are not good models for polymeric DNA, the results of these investigations may not be relevant to the interactions that take place between the compound and DNA in the cell [65]. In addition to DNA, the proteins kinase C, which is responsible for phosphorylating serine and threonine residues on proteins, and topoisomerase II, an enzyme that modifies the double-helical nature of DNA, also bind $[(Cp)_2TiCl_2]$ [65,78].

5.2.3 Structure, synthesis and properties of budotitane

Budotitane, which contains the β-diketone ligand 1-phenylbutane-1,3-dione, Hbzac, and two ethoxides (Figure 5.15), has also been in clinical trials for the treatment of cancer. As shown in Figure 5.16b and c, the compound can be made by reacting the β-diketone, which exists in a tautomeric, a keto and an enol form, with $TiCl_4$ in ethanol to give a mixture of compounds with the formula $[Ti(bzac)_2(OEt)_2]$. Since the proton involved in the tautomerization of Hbzac is acidic, $pK_a \sim 9$, it can easily be removed to form the β-diketonate ion, which, under the conditions of the synthesis, coordinates to the titanium ion. While the pK_a of proton removal from ethanol is ~ 16, the presence of the highly-charged Ti^{+4} cation facilitates the loss of the proton from the alcohol and it coordinates as ethoxide, EtO^-, to Ti^{+4}. Since the coordination geometry around the Ti^{+4} ion is octahedral, five isomers, shown in Figure 5.17, with the formula $[Ti(bzac)_2(OEt)_2]$, are possible. Three of these isomers, which have *cis* ethoxide ligands, have different relative spatial arrangements of the phenyl and methyl groups of the bzac ligand on the octahedron. Each of these isomers has a mirror image that is *not* superimposable on itself and thus each exists as an *enantiomeric* pair, but only the Λ-forms are shown in Figure 5.17, with the Δ form of one of the isomers shown in Figure 5.15 (Box 5.3). Since equal amounts of each enantiomer, Λ or Δ, are produced in the synthesis – that is, *racimates* are formed – there is no net optical activity in the products. Each of the remaining two isomers, which have *trans* ethoxide ligands, has a mirror image that *is* superimposable on itself; that is, these isomers cannot exist as enantiomers.

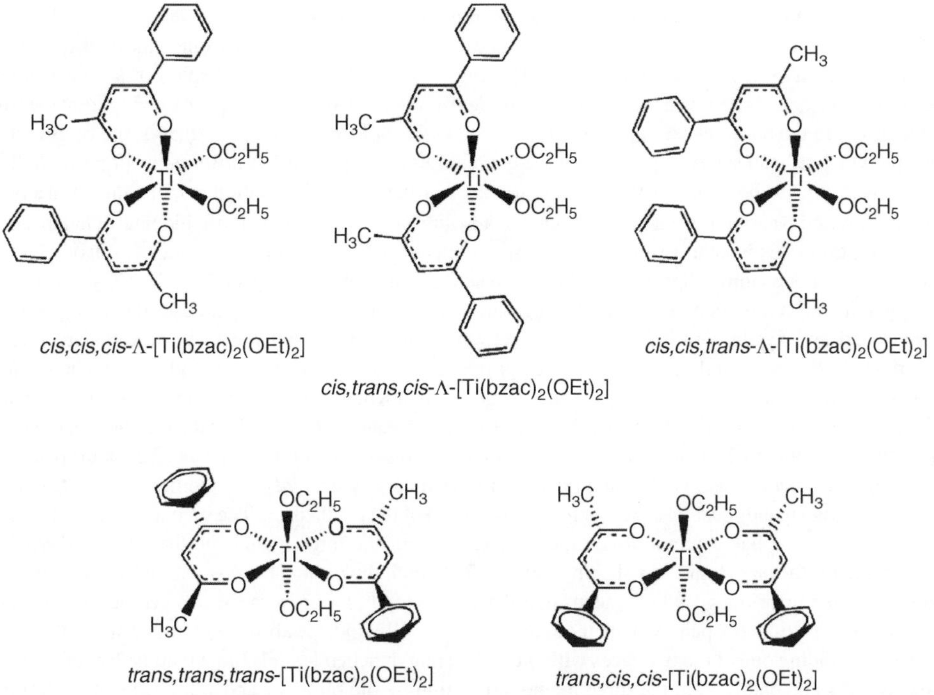

cis,cis,cis-Λ-[Ti(bzac)₂(OEt)₂]

cis,trans,cis-Λ-[Ti(bzac)₂(OEt)₂]

cis,cis,trans-Λ-[Ti(bzac)₂(OEt)₂]

trans,trans,trans-[Ti(bzac)₂(OEt)₂]

trans,cis,cis-[Ti(bzac)₂(OEt)₂]

Figure 5.17 *The isomers of budotitane, [Ti(bzac)₂(OEt)₂]. In the nomenclature used, the prefix cis,cis,cis, which can be abbreviated as c,c,c, refers to the dispositions of the ethoxide (OEt), β-diketonate phenyl and β-diketonate methyl groups, respectively. The three isomers with cis-OEt groups have mirror images that are not superimposable. They are optically-active and the structures shown have the Λ absolute configuration. The two isomers with trans-OEt have mirror images that are superimposable and are not optically-active*

Box 5.3 Geometric and optical isomerism of metal complexes [79]

The octahedron, which is the most common geometry of metal complexes, is made by joining eight identical equilateral triangles at their edges, which produces eight *faces* – the triangles – and six vertices – the points where four different triangles meet. If there are three identical bidentate ligands attached to the metal ion and the ligands have identical donor atoms on both 'ends', the complex formed is a *tris*-bidentate chelate compound containing three A–A-type ligands. An example of such an octahedral complex is [Co(en)₃]³⁺, which contains three coordinated ethylenediamine (en) ligands. When three A–A-type ligands are attached to the metal ion, *two* of the faces of the octahedron are unique in that they have *three* different attached ligands while the remaining six faces have only have *two* different attached ligands. The two unique faces, which are opposite each other on the octahedron, are shown as a dashed and a solid triangle in Figure 5.18.

An interesting property of a tris-bidentate chelate complex is that, even though the ligands themselves have no chiral centers (the ligands are not optically active), the arrangement of the three chelate rings on the octahedron produces a structure that has a nonsuperimposable mirror image; that is, the chelate ring system makes the compound chiral and there are two *enantiomers*. Inorganic chemists decided that the best way to specify the spatial arrangement of the bidentate ligands on the octahedron, which is called the *absolute configuration*, is to orient the enantiomer so that the eye of the observer is positioned above one of the unique faces; the view shown on the right of Figure 5.18. In this view, the observer is looking along an *axis of symmetry* of the enantiomer, called the *C₃ axis of rotation* (rotation by 120°), and the second unique face appears as a triangle on the 'backside' of the structure. If the structure is rotated so that the back face is positioned toward the observer and the face that was originally toward the observer is at the back, the view of the structure will be identical to the original view; that is, the enantiomer looks the same when viewed from either unique face. The assignment of absolute configuration of an enantiomer is made by noting the 'sense' of the chelate rings in passing from the front to the back of one of the unique faces. If when passing from the front to the back the chelate rings trace out motion in a *clockwise* direction – that is, the rings conform to a right-hand spiral or a right-hand screw or propeller – the enantiomer is assigned the Δ *absolute configuration*. If, on the other hand, in moving from the front to the back of the structure the observer sees the rings trace out a *counter-clockwise* motion – that is, the rings conform to a left-hand spiral or left-hand screw or propeller – the enantiomer is assigned the Λ *absolute configuration*. These two possibilities, with views above one of the two unique faces of the Λ and Δ enantiomers, are shown in Figure 5.18.

While the above convention is used to determine the absolute configuration of all tris-bidentate chelate complexes, ligands that have donor 'ends' that are not the same also produce structural (geometric) isomers, each of which exists in mirror-image pairs. For example, the anion of the amino acid glycine, glycinate (gly), $NH_2CH_2CO_2^-$, which is a good chelating agent, forms a five-membered chelate ring with a metal ion. Since this ligand uses the nitrogen atom of the amino group and an oxygen atom of the carboxyl group to bind to a metal ion to form a chelate ring, it is an unsymmetrical, A–B-type ligand. As a consequence of this *asymmetry*, the complex [M(gly)$_3$], and others like it that have three A–B-type ligands, exists in *two geometric forms*, both of which have nonsuperimposable mirror images. The name of the geometric isomer is determined by the location of donor atoms from each of the attached ligands. As shown in Figure 5.19, one geometric isomer has three 'A' donors from *three different ligands* on one face of the octahedron and three 'B' groups on a second face. This geometric isomer is designated the *facial* or *fac* isomer, indicating that common donors from the three ligands occupy a face (actually two faces) of the octahedron. The second geometric isomer (there are only two) is one that has common donor atoms from each of the three ligands on the *edge* or *meridian* of the octahedron, which is the *mer* isomer. As shown in Figure 5.19, both the *fac* and the *mer* geometric isomers also have a nonsuperimposable mirror image, which means that there are a total of four isomers: two geometric isomers each of which exists as an enantiomeric pair. As with the symmetrical A–A ligand, the absolute configurations of these isomers is determined by finding one of the two faces with three different attached ligands (note that the front face for the *mer* isomer has two As and one B from the three ligands), positioning the observer above the face and determining the 'sense' of the chelate rings – either right-hand, Δ, or left-hand, Λ – in passing from the front to the back of the structure.

Organic ligands that can be attached to a metal ion have a wide variety of different structures and while it is not possible to cover all of isomeric possibilities that might be encountered, some general comments concerning the sources of geometric and optical isomerism can be made. Some bidentate ligands have identical donor atoms, for example both atoms could be O or both could be N, but the organic structure connecting the two donor atoms in some way distinguishes the two 'ends' of the ligand. For example, one O-donor could be a deprotonated alcohol functional group and the other could be the oxygen atom of a deprotonated carboxylic acid functional group, as would occur with di-deprotonated glycolic acid, $^-OCH_2CO_2^-$. While both donor atoms are oxygen, the two ends of the ligand are clearly not the same. In this case two different orientations of the ligand on the metal are possible and, depending on the nature and number of other ligands bound to the metal, geometric isomers might be present. Sometimes octahedral complexes have only two chelate rings, with the remaining two sites being occupied by monodentate ligands, which do not generate a ring. If the monodentate ligands are *cis* to one another on the octahedron, the two chelate rings cause the resulting structure to have nonsuperimposable mirror images; that is, the compound exists as enantiomeric pairs. Determining the absolute configuration in this case is done by simply matching the sense of the chelate rings of the bis-bidentate compound with the tris-bidentate examples shown in Figures 5.18 and 5.19. Often the ligands attached to the metal ion contain chiral centers and are themselves optically active, for example amino acids. Placing ligands of this type in an octahedral environment around a metal ion can give rise to even more complex isomeric possibilities involving the relative arrangement of the donor atoms on the octahedron, the absolute configurations of the asymmetric atoms in the ligand and the absolute configuration of the ring system.

While all of this may seem to be on the fringe of what one needs to know about metal complexes that have found their way into medicine, it is important to point out that the biological system is a very chiral environment. Proteins are made up of amino acids, which have chiral centers; nucleic acids have chiral sugars; and the DNA double helix is chiral. Suppose, for example, a racemic metallo-drug is administered to the body and both enantiomers become bound to a target molecule to produce Δ-drug–target and Λ-drug–target. Since these two adducts are *diastereomeric* to one another (not exact mirror images of each other) they will not have the same physical properties, and most likely the binding constants of the Δ and Λ drug enantiomers toward the chiral target will not be the same. Since the strength of a drug–receptor interaction at the molecular level is often the basis for biological activity, choosing the correct optical isomer may mean the difference between high biological activity and none at all [80].

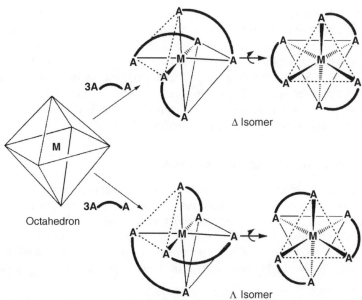

Figure 5.18 *Optical isomers and absolute configuration for an octahedral tris-bidentate chelate complex with three identical symmetrical, bidentate, A–A-type ligands. Examples of commonly found A–A-type ligands are ethylenediamine, $NH_2CH_2CH_2NH_2$, A = N, oxalate, $O_2CCO_2{}^{2-}$, A = O, and so on*

Because it is necessary to specify the spatial arrangement of the different groups on the octahedron, the nomenclature describing the budotitane isomers is rather complicated, but the convention given in Figure 5.17 is commonly employed [81–83]. Since budotitane is a *bis-bidentate* chelate, which is not exactly like the *tris-bidentate* chelate compound described in Figure 5.18, determining the absolute configuration of the budotitane isomers is done by simply matching the two chelate rings of a budotitane isomer with *any two of the three chelate rings* of the tris-bidentate chelate shown in Figure 5.18. When the rings match (overlap), the absolute configurations, indicated by Λ or Δ, are the same.

An interesting aspect of the synthesis of budotitane is that only the isomers with *cis*-OEt ligands are produced in the reaction and at room temperature in solution these isomers rapidly interconvert. This was determined by dissolving the budotitane in deuterated chloroform and measuring the proton NMR spectrum of the compound as a function of temperature. Since lowering temperature decreases the rate of a chemical process through (1.21) and (1.22), low temperature would be expected to slow the interconversion between the isomeric forms of budotitane. As shown in Figure 5.20, at 328 K (55 °C) only one resonance for the methyl group of the bzac ligand for the *c,c,c*, *c,t,c* and *c,c,t* isomers of budotitane can be observed. However, when the temperature of the solution is lowered, the proton NMR spectrum dramatically changes and at the lowest temperature of 243 K (−30 °C), four methyl resonances for bound bzac were observed (Figure 5.20). Inspection of the structure of the *c,c,t* isomer of budotitane (Figure 5.17) shows that two methyl groups of the coordinated bzac ligands are actually equivalent to one another in the sense that a methyl group on one ligand can be put into a position identical to that of the other by simply rotating the complex. From the standpoint of symmetry, this isomer is said to have a C_2 *axis of rotation* (rotation by 180°), which makes all of the atoms in one chelate ring equivalent to their counterparts in the other ring. The same equivalency also holds for the atoms of the two OEt groups. Since the methyl groups of both bound bzac ligands are in identical chemical environments, only *one* methyl resonance, at 2.26 ppm, is observed for the two methyl groups of this isomer (Figure 5.20). While a similar argument can made for the *c,t,c* isomer, which has only one methyl resonance, at 2.06 ppm (this isomer also has a C_2 axis), the methyl groups of the *c,c,c* isomer are *not* in identical chemical environments. Careful inspection of the *c,c,c* isomer shows that it is not possible to simply rotate the structure and have all of the atoms of the compound pass into equivalent positions. Thus there are *two* methyl resonances for this isomer and they appear at 2.25 and 2.02 ppm in the low-temperature spectrum of budotitane. Assigning the observed NMR peaks to specific isomers required the results of earlier studies, the synthesis and study of the corresponding exchange inert Co^{+3} complexes, and calculation of the expected distribution of isomers using computational methods [67,81,83].

5.2.4 Antitumor activity and clinical trials with budotitane

Colorectal cancer is one of the most frequent causes of death of patients suffering from cancer. In one study, rats with colon cancer were treated with budotitane, cisplatin and the anticancer drug 5-fluorouracil and it was found that budotitane was the most effective in reducing tumor volume [67].

In phase I clinical trials, doses of 100–230 mg m^{-2} of budotitane were given by IV infusion to patients who had solid tumors that were refractory to all other treatment modalities [84]. The maximum tolerated dose of budotitane administered twice weekly was 230 mg m^{-2}, with the dose-limiting toxicity being cardiac arrhythmia (abnormal heartbeat).

As may be surmised from the previous discussion, budotitane is quite reactive, and while chemical reactivity is certainly important for biological activity, in order for an agent to become a useful drug it must be stable enough over long periods of time to have a defined composition and exist in a form that is relatively

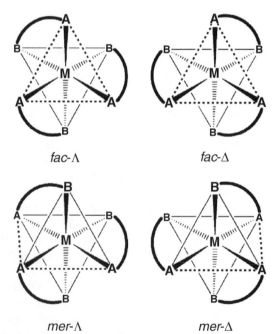

Figure 5.19 *Geometric and optical isomers for octahedral tris-bidentate chelate complexes with three identical, unsymmetrical, A–B-type ligands. Examples of commonly found ligands of this type are glycinate, $NH_2CHCO_2^-$, where $A = N$ and $B = O$, di-deprotonated glycolic acid, $^-OCH_2CO_2^-$, $A = O(OH)$, $B = O(CO_2H)$, and so on. The front unique face for the fac isomer and the edge, meridian, for the mer isomer are shown with dashed lines*

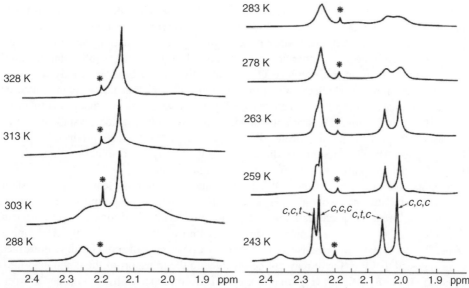

Figure 5.20   *Temperature dependence, in degrees Kelvin, K, of the 200 MHz ^{1}H NMR spectrum of budotitane in deuterated chloroform, CDCl$_3$. The singlets at 243 K are the methyl hydrogen resonances of the coordinated bzac ligands of budotitane and the resonance indicated by an asterisk is the methyl hydrogen resonance of uncoordinated bzac, a small amount of which is present in solution. The abbreviations are c = cis and t = trans. Adapted from B.K. Keppler et al., Struct. Bonding, 1991, 78, 97–127*

easy to administer by health care professionals. In attempting to prepare stable formulations of budotitane, Keppler and coworkers dissolved the agent in a medium containing the polymer glycerine polyethylene glycolericinoleate (CremophorEL), 1,2-propylene glycol and ethanol, and removed all of the solvent by evaporation to produce a solid called a co-precipitate [65,67,82]. When the co-precipitate was dissolved in water, micelles formed, which encapsulated the budotitane inside and protected it from hydrolysis. However, despite efforts to stabilize the agent, degradation still occurred, and compound instability and the fact that budotitane did not show a clear advantage over other drugs led to the abandonment of clinical trials.

5.2.5   Biological interactions of budotitane

With the high chemical reactivity of budotitane, using the compound in physiochemical studies involving DNA and proteins is challenging. However, early studies with budotitane reported that due to the presence of the phenyl groups on the β-diketonate ligands the compound may bind to DNA by intercalation [72,85].

5.2.6   Titanium anticancer drugs in development

The main lessons learned from titanocene dichloride and budotitane were that while the compounds have the desirable property of being less toxic than cisplatin, they are too unstable to formulate a solution that can practically be used in chemotherapy. In an effort to improve stability, research focused on ways to sequester the Ti^{+4} ion so that the resulting compound would be stable enough for a useful formulation, and at the same time to preserve enough reactivity that when the agent was introduced into the body it would react with compounds in the biological milieu.

Meléndez and coworkers recently synthesized an unusual tetranuclear complex, [Ti$_4$(ma)$_8$(μ-O)$_4$], containing Ti^{+4} ion bridged by oxide, O^{2-}, and containing the ligand 3-hydroxy-2-methyl-4-pyrone, maltol [86,87] Figure 5.21. Maltol is a commercially-available natural product that has low toxicity and is approved for use as an additive in cakes, breads, beer and other beverages to provide a 'malty' taste [88]. Maltol, Hma, has one acidic alcohol proton, which, in the presence of a metal cation and a base, is readily lost, forming the monovalent maltolate anion, ma, which can bind to a metal ion. Although both donors of maltolate are oxygen atoms – one is an alkoxide and the other is a ketone – the ligand is unsymmetrical; that is, maltolate is an A–B-type ligand, as shown in Figure 5.19. The compound, [Ti$_4$(ma)$_8$(μ-O)$_4$], which crystallizes with 18 water molecules, was made by reacting titanocene dichloride with maltol, which produced an intermediate mononuclear complex. Allowing the intermediate to stand in slightly basic solution ultimately resulted in [Ti$_4$(ma)$_8$(μ-O)$_4$], which crystallized from solution as an orange solid.

The investigators determined that in neutral aqueous solution the tetranuclear complex is stable over a period of 10 days, but that it may exist in a pH-dependent mono-tetramer equilibrium. In sharp contrast to titanocene dichloride, the tetranuclear compound does not transfer titanium ions to apo-transferrin, yet the compound is quite toxic toward HT-29 colon cancer cells, with values of IC$_{50}$ of $\sim 10^{-4}$ M during a 96 hour exposure. Since the cytotoxicity was insensitive to the presence or absence of apo-Tf in the culture medium, the investigators suggested that transferrin may not be involved in the transport of the compound into the cell.

The compound *cis*-[Ti(L^3)(OiPr)$_2$] (Figure 5.21), one of several reported by Tshuva and coworkers [89], contains a tetradentate N$_2$O$_2$ bis-phenolate ligand and two iso-propoxide (mono-deprotonated iso-propyl alcohol, $^-$OiPr) molecules in *cis*-coordination sites on Ti^{+4}. The compound was made by reacting Ti(OiPr)$_4$

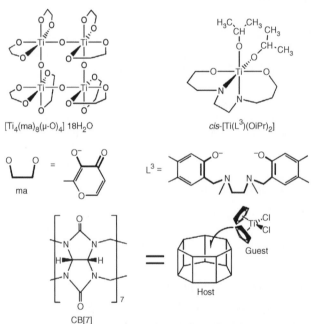

Figure 5.21 *Structures of titanium complexes which exhibit antitumor effects. The lower-case Greek letter μ is used to indicate a bridging ligand, which is in this case is O^{2-}, the ligand that is bridging the Ti^{+4} ions in [Ti$_4$(ma)$_8$(μ-O)$_4$] 18H$_2$O. The bold bonds in the ligands indicate the segment of the ligand that is involved in the chelate rings, shown in the structures of the complexes. Cucurbit[7]uril, CB[7], acting as a `host' for titanocene dichloride, the `guest', is shown at the bottom of the figure*

with the bis-phenolate ligand in an organic solvent. Cytotoxicity studies with ovarian OVCAR-1 and colon HT-29 cells showed that the titanium compound was more cytotoxic than cisplatin and it exhibited values of IC_{50} in the low micromolar range for a 72 hour exposure. Since the observed cytotoxicity was not affected by the presence or absence of apo-Tf in the culture medium, the investigators concluded that the compound is not transported by the transferrin system into the cell. Since one of the compounds studied lacked aromatic rings in the tetradentate ligand and was not cytotoxic, the investigators suggested that DNA intercalation by the flat aromatic parts of the compounds may be the basis for the biological activity of the complexes.

Harding and coworkers attempted to stabilize titanocene dichloride and thus make it more suitable as a possible drug by encapsulating it inside barrel-shaped molecules called cucurbit[*n*]urils (Figure 5.21) [90]. These molecules have 5–10 glycoluril units connected by methylene bridges and, since the structures have inside diameters ranging from 2 to 13 Å, are capable of binding small molecules in a host–guest relationship, where the cucurbit-uril is the 'host' and titanocene dichloride is the 'guest'. The investigators found that inclusion of titanocene dichloride into cucurbit[7]uril, CB[7], stabilized the titanium compound by blocking the protonation of the bound cyclopentadienyl ligands, raising the possibility that the approach might be useful for titanocene dichloride formulations. McGowan and Tacke and their coworkers have concentrated on modifying the Cp ring of titanocene dichloride [91,92]. Some of these compounds are cations with good water solubility and they show potent activity against a number of different cancer cell lines.

5.3   Gallium for treating cancer

The therapeutic properties of Ga^{+3} were recognized nearly 80 years ago when it was found that gallium tartrate could be used to treat syphilis [93]. In the years since, gallium compounds have been found to be useful for treating infectious disease, autoimmune disorders, accelerated bone resorption and cancer [93–99]. While the simple gallium salt gallium nitrate is approved for treating hypercalcemia, often associated with malignancies, the potential of gallium complexes for treating cancer will be summarized in this section.

5.3.1   Chemistry of gallium in biological media

Gallium, atomic number 31, is a semimetallic element that exhibits potent anticancer properties. The element has two common oxidation states, Ga^+ and Ga^{+3}, but only the latter is stable in the biological environment. The trivalent oxidation state of gallium, which has a $3d^{10}$ electronic configuration, has many properties in common with octahedral high-spin ($S = 5/2$) Fe^{+3}, which has a $3d^5$ configuration. Both ions have approximately the same octahedral ionic radius, 0.62 and 0.645 Å for Ga^{+3} and Fe^{+3}, respectively, and both have zero crystal field stabilization energy, CFSE. Since both Ga^{+3} and Fe^{+3} are hard acids, they form strong complexes with ligands that have N and O donor atoms. As is evident from Figure 1.20, both ions have water exchange rate constants of $\sim 10^3\,s^{-1}$, which indicates that the substitution kinetics involving metal complexes of the ions are likely to be similar. Not surprisingly, these similarities in physical and chemical properties cause Ga^{+3} to follow some of the biochemical pathways for Fe^{3+} in the body, often substituting for the latter in important iron-containing proteins [93,98,99]. An interesting difference between the two ions, and one that appears to be the basis for the antitumor effects of gallium compounds, is that unlike iron, gallium cannot easily change its oxidation state. This *redox* inertness can have serious consequences if Ga^{+3} is substituted for Fe^{+3} in a protein that requires the bound metal ion to change its oxidation state. This often leads to a nonfunctional protein, and if the protein is involved in an important catalytic function in the cell, the switch in metal ions can cause the cell to die.

One of the major drawbacks to working with simple gallium salts in aqueous media is that when the concentration of Ga^{+3} is relatively high (in the m*M* range) and the pH is near neutral, the ion reacts with water to produce polynuclear species, which precipitate from solution [94]. This aspect of gallium chemistry makes it difficult not only to study the ion in aqueous media but also to formulate stable solutions containing the ion for use in cancer chemotherapy.

The series of equilibrium reactions (5.1)–(5.4) illustrates the problems associated with aqueous gallium chemistry [94].

$$[Ga(H_2O)_6]^{3+} = [Ga(OH)(H_2O)_5]^{2+} + H^+, \quad pK_1 = 2.6 \qquad (5.1)$$

$$[Ga(OH)(H_2O)_5]^{2+} = [Ga(OH)_2(H_2O)_4]^+ + H^+, \quad pK_2 = 3.3 \qquad (5.2)$$

$$[Ga(OH)_2(H_2O)_4]^+ = [Ga(OH)_3(H_2O)_3] + H^+, \quad pK_3 = 4.4 \qquad (5.3)$$

$$[Ga(OH)_3(H_2O)_3] = [Ga(OH)_4(H_2O)]^- + H^+, \quad pK_4 = 6.3 \qquad (5.4)$$

As shown in (5.1)–(5.4), the protons of $[Ga(H_2O)_6]^{3+}$ are acidic and the fully protonated hexa-aqua species only exists at pH $< \sim 2$. The acidic nature of bound water is due to a Lewis acid effect, in which the Ga^{+3} ion removes electron density from the bound oxygen atom, which greatly polarizes and weakens the O–H bond of bound water. Also evident from the equilibria given is that as the charge on the complex ion decreases, it becomes progressively more difficult to remove a proton from the aqua complex in order to form a hydroxo ligand. Since the pK for proton removal depends on the charge on the metal ion to which the water is attached *and* the net charge on the metal complex from which the proton is being removed, this trend in the successive pKs (5.1)–(5.4) is expected.

A useful way to illustrate which species are present at different values of pH in a system with multiple equilibria is to construct a *species distribution plot*. Such a plot for Ga^{+3} in water [94], given in Figure 5.22, shows that at pH 7.4 the species in solution are the anionic complex $[Ga(OH)_4(H_2O)_2]^-$, $\sim 97\%$, and a small amount of the neutral complex $[Ga(OH)_3(H_2O)_3]$, $\sim 3\%$. The problem with aqueous solutions of Ga^{+3} near pH 7 is that the neutral complex $[Ga(OH)_3(H_2O)_3]$ has a strong tendency to self-associate to form di-, tri- and polynuclear complexes containing bridging hydroxo (OH^-) and oxo (O^{2-}) ligands, which are insoluble in

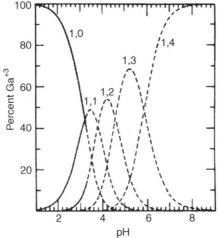

Figure 5.22 *Species distribution as a function of pH for 10 μM Ga^{+3}. The species are: $[Ga(H_2O)_6]^{3+}$, 1,0; $[Ga(OH)(H_2O)_5]^{2+}$, 1,1; $[Ga(OH)_2(H_2O)_4]^+$, 1,2; $[Ga(OH)_3(H_2O)_3]$, 1,3; $[Ga(OH)_4(H_2O)_2]^-$, 1,4. The number of water molecules was assumed to be six and it was assumed that all of the gallium remains in solution at all values of pH, which may not be the case. Reprinted from Baes and Mesmer, The Hydrolysis of Cations, 1976. Reprinted with permission of John Wiley & Sons Inc*

water. Since the amount of oligomerized product depends on the total concentration of $[Ga(OH)_3(H_2O)_3]$ in solution, keeping the Ga^{+3} concentration low (in the μM range) shifts the equilibrium in favor of the monomer, which prevents gallium precipitates from forming. Since working with systems that have both a solid and a solution phase is inherently difficult, high concentrations of Ga^{+3} should be avoided *or* the gallium should be sequestered by ligands that render it soluble in the medium being used. An additional complication with Ga^{+3}, indeed with all metal ions that are introduced into the biological system, is that biological fluids are replete with simple ions, for example Cl^-, CO_3^{2-}, PO_4^{3-} and so on, that can complex to the metal ion, thus changing the solubility of the ion in the medium. For example, phosphate can complex with Ga^{+3} to form $GaPO_4$, which is very insoluble in water [94].

5.3.2   Structures, synthesis and properties of gallium anticancer agents

In an effort to create stable solutions containing Ga^{+3} for use in cancer chemotherapy, investigators have focused on chelating the metal ion with ligands that will keep the ion in the complexed form, thus minimizing the concentration of free aquated Ga^{+3} in solution. Shown in Figure 5.23 are two gallium chelate compounds that are being evaluated for treatment of cancer [99–102]. Gallium maltolate, $[Ga(ma)_3]$, where ma is the deprotonated, anionic form of maltol (Hma), can be synthesized by the reaction of simple gallium salts with the ligand in aqueous and nonaqueous media in the presence of a base. When three maltolates are bound to Ga^{+3} ion in an octahedral geometry, the resulting tris-bidentate chelate complex exists in two isomeric forms, *fac* and *mer*, and both of these have a mirror image that is not superimposable on itself; that is, the *fac* and *mer* isomers exist as entiomeric pairs. The structure of one of these isomers, *fac-Δ-gallium maltolate*, is shown in Figure 5.23. While it is likely that the *fac* and *mer* isomers readily interconvert in solution, and resolution of the compound into enantiomers may not be possible, for the sake of completeness a specific structural form with the correct absolute configuration is shown [104]. A single crystal x-ray analysis of gallium maltolate confirmed the existence of the *mer* isomer but there have been no reports of the resolution of the compound into optically-active forms.

The compound KP46, $[GaQ_3]$, contains the well-known metal chelating agent 8-hydroxyquinoline, HQ, which itself has anticancer properties [103]. Similar to maltol, 8-hydroxyquinoline has an acidic phenolic proton, which in the presence of a metal ion and base can be lost to form the anion Q, which binds to Ga^{+3}. Since this ligand contains two different donor atoms – a phenylate oxygen and a pyridine-type nitrogen atom – formation of a tris-bidentate chelate complex with the ligand also leads to *fac* and *mer* isomers, both of which

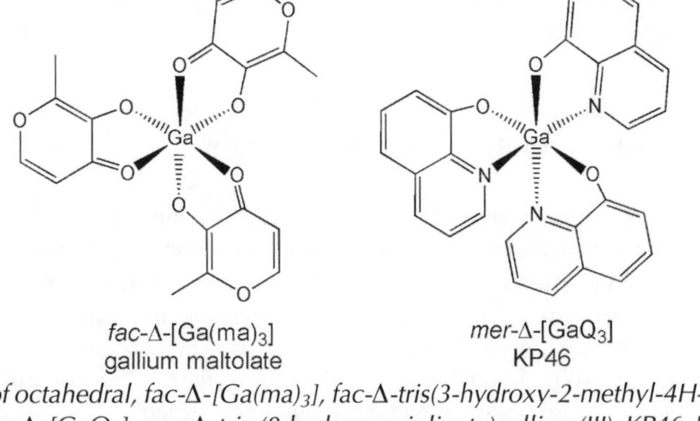

fac-Δ-[Ga(ma)₃]
gallium maltolate

mer-Δ-[GaQ₃]
KP46

Figure 5.23   *Structures of octahedral, fac-Δ-[Ga(ma)₃], fac-Δ-tris(3-hydroxy-2-methyl-4H-pyranonato)gallium(III), gallium maltolate, and mer-Δ-[GaQ₃], mer-Δ-tris-(8-hydroxyquiolinato)gallium(III), KP46. Both compounds exist as facial (fac) and meridinal (mer) isomers and both of these isomers can exist as enantiomeric pairs*

can exist in enantiomeric pairs. The structure of *mer*-Δ-[GaQ₃], one of the possible forms of KP46, is shown in Figure 5.23. Despite the fact that both gallium matolate and KP46 can exist in a number of isomeric forms, all of the biological studies with these compounds appear to have been carried out with substances with the correct molecular formula but of undefined stereochemical composition.

5.3.3 Uptake, cytotoxicity and reactivity of gallium agents in the biological system

Gallium maltolate exhibits several times more oral bioavailability (amount of Ga that reaches the blood) than orally-administered gallium chloride [101]. Doses for the compound are in the 100–500 mg range, the $t_{1/2}$ for elimination is 17–21 hours, and most of the gallium administered is ultimately excreted into the feces via the bile. Gallium maltolate was in phase II clinical trials for the treatment of bladder cancer, lymphoma, multiple myeloma and prostatic neoplasms but the trials were recently terminated.

The compound KP46 is superior to simple gallium salts in treating subcutaneous Walker carcinosarcoma 256, a cancer of the mammary gland, in rats [99]. The agent blocks the proliferation of multidrug-resistant non-small-cell lung cancer and shows synergistic effects with cisplatin, carboplatin and oxaliplatin *in vitro*. The acute toxicity of the compound in mice is somewhat higher than that of simple gallium salts and the highest concentrations of Ga^{+3} are found in bone, followed by the liver, spleen, kidneys and lung [99]. Phase I clinical trials with the compound suggest that KP46 may be useful for treating renal cancer.

Gallium in blood is mostly bound to the protein transferrin, Tf. Given that only ~30% of the total number of binding sites on transferrin in blood are normally occupied by Fe^{+3} ions, the protein has a significant capacity to accept other metal ions, including Ga^{+3} [99]. If orally-administered gallium maltolate and KP46 survive the acidic passage through the stomach and are adsorbed in the upper gastrointestinal tract, intact chelated Ga^{+3} complexes could circulate in the blood. While there is a paucity of information on the stability of these chelates, a recent report indicates that KP46 has pseudo first-order hydrolysis rate constants in water (pH 3.8, 37 °C) and physiological buffer (10 mM phosphate buffer, 100 mM NaCl, pH 7.4, 37 °C) of $k = 4.8 \pm 0.9 \times 10^{-5}\,s^{-1}$ and $k = 1.4 \pm 0.5 \times 10^{-5}\,s^{-1}$, respectively, showing the hydrolysis of the compound is about an order of magnitude slower than that of cisplatin [105,106].

The binding constants of Ga^{+3} toward the two lobes of apo-Tf in the presence of the co-ligand carbonate (27 mM) are very high, $\log K_1 = 20.3$ and $\log K_2 = 19.3$ [107]. In these studies, the metal-binding agent nitrilotriacetic acid, NTA, was present in the reaction medium to maintain a very low concentration of free Ga^{+3} so that at the neutral pH of the study, insoluble gallium-hydroxo/oxo species did not precipitate from solution. Having two phases, a solid and a solution phase, would compromise the equilibrium (thermodynamic) nature of the process being measured. NMR studies carried out in the presence of oxalate as a co-ligand show that the transferrin site in the C-lobe has a higher affinity for Ga^{+3} than the site in the N-lobe [108].

Box 5.4 Ribonucleotide reductase as a drug target

An enzyme that is considered a prime target for metal ions that are similar in size and properties to Fe^{+3} is ribonucleotide reductase (RNR), which is responsible for the reduction of ribonucleotides to the deoxyribonucleotides needed for the synthesis of DNA [109]. The R2 subunit of RNR is a homodimer of two 43 kDa monomers, with each monomer having a dinuclear iron complex located in the interior of the protein. In the oxidized state, the two high-spin ($S = 5/2$) Fe^{3+} ions, both of which have an octahedral coordination geometry, are connected by a bridging carboxylate of glutamic acid and a bridging oxo (O^{2-}) ligand (Figure 5.24a). During the catalytic cycle of the enzyme, the two Fe^{+3} ions are reduced to Fe^{+2} (a two-electron process) and the bridging oxo ligand is di-protonated to form water, which, with the two water molecules originally in the oxidized site, is lost to solvent. The reduced form contains two tetrahedral, high-spin ($S = 2$) Fe^{+2} ions bridged by the glutamic acid residue that was originally monodentate to one of the oxidized iron centers (Figure 5.24b). In the dinuclear iron site of the protein is a tyrosine residue (Tyr$_{122}$ in the *E. coli* enzyme) which exists as a radical, Y*, and which plays a key role in the catalytic cycle of RNR. Since this radical reacts with small-molecule reducing agents such as hydroxy urea and hydroxylamine to produce an inactive form of the enzyme, it, as well as the iron centers, is a potential target for drug action.

Studies by Chitambar and coworkers [110–114] have shown that gallium nitrate, which is approved for the treatment of hypercalcemia associated with malignancies, inhibits the enzyme ribonucleotide reductase, RNR, which plays a key role in DNA synthesis. While the mechanism by which this occurs in not known, blocking iron from reaching the enzyme or substituting iron with redox-inert Ga^{+3} would be expected to block the function of the enzyme. The effect of Ga^{+3} on the function of RNR is similar to that of the one-electron reducing agent hydroxyl urea, which is thought to react with the tyrosine radical in the iron site, thus inactivating the enzyme. Chitambar and coworkers [114] have also suggested that gallium maltolate inhibits cell proliferation and induces apoptosis through the mitochondrial pathway at lower concentrations and more rapidly than gallium nitrate, and that gallium-induced generation of reactive oxygen species, ROS, is likely involved.

5.3.4   Gallium anticancer agents in development

Proteasomes are large protein complexes inside cells that maintain intracellular homeostasis by degrading unwanted proteins. The proteins selected for elimination are modified at some of their lysine residues by the attachment of a small 76-amino-acid protein called *ubiquitin*, which tags the protein for 'death' by the proteasome. Since the proteasome has chymotrypsin-like activity, the interruption of which induces apoptosis in tumor cells, the proteasome is an ideal target for anticancer drug action.

Veriani and coworkers reported the synthesis, structure and anticancer properties of octahedral Ga^{+3} complexes containing the asymmetric tridentate N_2O ligands shown in Figure 5.25a [115]. The reaction of two equivalents of the chloro-substituted ligand as the mono anion with $GaCl_3$ in dry methanol produces the complex shown in Figure 5.25a. Examination of the cytotoxicity of the gallium complexes revealed that as a group they are very toxic toward human neuroblastoma (SK-N-BE(2)) cells that are highly resistant to cisplatin and other platinum drugs. In studying the mechanism of action of the compounds, Dou, Veriani and their coworkers found that the compounds target the proteasome in a variety of prostate cancer cell lines and

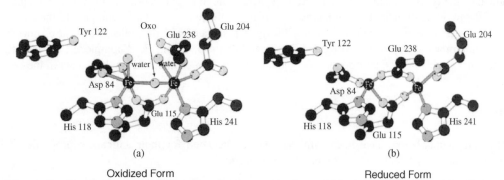

(a)   (b)

Oxidized Form   Reduced Form

Figure 5.24   *Oxidized, diferric (a), and reduced, diferrous (b), forms of the R2 subunit of E. coli, ribonucleotide reductase (RNR). Reprinted from Prog. Biophys. Mol. Biol., 77, H. Eklund et al., Structure and Function of the Radical Enzyme Ribonucleotide Reductase, 177–268. Copyright 2001, with permission from Elsevier*

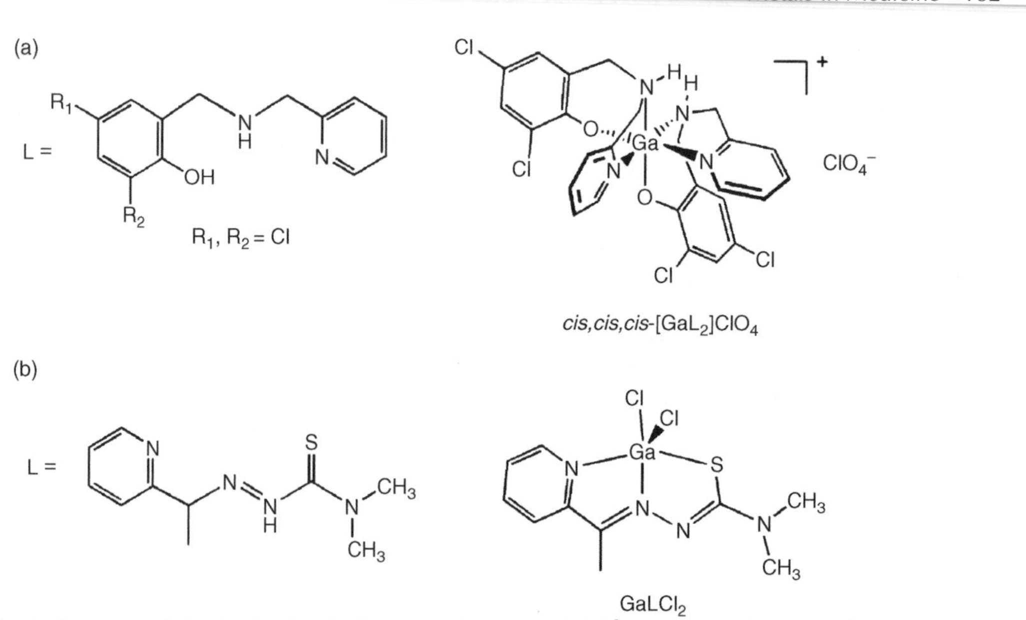

(a)

$R_1, R_2 = Cl$

cis,cis,cis-[GaL₂]ClO₄

(b)

GaLCl₂

Figure 5.25   *(a) Structure of cis,cis,cis-[GaL₂]ClO₄, one of a series of Ga⁺³ complexes that target the proteasome. (b) Structure of GaLCl₂, containing a pyridine-dimethylthiosemicarbazone ligand*

human prostate cancer xenografts [116,117]. These studies were carried out by measuring the ability of cells exposed to the gallium complexes to hydrolyze a small fluorogenic peptide via the chymotrypsin-like activity of the proteasome and by quantitating the build up of ubiquitinated proteins in the cell which could not be hydrolyzed by the inactivated proteasome. The ability of the gallium complexes to block the function of the proteasome is similar to that of the well-known boron-containing tripepetide anticancer drug *bortezomib* (*Velcade*), which is approved for use in treating multiple myeloma.

Dimethylthiosemicarbazone ligands of the type shown in Figure 5.25b exhibit anticancer properties that are enhanced when the ligands are incorporated into a Ga⁺³ complex [118]. Reaction of a pyridine-dimethylthio-semicarbazone ligand with GaCl₃ in dry ethanol produces the five-coordinate distorted square pyramidal complex, GaLCl₂ (Figure 5.25b), which exhibits potent toxicity toward human ovarian carcinoma (41M) and mammary carcinoma (SK-BR-3) cells. Investigations of the mechanism of action of the compound revealed that the radical on the tyrosine residue, Y*, located in the active site of ribonucleotide reductase may be the site of attack of the pyridine-dimethylthiosemicarbazone ligand delivered by the Ga⁺³ complex to the cancer cell. However, since measurements of the loss of the EPR signal associated with the radical with time do not correlate with the antitumor potency observed for the free ligand or its Ga⁺³ and Fe⁺³ complexes, the investigators suggested that targets other than ribonuclease reductase may also be attacked by the agents in the cell.

Problems

1. The compounds below have been synthesized and evaluated for their anticancer properties. Give a synthetic route by which each compound can be made.

    a. *trans*-[RuCl₄(Im)₂]HIm, where Im is imidazole.
    b. [(⁶η-*p*-cymene)Ru(pta)(cbdc)], where pta is 1,3,5-triaza-7-phospaadamantane, *p*-cymene is 1-methyl-4-(1-methylethyl)benzene and cbdca is 1,1-cyclobutane dicarboxylate.

2. The compound DW12, shown in Figure 5.2, exists in two enantiomeric forms. Draw both enantiomers.
3. Using crystal field arguments and expected relative values of the activation free energy, $\Delta G^{\ddagger}$ (see (1.22)), briefly explain why the hydrolysis rate of $[(\eta^6\text{-p-cymene})\text{Os(en)Cl}]^+$ is *slower* than that of the corresponding ruthenium compound, Ru^{+2}, $[Kr]4d^6$; Os^{+2}, $[Xe]5d^6$.
4. Draw the geometric and optical isomers of $[\text{Ti(bzac)}_2\text{(ox)}]$, where bzac is deprotonated 1-phenylbutane-1,3-dione and ox is oxalate.
5. The metallocene dichlorides, $[(^5\eta\text{-Cp})_2\text{MCl}_2]$, where $M = Mo$, V and Nb, are cytotoxic and are isostructural with titanocene dichloride.

    a. Determine the *formal oxidation state* for the metal in each of the compounds.
    b. Which compound(s) is (are) paramagnetic with one unpaired electron?

6. Shown below is the species distribution plot for the interaction of the bidentate ligand, isomaltol (HL), a structural relative of maltol, with Ga^{+3} in aqueous media as a function of pH. Isomaltol has an acidic alcohol proton which is lost to form L when the ligand reacts with a metal ion to form a complex. The curves were calculated for a solution containing 1 mM Ga^{+3} and 3 mM HL, in the presence of 0.15 M NaCl, at 25 °C. The dotted lines of the plot represent pH regions where the system may exist as two phases; that is, a solid and a solution phase. Answer the following, pertaining to the system:

    a. Assuming that the system remains entirely in the solution phase, give the approximate distribution of species in solution at pH 4.0.
    b. If GaL_3 is incorporated into a tablet to be taken orally, briefly comment on the nature of the Ga species expected to be present in the stomach at $pH = 2$.
    c. The $\log \beta_3$ for the overall reaction to form GaL_3 is 16.36 at 25 °C. Calculate the equilibrium constant, K, and the value of ΔG at 25 °C for this reaction. $R = 8.314\,J\,°K^{-1}\,mol^{-1}$
    d. Draw the structure of *fac*-Λ-GaL_3.

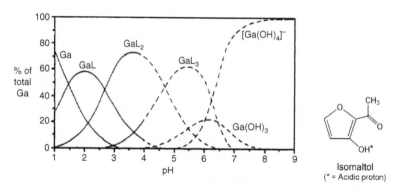

Figure reprinted with permission from T.G. Lutz et al., Metal Chelation with Natural Products: Isomaltol Complexes of Aluminum, Gallium and Indium, Inorg. Chem. 28, 715–719. Copyright 1989 American Chemical Society

7. The UV absorption data for Al(ma)_3 and Ga(ma)_3, where ma is maltolate, are 305 (18 800), 250 (8900) and 305 (20 130), 250 (9030) respectively for λ_{max}, nm and ε in $M^{-1}\,cm^{-1}$ in parenthesis. The octahedral ionic radii for Al^{+3} and Ga^{+3} are 0.535 and 0.62 Å, respectively.

    a. Assign the absorption bands observed in the spectrum of the compounds. Consider the following possibilities: *d-d* absorptions, charge transfer absorptions and ligand transitions. Briefly explain your choice.

b. Which complex, Al(ma)$_3$ or Ga(ma)$_3$, would be expected to have the *smaller* rate constant for hydrolysis; that is, reaction with water? Briefly explain your choice.

8. The two Fe^{+3} ions in the R2 subunit of ribonucleotide reductase, RNR, can be removed from the protein to form apo-R2 using an excess of the metal-chelating agent lithium 8-hydroxyquinoline-5-sulfonate. Similar to 8-hydroxyquinoline, lithium 8-hydroxyquinoline-5-sulfonate loses a phenolic proton to form SQ and reacts with metal ions to produce tris-bidentate chelates.

a. Given that the gallium-containing anticancer agent KP46, GaQ$_3$, contains the 8-hydroxyquinolate anion (Q), which is similar in structure to SQ, write the *overall equilibrium expression* in terms of the concentrations and the equilibrium constant, K, for the reaction given below:

$$Fe_2 R2 + 2\,GaQ_3 \leftrightarrow Ga_2R2 + 2\,FeQ_3$$

b. If the overall equilibrium constant for the binding of two aquated Fe^{+3} ions to apo-R2 to form Fe$_2$R2 is K_1, the overall equilibrium constant for the binding of two aquated Ga^{+3} ions to apo-R2 to form Ga$_2$R2 is K_2, and the equilibrium constants for the formation of FeQ$_3$ and GaQ$_3$ from the aquated ions and Q are K_3 and K_4, respectively, write the equilibrium expression in part (a) in terms of K and K_{1-4}.

HSQ

References

1. Alessio, E., Mestroni, G., Bergamo, A., and Sava, G. (2004) Ruthenium anticancer drugs. *Metal Ions in Biological Systems*, **42**, 323–347.
2. Bergamo, A. and Sava, G. (2007) Ruthenium complexes can target determinants of tumor malignancy. *Dalton Transactions*, 1267–1272.
3. Jakupec, M.A., Galanski, M., Arion, V.B., and Hartinger, C.G. (2008) Keppler, Antitumor metal compounds: More than theme and variations. *Dalton Transactions*, 183–194.
4. Peacock, A.F.A. and Sadler, P.J. (2008) Medicinal organometallic chemistry: Designing metal arene complexes as anticancer agents. *Chemistry, an Asian Journal*, **3**, 1890–1899.
5. Clark, M.J. (2003) Ruthenium metallopharmaceuticals. *Coordination Chemistry Reviews*, **236**, 209–233.
6. Clarke, M.J., Zhu, F., and Frasca, D.R. (1999) Non-platinum chemotherapeutic metallopharmaceuticals. *Chemical Reviews*, **99**, 2511–2533.
7. Czap, A. and van Eldik, R. (2003) The unusually fast reactions between ruthenium(III)-ammine complexes and NO revisited. *Dalton Transactions*, 665–671.
8. Rademaker-Lakhai, J.M., van den Bongard, D., Pluim, D. *et al.* (2004) A phase I and pharmacological study with imidazolium-trans-DMSO-imidazole-tetrachlororuthenate, a novel ruthenium anticancer agent. *Clinical Cancer Research*, **10**, 3717–3727.
9. Hartinger, C.G., Zorbas-Seifried, S., Jakupec, M.A. *et al.* (2006) From bench to bedside-preclinical and early clinical development of the anticancer agent indazolium trans-[tetrachlorobis(1H-indazole)ruthenate(III)](KP1019 or FFC14A). *Journal of Inorganic Biochemistry*, **100**, 891–904.
10. Morris, R.E., Aird, R.E., del Socorro Murdoch, P. *et al.* (2001) Inhibition of cancer cell growth by ruthenium (II) arene complexes. *Journal of Medicinal Chemistry*, **44**, 3616–3621.

11. Bugarcic, T., Nováková, O., Halámiková, A. *et al.* (2008) Cytotoxicity, cellular uptake, and DNA interactions of new monodentate ruthenium (II) complexes containing terphenyl arenes. *Journal of Medicinal Chemistry*, **51**, 5310–5319.

12. Bugarcic, T., Habtemariam, A., Stepankova, J. *et al.* (2008) The contrasting chemistry and cancer cell cytotoxicity of bipyridine and bipyridinediol ruthenium (II) arene complexes. *Inorganic Chemistry*, **47**, 11470–11486.

13. Bruijnincx, P.C.A. and Sadler, P.J. (1008) New trends for metal complexes with anticancer activity. *Current Opinion in Chemical Biology*, **12**, 197–206.

14. Ang, W.H. and Dyson, P.J. (2006) Classical and non-classical ruthenium-based anticancer drugs: Towards targeted chemotherapy. *European Journal of Inorganic Chemistry*, 4003–4018.

15. Scolaro, C., Chaplin, A.B., Hartinger, C.G. *et al.* (2007) Tuning the hydrophobicity of ruthenium (II)-arene (RAPTA) drugs to modify uptake, biomolecular interactions and efficacy. *Dalton Transactions*, 5065–5072.

16. Dougan, S.J., Habtemariam, A., McHale, S.E. *et al.* (2008) Catalytic organometallic anticancer complexes. *Proceedings of the National Academy of Sciences of the United States of America*, **105**, 11628–11633.

17. Pagano, N., Maksimoska, J., Bregman, H. *et al.* (2007) Ruthenium half-sandwich complexes as protein kinase inhibitors: Derivatization of the pyridocarbazole pharmacophore ligand. *Organic and Biomolecular Chemistry*, **5**, 1218–1227.

18. Smalley, K.S.M. *et al.* (2007) An organometallic protein kinase inhibitor pharmacologically activates p53 and induces apoptosis in human melanoma cells. *Cancer Research*, **67**, 209–217.

19. Mestroni, G., Alessio, E., and Sava, G. (1998) New salts of anionic complexes of Ru(III) as antimetastatic and antineoplastic agents. Int. Patent WO 98/000431.

20. Bacac, M., Hotze, A.C.G., van der Schilden, K. *et al.* (2004) The hydrolysis of the anti-cancer ruthenium complex NAMI-A affects its DNA binding and antimetastatic activity: An NMR evaluation. *Journal of Inorganic Biochemistry*, **98**, 402–412.

21. Brindell, M., Piotrowska, D., Shoukry, A.A. *et al.* (2007) Kinetics and mechanism of the reduction of (ImH)[*trans*-RuCl$_4$(dmso)(Im)] by ascorbic acid in acidic aqueous solution. *Journal of Biological Inorganic Chemistry: JBIC: a Publication of the Society of Biological Inorganic Chemistry*, **12**, 809–818.

22. Gava, B., Zoret, S., Spessotto, P. *et al.* (2006) Inhibition of B16 melanoma metastasis with the ruthenium complex imidazolium trans-imidazoledimethylsulfoxidetetrachlororuthenate and down regulation of tumor cell invasion. *Journal of Pharmacology and Experimental Therapeutics*, **317**, 284–291.

23. Bouma, M., Nuijen, B., Jansen, M.T. *et al.* (2002) A kinetic study of the chemical stability of the antimetastatic ruthenium complex NAMI-A. *International Journal of Pharmaceutics*, **248**, 239–246.

24. Messori, L., Orioli, P., Vullo, D. *et al.* (2000) A spectroscopic study of the reaction of NAMI, a novel ruthenium (III) anti-neoplastic complex, with bovine serum albumin. *European Journal of Biochemistry*, **267**, 1206–1213.

25. Messori, L., Vilchez, F.G., Vilaplana, R. *et al.* (2000) Binding of antitumor ruthenium(III) complexes to plasma proteins. *Metal Based Drugs*, **7**, 335–342.

26. Bergamo, A., Messori, L., Piccioli, F. *et al.* (2003) Biological role of adduct formation of the ruthenium (III) complex NAMI-A with serum albumin and serum transferrin. *Investigational New Drugs*, **21**, 401–411.

27. Ravera, M., Baracco, S., Cassino, C. *et al.* (2004) Electrochemical measurements confirm the preferential bonding of the antimetastatic complex [ImH][RuCl$_4$(dmso)(Im)] (NAMI-a) with proteins and the weak interaction with nucleobases. *Journal of Inorganic Chemistry*, **98**, 984–990.

28. Messori, L., Kratz, F., and Alessio, E. (1996) The interaction of the antitumor complexes Na[trans-RuCl$_4$(dmso)(Im)] and Na[trans-RuCl$_4$(dmso)(Ind)] with apotransferrin: A spectroscopic study. *Metal Based Drugs*, **3**, 1–9.

29. Casini, A., Mastrobuoni, G., Terenghi, M. *et al.* (2007) Ruthenium anticancer drugs and proteins: a study of the interactions of the Ruthenium(III) complex imidazolium trans-[tetrachloro(dimethyl sulfoxide)(imidazole)ruthenate(III)] with hen egg white lysozyme and horse heart cytochrome c. *Journal of Biological Inorganic Chemistry: JBIC: a Publication of the Society of Biological Inorganic Chemistry*, **12**, 1107–1117.

30. Gallori, E., Vettori, C., Alessio, E. *et al.* (2000) DNA as a possible target for antitumor ruthenium (III) complexes. *Archives of Biochemistry and Biophysics*, **376**, 156–162.

31. Brindell, M., Stawoska, I., Supel, J. *et al.* (2008) The reduction of (ImH)[*trans*-Ru^{III}Cl$_4$(dmso)(Im)] under physiological conditions: Preferential reaction of the reduced complex with human serum albumin. *Journal of Biological Inorganic Chemistry: JBIC: a Publication of the Society of Biological Inorganic Chemistry*, **13**, 909–918.

32. Hartinger, C.G., Jakupec, M.A., Zorbas-Seifried, S. *et al.* (2008) KP1019, a new redox-active anticancer agent-preclinical development and results of a clinical phase I study in tumor patients. *Chemistry & Biodiversity*, **5**, 2140–2155.

33. Küng, A., Pieper, T., Wissiack, R. *et al.* (2001) Hydrolysis of the tumor-inhibiting Ruthenium(III) complexes HIM trans-[RuCl$_4$(im)$_2$] and Hind trans-[RuCl$_4$(ind)$_2$] investigated by means of HPCE and HPLC-MS. *Journal of Biological Inorganic Chemistry: JBIC: a Publication of the Society of Biological Inorganic Chemistry*, **6**, 292–299.

34. Schluga, P., Hartinger, C.G., Egger, A. *et al.* (2006) Redox behavior of tumor-inhibiting ruthenium(III) complexes and effects of physiological reductants on their binding to GMP. *Dalton Transactions*, 1796–1802.

35. Heffeter, P., Pongratz, M., Steiner, E. *et al.* (2005) Intrinsic and acquired forms of resistance against the anticancer ruthenium compound KP1019 [Indazolium][*trans*-tetrachlorobis(1*H*-indazole)ruthenate (III)] (FFC14C). *The Journal of Pharmacology and Experimental Therapeutics*, **312**, 281–289.

36. Pongratz, M., Schluga, P., Jakupec, M.A. *et al.* (2004) Transferrin binding and transferrin-mediated uptake of the ruthenium coordination compound KP1019, studied by means of AAS, ESI-MS and CD spectroscopy. *Journal of Analytical Atomic Spectrometry*, **19**, 46–51.

37. Kratz, F., Hartman, M., Keppler, B., and Messori, L. (1994) The binding properties of two antitumor ruthenium (III) complexes to apotransferrin. *The Journal of Biological Chemistry*, **269**, 2581–2588.

38. Kratz, F., Keppler, B.K., Messori, L. *et al.* (1994) Protein-binding properties of two antitumor Ru(III) complexes to human apotransferrin and apolactoferrin. *Metal-Based Drugs*, **1**, 169–173.

39. Esposito, B. and Najjar, R. (2002) Interaction with antitumoral platinum-group metallodrugs with albumin. *Coordination Chemistry Reviews*, **232**, 137–149.

40. Timerbaev, A.R., Hartinger, C.G., Aleksenko, S., and Keppler, B.K. (2006) Interactions of antitumor metallodrugs with serum proteins: Advances in characterization using modern analytical methodology. *Chemical Reviews*, **106**, 2224–2248.

41. Timerbaev, A.R., Rudnev, A.V., Semenova, O. *et al.* (2005) Comparative binding of antitumor indazolium [trans-tetrachlorobis(1H-indazole)ruthenate(III)] to serum transport proteins assayed by capillary zone electrophoresis. *Analytical Biochemistry*, **341**, 326–333.

42. Baker, H.M., He, Q.-Y., Briggs, S.K. *et al.* (2003) Structural and functional consequences of binding site mutations in transferrin: Crystal structures of the Asp$_{63}$Glu and Arg$_{124}$Ala mutants of the N-lobe of human transferring. *Biochemistry*, **42**, 7084–7089.

43. Kurokawa, H., Mikami, B., and Hirose, M. (1995) Crystal structure of diferric hen ovotransferrin at 2.4 Å resolution. *Journal of Molecular Biology*, **254**, 196–207.

44. Chen, D.C., Newman, B., Turkall, R.M., and Tsan, M.F. (1982) Transferrin receptors and Gallium-67 uptake *in vitro*. *European Journal of Nuclear Medicine*, **7**, 536–540.

45. Plamer, D.A. and van Eldik, R. (1983) The chemistry of metal carbonato and carbon dioxide complexes. *Chemical Reviews*, **83**, 651–731.

46. Wang, F., Chen, H., Parsons, S. *et al.* (2003) Kinetics of aquation and anation of ruthenium(II) arene anticancer complexes, acidity and x-ray structures of aqua adducts. *Chemistry – A European Journal*, **9**, 5810–5820.

47. Berger, I. *et al.* (2008) *In vitro* anticancer activity and biologically relevant metabolism of organometallic ruthenium complexes with carbohydrate-based ligands. *Chemistry – A European Journal*, **14**, 9046–9057.

48. Habtemariam, A. *et al.* (2006) Structure-activity relationships for cytotoxic ruthenium(II) complexes containing N,N-, N,O- and O,O chelating ligands. *Journal of Medicinal Chemistry*, **49**, 6858–6868.

49. Aird, R.E., Cummings, J., Ritchie, A.A. *et al.* (2002) *In vitro* and *in vivo* activity and cross resistance profiles of novel ruthenium (II) organometallic arene complexes in human ovarian cancer. *British Journal of Cancer*, **86**, 1652–1657.

50. Scolaro, C., Bergamo, A., Brescacin, L. *et al.* (2005) *In vitro* and *in vivo* evaluation of ruthenium(II)-arene PTA complexes. *Journal of Medicinal Chemistry*, **48**, 4161–4171.

51. Maksimoska, J., Feng, L., Harms, K. *et al.* (2008) Targeting large kinase active site with rigid, bulky octahedral ruthenium complexes. *Journal of the American Chemical Society*, **130**, 15764–15765.

52. Castellano-Castillo, M., Kostrhunova, H., Marini, V. *et al.* (2008) Binding of mismatch repair protein MutS to mismatch DNA adducts of intercalating ruthenium (II) arene complexes. *Journal of Biological Inorganic Chemistry: JBIC: A Publication of the Society of Biological Inorganic Chemistry*, **13**, 993–999.

53. Melchart, M., Habtemariam, A., Novakova, O. *et al.* (2007) Bifunctional amine-tethered ruthenium(II) arene complexes form monofunctional adducts on DNA. *Inorganic Chemistry*, **46**, 8950–8962.

54. Casini, A. *et al.* (2008) Emerging protein targets for anticancer metallodrugs: inhibition of thioredoxin reductase and cathepsin B by antitumor ruthenium(II)-arene compounds. *Journal of Medicinal Chemistry*, **51**, 6773–6781.

55. Schatzschneider, U., Niessel, J., Ott, I. *et al.* (2008) Cellular uptake, cytotoxicity, and metabolic profiling of human cancer cells treated with ruthenium(II) polypyridyl complexes [Ru(bpy)$_2$(N-N)]Cl$_2$ with N=bpy, phen, dpq, dppz, and dppn. *ChemMedChem*, **3**, 1104–1109.

56. Liu, J., Zheng, W., Shi, S. *et al.* (2008) Synthesis, antitumor activity and structure-activity relationships of a series of Ru(II) complexes. *Journal of Inorganic Biochemistry*, **102**, 193–202.

57. Tan, C., Liu, J., Chen, L. *et al.* (2008) Synthesis, structure and characterization, DNA binding properties and cytotoxicity studies of a series of Ru(III) complexes. *Journal of Inorganic Biochemistry*, **102**, 1644–1653.

58. Mishra, L., Yadaw, A.K., Bhattacharya, S., and Dubey, S.K. (2005) Mixed-ligand Ru(II) complexes with 2,2′-bipyridine and aryldiazo-beta-diketonato auxiliary ligands: Synthesis, physico-chemical study and antitumor properties. *Journal of Inorganic Biochemistry*, **99**, 1113–1118.

59. Karki, S.S., Thota, S., Darj, S.Y. *et al.* (2007) Synthesis, anticancer, and cytotoxic activities of some mononuclear Ru(II) compounds. *Bioorganic and Medicinal Chemistry*, **15**, 6632–6641.

60. Grguric-Sipka, S.R., Vilaplana, R.A., Pérez, J.M. *et al.* (2003) Synthesis, characterization, interaction with DNA and cytotoxicity of the new potential antitumor drug cis-K[Ru(eddp)Cl$_2$]. *Journal of Inorganic Biochemistry*, **97**, 215–220.

61. Chatterjee, D., Mitra, A., Levina, A., and Lay, P.A. (2008) A potential role for protein tyrosine phosphatase inhibition by a Ru(III)-edta complex (edta=ethylenediaminetetraacetate) in its biological activity. *Chemical Communications (Cambridge, England)*, 2864–2866.

62. Chifotides, H.T. and Dunbar, K.R. (2005) Interactions of metal–metal-bonded antitumor active complexes with DNA fragments and DNA. *Accounts of Chemical Research*, **38**, 146–156.

63. Fricker, S.P. (2007) Metal based drugs: From serendipity to design. *Dalton Transactions*, 4903–4917.

64. Köpf, H. and Köpf-Maier, P. (1979) Titanocene dichloride – the first metallocene with cancerostatic activity. *Angewandte Chemie – International Edition in English*, **18**, 477–478.

65. Meléndez, E. (2002) Titanium complexes in cancer treatment. *Critical Reviews in Oncology/Hematology*, **42**, 309–315.

66. Harding, M.M. and Waern, J.B. (2004) Coordination chemistry of the antitumor metallocene molybdocene dichloride with biological ligands. *Inorganic Chemistry*, **43**, 206–213.

67. Keppler, B.K., Friesen, C., Moritz, H.G. *et al.* (1991) Tumor-inhibiting bis(β-diketonato) metal complexes. Budotitane, *cis*-diethoxybis(1-phenylbutane-1,3-dionato)titanium(IV). *Struct Bonding*, **78**, 97–127.

68. Wilkinson, G. and Birmingham, J.M. (1954) Bis-cyclopentadienyl compounds of Ti, Zr, V, Nb, and Ta. *Journal of the American Chemical Society*, **76**, 4281–4284.

69. Toney, J.H. and Marks, T.J. (1985) Hydrolysis chemistry of the metallocene dichlorides M(η^5-C$_5$H$_5$)$_2$Cl$_2$, M = Ti, V, Zr. Aqueous kinetics, equilibria, and mechanistic implications for a new class of antitumor agents. *Journal of the American Chemical Society*, **107**, 947–953.

70. Abeysinghe, P.M. and Harding, M.M. (2007) Antitumor bis(cyclopentadienyl) metal complexes: Titanocene and molydocene dichloride and derivatives. *Dalton Transactions*, 3474–3482.

71. Ravera, M., Cassino, C., Monti, E. *et al.* (2005) Enhancement of the cytotoxicity of titanocene dichloride by aging in organic co-solvent. *Journal of Inorganic Biochemistry*, **99**, 2264–2269.

72. Caruso, F. and Rossi, M. (2004) Antitumor titanium compounds and related metallocenes *Metal Ions in Biological Systems*, **42**, 353–384.

73. Guo, M., Sun, H., McArdle, H.J. *et al.* (2000) Ti^{IV} uptake and release by human serum transferrin and recognition of Ti^{IV}-transferrin by cancer cells: Understanding the mechanism of action of the anticancer drug titanocene dichloride. *Biochemistry*, **39**, 10023–10033.

74. Tinoco, A.D., Eames, E.V., and Valentine, A.M. (2008) Reconsideration of serum Ti(IV) transport: Albumin and transferrin trafficking of Ti(IV) and its complexes. *Journal of the American Chemical Society*, **130**, 2262–2270.

75. Campbell, K.S., Dillon, C.T., Smith, S.V., and Harding, M.M. (2007) Radiotracer studies of the antitumor metallocene molybdocene dichloride with biomolecules. *Polyhedron*, **26**, 456–459.

76. Köpf-Maier, P. (1990) Intracellular localization of titanium within xenografted sensitive human tumors after treatment with the antitumor agent titanocene dichloride. *Journal of Structural Biology*, **105**, 35–45.

77. McLaughlin, M.L., Cronan, J.M., Schaller, T.R., and Snelling, R.D. (1990) DNA–metal binding by antitumor-active metallocene dichlorides from inductively coupled plasma spectroscopy analysis: Titanocene dichloride forms

DNA-Cp$_2$Ti or DNA-CpTi adducts depending on the pH. *Journal of the American Chemical Society*, **112**, 8949–8952.

78. Harding, M.M. and Mokdsi, G. (2000) Antitumor metallocenes: Structure–activity studies and interactions with biomolecules. *Current Medicinal Chemistry*, **7**, 1289–1303.

79. Nakamoto, K. and McCarthy, S.J. (1968) *Spectroscopy and Structure of Metal Chelate Compounds*, John Wiley & Sons Inc., New York.

80. Waldeck, B. (1993) Biological significance of the enantiomeric purity of drugs. *Chirality*, **5**, 350–355.

81. Serpone, N. and Fay, R.C. (1967) Stereochemistry and lability of dihalobis(β-diketonato)titanium(IV) complexes. II. Benzoylacetonates and dibenzoylmethanates. *Inorganic Chemistry*, **6**, 1835–1843.

82. Dubler, E., Buschmann, R., and Schmalle, H.W. (2003) Isomer abundance of bis(β-diketonato) complexes of titanium(IV). Crystal structures of the antitumor compound budotitane [Ti^{IV}(bzac)$_2$(OEt)$_2$] and of its dichloro-derivative [Ti^{IV}(bzac)$_2$Cl$_2$] (bzac = 1-phenylbutane-1,3-dionate). *Journal of Inorganic Biochemistry*, **95**, 97–104.

83. Comba, P., Jakob, H., Nuber, B., and Keppler, B.K. (1994) Solution structures and isomer distribution of bis (β-diketonato) complexes of titanium(IV) and cobalt(III). *Inorganic Chemistry*, **33**, 3396–3400.

84. Schilling, T., Keppler, K.B., Heim, M.E. *et al.* (1996) Clinical phase I and pharmacokinetic trial of the new titanium complex budotitane. *Investigational New Drugs*, **13**, 327–332.

85. Frühauf, S. and Zeller, W.J. (1991) New platinum, titanium, and ruthenium complexes with different patterns of DNA damage in rat ovarian tumor cells. *Cancer Research*, **51**, 2943–2948.

86. Lamboy, J.L., Pasquale, A., Rheingold, A.L., and Meléndez, E. (2007) Synthesis, solution and solid state structure of titanium–maltol complex. *Inorganica Chimica Acta*, **360**, 2115–2120.

87. Hernández, R., Lamboy, J., Gao, L.M. *et al.* (2008) Structure–activity studies of Ti(IV) complexes: Aqueous stability and cytotoxic properties in colon cancer HT-29 cells. *Journal of Biological Inorganic Chemistry*, **13**, 685–692.

88. Thompson, K.H., Barta, C.A., and Orvig, C. (2006) Metal complexes of maltol and close analogues in medicinal inorganic chemistry. *Chemical Society Reviews*, **35**, 545–556.

89. Shavit, M., Peri, D., Manna, C.M. *et al.* (2007) Active cytotoxic reagents based on non-metallocene non-diketonato well-defined C$_2$-symmetrical titanium complexes of tetradentate Bis(phenolate) ligands. *Journal of the American Chemical Society*, **129**, 12098–12099.

90. Buck, D.P., Abeysinghe, P.M., Cullinane, C. *et al.* (2008) Inclusion complexes of the antitumor metallocenes Cp$_2$MCl$_2$ (M = Mo, Ti) with cucurbit[n]urils. *Dalton Transactions*, 2328–2334.

91. Allen, O.A., Gott, A.L., Hartley, J.A. *et al.* (2007) Functionalized cyclopentadienyl titanium compounds as potential anticancer drugs. *Dalton Transactions*, 5082–5090.

92. Strohfeldt, K. and Tacke, M. (2008) Bioorganometallic fulvene-derived titanocene anti-cancer drugs. *Chemical Society Reviews*, **37**, 1174–1187.

93. Bernstein, L.R. (1998) Mechanism of therapeutic activity for gallium. *Pharmacological Reviews*, **50**, 665–680.

94. Base, C.F. and Mesmer, R.E. (1976) *The Hydrolysis of Cations*, John Wiley & Sons, New York, NY.

95. Jakupec, M.A., Galanski, M., Arion, V.B. *et al.* (2008) Antitumor metal compounds: More than theme and variations. *Dalton Transactions*, 183–194.

96. Apseloff, G. (1999) Therapeutic uses of gallium nitrate: Past, present, and future. *American Journal of Therapeutics*, **6**, 327–339.

97. Bockman, R. (2003) The effects of gallium nitrate on bone resorption. *Seminars in Oncology*, **30**, 5–12.

98. Jakupec, M.A. and Keppler, B.K. (2004) Gallium in cancer treatment. *Current Topics in Medicinal Chemistry*, **4**, 1575–1583.

99. Jakupec, M.A. and Keppler, B.K. (2004) Gallium and other main group metal compounds as antitumor agents. *Metal Ions in Biological Systems*, **42**, 425–462.

100. Bernstein, L.R. (Nov. 12. (1996)) United States Patent 5,574,027.

101. Bernstein, L.R., Tanner, T., Godfrey, C., and Noll, B. (2000) Chemistry and pharmacokinetics of gallium maltolate, a compound with high oral gallium bioavailability. *Metal-Based Drugs*, **7**, 33–47.

102. Collery, P., Domingo, J.L., and Keppler, B.K. (1996) Preclinical toxicology and tissue distribution of a novel antitumor gallium compound: Tris(8-quinolinato)gallium(III). *Anticancer Research*, **16**, 687–691.

103. Shen, A.Y., Wu, S.N., and Chiu, C.T. (1999) Synthesis and cytotoxicity evaluation of some 8-hydroxyquinoline derivatives. *The Journal of Pharmacy and Pharmacology*, **51**, 543–548.

104. Matsuba, C.A., Nelson, W.O., Rettig, S.J., and Orvig, C. (1988) Neutral water-soluble indium complexes of 3-hydroxy-4-pyrones and 3-hydroxy-4-pyridinones. *Inorganic Chemistry*, **27**, 3935–3939.
105. Rudnev, A.V., Foteeva, L.S., Kowol, C. *et al.* (2006) Preclinical characterization of anticancer gallium(III) complexes: Solubility, stability, lipohilicity and binding to serum proteins. *Journal of Inorganic Biochemistry*, **100**, 1819–1826.
106. Miller, S.E., Gerard, K.J., and House, D.A. (1991) The hydrolysis of *cis*-diamminedichloroplatinum(II) 6. A kinetic comparison of the cis- and trans-isomers and other cis-di(amine)di(chloro)platinum(II) compounds. *Inorganica Chimica Acta*, **190**, 135–144.
107. Harris, W.R. and Pecoraro, V.L. (1983) Thermodynamic binding constants for gallium transferrin. *Biochemistry*, **22**, 292–299.
108. Beatty, E.J., Cox, M.C., Frenkiel, T.A. *et al.* (1996) Interlobe communication in ^{13}C-methionine-labeled human transferrin. *Biochemistry*, **35**, 7635–7642.
109. Eklund, H., Uhlin, U., Färnegårdh, M. *et al.* (2001) Structure and function of the radical enzyme ribonucleotide reductase. *Progress in Biophysics and Molecular Biology*, **77**, 177–268.
110. Chitambar, C.R., Narasimhan, J., Guy, J. *et al.* (1991) Inhibition of ribonucleotide reductase by gallium in murine leukemia L1210 cells. *Cancer Research*, **51**, 6199–6201.
111. Chitambar, C.R. (2004) Apoptotic mechanisms of gallium nitrate: Basic and clinical investigations. *Oncology (Williston Park)*, **18** (13 Suppl 10), 39–44.
112. Narasimhan, J., Antholine, W.E., and Chitambar, C.R. (1992) Effect of gallium on the tyrosyl radical of the iron-dependent M2 subunit of ribonucleotide reductase. *Biochemical Pharmacology*, **44**, 2403–2408.
113. Chitambar, C.R. (2004) Gallium compounds as antineoplastic agents. *Current Opinion in Oncology*, **16**, 547–552.
114. Chitambar, C.R., Purpi, D.P., Woodliff, J. *et al.* (2007) Development of gallium compounds for treatment of lymphoma: Gallium maltolate, a novel hydroxypyrone gallium compound, induces apoptosis and circumvents lymphoma cell resistance to gallium nitrate. *The Journal of Pharmacology and Experimental Therapeutics*, **322**, 1228–1236.
115. Shakya, R., Peng, F., Liu, J. *et al.* (2006) Synthesis, structure, and anticancer activity of gallium(III) complexes with asymmetric tridentate ligands: Growth inhibition and apoptosis induction of cisplatin-resistance neuroblastoma cells. *Inorganic Chemistry*, **45**, 6263–6268.
116. Chen, D., Frezza, M., Shakya, R. *et al.* (2007) Inhibition of the proteasome activity by gallium(III) complexes contributes to their anti-prostate tumor effects. *Cancer Research*, **67**, 9258–9265.
117. Chen, D., Milacic, V., Frezza, M., and Dou, Q.P. (2009) Metal complexes, their cellular targets and potential for cancer therapy. *Current Pharmaceutical Design*, **15**, 777–791.
118. Kowol, C.R. *et al.* (2007) Gallium(III) and iron(III) complexes of α-N-heterocyclic thiosemicarbazones: Synthesis, characterization, cytotoxicity, and interaction with ribonucleotide reductase. *Journal of Medicinal Chemistry*, **50**, 1254–1265

Further reading

Baes, C.F. and Mesmer, R.E. (1976) *The Hydrolysis of Cations*, John Wiley & Sons, New York, NY.
Ebbing, D.D. and Gammon, S.D. (2009) *General Chemistry*, 9th edn, Houghton Mifflin Co., Boston, MA.
Housecroft, C.E. and Sharpe, A.G. (2007) *Inorganic Chemsitry*, 3rd edn, Pearson Education Limited, Edinburgh Gate, UK.
Huheey, J.E., Keiter, E.A., and Keiter, R.L. (1993) *Inorganic Chemistry. Principles of Structure and Reactivity*, 4th edn, Benjamin-Cummings Publishing Co., San Francisco, CA.
Lever, A.B.P. (1984) *Inorganic Electronic Spectroscopy*, 2nd edn, Elsevier Science Publishers B. V., Amsterdam, The Netherlands.
Miessler, G.L. and Tarr, D.A. (2004) *Inorganic Chemistry*, 3rd edn, Pearson Prentice Hall, Upper Saddle River, NJ.
West, J.B. (1990) *Physiological Basis of Medical Practice*, 12th edn, Williams & Wilkins, Baltimore, MD.
Tinoco, I., Sauer, K., Wang, J.C., and Puglisi, J.D. (2002) *Physical Chemistry. Principles and Applications in Biological Chemistry*, 4th edn, Prentice Hall, Upper Saddle River, NJ.

6

Gold Compounds for Treating Arthritis, Cancer and Other Diseases

Elemental gold is a low-melting lustrous metal that can easily be extracted from the earth and shaped into a variety of different forms. Since the metal is relatively rare and chemically inert it has been a fascinating substance throughout history and has found its way into nearly every human culture in the form of jewelry, ornaments and monetary exchange. While gold compounds carry none of the mystic and allure of the element, they are much more chemically reactive, which makes them useful for applications in medicine. While this chapter is primarily on the use of gold compounds for treating arthritis and cancer, their potential for treating AIDS, malaria and Chagas disease will also be briefly discussed. For additional information on the therapeutic effects of gold, the reader is directed to excellent reviews by Sutton [1], Shaw [2], Messori and Giordana [3,4], Tiekink [5], Kostova [6] and Howard-Lock [7].

6.1 Chemistry of gold in biological media

Gold, atomic number 79, has two oxidation states that are found under biological conditions, Au^+, $[Xe]4f^{14}5d^{10}$, and Au^{3+}, $[Xe]4f^{14}5d^8$. The monovalent oxidation state has a completely filled outer electronic shell, $5d^{10}$, is diamagnetic with $S = 0$, and thus has zero crystal field stabilization energy. With no CFSE, the structures of Au^+ complexes are dominated by steric effects of the ligands and simple electrostatic interactions associated with the formation of the complex. The three principal coordination geometries for the ion are linear two-coordinate, trigonal three-coordinate and tetrahedral four-coordinate (Figure 1.1) [2,3]. Since Au^+ is a large cation with a low oxidation state, its outer electronic distribution is easily polarized, making it a *soft acid*, which prefers to bind to a *soft base*. If the ligands attached to Au^+ are *anionic* and soft bases, such as cyanide, CN^-, or thiolate, RS^-, the geometry with the highest stability is linear two-coordinate – that is, $[Au(CN)_2]^-$ and $[Au(RS)_2]^-$ – where R is an alkyl or aryl group. Linear two-coordination is favored because in order to form a three-coordinate structure, a third *anionic* ligand would need to attack the *anionic* complex, which, for electrostatic reasons, is not less probable. If the ligands are *neutral* and soft bases, such as phosphines, PR_3, or arsines, AsR_3, the Au^+ ion accepts the ligands in a stepwise manner to form linear, $[AuL_2]^+$, trigonal, $[AuL_3]^+$, and tetrahedral, $[AuL_4]^+$, complexes (Figure 1.1), all of which are in equilibrium with each other according to (6.1)–(6.3) [8]. Since the group being added in the metal ion is uncharged, there is no electrostatic impediment for successive ligand addition, except that at some point the coordination environment about the cation becomes sterically crowded, thus making it more difficult to produce species with higher coordination numbers.

$$AuX(PR_3) + PR_3 = [Au(PR_3)_2]^+ + X^- \text{(monovalent anion)} \tag{6.1}$$

$$[Au(PR_3)_2]^+ + PR_3 = [Au(PR_3)_3]^+ \tag{6.2}$$

$$[Au(PR_3)_3]^+ + PR_3 = [Au(PR_3)_4]^+ \tag{6.3}$$

Phosphorous-31 NMR studies show that simple phosphine ligands bound to Au^+ are in rapid exchange with unbound ligand. Since the naturally abundant phosphorous isotope ^{31}P has the same *nuclear spin quantum number*, I, as the proton, $I = \frac{1}{2}$, the gold–phosphine interaction is easily detected and studied using ^{31}P-NMR. With this technique, individual NMR peaks for the various complexes in (6.1)–(6.3) are not found, but rather a single ^{31}P-NMR resonance is observed to shift to new positions as the Au-P ratio is changed [7]. This indicates that the phosphines are in *fast chemical exchange* with the gold ion on the NMR time scale and that the exchange rate constant for the system is $k > \sim 10\,s^{-1}$. Since there is generally ample room for an exchanging (attacking) ligand to approach the metal ion in the complex, exchange processes involving Au^+ are believed to be *associative* in nature.

The outer electronic configuration of Au^{3+} is $5d^8$, which is isoelectronic with Pt^{+2}. Thus, complexes of Au^{+3} are square planar and the occupancy of the d-orbitals is d_{xz}^2, d_{yz}^2 (degenerate), $d_{z^2}^2$, d_{xy}^2, $d_{x^2-y^2}^0$, which makes the complexes diamagnetic; that is, $S = 0$ (Figure 1.5). Since the charge on Au^{+3} is one unit higher than that on Pt^{+2}, crystal field theory predicts that CFSE (Au^{+3}) > CFSE (Pt^{+2}), which is supported by the few cases in which comparisons can be made. While UV–visible absorption spectra would be ideally suited to examining crystal field effects, crystal field transitions for Au^{+3} are usually obscured by the much stronger LMCT bands exhibited by many Au^{+3} complexes. Because of its trivalent charge, Au^{+3} is an *intermediate acid*, so it would be expected to form strong complexes not just with soft bases like thiolates on proteins, but also with bases on the 'harder' end of the HSAB scale, like the nitrogen atoms in the heterocyclic bases of DNA. As is the case with Pt^{+2}, complexes of Au^{+3} are believed to undergo substitution reaction via an *associative* mechanism (1.23).

A property of Au^{+3} that greatly complicates its chemistry is that many of its simple complexes can easily be reduced to Au^+ by a variety of ligands, including thiols and thioethers found on cysteine and methionine residues of peptides and proteins [2,9]. Even the disulfide linkage, RSSR, which is generally considered a poorer ligand than a thiol or a thioether, binds to and reduces Au^{+3} to Au^+ [10]. Since there are agents in the biological system that can oxidize Au^+ to Au^{+3} [2], gold compounds can, in principle, exist in a variety of different coordination states in the biological system. These properties, and the fact that the concentrations of gold compounds normally encountered in therapeutic situations are very low, make it difficult to determine the chemistry of gold in the biological environment.

6.2   Gold compounds for treating arthritis

Rheumatoid arthritis (RA) is a systemic disorder that causes the immune system to attack the joints of the body as well as other organs [11]. The attack on a joint, which ultimately results in the loss of the lubricating fluid (synovial fluid) in the bone–bone interface, causes the joint to swell, deform and lose its articulation. The disease affects about 1% of the population in developed countries, is more common in women than men and becomes worse with age. In its later stages, RA is quite debilitating, causing changes in working and living habits, and ultimately produces functional disability [11].

Medical treatments of RA rely on two types of drug, those that relieve the symptoms without treating the disease, which include aspirin and steroids, and those that control the progression of the disease, which include gold-containing agents, antimalarials, D-penicillamine, sulfasalzine and immunosuppressive drugs such as methotrexate. The second class of drug, often referred to as *disease-modifying arthritis rheumatoid drugs* is given the acronym DMARD [2,3].

The first use of gold compounds for treating RA, a therapy which has now become known as *chrysotherapy*, was described by a French physician, Jacque Forestier [12]. Forestier was aware that sodium dicyanoaurate, $Na[Au(CN)_2]$, was effective in treating tuberculosis and since he believed that rheumatoid arthritis and tuberculosis were in some way connected, he decided to treat RA patients with gold compounds. Since the patients showed a positive response to dicyanoaurate, many related compounds were tested and it was ultimately found that gold thiolate complexes were especially effective at slowing the progression of RA [2,3].

6.2.1 Structures, synthesis and properties of gold antiarthritic drugs

Some gold compounds that are in current use for the treatment of RA are shown in Figure 6.1 and their synthesis is outlined in Figure 6.2. Myochrysine and solganol can be synthesized in water from AuCl and the ligand in the presence of base. The function of the base, for example NaOH, is to deprotonate the carboxylic acid and thiol functional groups of thiomalic acid in the synthesis of myochrysine and to deprotonate the thiol group of thioglucose in the synthesis of solgonal. Since AuCl is prone to oxidation, and obtaining it in a pure state is sometimes difficult, it is also possible to use Au^{+3}, for example $Na[AuCl_4]$, as a starting material in the reaction [13]. In this case, excess thiol ligand is added, which serves as a reducing agent to reduce Au^{+3} to Au^+ *in situ*, with the thiol being oxidized to the disulfide, RSSR. Once Au^+ is formed, the excess thiol in solution reacts with the gold ion to form the drug.

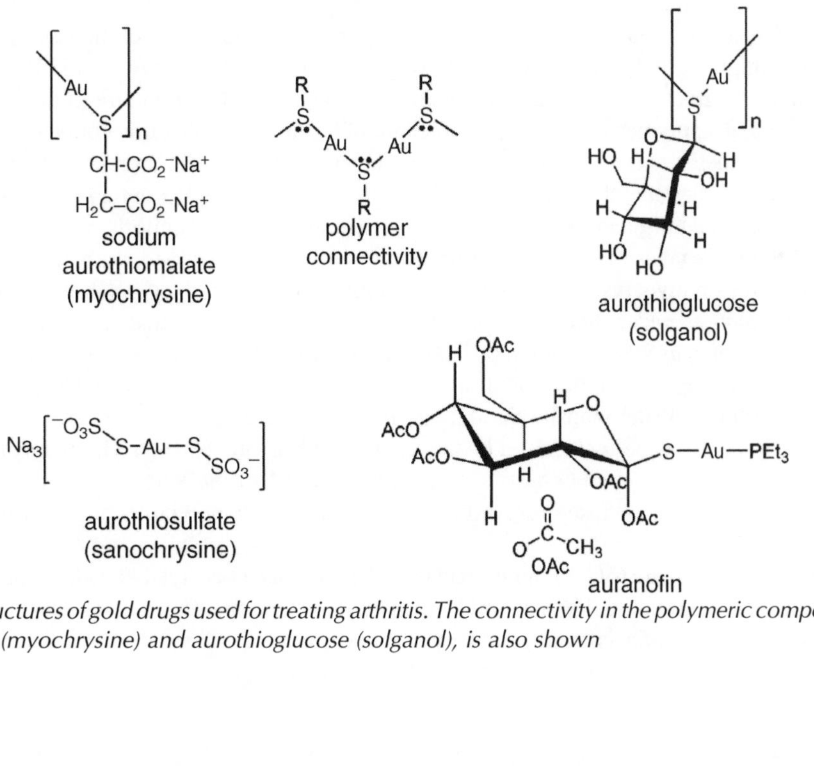

Figure 6.1 *Structures of gold drugs used for treating arthritis. The connectivity in the polymeric compounds, sodium aurothiomalate (myochrysine) and aurothioglucose (solganol), is also shown*

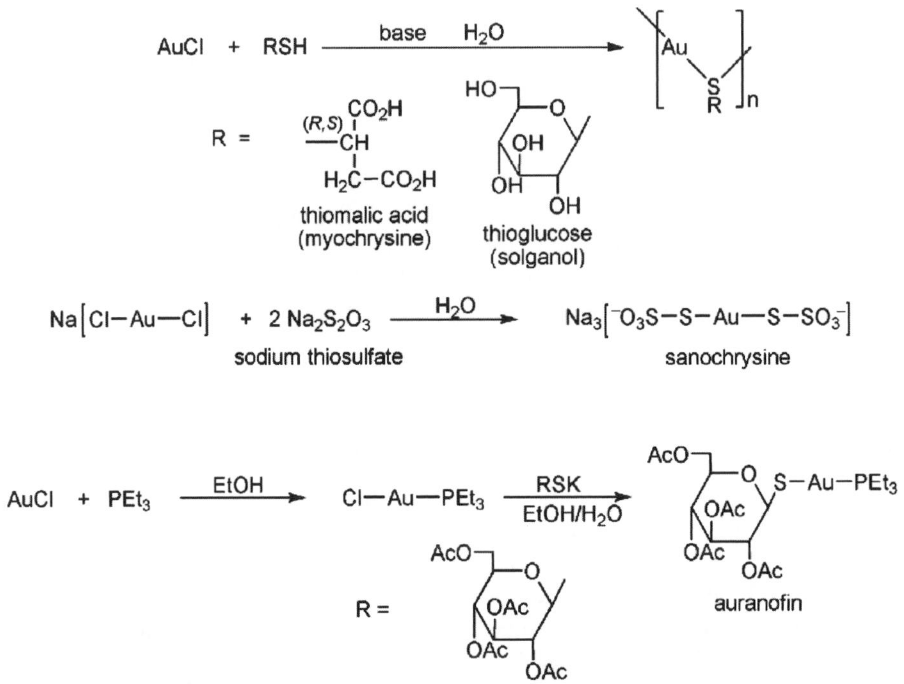

Figure 6.2 *Synthesis of gold drugs that exhibit antiarthritic properties*

As is indicated in Figure 6.1, myochrysine and solganol are actually polymers that are formed by the thiolate ligand bridging (through two of its lone pairs) two Au$^+$ ions to form extended chains. Despite the fact that both drugs are polymers, the deprotonated carboxyl groups of myochrysine and the alcohol function groups on glucose of solgonal allow both drugs to have solubility in water, with solganol the less soluble of the two. While characterizing these polymers proved to be very difficult, in 1998 Robert Bau [14] was able to grow crystals and solve the structure of myochrysine using x-ray analysis. The structure of the drug (Figure 6.3) shows that it is indeed a polymer with an extended chain of S-Au-S units that trace out a helix-type structure with a fourfold symmetry axis (rotation by 90°) (Figure 6.3a). When the structure is turned on its side (Figure 6.3b) it can be seen that the polymerized S-Au-S units actually trace out a *left-hand* helix, which is intertwined with a second left-hand helix (*S* enantiomer). While the S-Au-S units themselves are nearly linear, the 'turning' of the helix takes place at the sulfur atoms. While the crystal structure provides very detailed structural information on myocrysine, it is not clear whether polymers of the type observed in the solid state actually exist in solution under therapeutic conditions. It is more likely that the low concentrations of myochrysine and solganol used in crysotherapy (μM range) cause polymers to break up into smaller *oligomers* or perhaps even monomers. The latter could have a bio-ligand, for example H_2O, Cl^-, or a second thiol, occupying the coordination site opposite the thiolate ligand of the drug [2].

The drug sanochrysine can be synthesized by the reaction of Na[AuCl$_2$] with sodium thiosulfate, as shown in Figure 6.2 [15]. While the thiolate-type sulfur atoms bonded to the Au$^+$ ion could in principle also participate in bridges similar to those observed for myochrysine, polymerization apparently does not occur with this drug [16,17].

The last drug shown in Figure 6.1, auranofin (Ridaura; 2,3,4,6-tetra-*O*-acetyl-1-thio-β-D-glucopyranosato-*S*; triethylphosphine gold(I)) is unusual in that, unlike the previous three agents, it is not very soluble water. As is shown in Figure 6.2, auranofin is synthesized by the reaction of the liner two-coordinate Au$^+$ complex

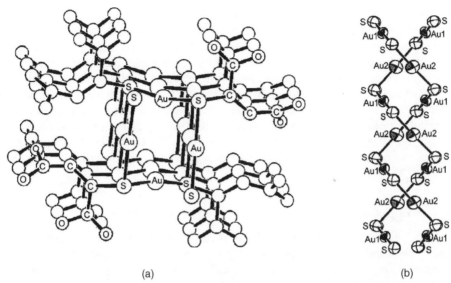

(a)   (b)

Figure 6.3 *(a) View approximately down the fourfold helical symmetry axis of gold(I) thiomalate. A fourfold symmetry axis is one in which the structure can be rotated by 90° (one quarter of a circle (360°)) to obtain a structure that is equivalent to the original orientation. (b) Side view, approximately perpendicular to the fourfold symmetry axis. This view shows two left-hand (left-hand screw) intertwined helices produced by the S-Au-S network when S-thiomalate is used as the bridging ligand. The enantiomer, R-thiomalate, produces a structure identical to the one shown except that the S-Au-S network exists as two right-handed (right-hand screw) intertwined helices. Since the racemic form of malonate was used in the synthesis, there are an equal number of left- (from S-malonate) and right- (from R-malonate) hand intertwined helices in the crystal lattice. Reprinted with permission from R. Bau, Crystal Structure of the Antiarthritic Drug Gold Thiomalate (Myochrysine): A Double-Helical Geometry in the Solid State, J. Am. Chem. Soc. 120, 9380–81. Copyright 1998 American Chemical Society*

[AuCl(PEt$_3$)] with tetra-acylated thiol-glucose in ethanol [18]. Unlike myochrysine and solganol, one side of the two-coordinate Au$^+$ complex is 'blocked' by a strongly-binding triethylphosphine ligand, so the acylated thioglucose can only occupy one of the coordination sites, producing a monomeric linear two-coordinate Au$^+$ complex.

6.2.2   Formulation, administration and pharmacokinetics of gold antiarthritic drugs

Sodium aurothiomalate (myochrysine), aurothioglucose (solganol) and aurothiosulfate (sanochrysine) are water-soluble polymeric compounds that are administered to the patient by injection, so-called *injectable* or *parenteral drugs*, while auranofin, which is only slightly soluble in water, is given to the patient *orally* in capsule form [19–21].

The injectable gold drugs are given in weekly or biweekly dose ($\sim$50 mg) over a long period of time, 4–6 months, and the patient is monitored for remission and/or toxic effects. The result of the monitoring is that the dose is either changed or, if there are toxic side effects, the therapy is discontinued [19]. The initial serum half life of the parenteral drugs is $\sim$7 days but as therapy is continued, Au slowly accumulates in all tissues of the body, with the sites of inflammation seeming to have the highest levels of gold. Studies with myochrysine and solganol, which have been labeled with two radioactive isotopes, ^{195}Au and ^{35}S, show that in mice and rats the thiol ligands are rapidly displaced from the Au$^+$ ion, with most of the gold being excreted in the urine.

Auranofin is administered once, twice or three times daily (typically a total of ~6 mg/day in capsules) and the patient is monitored for signs of toxicity or therapeutic response. Since the effects of the drug are slow to become apparent, auranofin chrysotherapy can last for a considerable period, for example 1 year [20,21]. While 20–30% of the oral dose of the drug or its biotransformation products are absorbed through the GI tract, studies with a triple radioactively-labeled auranofin, ^{195}Au, ^{35}S (acylthioglucose) and ^{32}P (phosphine) show that components of the drug are separated at different rates *in vivo*. The acylthioglucose group is quickly separated from the gold ion and, while the gold–phosphine fragment is longer-lived, the triethylphosphine ligand is ultimately displaced from the metal ion and is oxidized by biologically-occurring oxygen to the phosphine oxide, $Et_3^{32}P{=}O$. The half lives for excretion of $Et_3^{32}P{=}O$, ^{35}S and ^{195}Au from the urine are, 8 hours, 16 hours and 20 days, respectively, indicating that each of the fragments of the drug has a different biological fate [2,19].

6.2.3   Reactions in biological media, uptake and cytotoxicity of gold drugs

The reaction of the gold drugs in biological media has been studied by a number of investigators [2,3]. In the case of auranofin, the stability of the compound with pH is a critical issue because the pH of the stomach, as HCl, is in the acidic range, pH 1–3. Hempel, Pasternack and their coworkers [22] studied the kinetics of reaction of auranofin under conditions that simulated the environment of the stomach using a *stopped-flow* kinetic technique coupled with detection via optical spectroscopy. With this technique, it is possible to measure the kinetics of fast chemical reactions that have half lives in the millisecond range, $\sim 10^{-3}$ s. The investigators found that the degradation of auranofin depends on both the hydrogen ion and chloride ion concentration in the medium and they proposed the mechanism shown in Figure 6.4. In the first step, which is an equilibrium reaction (Figure 6.4a), the chloride ion attacks the gold ion to produce a three-coordinate Au$^+$ intermediate. In acidic media this complex is protonated, probably at the thiol sulfur, which results in the formation of the protonated thiol sugar and the linear two-coordinate complex [AuCl(PEt$_3$)] (Figure 6.4b). At pH 1.6 with $[Cl^-]=0.5\,M$ the rate constant for the degradation at 25 °C is $\sim 70\,s^{-1}$, $t_{1/2} \sim 1 \times 10^{-2}\,s$, showing that the compound rapidly degrades under these conditions. Since auranofin is not very soluble in water, it is not clear how much of the drug is degraded in the stomach but about 25% of the oral dose is believed to be absorbed from the gastrointestinal tract for distribution throughout the body [19,23].

Albumin, the most abundant protein in blood, $\sim 600\,\mu M$, has a molecular weight of 66 kDa. The protein consists of a single polypeptide chain of 585 amino acids with 17 disulfide bridges and only one free cysteine residue, Cys$_{34}$. Since the thiolate functional group of cysteine is a soft base and Au$^+$ is a soft acid, Cys$_{34}$

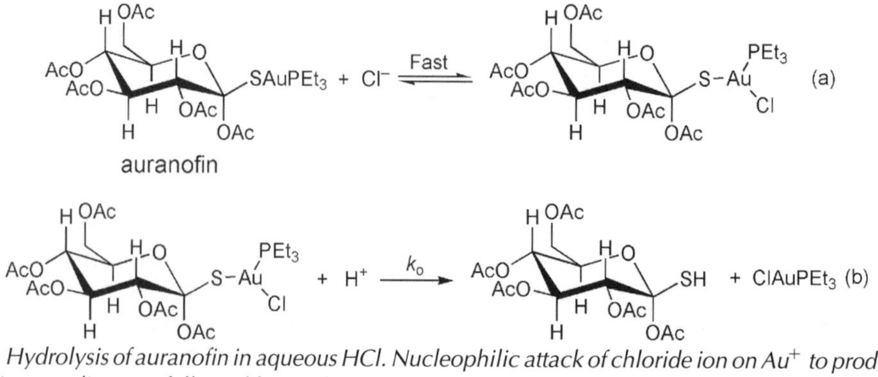

Figure 6.4 *Hydrolysis of auranofin in aqueous HCl. Nucleophilic attack of chloride ion on Au$^+$ to produce a three-coordinate intermediate (a), followed by acid-catalyzed cleavage of the gold thiolate bond (b)*

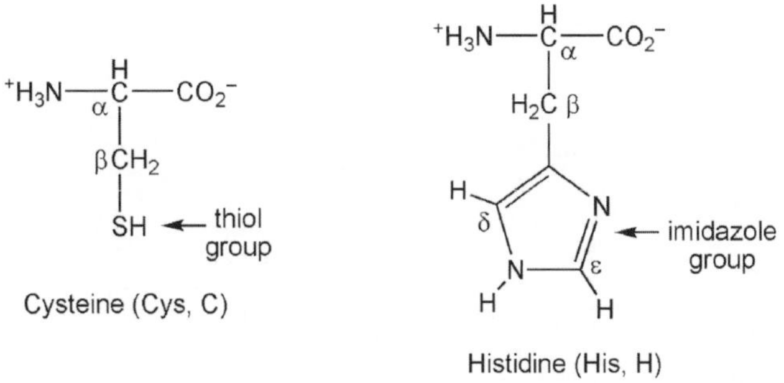

Figure 6.5   *Structures of the zwitter ionic forms of the amino acids cysteine and histidine*

of albumin is an ideal binding site for the gold ion. Shaw and coworkers [24] studied the reaction of auranofin with bovine serum albumin, BSA, and concluded that the drug reacts with Cys_{34} of BSA according to (6.4), where $[Au(SATg)(Et_3P)]$ is auranofin and HSATg is the protonated form of the acylated thioglucose group.

$$BSA + [Au(SATg)(Et_3P)] = [Au(BSA)(PEt_3)] + HSATg \qquad (6.4)$$

The second-order rate constant for the reaction at pH 7.2 (21 °C) is $k_2 = 8 \pm 2 \times 10^2 \, M^{-1} \, s^{-1}$ and since the concentration of orally-administered auranofin in blood is $\sim$10–25 μM, the pseudo first-order rate constant (where the protein is in excess relative to the drug) for the reaction is $k_1 = \sim 0.48 \, s^{-1}$, $t_{1/2} \sim 1$ s. The rapid reaction with BSA suggests that human serum albumin, HSA, might be the means by which the gold ion is distributed throughout the body in chrysotherapy with auranofin.

The interaction of auranofin with HSA, which is present in blood plasma in high concentration (600 μM), was studied by Sadler and coworkers using 1D ^{1}H NMR [25]. Since blood plasma contains many proteins and other molecules with protons with similar chemical shifts, its 1D ^{1}H NMR spectrum is very complicated, with many overlapped resonances. However, since the imidazole group of the amino acid histidine is aromatic, its ^{1}H NMR proton resonances are strongly deshielded and they appear in the low-field region of the NMR spectrum, 6–8 ppm, which allows them to be observed in blood plasma. As shown in Figure 6.5, the imidazole group of histidine has two aromatic protons, one on the δ-carbon atom (Hδ2) and the other on the ε-carbon atom (Hε1), and since the protons are not spin-coupled to each other or other protons in the structure, they appear as singlets in an NMR spectrum.

In studying the interaction of auranofin with HSA, Sadler and coworkers discovered that binding of the $Au(PEt_3)$ fragment of the gold drug to Cys_{34} of the protein caused a resonance associated with Hε1of His$_3$ of HSA (the protein has 10 imidazole groups) to shift to a new position. This observation, which they called the 'n$'$ → n switch', made it possible to detect chemical changes at Cys_{34} of HSA by observing the proton NMR resonance of His$_3$, which is remote from the actual site of gold binding. As shown in Figure 6.5, cysteine has methylene protons adjacent to the thiol functional group, but since there are many methylene-type protons present in HSA and even more in plasma, directly observing the methylene resonances of Cys_{34} to detect chemical changes at the adjacent thiol is not possible.

Shown in Figure 6.6 is the low-field portion of the ^{1}H NMR spectrum of human blood plasma in the absence (A) and presence (B) of 0.4 mol equivalents of auranofin [25]. In addition to numerous proteins, human blood plasma contains small amounts of all of the amino acids and the resonance for Hε1 of free histidine in plasma is at $\sim$7.8 ppm, while the corresponding proton on His$_3$ of HSA is at $\sim$7.7 ppm. When 0.4 mol equivalents of auranofin are added to the solution, the original NMR peak for Hε1 of His$_3$ in HSA, indicated as n$'$, *decreases* in intensity, and a new peak, indicated as n, at slightly lower field appears (B); that is, the n$'$ → n switch occurs. Since the researchers established that the same shift was observed when Cys_{34} was converted into a disulfide, they proposed that the reaction of Cys_{34} with auranofin moved the cysteine

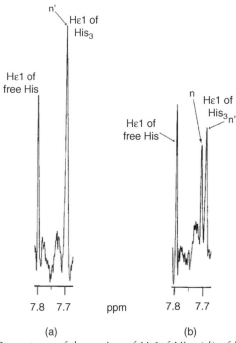

Figure 6.6 *600 MHz ¹H NMR spectrum of the region of Hε1of His₃ (n′) of human blood plasma at pH 7.4. (a) Normal plasma in D₂O. (b) Plasma plus 0.4 mol equivalents (with respect to albumin) of the oral antiarthritic drug, auranofin. The n′ → n 'switch' is produced by Au-PEt₃ of auranofin binding to Cys₃₄ of the protein, which causes a structural change in albumin. This structural change causes His₃ to be relocated to a new chemical environment, which results in a change of the chemical shift of the Hε1of His₃ (n). Reprinted from FEBS Lett. 376, J. Christodoulou et al., 1H NMR of Albumin in Human Blood Plasma: Drug Binding and Redox Reactions at Cys34, 1–5. Copyright 1995, with permission from Elsevier*

residue from a crevice on the inside of a solvent-exposed unstructured loop of HSA to an exposed environment. This is shown in a schematic form in Figure 6.7 [25]. This movement was coupled with the movement of His₃ near the N-terminus of the protein to a new environment (labeled n), causing its Hε1 proton to be slightly deshielded relative to its original location. In the reaction, Cys₃₄ very likely attacks the gold ion

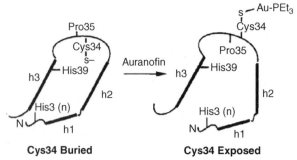

Figure 6.7 *Diagram of the His₃-Cys₃₄ 'switch' of albumin. Binding of Au-PEt₃ to Cys₃₄ causes a structural change in the albumin protein that changes the chemical environment of His₃ near the N-terminus of the protein from n′ (Cys₃₄ buried) to n (Cys₃₄ exposed). Reprinted from FEBS Lett. 376, J. Christodoulou et al., 1H NMR of Albumin in Human Blood Plasma: Drug Binding and Redox Reactions at Cys34, 1–5. Copyright 1995, with permission from Elsevier*

of auranofin to form an unstable three-coordinate Au^+ intermediate, which loses the acyl thioglucose group of the drug to form a two-coordinate, Cys_{34}-$AuPEt_3$ HSA adduct. The investigators pointed out that the $n' \rightarrow n$ switch could be useful for studying disulfide formation at Cys_{34} as well as the binding of other metal ions to this site on HSA in blood plasma.

While it seems unlikely that significant amounts of the gold-based drugs enter the cell intact, at least some fragments of the agents, for example $[Au(PEt_3)]^+$, or their biotransformation products, for example $[Au(CN)_2]^-$, are taken up by cells [2]. As was pointed out above, $[Au(PEt_3)]^+$ becomes attached to Cys_{34} when auranofin reacts with albumin. Although it does not seem likely under biological conditions in which the concentration of the drug is very dilute, if $[Au(HSA)(PEt_3)]$ is attacked by a second gold–phosphine fragment, $[Au(PEt_3)]^+$, an albumin adduct with two $[Au(PEt_3)]^+$ groups attached to Cys_{34} can form, producing [HSA-S $(AuPEt_3)_2]^+$. This chemistry is similar to that observed in the acid hydrolysis of auranofin, in which the initially-released linear two-coordinate complex $[AuCl(PEt_3)]$ (Figure 6.4) attacks auranofin to produce the *bis* gold adduct, $[auranofin-SAu(PEt_3)_2]^+$ [22]. Exposing red blood cells to $[AuCl(PEt_3)]$ shows that $[Au(PEt_3)]^+$ can in fact be internalized by the cells but it is not clear if this is important in the biochemical mechanism of action of the drug.

An interesting biotransformation product of chrysotherapy is the linear two-coordinate complex dicyanogold(I), $[Au(CN)_2]^-$ [26,27]. Cyanide ion, which is one of the strongest metal-binding ligands known, forms a strong complex with Au^+, $\log \beta_2 = 36.6$ [28]. While CN^- is quite toxic, small amounts of cyanide naturally occur in the body, and much higher levels are found in the blood of individuals who smoke tobacco products, which release HCN in the lungs. Dicyanogold(I) is taken up by red blood cells, binds to serum albumin and is found in the urine of patients, independent of their smoking habits, receiving chrysotherapy [2,3,27].

Dicyanogold(I) is formed by the interaction of Au^+ biotransformation products with polymorphonuclear leukocytes (white blood cells with multi-lobed nuclei) but the exact mechanism of its formation is not known. The key enzyme in the production of $[Au(CN)_2]^-$ is myeloperoxidase, which consumes hydrogen peroxide (H_2O_2) to form the oxidizing agents hypochlorite (OCl^-), hypobromite (OBr^-) and hypothiocyanite $(OSCN^-)$ from Cl^-, Br^- and SCN^- (thiocyanate), respectively. Cells in an inflamed area can experience an *oxidative burst*, during which there is rapid but transient production of ROS, for example $O_2^{-\bullet}$ (superoxide), $HO^\bullet$ (hydroxyl radical) and H_2O_2 (hydrogen peroxide). While the detailed chemistry is not known, oxidation of SCN^- can produce HCN, which is believed to be the source of CN^- for the formation of $[Au(CN)_2]^-$. Since $[Au(CN)_2]^-$ electrostatically binds in an equilibrium manner to the anion-binding site of serum albumin, not Cys_{34}, the chemical composition of the gold complex is not changed. Equilibrium binding to albumin allows the compound to be carried to all parts of the body, which explains why dicyanogold(I) is widely distributed in chrysotherapy patients.

An interesting finding by Shaw and coworkers [29] is that $[Au(CN)_2]^-$, in the presence of CN^- and Cl^-, can be oxidized by OCl^- to the square planar compound $[Au(CN)_4]^-$. Since regions of inflammation have an abundance of oxidizing agents, including OCl^-, Au^+ species, which are the result of the administration of antiarthritic drugs, can be biotransformed to Au^{3+} complexes. This finding suggests that the biological effects of chrysotherapy could involve two different oxidation states of gold, which significantly expands mechanistic options for the metal ion in the biological milieu. Since Au^{3+} can be reduced to Au^+ by naturally-occurring thiols such as glutathione (GSH), redox chemistry may also have a role in the mechanism of action of gold-containing antiarthritic drugs [29].

6.2.4   Interactions with cellular targets

Treatment of RA with gold compounds is typically carried out over an extended period of time. In chrysotherapy the patient receives regular doses of the drugs over a six month to one year period and eventually all parts of the body slowly accumulate small amounts of gold. Uncovering the mechanism of action of this type of agent is much more challenging than doing so for, for example, a drug for treating cancer in which relatively few carefully-controlled doses are given and the effects of the drug, at least on the 'chrysotherapy time scale', are 'immediate'. Fast response always makes it easier to establish *cause and effect* or the *causality* associated with the agent and ultimately leads to a clearer picture of how the agent works in the body. In addition to a long treatment period, gold compounds for treating RA have their chemical compositions rapidly modified soon after they enter the biological system. When rapid changes in chemical composition are combined with the possibility of changes in the oxidation state, it is not surprising that, despite many years of research, a clear mechanistic picture of how the gold antiarthritic drugs work in the body has not evolved.

In trying to understand how the agents work, at the molecular level most investigators have studied the interaction of the drugs or their known biotransformation products with potential target molecules, usually in a simple environment that involves the target, the drug and a buffer. In 1979 Lorber and coworkers [30] reported that auranofin inhibited the growth of HeLa cells, thus showing that, besides antiarthritic properties, the drug may have useful anticancer properties. This early report prompted Blank and Dabrowiak [31] to use absorption and circular dichroism spectroscopies to study the interaction of auranofin, myochrysine, solganol $[Au(PEt_3)_2]Cl$ and $[AuX(PEt_3)]$, where X is Cl^- and Br^-, with calf thymus DNA. Since only $[AuCl(PEt_3)]$ and $[AuBr(PEt_3)]$ bound in a nondenaturating fashion to guanine and cytosine sites on DNA, the investigators concluded that only Au^+ complexes with a weakly-bound, displaceable anion, for example Cl^-, which is a hard base, and Br^-, which is an intermediate base, bound to the polymer. A later study by Mirabelli *et al.* [32] used agarose gel electrophoresis to study the binding of auranofin and other gold compounds to pBR322 DNA. While this study also found that auranofin does not bind to DNA – that is, it did not change the mobility of the closed circular form of pBR322 DNA in a gel electrophoresis experiment – it showed that the linear two-coordinate complexes $[AuX(PEt_3)]$, where X is Cl^-, Br^- and SCN^- do indeed bind to and unwind closed circular DNA. Addition of acyl thioglucose, the thiol ligand in auranofin, to the reaction medium inhibited binding of the linear two-coordinate gold complexes to DNA. Since this thiol would be expected to react with the linear two-coordinate gold complexes *in situ* to form auranofin, the gel experiments confirmed the conclusions of the earlier absorption and CD experiments: auranofin does not bind to DNA.

By far the most likely targets for Au^+-containing drugs in the cell are proteins and other small molecules that have thiol or thioether functional groups. This of course was realized early in chrysotherapy, which led to the study of a number of different proteins as potential targets for the gold drugs. One important target may be the *cathepsins*, which are a group of cysteine proteases that hydrolyze elastin in cartilage and collagen in bone and which are believed to play an important role in the resorption (degrading) of bone which occurs in RA. The enzymes are found inside lysomes in osteoclast cells, where gold from chrysotherapy is known to accumulate. Recently, Brömme and coworkers [33] solved the crystal structure of myochrysine bound to the active site of cathepsin K. The structure (Figure 6.8) shows the thiomalate-Au fragment of the drug bound to Cys_{25} of the protein, a residue that is important for the catalytic function of the enzyme. As expected, the geometry around the gold ion is nearly linear (173.1°) and the *S*-enantiomer of thiolmalate (the drug is a racemic mixture of both enantiomers) appears bound to the active site. The selection of only one enantiomer by the protein, a form of *enantiomeric recognition*, is due to the fact that the active site of cathepsin K is a highly chiral environment, which apparently allows one of the enantiomers (*S*) of the drug to preferentially bind to the site (Box 5.3). Using a fluorescence assay, the investigators also showed that both myochrysine and auranofin inhibit the catalytic function of cathepsin K inside human RA synovial fibroblasts (extracellular matrix cells) and monocyte-derived macrophages (immune system cells), with myochrysine being the more effective of the two agents. Since free Cys_{25} is known to be important in the catalytic cycle of the enzyme, the gold drugs appear to inhibit cathepsin K by binding directly to the thiolate ion of Cys_{25}, thereby blocking its ability to act as a nucleophile in protein degradation. Since cathepsins are implicated in the pathogenesis of rheumatoid arthritis, inhibition of their function by gold drugs may be important in the mechanism of action of the gold agents used in chrysotherapy.

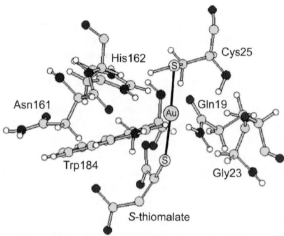

Figure 6.8   *Binding region of myochrysine–cathepsin K adduct. The coordination geometry about the Au⁺ ion is linear (S-Au-S angle = 173.1°), with the thiolate from S-thiomalate and the thiolate from cysteine 25 of the protein as donor atoms to the gold ion. The carboxylates of the thiomalate are involved in electrostatic interactions with nearby residues on the protein. PDB ID: 2ATO, E. Weidauer et al., Effects of Disease-Modifying Anti-Arthritic Drugs (DMARDs) on the Activities of Rheumatoid Arthritis-Associated Cathepsins K and S., Biol. Chem. 2007, 388, 331–336*

Barrios and coworkers [34] also focused on the cathepsins as targets for the gold antiarthritic drugs. Since the Au-P bond of auranofin remains intact longer that the Au-S bond of the drug in the biological system [2], these investigators modified the phosphine portion of auranofin and studied the effects of a series of auranofin analogs on the catalytic activity of cathepsin B. They found that changing the steric and electronic properties of the phosphine ligand affected the catalytic activity of the enzyme by more that three orders of magnitude, suggesting that, at least from the standpoint of inhibiting cathepsin B with auranofin analogs, agents more potent than the parent drug are possible.

Box 6.1 Zinc finger proteins

Zinc finger proteins are a large class of Zn^{+2}-containing metallo-proteins, many of which are *transcription factors* that regulate the expression of genes in the cell. The proteins have two to thirteen 'finger'-type polypeptide structures, each with a coordinated Zn^{+2} ion, that are small enough to bind within the major groove of double-helical DNA. The function of the Zn^{+2} ion in each of the fingers is to organize the amino acids in the finger so that when the structure is bound to DNA, other amino acid residues in the finger can make hydrogen bonds and electrostatic contact with sites on DNA. This allows the finger to 'read' the DNA sequence, and when it is bound it initiates a series of events that ultimately result in the synthesis of messenger RNA from a specific gene and the production of a protein.

Perhaps the most well-studied zinc finger protein is the transcription factor TFIIIA. This protein, which controls the production of a large RNA molecule that is part of the ribosome on which protein synthesis takes place, has nine zinc fingers, each with a Cys_2His_2 donor set of amino acid residues bound to a tetrahedral Zn^{+2} ion. TFIIIA binds to a 50-base-pair section of double-stranded DNA, with its nine zinc fingers 'draped' along the DNA molecule and lying in the major groove. The structure of the complex formed between six of the nine fingers of TFIIIA and a 31-base-pair segment of double-stranded DNA is shown in Figure 6.9 and an enlarged view of one of the fingers, Finger 6, is shown in Figure 6.10 [35,36]. As is evident in the figures, each of the fingers has a protein 'scaffold' that consists of an α-helical and β-sheet structure from which two imidazole residues of histidine (from the α-helix portion) and two cysteine thiolate ions (from the β-sheet portion) are donated to the Zn^{2+} ion.

> The fact that zinc fingers are metalloproteins makes them ideal targets for metallo-drugs that contain metal ions with high affinity for the protein residues normally bound to the Zn^{+2} ion in the finger. Since substitution of the zinc ion in the finger often leads to a change in coordination geometry around the metal ion, which affects the structure of the finger, metallo-drug modification of zinc fingers is an effective way to block transcription, which often leads to the death of the cell.

Zinc finger proteins, which are DNA binding proteins important in gene expression, contain Zn^{+2} bound to one or more thiolate residue of cysteine. Since thiolate is a soft base, its most stable complexes are formed with soft acids. On the HSAB scale, Zn^{+2} is an *intermediate acid*, which means the zinc ion in a zinc finger is thermodynamically susceptible to replacement by a metal ion that is 'softer' than zinc. For example cisplatin, which contains the soft acid Pt^{+2}, can readily displace the Zn^{+2} ion from a zinc finger peptide [37] and there is also evidence that Cd^{+2}, a soft acid, can displace the zinc ion from zinc fingers of DNA repair enzymes [38]. Realizing that a softer metal ion may be able to displace the Zn^{+2} ion from a zinc finger, Larabee *et al.* [39] studied the interaction of the gold-containing antiarthritic drug myochrysine with a 28-mer peptide that is a model of the third finger in the zinc finger transcription factor Sp1. The sequence of the peptide used in the study was KKFACPECPKRFMSDHLSKHIKTHQNKK (Sp1-3), which uses two cysteine thiolates and two histidine nitrogen atoms (imidazole) to coordinate Zn^{+2}. After complexing Zn^{+2} to the peptide, the researchers used mass spectrometry, gel techniques and circular dichroism (CD) to show that the Au^+ ion of myochrysine can displace the Zn^{+2} ion in the Zn-Sp1-3 complex. The CD spectra in Figure 6.11 show that titration of Zn-Sp1-3 with myochrysine causes the intensity of the CD band at $\sim$203 nm to *decrease* while at the same time the CD intensity at $\sim$190 nm *increases*. The spectra also show that all of the curves pass through a single point at $\sim$197 nm, which is called an *isoelliptic point*, where the value of the molar ellipticity, $[\theta]$, for all of the species in solution is the same. This classic set of CD curves most often indicates that there are only two compounds in solution and that one compound, A, in this case Zn-Sp1-3, is being converted into the second compound, B, Au-Sp1-3, and also that both compounds have the same value of $[\theta]$ at $\sim$197 nm. Also evident from the spectra is that the intensity of the CD band at $\sim$203 nm, which is known to be a measure of the α-helical content of a peptide, is *less* for [Au-Sp1-3] (the value of $[\theta]$ is less negative) than it is for the native zinc finger, [Zn-Sp1-3]. This indicates that Au^+ is not as efficient as Zn^{2+} in maintaining the α-helical content

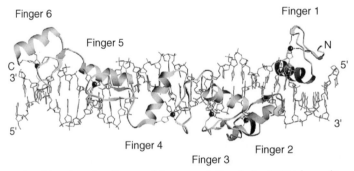

Figure 6.9 *Shown are six of the nine zinc fingers of the transcription factor TFIIIA bound to a 31-base-pair segment of DNA. Each finger consists of a Zn^{+2} ion (black sphere), about which is arranged an α-helix, which donates two imidazole residues of histidine to the metal ion, and a β-sheet-type protein structure, which donates two thiolate ions from two cysteine residues to the Zn^{+2} ion; that is, a Cys_2His_2 type zinc finger. The 3' and 5' ends of the two strands of the DNA double helix and the N- and C-termini of the protein are also indicated. Amino acids in the α-helical portion of the fingers make specific hydrogen bonds and electrostatic contacts with sites located in the major groove of DNA. From R.T. Nolte et al., Differing Roles for Zinc Fingers in DNA Recognition: Structure of a Six-Finger Transcription Factor IIIA Complex, Proc. Natl. Acad. Sci., 95, 2938-43. Copyright 1998 National Academy of Sciences, U.S.A.*

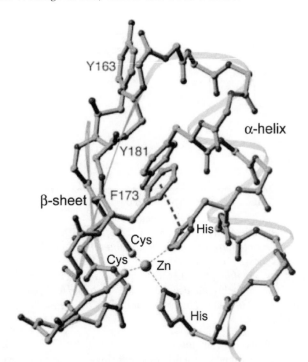

Figure 6.10   *Structure of finger 6 of TFIIIA. The Zn^{+2} ion is in a tetrahedral environment coordinated by N-1 of the imidazole residues of two histidines and the thiolate ions of two cysteine residues to produce a Cys$_2$His$_2$-type coordination environment. For this finger of TFIIIA there is a stacking interaction between one of the histidine donors and a phenylalanine residue (F173), which, in addition to the coordinated Zn^{+2} ion, stabilizes the finger. The α-helix and β-sheet structures and other amino acid residues are also indicated. From D. Lu and A. Klug, Invariance of the Zinc Finger Model: A Comparison of the Free Structure With Those in Nucleic-Acid Complexes, Proteins: Structure, Function and Bioinformatics, 2007, 67, 508–512. Reprinted with permission of Wiley-Liss Inc. a subsidiary of John Wiley & Sons Inc.*

of the peptide, which may be the reason why the gold finger does not bind very well to the interaction site for Zn-Sp1-3 on double-stranded DNA.

In a related study, Franzman and Barrios [40] investigated the interaction of the auranofin biotransformation product, [AuCl(PEt$_3$)], with the zinc finger-type peptides shown in Figure 6.12. When Zn^{+2} is bound to these peptides, the finger structures indicated by the schematic at the bottom of the figure are produced. Titrations of the unmetallated peptides with [AuCl(PEt$_3$)] resulted in the absorption spectra shown in Figure 6.13, showing that gold interacts with the peptides to produce 'gold fingers'. Since the formation of an Au-S bond in the gold finger results in the appearance of a strong LMCT band at ~310 nm, the investigators used this wavelength to measure the amount of gold finger produced in the titration. This was done by plotting the absorbance of the solution at 310 nm, which is proportional to the concentration of Au-S bonds formed in solution, as a function of the concentration of [AuCl(PEt$_3$)] present in solution. By constructing plots of this type, the researchers were able to show that the gold–peptide ratios for all three peptides were not the same. While the CCHH, Cys$_2$His$_2$, peptide binds one Au$^+$ ion per peptide – that is, Au$^+$/CCHH = 1 (Figure 6.13a) – the binding stoichiometries for the remaining two peptides are Au$^+$/CCCC = 2 (Figure 6.13b) and Au$^+$/CCHC = 1.5 (Figure 6.13c). The researchers suggested that the non-integer stoichiometry of CCHC can be explained if two Au$^+$ ions are bound to three peptides. One gold ion could be involved in an *intramolecular* Cys-Au-Cys-type

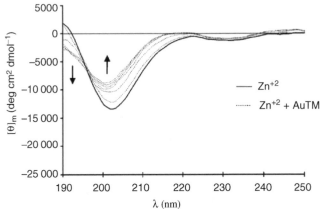

Figure 6.11 *Addition of myochrysine [Au(TM)] to 25 μM [Zn-Sp1-3]. The zinc finger was formed in a 5 mM ammonium acetate buffer, pH 6.8, containing 90 μM DTT (dithiothretol), 38 μM Zn(OAc)$_2$, where OAc is the acetate ion, and 25 μM Sp1-3. The curves shown are for solutions with 5, 15, 25, 30, 38 and 48 μM myochrysine [Au(TM)] in solution, waiting for 5 minutes, and obtaining the CD spectrum. The arrows indicate the direction of the shift in the CD spectrum as the concentration of myochrysine was increased. Reprinted with permission from J.L. Larabee et al., Mechanisms of Aurothiomalate-Cys2His2 Zinc Finger Interactions, Chem. Res. Toxicol, 18, 1943–54. Copyright 2005 American Chemical Society*

bond and the other could be involved in an *intermolecular* Cys$_A$-Au-Cys$_B$ bond, which occurs between two peptides. Using CD, it was determined that while the gold fingers have more ordered secondary structure than the metal-free peptides, they are all less ordered – that is, they produce less α-helical character – than the corresponding zinc fingers. The fact that the metal ions had different effects on the secondary structure of the peptide appears directly related to the fact that Au$^+$ prefers linear two-coordination or possibly trigonal three-coordination while Zn^{+2} prefers tetrahedral coordination.

Bindoli and coworkers [41] showed that auranofin, myochrysine and other complexes of Au$^+$ and Au^{+3} bind to thioredoxin reductase, TrxR, an enzyme responsible for regulating the redox state of the cell (Box 4.3). This enzyme contains a selenocysteine, which in its deprotonated form as selenate ion, R-Se$^-$, is an even softer base than the thiolate ion, R-S$^-$, of cysteine. Gold complexes, as well as platinum, appear to bind to TrxR in mitochondria, which blocks the function of the enzyme and may also impair cellular respiration [42,43]. However, a recent study suggests that different types of Au$^+$ and Au^{+3} complexes can inhibit TrxR at concentrations that do not significantly inhibit mitochondrial function [44].

Since a large number of biological macromolecules have cysteine and, in some cases, selenocysteine residues, it is not surprising that many different potential protein targets for the gold antiarthritic drugs have

CCHH   P Y K C P E C K S F S Q K S D L V K H Q R T H G
CCCC   P Y K C P E C K S F S Q K S D L V K C Q R T C G
CCHC   P Y K C P E C K S F S Q K S D L V K H Q R T C G

Figure 6.12 *Sequences of the three zinc finger model peptides CCHH, CCCC and CCHC, along with a schematic that indicates the secondary structure in the canonical zinc fingers, where a straight arrow indicates a β-sheet, a curved arrow indicates an α-helix and the curved line indicates a flexible loop. Reprinted from M.A. Franzman and A. M. Barrios, Spectroscopic Evidence for the Formation of Goldfingers, Inorg. Chem, 47, 3928–30. Copyright 2008 American Chemical Society*

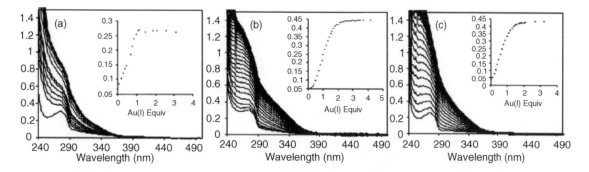

Figure 6.13 *Titartions of reduced zinc finger peptides with [AuCl(PEt₃)] monitored by UV–visible absorption spectroscopy. The insets show increases in A_{310} as a function of added [AuCl(PEt₃)]. (a) Solution of 1.34×10^{-4} M CCHH, coordinates 1 mol equivalent of [AuCl(PEt₃)]; (b) CCCC (1.42×10^{-4} M) and (c) (1.50×10^{-4} M) bind 2 and 1.5 mol equivalent of [AuCl(PEt₃)], respectively. Reprinted with permission from M.A. Franzman and A.M. Barrios, Spectroscopic Evidence for the Formation of Goldfingers, Inorg. Chem. 47, 3928–30. Copyright 2008 American Chemical Society*

been identified and studied. For a more comprehensive discussion of other possible targets, the reader is directed to the excellent review article by Messori and Marcon [3].

6.3   Gold complexes for treating cancer

The connection between cancer and arthritis was made years ago when it was found that the important anticancer drugs 6-mercaptopurine and cyclophosphamide have immunosuppressive and anti-inflammatory properties, which are the same effects exhibited by antiarthritic agents [5]. It was also learned that patients on chrysotherapy had statistically lower incidences of cancer, suggesting that agents used for treating RA may actually be useful for treating cancer. This connection between the two diseases, and the success of the platinum drugs, stimulated considerable research on gold compounds for the treatment of cancer, a brief summary of which will be presented in this section [2,4–6].

6.3.1   Structures, synthesis and properties of gold anticancer agents

The early work on auranofin and its analogs revealed that Au^+ complexes that have phosphine and thioglucose ligands were effective in killing B16 melanoma and P388 leukemia cells in culture [30,45]. One compound that showed a significantly broader range of activity than auranofin and its analogs against a number of different tumor models implanted in mice was the tetrahedral cation $[Au(dppe)_2]^+$, *bis*[1,2-*bis*(diphenyl-phosphino)ethane]gold(I). As shown in Figure 6.14, this complex can be made by the reaction of AuCl with two equivalents of the phosphine ligand in an organic solvent [46,47]. The finding that $[Au(dppe)_2]^+$ is active against cancer led to the synthesis and testing of a number of gold phosphine analogs, some of which are now showing considerable potential as useful antitumor agents [48].

   In the period following the initial reports that Pt^{+2} complexes have potent antitumor properties, researchers quickly realized that complexes of Au^{+3}, which are isoelectronic and isostructural with those of Pt^{+2}, may also be useful for treating cancer. However, unlike complexes of Pt^{+2}, which are stable toward reduction, many Au^{+3} complexes, for example $[AuCl_4]^-$, are easily reduced, especially by thiols, which are commonly found in the biological system. This meant that if the reduction were not blocked, many potential Au^{+3} antitumor complexes

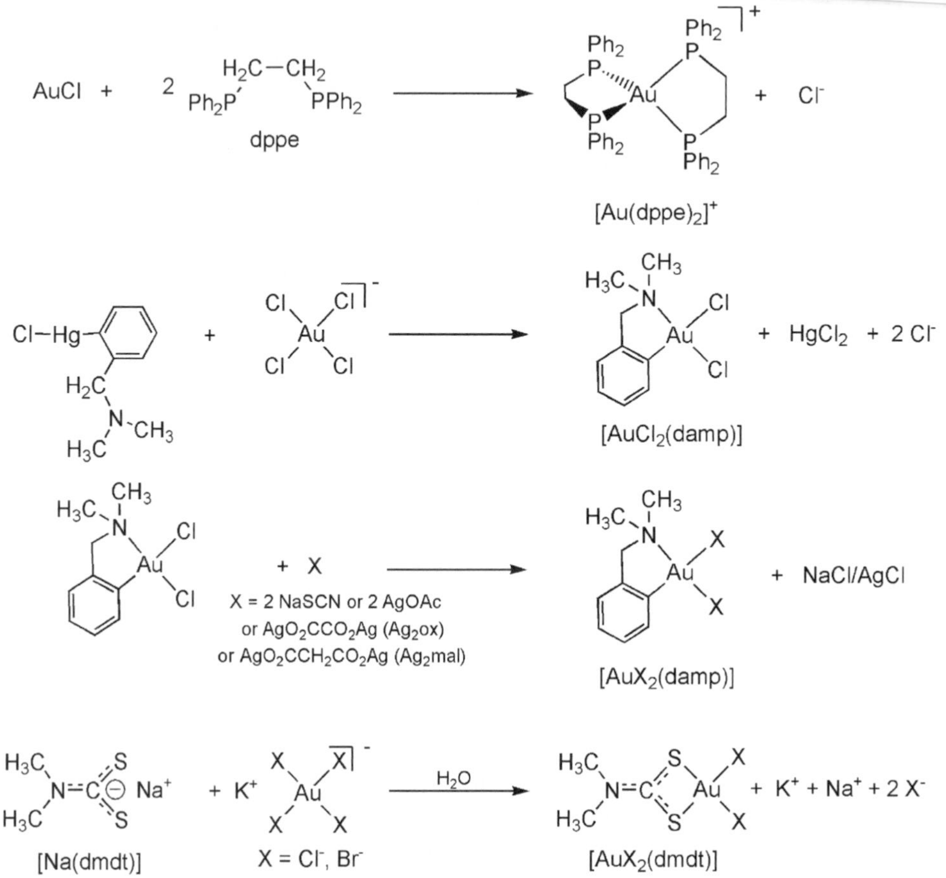

Figure 6.14 Synthesis and structure of gold complexes exhibiting anticancer properties

would quickly be reduced, most likely in the blood, before they had a chance to reach target molecules inside the cell as Au^{+3} compounds. While it is not clear that bioreduction would be detrimental to the efficacy of these agents, investigators worked on ways to stabilize the trivalent oxidation state of the metal ion and at the same time incorporate labile sites that would be essential for the interaction of the compound with biological targets such as proteins and DNA. An early success was the complex [Au(damp)X$_2$], in which X = Cl$^-$, SCN$^-$, OAc$^-$, $\frac{1}{2}$ oxalate (ox) or $\frac{1}{2}$ malonate (mal), dichloro(2-((dimethylamino)methyl)phenyl)gold(III) (Figure 6.14). The dichloro complex can be made by reacting a linear two-coordinate Hg^{+2} precursor complex with [AuCl$_4$]$^-$ in an organic solvent. Once this complex is obtained, the two chloro ligands can easily be exchanged using a *metathesis reaction*, in which one ligand is exchanged for another. This can be done by adding an excess of the sodium salt of the ligand, for example NaSCN, or by adding stoichiometric amounts of an Ag$^+$ salt of the ligand, resulting in the formation of insoluble AgCl, which precipitates from solution, thus driving the reaction to product through a phase change [49]. Since the 'damp'-type complexes have a carbon–metal bond that strongly donates electrons to the metal ion via a σ-bond, thus making the ion more difficult to reduce, the damp ligand effectively stabilizes Au^{+3}. Another attractive feature of the damp complexes is the presence of two easily-displaced ligands, X = Cl$^-$, SCN$^-$ or OAc$^-$, in *cis* sites of the compound. This of course follows the paradigm established by cisplatin, indicating that the presence of *cis*-leaving ligands is important for anticancer activity. While the bidentate chelates, oxalate (ox) and malonate (mal), are more difficult to displace than the monodentate ligands, previous work on cisplatin and its analogs indicates that these derivative would likely be biologically active as well. The damp complexes show activity against bladder and ovarian cancer cell lines and the acetato and malonato damp complexes exhibit activity comparable to cisplatin itself against human HT1376 bladder cancer grafted to animals (as *xenografts*) [3,50,51].

The successes with the damp compounds led to the synthesis and testing of other Au^{+3} complexes that were also stabilized by electron-donating ligands. Fregona and coworkers [52,53] synthesized an interesting group of square planar Au^{3+} complexes containing the dithiocarbamate ligand, one of which, $[Au(dmdt)X_2]$, where dmdt is N,N-dimethyldithiocarbamate and $X = Cl^-, Br^-$, is shown in Figure 6.14. These compounds were made by reacting the sodium salt of the dithiocarbamic acid with the potassium salt of Au^{+3} halides in water. Dithiocarbamate is a negatively-charged ligand which donates two sulfur atoms to form a four-membered chelate ring with a metal ion. The electronic structure of the ligand is such that it strongly donates electron density to the metal ion via a σ mechanism, so it, like the damp ligand, stabilizes the trivalent oxidation state of gold. The structures of these compounds also follow the cisplatin paradigm in that they have ligands in *cis* positions that are easily displaced by reaction with biological molecules. The compounds were found to be much more cytotoxic than cisplatin and, since they have activity against cisplatin-resistant lines, their mechanism of action may be different from those of platinum drugs.

In order to increase the stability of Au^{+3} in the biological milieu, Che and coworkers synthesized the square planar complex, [Au(TTP)]Cl, with the well-known tetraphenylporphyrin ligand bound to Au^{+3} (Figure 6.15) [54–56]. Due to the macrocyclic effect, the complex is very stable, and since Au^+ would prefer linear two-coordination, trigonal three-coordination or tetrahedral four-coordination, and the porphyrin ligand cannot easily accommodate any of these geometries, [Au(TTP)]Cl resists reduction to Au^+ and is stable in the biological environment. *In vitro* screening of [Au(TTP)]Cl showed that the compound has high anticancer activity with values of IC_{50} in the submicromolar range for a 48 hour incubation against cisplatin- and multidrug-resistant cells with the highest activity (greater than 100 times higher than cisplatin) against nasopharyngeal carcinoma.

6.3.2   Reactions in biological media, uptake and cytotoxicity of gold anticancer agents

Using ^{31}P NMR, Berners-Price and Sadler studied the thermodynamic and kinetic stability of $[Au(dppe)_2]^+$ in solution, finding that the compound has unusually high thermodynamic stability and that it resists ligand

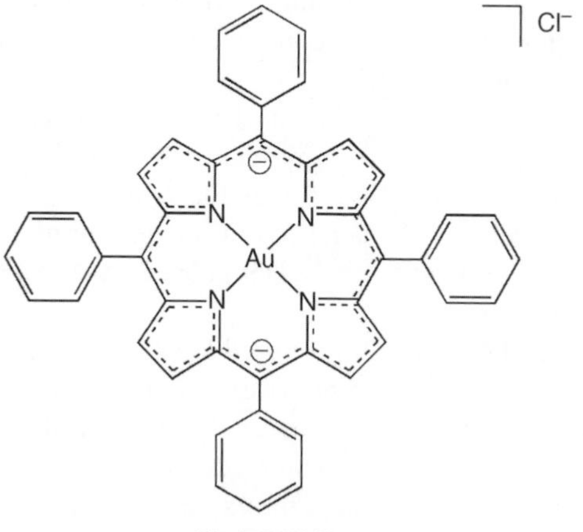

[Au(TPP)]Cl

Figure 6.15   *Structure of [Au(TPP)]Cl, which has anticancer properties*

exchange [47]. These properties appear related to the presence of five-membered chelate rings and the fact that the phosphorous atoms have bulky phenyl groups which block nucleophiles from attacking the gold ion. This makes the compound quite resistant to attack by thiols and it is able to survive the harsh nucleophilic environment of blood [57,58]. Studies using inductively-coupled plasma mass spectrometry (ICP-MS) show that $[Au(dppe)_2]^+$ is readily taken up by CH-1 cells and that uptake is related to the high lipophilicity of the compound. Although the compound is a monovalent cation, its exterior surface is covered by hydrophobic phenyl and methylene groups, which apparently allows it cross the lipid bilayer of the cellular membrane [57]. The compound is active alone, and in combination with cisplatin, against P388 leukemia in mice, and it is also active against various sarcomas in mice [59].

The bulk of the mechanistic studies on the damp complexes have focused on the bis-acetato analog, $[Au(OAc)_2(damp)]$ [60]. This compound, which is the most water-soluble of the complexes shown in Figure 6.14 ($>5.2\,mM$, $25\,°C$), readily loses the two acetate ligands in water to form the monoaqua and diaqua complex. While no detailed studies have been done on the reaction of this compound with components in blood, it would probably react with albumin and other thiol-containing proteins in serum, for example the immunoglobins. In the early studies of the antitumor properties of the damp complexes, the acetato analog $[Au(OAc)_2(damp)]$ and the malonato derivative $[Au(mal)(damp)]$, administered i.p. (intraperitoneally), were found to have significant activity against bladder and ovarian cancers that were subcutaneously implanted (as xenografts) in mice [50].

6.3.3 Reactions with cellular targets

Biological targets for $[Au(dppe)_2]^+$ and related Au^+-phosphine compounds are not known but DNA as a target seems unlikely [31]. Since Au^+ is a soft acid, it would prefer to bind to a soft base like a free thiol in a protein, but since $[Au(dppe)_2]^+$ is quite stable in blood, which contains many potential thiol nucleophiles, these dppe type compounds seem to be relatively resistant to attack by thiols. While no specific target molecules for $[Au(dppe)_2]^+$ have been identified, most researchers believe that the site of action for the compound is the mitochondria in the cell, but the exact mechanism by which the compound attacks this organelle is unknown [57,58]. Early studies with gold–phosphine complexes suggested that they produce breaks in DNA and that they can serve as a link (bridge) that allows the attachment of a protein molecule to DNA [58].

Since Au^{+3} is an *intermediate acid* on the HSAB scale, a number of studies have focused on whether and to what extent it can bind to DNA that contains *intermediate bases* – that is, nitrogen atoms on the heterocyclic bases – and *hard bases* – phosphate oxygen atoms on the backbone of the polymer. While most of these binding studies have been carried out in simple buffer media containing only a gold complex and DNA, they outline the types of interaction that may take place in a cell.

Shown in Figure 6.16 is a group of Au^{+3} complexes that have been studied in connection with their ability to bind to DNA. The complex $[AuCl_3(py)]$, where py is pyridine, was found by Mirabelli *et al.* [32] to bind to and unwind supercoiled pBR322 DNA, demonstrating that the complex can clearly change the degree of supercoiling of DNA. Since the altered DNA persisted in the gel during migration in the electrophoresis experiment, the 'off' rate of the gold complex from the DNA appeared to be slow enough that DNA unwinding could easily be detected in the gel experiment.

A year after that report, Ward and Dabrowiak [61] used DNA sequencing methodology to show that the complex $[AuBr_3(PEt_3)]$ binds to N-7 of the guanine bases of double-stranded DNA. This was done by first binding the gold complex to a DNA fragment that was labeled at one end of one strand with radioactive ^{32}P. This isotope of phosphorous emits a β-particle, which can easily be detected in a sequencing gel using photographic and other techniques. In the next step, the experimenters added the alkylating agent dimethyl sulfate (DMS), which was known to specifically attack N-7 of G and produce a strand break at all G sites on double-stranded DNA. By comparing the DNA breakage patterns in the presence and absence of the gold

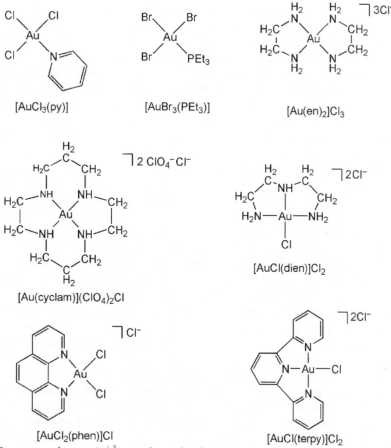

[AuCl₃(py)]   [AuBr₃(PEt₃)]   [Au(en)₂]Cl₃

[Au(cyclam)](ClO₄)₂Cl   [AuCl(dien)]Cl₂

[AuCl₂(phen)]Cl   [AuCl(terpy)]Cl₂

Figure 6.16   *Structures of some Au⁺³ complexes that have been studied in connection with binding to DNA*

complex using a DNA sequencing gel, the researchers concluded that the gold complex blocked alkylation at all G sites, which meant that the gold complex was most likely directly bonded to the N-7 of G, the site on the guanine base that is being attacked by DMS in the DNA breakage reaction. Since most of the G sites were isolated – that is, not adjacent to another G – the study indicated that monoadducts between the gold compound and DNA were the major lesions formed in the interaction.

More recently, Messori *et al.* [62] used absorption and circular dichroism spectroscopy plus various filtration techniques to study the interaction of five Au⁺³ complexes with DNA. The basic approach was to first incubate polymeric calf thymus DNA in a buffer with the gold compound using different ratios of gold to DNA base pairs and, at various times after mixing the gold complex and DNA, to collect absorption and CD data. After examining the data, the authors concluded that the complexes studied did indeed bind to DNA (the site of binding was not specified), and while binding constants were not measured, that the order of the affinities was [AuCl(terpy)]Cl₂, ~ [AuCl(dien)]Cl₂ ≫ [AuCl₂(phen)]Cl > [Au(en)₂]Cl₃ > [Au(cyclam)](ClO₄)₂Cl. By filtering solutions of the gold complex plus DNA several times through a filter that only allowed a small molecule like the gold complex to pass through its holes, the authors concluded that the gold complex slowly became unbound from the calf thymus DNA. By continually removing unbound gold complex from solution, the equilibrium is driven in favor of free complex and free DNA. Since more gold was removed from DNA than platinum (as cisplatin) using this procedure, the 'off rate' of the gold complex from DNA appears to be faster than the corresponding off rate for platinum.

The binding of the dithiocarbamate complex [AuCl$_2$(dmdt)] to purified calf thymus DNA in a buffer was studied by Fregona and coworkers [52,53]. In this case the authors used UV–visible spectroscopy to measure the absorbance of a solution at 311 nm, which is due to a $\pi^* \leftarrow \pi$ electronic transition associated with the CSS group of the dithiocarbamate ligand. Since this transition is quite strong, $\varepsilon \sim 2300\,M^{-1}\,cm^{-1}$, and it is separated from the $\pi^* \leftarrow n$ and the $\pi^* \leftarrow \pi$ transitions of the heterocyclic bases of DNA at $\sim$260 nm, the intensity of the dithiocarbamate absorption can be used as a measure of the amount of the dithiocarbamate ligand (which is attached to the gold ion) that is present in the solution. After separating free gold complex from DNA-bound gold complex at various times after mixing, the authors concluded that [AuCl$_2$(dmdt)] binds to DNA with an 'on rate' that is much faster than cisplatin, but no rate constants for the interaction were given.

Fregona and coworkers [53] also compared the number of *interstrand crosslinks, ISC*, which are created by the gold ion bridging two opposing DNA strands, that are produced by [AuCl$_2$(dmdt)] vs. cisplatin. This was done by exposing calf thymus DNA to either the gold complex or cisplatin for various times, heating the metallated DNA to 100 °C for a brief period of time, which separates the two DNA strands – that is, melts the DNA – followed by cooling the DNA solution to room temperature. If the metal ion links two DNA strands, melting will never completely separate one strand from its compliment (the crosslink will keep the strands connected). When the system is cooled, the bases on the two opposing strands will easily find their compliments to reform duplex DNA. If there is no crosslink between the strands, melting will completely separate the strands in solution and the possibility of one strand finding its exact complement when the solution is cooled is remote. The easy way to detect when two strands of DNA are in a duplex form is to add the dye ethidium bromide, which intercalates between the base pairs of DNA and produces a strong fluorescence signal that can easily be measured. Since the intensity of the signal is a measure of the amount of double-stranded character that is present in the DNA molecules in solution, the percentage of ISC (compared to a control in which no metal complex is present) can easily be measured. Figure 6.17 shows that [AuCl$_2$(dmdt)] and four other Au^{+3} dithiocarbamates form ISC, and do so at a faster rate than that of cisplatin [53]. Since DNA polymerases in the cell require that the two stands of DNA be easily separated in order to synthesize RNA in transcription, and the gold compounds do not allow the strands to be completely separated, ISC formed by the gold compounds would be lethal to the cell.

Fregona and coworkers [53] also found that the gold complexes used in the study rapidly disrupted the membrane of red blood cells, causing the release of cytosolic components to the surrounding environment; that is, they caused cell lysis. Since this behavior is not observed with cisplatin, the authors suggested that the gold complexes may not target DNA, and that they may have a mechanism of action that is very different than that of cisplatin.

While the mechanism of action of [Au(TTP)]Cl is not known, the compound alters the biosynthesis of enzymes involved in energy production and proteins responsible for maintaining the redox balance in the cell, which suggests that mitochondria may be the site of attack [54]. This is also consistent with the observed compound-induced release of cytochrome c from mitochondria and the activation of caspase-9 and caspase-3, enzymes in the apoptotic pathway. The gold–porphyrin complex also binds to calf thymus DNA with a relatively high binding constant of $\sim$10$^6\,M^{-1}$, but the mode of binding to the polymer is unknown.

6.4   Gold complexes for treating AIDS and other diseases

As if activity against two major diseases were not impressive enough, gold compounds may also be useful for treating human immunodeficiency virus (HIV), which causes AIDS. The compound aurocyanide, [Au(CN)$_2$]$^-$, which is a biotransformation product in chrysotherapy [63], has been found to inhibit

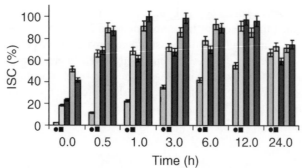

Figure 6.17   *Percentage of interstrand crosslinks (ISC) on calf thymus DNA as a function of time at 37°C, in a pH 7.4 phosphate buffer, in the presence of either cisplatin (solid circles under bars) or [AuCl$_2$(dmdt)] (solid squares under bars) using r = 0.05, where r = [Agent]/[DNA-nt]. The additional bars in the plot are associated with other compounds in the study. Adapted with permission from L. Ronconi et al., Gold(III) Dithiocarbamate Derivatives for the Treatment of Cancer: Solution Chemistry, DNA Binding, and Hemolytic Properties, J. Med. Chem. 49, 1648–1657. Copyright 2006 American Chemical Society*

proliferation of HIV in a strain of CD4$^+$ T cells, a type of white blood cell (lymphocyte) that plays an important role in the immune system. The concentrations of [Au(CN)$_2$]$^-$ at which inhibition occurs are in the 20 n*M* range, which is similar to the concentration of this complex produced in chrysotherapy [2]. A gold complex with two attached thioglucose ligands has been shown to protect MT-4 cells from the HIV virus by binding to a specific cysteine residue on a 120 kDa protein, gp120, which is part of the outer envelope of the virus [64]. The gold compound may work by binding to a cysteine residue on the protein, causing the release of the gp120 protein from the surface of the virus and thereby preventing HIV from recognizing its target cell.

The compound [bpza][AuCl$_4$], where bpza is the diprotonated-chloride form of a bis-pyrazole ligand, inhibits both reverse transcriptase, RT, and HIV-1 protease, PR [65]. Since these enzymes function differently in the life cycle of the HIV virus, inhibiting both with a single compound is unusual. Reverse transcriptase is responsible for converting viral RNA into double-stranded DNA prior to the integration of the latter into genomic DNA of the T cell, and HIV-1 protease controls the maturation and production of infectious virons (virus particles). Although the compound contains both [AuCl$_4$]$^-$ and [bpza] as separate, uncomplexed entities, the presence of both is required for biological activity against RT and PR. Since [bpza][AuCl$_4$] is nontoxic toward peripheral blood mononuclear cells in the immune system, the compound may have potential as an anti-HIV agent.

Malaria is a major challenge on the international scale and, owing to the frequent appearance of resistant strains, there is a continuing need for new drugs and agents for treatment of the disease. A recent report by Sannella *et al.* [66] showed that auranofin and other gold compounds inhibit the growth of *Plasmodium falciparum*, a protozoan parasite carried by *Anopheles* mosquitoes that causes malaria. The researchers suggested that the mechanism by which the gold compounds inhibit the growth of *P. falciparum* is related to the ability of the complexes to block the function of the enzyme thioredoxin reductase, TrxR (Box 4.3). Since *P. falciparium* is known to be sensitive to oxidants released in the cell during *oxidative stress* and TrxR helps to reduce the level of these oxidants, inhibiting the enzyme with gold compounds would alter the *redox balance* of the cell, which could be the reason the gold compounds are effective in killing the parasite.

Chagas disease is a disease common to tropical South America and other tropical regions of the world that is transmitted to humans by certain blood-sucking insects that are able to inject a flagellated protozoan called *Trypanosoma cruzi* into the body [66]. The manifestation of the disease, which takes years to develop, includes heart disease, deformation of intestines and other disorders, and if the disease is left untreated, those afflicted with it often die. A recent study by Fricker *et al.* [67] examined the effects of a series of Au^{+3}, Pd^{+2} and Re^{+5}

complexes on the enzymatic activity of cathepsin B and cysteine proteases from *T. crizi* and a related parasite, *L. major*. They also examined the effects of the compounds on macrophage cells that were infected with *T. crizi*. While the lone Au^{+3} compound tested against infected macrophage cells had little effect on the life cycle of the parasite, all of the gold compounds, including three 'damp'-type complexes (Figure 6.14), inhibited the catalytic function of the cysteine proteases. The researchers concluded that metal complexes of Au^{+3}, Pd^{+2} and Re^{+5} that target parasite cysteine proteases show promise for the treatment of Chagas disease and the related disorder, *leishmaniasis*.

Problems

1. The first step in the release of the acylthioglucose group from auranofin is the rapid formation of a three-coordinate Au^+ intermediate, reaction (a), followed by a rate-limiting proton-assisted scission of the gold–sulfur bond, reaction (b). The equilibrium constant, K, for reaction (a) at $31\,°C$ is $0.30\,M^{-1}$.

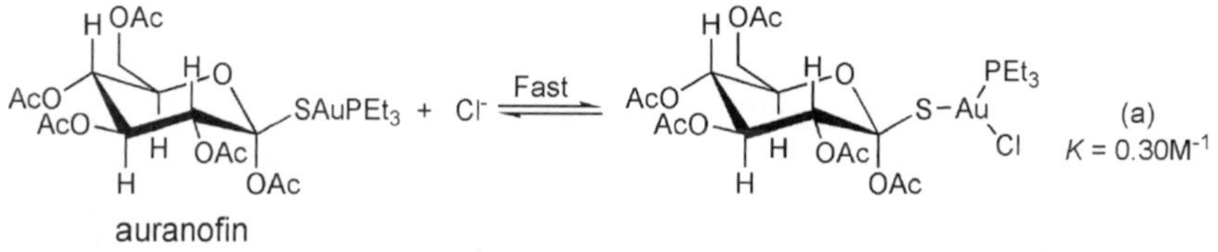

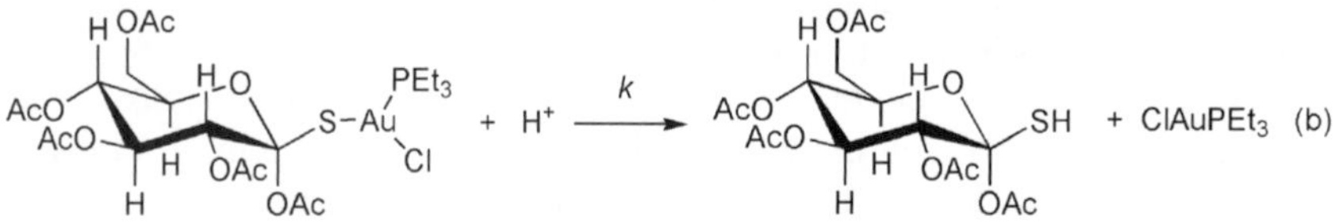

    a. An oral dose of auranofin produces a drug concentration in the stomach of $\sim 10^{-5}\,M$. If, in the stomach, $pH = 1$ in hydrochloric acid, calculate the equilibrium concentration of the three-coordinate intermediate from the information given.
    b. If $\Delta H°$ for reaction (a) is $-8\,kJ\,mol^{-1}$, calculate $\Delta S°$ for the reaction at $31\,°C$. Is the reaction favored or disfavored by entropy?
    c. The degradation of auranofin in perchloric acid, $HClO_4$, is considerably slower than in hydrochloric acid, HCl, in water. Based on the proposed degradation mechanism, briefly explain why this result may be expected.

2. The linear two-coordinate Au^+ complex $[AuCl(PEt_3)]$, where PEt_3 is triethylphosphine, is implicated in the reactions of the antiarthritic drug *auranofin*. When this complex reacts with nucleophiles by losing the chloride ligand, will it form a more stable complex with the ε-amino group (NH_2-R) of the amino acid lysine or the thiol group (HS-R) of the amino acid cysteine? Briefly explain your choice.
3. Using Table 1.1, which gives the one-electron energies of the *d*-orbitals in different crystal field environments, calculate the approximate CFSE, in terms of Δ_o and P the pairing energy, for the square planar complex $[Au(en)_2](ClO_4)_3$, where en is 1,2-diaminoethane, $H_2NCH_2CH_2NH_2$.
4. Shown below is the amount of bromide ion released to solution from the anticancer agent $[Au(dmdt)Br_2]$ as a function of time due to hydrolysis in a pH 7.4 phosphate buffer. The amount of Br^- released, given on the y-axis of the plot, is the ratio of the concentration of free bromide ion in solution divided by the initial concentration of $[Au(dmdt)Br_2]$. The released Br^- ion was measured using a bromide-sensitive electrode

that is quantitative for the ion. From the curve shown, *estimate* the observed first-order *rate constant*, k_1, for the first step of the hydrolysis of the complex. (Reprinted with permission from L. Ronconi *et al.*, Gold(III) Dithiocarbamate Derivatives for the Treatment of Cancer: Solution Chemistry, DNA Binding, and Hemolytic Properties, *J. Med. Chem.* 49, 1648–1657. Copyright 2006 American Chemical Society.)

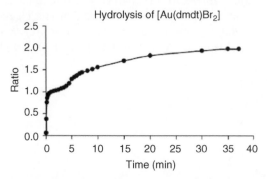

5. Shown below is the amount of halide ion released to solution from two anticancer agents, [Au(ESDT)Br$_2$] and [Au(DMDT)Cl$_2$] (Au^{+3}, 5d^8), as a function of time due to hydrolysis in a 0.01 *M*, pH 7.4 phosphate buffer. The amount of halide ion released, given on the y-axis of the plot, is the ratio of the concentration of free halide ion in solution divided by the initial concentration of complex. Both complexes undergo a two-step hydrolysis, which is characterized by forward and reverse rate constants, k_1, k_{-1} and k_2, k_{-2}.

    a. Which compound likely has the larger value of k_1? Which has the larger value of k_2? Briefly explain your choices.

    b. Propose a transition state (draw the structure) associated with the conversion of [Au(DMDT)Cl$_2$] to [Au(DMDT)Cl(H$_2$O)]$^+$.

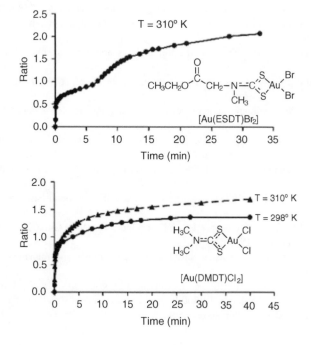

6. Shown below is the absorption spectrum of $4.8 \times 10^{-4}\,M$ [Au(en)$_2$]Cl$_3$ in the buffer (50 mM phosphate, 4 mM NaCl, pH 7.4).

   a. Assuming that the absorption spectrum was collected in a 1 cm path-length cell, calculate the molar extinction coefficient, ε ($M^{-1}\,cm^{-1}$), for the band at $\sim$340 nm in the spectrum of the compound.
   b. Assign the observed transition, for example 'the transition is a charge transfer band' and so on. (Reprinted from *Biochem. Biophys. Res. Commun.* 281, L. Messori *et al.*, Interactions of Selected Gold (III) Complexes with Calf Thymus DNA, 352–360. Copyright 2001, with permission from Elsevier.)

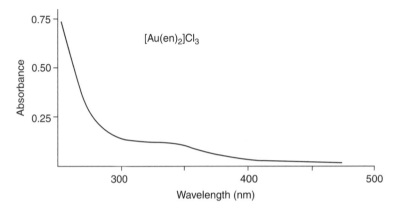

References

1. Sutton, B.M. (1983) Overview and current status of gold-containing antiarthritic drugs. Platinum, Gold, and Other Metal Chemotherapeutic Agents, ACS Symposium Series (ed. S.J. Lippard), 209, pp. 355–369.
2. Shaw, C.F. III (1999) Gold-based therapeutic agents. *Chemical Reviews*, **99**, 2589–2600.
3. Messori, L. and Giordana, M. (2004) Gold complexes in the treatment of rheumatoid arthritis. *Metal Ions in Biological Systems*, **41**, 279–304.
4. Messori, L. and Giordana, M. (2004) Gold complexes as antitumor agents. *Metal Ions in Biological Systems*, **42**, 385–424.
5. Tiekink, E.R.T. (2002) Gold derivatives for the treatment of cancer. *Critical Reviews in Oncology/Hematology*, **42**, 225–248.
6. Kostova, I. (2006) Gold coordination complexes as anticancer agents. *Anti-Cancer Agents in Medicinal Chemistry*, **6**, 19–32.
7. Howard-Lock, H.E. (1999) Structures of gold(I) and silver(I) thiolate complexes of medicinal interest: a review and recent results. *Metal-Based Drugs*, **6**, 201–209.
8. Concepción, G. and Laguna, A. (1997) Three- and four-coordinate gold(I) complexes. *Chemical Reviews*, **97**, 511–522.
9. Elder, R.C. and Eidsness, M.K. (1987) Synchrotron x-ray studies of metal-based drugs and metabolites. *Chemical Reviews*, **87**, 1027–1046.
10. Shaw, C.F., Cancro, M.P. III, Witkiewicz, P.L., and Eldridge, J.E. (1980) Gold(III) oxidation of disulfides in aqueous solution. *Inorganic Chemistry*, **19**, 3198–3201.
11. Majithia, V. and Geraci, S. (2007) Rheumatoid arthritis: diagnosis and management. *The American Journal of Medicine*, **120**, 936–939.
12. Forestier, J. (1934) Rheumatoid arthritis and its treatment by gold salts. *Lancet*, **2**, 646–648.
13. Nomiya, K., Noguchi, R., and Sakurai, T. (2000) Synthesis and crystal structure of a water-soluble gold(I) complex, {K$_3$[Au(mba)$_2$]}$_2$ formed by 2-mercaptobenzoic acid (H$_2$mba), with aurophilic interaction in the solid state. *Chemistry Letters*, 274–275.

14. Bau, R. (1998) Crystal structure of the antiarthritic drug gold thiomalate (Myochrysine): A double-helical geometry in the solid state. *Journal of the American Chemical Society*, **120**, 9380–9381.

15. McClusky, K.L. and Eichelberger, L. (1926) New methods for the preparation of sodium aurothiosulfate. *Journal of the American Chemical Society*, **48**, 136–139.

16. Hill, D.T., Sutton, B.M., Isab, A.A. *et al.* (1983) Gold-197 mössbauer studies of some gold(I) thiolates and their phosphine complexes including certain antiarthritic gold drugs. *Inorganic Chemistry*, **22**, 2936–2942.

17. Ruben, H., Zalkin, A., Faltens, M.O., and Templeton, D.H. (1974) Crystal strcuture of sodium gold(I) thiosulfate dihydrate, $Na_3Au(S_2O_3)_2 \cdot 2H_2O$. *Inorganic Chemistry*, **13**, 1836–1839.

18. Sutton, B.L., McGusty, E., Walz, D.T., and DiMartino, M.J. (1972) Oral Gold. Antiarthritic properties of alkylphosphinegold coordination complexes. *Journal of Medicinal Chemistry*, **15**, 1095–1098.

19. Papp, K.A. and Shear, N.H. (1991) Systemic gold therapy. *Clinics in Dermatology*, **9**, 535–551.

20. Blocka, K. (1983) Auranofin versus injectable gold. Comparison of pharmacokinetic properties. *The American Journal of Medicine*, **75**, 114–122.

21. Tett, S.E. (1993) Clinical pharmacokinetics of slow acting antiheumatic drugs. *Clinical Pharmacokinetics*, **25**, 392–407.

22. Boles Bryan, D.L., Mikuriya, Y., Hemple, J.C. *et al.* (1987) Reactions of auranofin ((1-thio-β-D-glucopyranose-2,3,4,6-tetraacetato-S)(triethylphosphine)gold(I)) in aqueous hydrochloric acid. *Inorganic Chemistry*, **26**, 4180–4185.

23. Intoccia, A.P., Flanagan, T.L., Walz, D.T. *et al.* (1982) Pharmacokinetics of auranofin in animals. *Journal of Rheumatology. Supplement*, **8**, 90–98.

24. Roberts, J.R., Xiao, J., Schliesman, B. *et al.* (1996) Kinetics and mechanism of the reaction between serum albumin and auranofin (and its isopropyl analog) *in vitro*. *Inorganic Chemistry*, **35**, 424–433.

25. Christodoulou, J., Sadler, P.J., and Tucker, A. (1995) ^1H NMR of albumin in human blood plasma: drug binding and redox reactions at Cys^{34}. *FEBS Letters*, **376**, 1–5.

26. Graham, G.G., Whitehouse, M.W., and Bushell, G.R. (2008) Aurocyanide, dicyano-aurate (I), a pharmacologically active metabolite of medicinal gold complexes. *Inflammopharmacology*, **16**, 126–132.

27. Elder, R.C., Zhao, Z., Zhang, Y. *et al.* (1993) Dicyanogold (I) is a common human metabolite of different gold drugs. *The Journal of Rheumatology*, **20**, 268–272.

28. Skibsted, L.H. and Bjerrum, J. (1977) Studies on gold complexes. III. The standard electrode potentials of aqua gold ions. *Acta Chemica Scandinavica (Copenhagen, Denmark: 1989)*, **A31**, 155–156.

29. Canumalla, A.J., Al-Zamil, N., Phillips, M. *et al.* (2001) Redox and ligand exchange reactions of potential gold(I) and gold(III)-cyanide metabolites under biometic conditions. *Inorganic Chemistry*, **85**, 67–76.

30. Simon, T.M., Kunishima, D.H., Vibert, G.J., and Lorber, A. (1979) Inhibitory effects of a new oral gold compound on HeLa cells. *Cancer*, **44**, 1965.

31. Blank, C.E. and Dabrowiak, J.C. (1984) Absorption and circular dichroism studies of a gold(I)-DNA complex. *Journal of Inorganic Biochemistry*, **21**, 21–29.

32. Mirabelli, C.K., Sung, C.-M., Zimmerman, J.P. *et al.* (1986) Interactions of gold coordination complexes with DNA. *Biochemical Pharmacology*, **35**, 1427–1433.

33. Weidauer, E., Yasuda, Y., Biswal, B.K. *et al.* (2007) Effects of Disease-Modifying Anti-Arthritic Drugs (DMARDs) on the Activities of Rheumatoid Arthritis-Associated Cathepsins K and S. *Biological Chemistry*, **388**, 331–336.

34. Gunatilleke, S.S. and Barrios, A.M. (2008) Tuning the Au(I)-medaiated inhibition of cathepsin B through ligand substitutions. *Journal of Inorganic Chemistry*, **102**, 555–563.

35. Nolte, R.T., Conlin, R.M., Harrison, S.C., and Brown, R.S. (1998) Differing roles for zinc fingers in DNA recognition: structure of a six-finger transcription factor IIIA complex. *Proceedings of the National Academy of Sciences of the United States of America*, **95**, 2938–2943.

36. Lu, D. and Klug, A. (2007) Invariance of the zinc finger module: a comparison of the free structure with those in nucleic-acid complexes. *Proteins*, **67**, 508–512.

37. Bose, R.N., Yang, W.W., and Evanics, F. (2005) Structural perturbation of a C4 Zinc-finger module by cis-daimminedichloroplatinum(II): insights into the inhibition of transcription processes by the antitumor drug. *Inorganica Chimica Acta*, **358**, 2844–2854.

38. Jin, Y.H., Clark, A.B., Slebos, R.J. *et al.* (2003) Cadmium is a mutagen that acts by inhibiting mismatch repair. *Nature Genetics*, **34**, 326–341.

39. Larabee, J.L., Hocker, J.R., and Hanas, J.S. (2005) Mechanisms of aurothiomalate-Cys$_2$His$_2$ Zinc finger interactions. *Chemical Research in Toxicology*, **18**, 1943–1954.

40. Franzman, M.A. and Barrios, A.M. (2008) Spectroscopic evidence for the formation of goldfingers. *Inorganic Chemistry*, **47**, 3928–3930.

41. Rigobello, M.P., Messori, L., Marcon, G. *et al.* (2004) Gold complexes inhibit mitochondrial thioedoxin reductase: consequences on mitochondrial functions. *Journal of Inorganic Biochemistry*, **98**, 1634–1641.

42. Witte, A.-B., Anestal, K., Jerremalm, E. *et al.* (2005) Inhibition of thioredoxin reductase but not of glutathione reductase by the major classes of alkylating and platinum-containing anticancer compounds. *Free Radical Biology & Medicine*, **39**, 696–703.

43. Urig, S., Fritz-Wolf, K., Réau, R. *et al.* (2006) Undressing of phosphine gold(I) complexes as irreversibel inhibitors of human disulfide reductases. *Angewandte Chemie-International Edition*, **45**, 1881–1886.

44. Omata, Y., Folan, M., Shaw, M. *et al.* (2006) Sublethal concentration of diverse gold compounds inhibit mammalian cytosolic thioredoxin reductase (TrxR1). *Toxicology In Vitro*, **20**, 882–890.

45. Mirabelli, C.K., Johnason, R.K., Hill, D.T. *et al.* (1986) Correlation of the *in vitro* cytotoxic and antitumor activities of gold(I) coordination complexes. *Journal of Medicinal Chemistry*, **29**, 218–223.

46. Mirabelli, C.K., Hill, D.T., Faucette, L.F. *et al.* (1987) Antitumor activity of the bis(diphenylphosphino)alkanes, their gold(I) coordination complexes, and related compounds. *Journal of Medicinal Chemistry*, **30**, 2181–2190.

47. Berners-Price, S.J. and Sadler, P.J. (1986) Gold(I) complexes with bidentate tertiary phosphine ligands: formation of annular vs. tetrahedral chelated complexes. *Inorganic Chemistry*, **25**, 3822–3827.

48. Liu, J.J., Galettis, P., Farr, A. *et al.* (2008) *In vitro* antitumor and hepatotoxicity profiles of Au(I) and Ag(I) bidentate pyridyl phosphine complexes and relationships to cellular uptake. *Journal of Inorganic Biochemistry*, **102**, 303–310.

49. Parish, R.V., Howe, B.P., Wright, J.P. *et al.* (1996) Chemical and biological studies of dichloro(2-((dimethylamino)methyl)phenyl)gold(III). *Inorganic Chemistry*, **35**, 1659–1666.

50. Buckley, R.G., Elsome, A.M., Fricker, S.P. *et al.* (1996) Antitumor properties of some 2-[(dimethylamino)methyl]phenylgold(III) complexes. *Journal of Medicinal Chemistry*, **39**, 5208–5214.

51. Milacic, V., Fregona, D., and Dou, Q.P. (2008) Gold complexes as prospective metal-based anticancer drugs. *Histology and Histopathology*, **23**, 101–108.

52. Ronconi, L., Giovagnini, L., Marzano, C. *et al.* (2005) Gold dithiocarbamate derivatives as potential antineoplastic agents: Design, spectroscopic properties, and *in vitro* antitumor activity. *Inorganic Chemistry*, **44**, 1867–1881.

53. Ronconi, L., Marzano, C., Zanello, P. *et al.* (2006) Gold(III) dithiocarbamate derivatives for the treatment of cancer: Solution chemistry, DNA binding, and hemolytic properties. *Journal of Medicinal Chemistry*, **49**, 1648–1657.

54. Sun, R., Ma, D.-L., Wong, E., and Che, C.-M. (2007) Some uses of transition metal complexes as anti-cancer and anti-HIV agents. *Dalton Trans.*, 4884–4892.

55. Che, C.-M., Sun, R., Yu, W.-Y. *et al.* (2003) Gold(III) porphyrins as a new class of anticancer drugs: cytotoxicity, DNA binding and induction of apoptosis in human cervix epitheloid cancer cells. *Chem. Commun.*, 1718–1719.

56. To, Y.F., Sun, R.W., Chen, Y. *et al.* (2009) Gold(III) porphyrin compelx is more potent than cisplatin in inhibiting growth of nasopharyngeal carcinoma *in vitro* and *in vivo*. *International Journal of Cancer*, **124**, 1971–1979.

57. Berners-Price, S.J., Jarrett, P.S. and Sadler, P.J. (1987) [^{31}P] NMR studies of [Au$_2$(μ-dppe)]$^+$ antitumor complexes. Conversion into [Au(dppe)$_2$]$^+$ induced by thiols and blood. *Inorganic Chemistry*, **26**, 3074–3077.

58. Berners-Price, S.J., Mirabelli, C.K., Johnson, R.K. *et al.* (1986) *In vivo* antitumor activity and *in vitro* cytotoxic properties of bis[1,2-bis(diphenylphosphino)ethane]gold(I) chloride. *Cancer Research*, **46**, 5486–5493.

59. McKeage, M.J., Berners-Price, S.J., Galettis, P. *et al.* (2000) Role of lipophilicity in determineing cellular uptake and antitumor activity of gold phosphine complexes. *Cancer Chemotherapy and Pharmacology*, **46**, 343–350.

60. Fricker, S.P. (1999) A screening stragety for metal antitumor agents as exemplified by Gold(III) complexes. *Metal Based Drugs*, **6**, 291–300.

61. Ward, B. and Dabrowiak, J.C. (1987) DNA binding specificity of the gold(III) complex (C$_2$H$_5$)$_3$PAuBr$_3$. *Journal of the American Chemical Society*, **109**, 3810–3811.

62. Messori, L., Orioli, P., Tempi, C., and Marcon, G. (2001) Interactions of selected gold(III) complexes with calf thymus DNA. *Biochemical and Biophysical Research Communications*, **281**, 352–360.

63. Tepperman, K., Shinju, Y., Roy, P.W. *et al.* (1994) Transport of dicyanogold(II) anion. *Metal-Based Drugs*, **1**, 433–444.

64. Okada, T., Patterson, B.K., Ye, S.Q., and Gurney, M.E. (1993) Aurothiolates inhibit HIV-1 infectivity by gold(I) ligand exchange with a component of the viral surfacer. *Virology*, **192**, 631–642.
65. Fonteh, P.N., Keter, F.K., Meyer, D. *et al.* (2009) Tetra-chloro(bis(3,5-dimethylpyrazolyl)methane)gold(III) chloride: An HIV-1 reverse transcriptase and protease inhibitor. *Journal of Inorganic Biochemistry*, **103**, 190–194.
66. Sannella, A.R., Casini, A., Gabbiani, C. *et al.* (2008) New uses for old drugs. Auranofin, a clinically established antiarthritic metallodrug, exhibits potent antimalarial effects *in vitro*: mechanistic and pharmacological implications. *FEBS Letters*, **582**, 844–847.
67. Fricker, S.P. *et al.* (2008) Metal compounds for the treatment of parasitic diseases. *Journal of Inorganic Biochemistry*, **102**, 1839–1845.

Further reading

Huheey, J.E., Keiter, E.A., and Keiter, R.L. (1993) Inorganic Chemistry, *Principles of Structure and Reactivity*, 4th edn, Benjamin-Cummings Publishing Co., San Francisco.
Puddephatt, R. (1978) *The Chemistry of Gold*, Elsevier Scientific Publishing Co., Amsterdam.

7

Vanadium, Copper and Zinc in Medicine

7.1 Vanadium for treating diabetes

The first report of the use of vanadium for the treatment of diabetes mellitus (DM) was published by the French physician Dr. B. Lyonnet, who in 1899 described the ability of the simple vanadium compound sodium vanadate, Na_3VO_4 (Figure 7.1), to lower the blood sugar levels of diabetic patients [1]. More than 80 years later, both vanadate and the related ion, vanadyl, VO^{2+} (Figure 7.1), were shown to exhibit insulin-mimetic effects on the oxidation of glucose in adipocytes (fat cells) of rats [2–4], and sodium vanadate was found to be effective in controlling blood glucose levels and reducing cardiac decline in diabetic rats [5]. These early reports led to the synthesis and study of a large number of vanadium-based insulin-mimetic agents, a few of which are shown in Figure 7.1 [6–14]. One of these compounds, bis(ethylmaltolato)oxovanadium(IV), BEOV, is currently in phase II clinical trials in the United States as a potential new insulin-mimetic agent for treating diabetes.

Vanadium, element 23, is found in low concentrations in the body ($\sim$100 µg total) and its function is not known. While vanadium has many oxidation states, those commonly found in the biological system and incorporated into vanadium complexes for treating diabetes are V^{+5}, which has the electronic configuration $[Ar]3d^0$, and V^{+4}, with the configuration $[Ar]3d^1$. The formula for vanadate ion is VO_4^{-3}, but since the pK_as for the first and second protonation of the ion are $\sim$13 and $\sim$8 respectively, at neutral pH in dilute solution the ion exists mainly in its diprotonated form, $[H_2VO_4]^-$ (Figure 7.1). Since vanadate is isostructural and isoelectronic with phosphate but cannot undergo the same chemical reactions as phosohate, vanadate is often used as an inhibitor in mechanistic studies of biochemical reactions that require phosphate. Like phosphate, vanadate can also form higher-order structures, but vanadate has a stronger tendency to nucleate and can form species that contain up to 10 vanadium ions, such as decavanadate, $[V_{10}O_{28}]^{6-}$ [15].

The simplest ion in water at neutral pH for aquated V^{+4} is $[VO(OH)(H_2O)_4]^+$, which since it contains a VO^{2+} species is often referred to simply as *vanadyl*. This octahedral complex contains an oxo ligand (O^{2-}), a hydroxo group (OH^-) and four bound water molecules. Depending on the concentration of vanadium and the pH of the medium, vanadyl can also form higher-order multinuclear structures in solution. The water exchange rate constant for $[VO(OH)(H_2O)_4]^+$ at 25 °C is $k \sim 5 \times 10^2\,s^{-1}$ (Figure 1.20) [16].

Complexes containing V^{+5} are diamagnetic ($S = 0$). Since the main isotope, ^{51}V, is NMR 'active' ($I = 7/2$) and has high natural abundance (>99%), compounds containing V^{+5} can easily be studied using NMR. Mononuclear vanadyl complexes are paramagnetic with one unpaired electron – that is, $S = {}^1/_2$ – and, although

Metals in Medicine James C. Dabrowiak
© 2009 John Wiley & Sons, Ltd

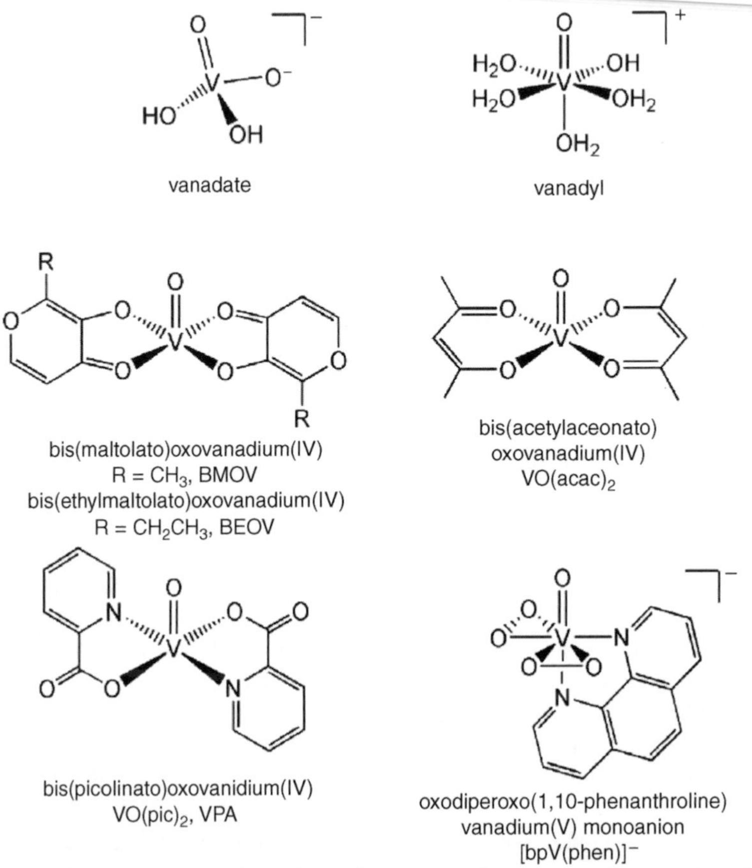

Figure 7.1 *Structures of some vanadium complexes that are insulin mimics. The prevalent forms of vanadate, which contains V^{+5}, and vanadyl, VO^{2+}, which contains V^{+4}, at neutral pH in dilute solution are shown at the top of the figure. Three of the indicated metal complexes with antidiabetic properties contain five-coordinate V^{+4}, while [bpV(phen)]$^-$, which is seven-coordinate, contains V^{+5}. In aqueous solution, it is possible for the five-coordinate complexes to bind a water molecule as the sixth ligand*

not studied with NMR, they can be investigated using electron paramagnetic resonance (EPR) see (Box 7.1). Both V^{+5} and V^{+4} are hard acids and as such they form stable complexes with ligands with hard-base donors such as oxygen and nitrogen. Because vanadyl complexes are less toxic than vanadate and the latter is thought to be reduced to V^{+4} in the biological system, vanadyl compounds have most often been studied in connection with diabetes. Vanadyl complexes often have a square pyramidal structure, with the oxo ligand occupying the axial coordination site and the remaining four sites in the base of the pyramid occupied by the organic ligands attached to the V^{+4} ion. However, in some cases vanadyl complexes have been shown to be six-coordinate, octahedral, with a water molecule occupying one of the coordination sites of the octahedron [17,18].

7.1.1 Diabetes mellitus (DM)

Diabetes mellitus, commonly called diabetes, is a disease characterized by the overproduction of glucose in the liver and underutilization of glucose by other organs in the body. There are two types of diabetes: type 1, in which the pancreas fails to produce the key peptide hormone *insulin*, and type 2, in which insulin is produced by the pancreas but the body is unresponsive to the hormone. The normal level of glucose in blood is $\sim$5 mM but when food is ingested the level rises, which if not restored produces a condition called *hyperglycemia*. In order to readjust the level to the optimal value, insulin, which is produced in the islets of Langerhans in the pancreas, is released to the blood and binds to receptors mainly found on cells in skeletal muscle and adipose tissue. This signals the cells to take up glucose and convert it to glycogen, which lowers the glucose level in blood to the optimum value. During periods of fasting, when there is no intake of food, or during strenuous exercise, when blood glucose is consumed, the concentration of glucose in blood falls below $\sim$5 mM. In order to increase the concentration, α-cells located in the pancreas release the peptide hormone *glucagon*, which causes the liver cells to stimulate the production of glucose and release it into the blood. Clearly, this system of glucose *homeostasis* is subject to many controls and while people with type 2 diabetes usually have insulin in their blood, the system does not respond to the hormone to maintain proper glucose levels.

Type 2 diabetes afflicts older people. It can to a certain extent be suppressed with the proper diet and exercise, but genetics, which cannot be altered, probably also plays a role. While injections of insulin may help, increasing lack of response usually causes this method of treatment to become ineffective. Some orally-applied drugs are available which stimulate insulin production or uptake of glucose by fat cells, but they often have undesirable side effects. Vanadyl compounds appear to exhibit few side effects and since certain complexes help to maintain proper glucose levels for individuals with type 2 diabetes, vanadium compounds are being developed as new anti-diabetic agents for treating type 2 DM.

7.1.2 Bis(ethylmaltolato)oxovanadium(IV), BEOV and bis(maltolate)oxovanadium(IV), BMOV

The compound bis(ethylmaltolato)oxovanadium(IV), BEOV (Figure 7.1), is currently in phase II clinical trials as an orally-administered antidiabetic agent [10]. As was initially reported by Thompson *et al.* [20], BEOV can be synthesized by reacting two equivalents of the maltol ligand with one equivalent of vanadyl sulfate, $VOSO_4$, under an argon gas atmosphere (BEOV is sensitive to oxidation by oxygen). After slowly adjusting the pH of the solution to 8.5 and heating to reflux for 2 hour, BEOV, which is only sparingly soluble in water, crystallizes as a dark blue-gray solid from solution. The compound has a magnetic moment of $\mu_{eff} = 1.72$ BM, indicative of one unpaired electron, consistent with the electronic configuration [Ar]$3d^1$. Potentiometric titrations show that at 25 °C in an aqueous medium containing 0.16 M NaCl, log K_1 and log β_2 for BEOV are 8.79 and 16.43, respectively. While the single crystal x-ray structure of BEOV has not been reported, the structure of the related compound, BMOV (Figure 7.1), suggests that the coordination geometry about the V^{+4} ion in BEOV is likely five-coordinate square pyramidal, with the two maltol ligands in the *trans* geometry [21]. The EPR parameters of BEOV, with g_x, g_y and g_z of 1.988, 1.976 and 1.935, respectively, and A_x, A_y and A_z of 60.5, 60.0 and 170.0 $\times 10^{-4}$ cm^{-1}, respectively, indicate that the compound has rhombic symmetry, clearly showing that the crystal field effects of the two maltolate oxygen atoms on the V^{+4} ion are not the same.

The antidiabetic effects of both BEOV and BMOV were studied using diabetic rats in which diabetes was induced by administration of the natural product, streptozotocin (STZ), which severely impairs the β-cells in the pancreas from producing insulin [19,20]. In a study by Thompson *et al.* [20], the STZ-diabetic rats were given either BMOV or BEOV by gavage (force feeding) and the concentration of glucose in plasma was determined 20 hours after administration. As shown in Figure 7.2, rats that were not given a dose of either BMOV or BEOV (both compounds exhibit identical effects) were quite hyperglycemic, having a plasma glucose concentration of $\sim$20 mM, but those given the vanadium complex exhibited reduced glucose concentrations in a dose-dependent manner. At $\sim$0.8 mmol V kg^{-1} the plasma glucose concentration in the animals was brought into the normal range by both BEOV and BMOV.

In order for BEOV and BMOV to express their glucose-lowering effects, the intact complexes or bioproduced reaction products containing vanadium must pass from the stomach to the upper GI tract and into the blood. Pharmacokinetics studies of BMOV and BEOV by labeling the compounds with radioactive [48]V

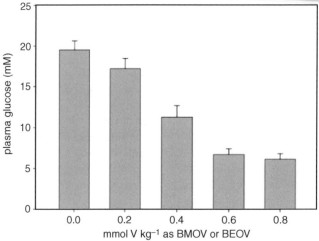

Figure 7.2   *Concentration of glucose in plasma of STZ-diabetic rats 20 hours after feeding the rats with a single dose with either BMOV or BEOV. Adapted with kind permission from Springer Science + Business Media: J. Biol. Inorg. Chem., Preparation and Characterization of Vanadyl Complexes with Bidentate Maltol-Type Ligands; In Vivo Compairson of Anti-Diabetic Therapeutic Potential, 8, 2003, 66–74, K.H. Thompson et al., Fig. 2*

or ^{14}C showed that vanadium appears in the blood before the maltol ligand, clearly indicating that the compounds degrade during the absorption process associated with oral administration [9,10,20]. While the site of degradation is not known, BMOV readily reacts with human serum albumin and apo-transferrin found in blood, and modeling and kinetic measurements show that the compound can react with citrate, ascorbate and other small ligands that are also found in blood [22–26]. Calorimetric measurements, which can be used to determine the enthalpy of a metal–ligand interaction, were used to measure the binding constant of BMOV toward the C- and N-lobes of apo-transferrin [26]. Since the binding constants of BMOV and vanadyl sulfate toward both transferrin sites were about the same, $\sim 10^5 M^{-1}$, investigators suggested that solvated VO^{2+} released from BMOV is binding to the protein, but that due to the presence of dissolved oxygen, V^{+4} is oxidized to V^{+5} and the latter is bound to the protein [26].

An important shortcoming of many vanadium compounds studied in connection with diabetes is that they have poor bioavailability and not much of the administered vanadium is absorbed by the body (5–10%), with most of the vanadium being eliminated in the feces and urine [27,28]. However, absorption of vanadium from administered BMOV and BEOV into most tissues, with bone being the most absorbant, is two to three times greater than with $^{48}VOSO_4$, showing that the maltolate ligands must influence some aspect of the absorption process.

Mechanistic studies on BEOV and BMOV, indeed on all vanadium insulin-mimetic agents, are hampered by the complex nature of the glucose regulatory system. Pharmacokinetic studies with BMOV and BEOV clearly show that while the vanadium concentration in blood peaks within a few hours after administration, the maximum glucose lowering occurs $\sim$24 hours after administration, showing that the biological response to vanadium is slow [9,20,28,29]. Since bone and other tissue incorporates vanadium, it is also possible that these sites can release vanadium, which may be the reason why in some cases the glucose-lowering effects of the ion persist long after treatment is terminated. In view of the complicated nature of the signaling system and its apparent slow response to a perturbation (the vanadium), it is not surprising that biological targets for vanadium-induced glucose lowering have not yet been identified.

The compound BEOV is currently in phase IIa clinical trials for the treatment of type 2 diabetes. In phase I clinical trials, doses of 10–90 mg of BEOV were orally given to nondiabetic volunteers and produced no

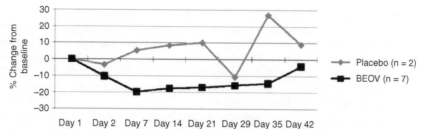

Figure 7.3 *Fasting plasma glucose following 28 days of administration of bis(ethylmaltolato)oxovanadium(IV), BEOV, 20 mg day⁻¹ orally to seven type 2 diabetic human subjects, compared to two type 2 diabetic individuals receiving a placebo. Reprinted from J. Inorg. Biochem. 103, K.H. Thompson et al., Vanadium Treatment of Type 2 Diabetes: A View to the Future, 554–8. Copyright 2009, with permission from Elsevier*

adverse side effects, with all biochemical parameters of the subjects remaining within normal limits. In the phase IIa trial involving nine type 2 diabetic patients, BEOV was given – 20 mg day⁻¹ for 28 days during periods of fasting – to seven individuals and placebos were given to the remaining two. As shown in Figure 7.3, blood glucose of the seven patients treated with BEOV declined an average of ∼15% relative to the initial glucose level, with the reduced glucose concentration persisting for about a week after termination of treatment. In addition to exhibiting a reduction in blood glucose levels, BEOV also enhanced the ability of the treated group to handle a glucose challenge (like eating a candy bar) during glucose tolerance tests. These encouraging results are allowing BEOV to undergo further testing as a potential new drug for treating diabetes.

Box 7.1 Electron paramagnetic resonance

Electron paramagnetic resonance, EPR, or electron spin resonance, ESR, is a type of spectroscopy that utilizes radiation in the microwave range to study molecules, ions or atoms with unpaired electrons [30–32]. If a molecule with an unpaired electron is placed in a magnetic field, quantum mechanics specifies that there are only two possible orientations of the magnetic moment of the unpaired electron relative to the applied field. One orientation, which is lowest-energy (most stable) and has the magnetic spin quantum number $m_s = -1/2$, is referred to as the β-spin state (Figure 7.4). For this state, the magnetic moment of the electron is aligned with the applied field. The second orientation, which is highest in energy (least stable) and has $m_s = +1/2$, is referred to as the α-spin state. This state has the magnetic moment of the electron opposing the field. The energies of the two possibilities are given by (7.1), where E_{ms} is the energy of the state, g_e is the g-factor for the electron (2.0023), μ_B is the Bohr magneton (9.274 × 10⁻²⁴ J T⁻¹) and H is the strength of the magnetic field in tesla, T.

$$E_{m_s} = g_e \mu_B H m_s m_s = \pm 1/2 \tag{7.1}$$

The energy separation between the two states is given by (7.2). When a sample is exposed to electromagnetic radiation of frequency v, absorption of energy occurs – a condition called *resonance* – when (7.3) is satisfied.

$$\Delta E = E_\beta - E_\alpha = g_e \mu_B H \tag{7.2}$$

$$\Delta E = h v = g_e \mu_B H \tag{7.3}$$

The magnetic field used in a type of EPR instrument called an *X-band spectrometer* is ∼0.34 T, where 1 T = 10⁴ Gauss (G). Since the energy separation between the two states in a field of this strength is ∼2 × 10⁻²⁴ J or ∼0.1 cm⁻¹, the frequency, v, corresponding to ΔE ($h = 6.626 × 10^{-34}$ J s) for this field strength is ∼9500 MHz or ∼9.5 GHz, which is radiation in the microwave range. By comparison, the crystal field splitting parameter, Δ, for many transition metal complexes is ∼10⁴ cm⁻¹, which is about five orders of magnitude *larger* than the splitting between the two sates in Figure 7.4.

The EPR experiment can be carried out on solutions at room temperature or on frozen 'glasses', in which the paramagnetic compound is randomly oriented in a ridged glass-like matrix. An EPR spectrum is obtained by placing the sample in the field of an electromagnet and irradiating it with microwave radiation of a fixed frequency, v, which corresponds to a fixed value of ΔE. In order to produce resonance, the current to the electromagnet is varied, which changes the strength of the magnetic field and thus the energy difference between the two states in Figure 7.4. When the energy difference between the two levels exactly matches the energy being supplied by the microwave source, radiation will be absorbed by the sample and resonance will occur.

If a large number of molecules, each with an unpaired electron, are placed in the field of the magnet, not all of the electron spins will be aligned with the field; that is, in the lowest-energy, E_β, state. This is because at room temperature the amount of thermal energy available in terms of wavenumbers is $\sim 200 \, \mathrm{cm}^{-1}$, which, since the spacing between E_α and E_β is only $\sim 0.1 \, \mathrm{cm}^{-1}$, allows *both states* to be populated, with slightly more spins in the lower state, E_β. The expression that addresses the number of spins in each state, denoted by the population, N, in the state, is the *Boltzmann distribution equation* (see Box 9.2), which when applied to two states separated by $\sim 0.1 \, \mathrm{cm}^{-1}$ shows that the ratio $N(E_\alpha)/N(E_\beta) \approx 0.999$. While this difference is small, it is large enough to detect a change in the populations of the states when resonance occurs.

Shown in Figure 7.5a is the absorption curve associated with the transition $E_\alpha \leftarrow E_\beta$ for a 'free electron', $g_e = 2.0023$, in a magnetic field with $H = 0.335 \, \mathrm{T}$, for which, from (7.3), the resonance frequency $v = 9388.2 \, \mathrm{MHz}$. In order to uncover more detail in the absorption/resonance effect, EPR spectroscopists usually plot the first derivative of the absorption curve, dI/dG, where I is the intensity associated with the absorption, which is shown in Figure 7.5b. Unlike the absorption curve, the first derivative curve is very sensitive to subtle features in the absorption envelope that might otherwise go undetected in the experiment. The number of maxima in the first derivative spectrum always equals the number of absorptions, which, in the example given in Figure 7.5, is one.

Since the magnetic moment associated with the electron can couple quantum mechanically with the magnetic moment of the *nucleus* of an atom or ion that is near the electron, the simple two-level diagram in Figure 7.4 can change into a more complicated diagram. This coupling, which gives rise to the *hyperfine structure* often observed in EPR spectra, increases the number of states (levels) in the diagram, which results in a number of absorption bands. As an example of hyperfine coupling, consider the EPR spectrum of a five-coordinate V^{+4}, $[\mathrm{Ar}]3d^1$, $S = {}^1/_2$, complex shown in Figure 7.6. The most abundant naturally-occurring isotope of vanadium, $>99\%$, is ${}^{51}\mathrm{V}$, which has a nuclear spin quantum number of $I = 7/2$. The spectrum in Figure 7.6 shows that the typical V^{+4} complex exhibits eight absorptions, eight peaks in the first derivative curve, which are due to the coupling between the magnetic moments of the electron and the ${}^{51}\mathrm{V}$ nucleus. The number of absorptions due to hyperfine coupling is given by (7.4), where n is the number of nuclei coupling with the electron, in this case one, and I is the nuclear spin quantum number of the coupling nucleus, which for ${}^{51}\mathrm{V}$ is 7/2.

$$2nI + 1 \tag{7.4}$$

In addition to complexes of V^{+4}, metal compounds of Cu^{+2}, $[\mathrm{Ar}]3d^9$, $S = {}^1/_2$, are also commonly found in medicine. Naturally-abundant copper mainly exists as two isotopes, ${}^{63}\mathrm{Cu}$ and ${}^{65}\mathrm{Cu}$, in the approximate ratio $70 : 30$, and both isotopes have a nuclear spin quantum number of $I = 3/2$. If it is assumed that coupling effects of both isotopes are identical, coupling of the electron and the nuclear magnetic moments produces a four-line hyperfine pattern for a Cu^{+2} complex through (7.4). In addition to coupling to the nuclear spin of the metal ion, the unpaired electron can also couple to nuclear spins of atoms in the ligands if the coupling atom has $I \neq 0$. However, in order to do so, the unpaired electron must spend time in a molecular orbital that has a significant contribution from the atomic wavefunction of the atom with the non-zero nuclear spin. The parameter that is typically reported is the *hyperfine coupling constant*, A, which is the separation, usually in wavenumbers, cm^{-1}, between the bands in the hyperfine pattern.

In addition to hyperfine coupling, spectra of transition metal complexes usually exhibit more than one value of the g-factor. This is because g is sensitive to the coupling between the spin and orbital angular momentum of the electron, which depends on the structure of the complex. If a complex with an unpaired electron on the metal ion has high symmetry, for example an *octahedral* compound, only one value of g will be observed and g is said to be *isotropic*. For an octahedral complex, the crystal fields on the x-, y- and z-axes are identical, but if the crystal field on the z-axis is different than that on x and y, for example in a *tetragonal* complex (Figure 1.5), the value of g associated with the z-axis will be different than the value of g in the x,y-plane. Such a system has an *anisotropic* g-factor and the compound produces separate absorptions for g_x and g_y ($g_\perp$) and g_z ($g_{\parallel}$) and, if hyperfine coupling is present, A_x, A_y ($A_\perp$) and A_z ($A_{\parallel}$) in the EPR spectrum. If the crystal field is such that x $\neq$ y $\neq$ z, the complex is said to be *rhombic* and separate features for g_x, g_y and g_z, and A_x, A_y and A_z are expected in the spectrum. In this way EPR can be used to uncover some structural features of the complex, but since EPR absorptions are usually highly overlapped, extracting values of the anisotropic g-factors and hyperfine coupling constants from spectra can be challenging. Fortunately, computer software can be used to simulate spectra, which allows the extraction of the various EPR parameters from the data.

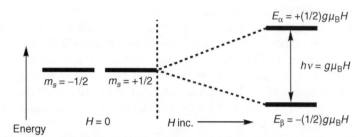

Figure 7.4   *Energy states for and electrons in an applied magnetic field, H*

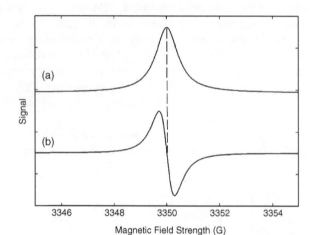

Figure 7.5   *EPR absorption curve (a) and the first derivative of the absorption curve (b) at 3350 G (0.3350 T), v = 9388.2 MHz*

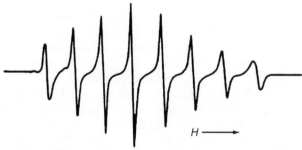

Figure 7.6   *EPR spectrum of a complex containing vanadyl, VO^{2+}*

7.1.3 Bis(acetylacetonato)oxovanadium(IV), VO(acac)₂, bis(picolinato)oxovanadium(IV), VO(pic)₂ and oxodiperoxo(1,10-phenanthroline)vanadium(V), [bpV(phen)]⁻

The β-diketone, 2,4-pentanedione or acetylacetone, is a common metal-binding ligand. Since one of the methylene protons of acetylacetone is acidic and is easily lost upon complexation to produce the acetylacetonate ion, *acac*, the reaction of two equivalents of acetylacetone with vanadyl sulfate produces the five-coordinate compound VO(acac)₂ (Figure 7.1), which has insulin-mimetic properties. The acac complex slightly lowers plasma glucose levels in STZ-rats and stimulates the uptake and conversion of glucose to fatty acids by fat cells, which is one of the properties of the antidiabetic vanadium compounds. In one study, Mekinen and Brady [33] focused on the importance of albumin as a possible transport vehicle for VO(acac)₂ by studying whether and to what extent albumin in the culture medium with the vanadium compound affects the ability of the to 3T3-L1 adipocytes to remove glucose from the culture medium. As shown in Figure 7.7, the presence of bovine serum albumin (BSA) in the culture medium dramatically affects the ability of the cell to take up a radiolabeled form of glucose from the medium. Using EPR, the investigators showed that BSA bound one VO(acac)₂ complex, and they suggested that formation of protein-complex adducts of this type may be important not just for the insulin-mimetic effects of VO(acac)₂, but for those of other vanadium complexes as well.

The insulin-mimetic complex bis(picolinato)oxovanadium(IV), VO(pic)₂, VPA, can be made by the reaction of two equivalents of the pyridine carboxylic acid with vanadyl sulfate. A single crystal x-ray structural analysis of an analog of the compound shows that VPA is most likely a distorted six-coordiante structure with a water molecule occupying the coordination site *trans* to the oxo-ligand [18]. When administered to STZ-rats either intraperitoneally or orally for a period of 14 days, VPA was reported to maintain the serum glucose levels in the normal range for ~30 days [34]. Modeling studies using the stability constants of the complex and the affinities of its ternary complexes formed with molecules found in serum, oxalic, lactic and citric acids, phosphate, HSA and transferrin suggest that VPA very likely forms ternary complexes with ions and proteins in blood, and EPR measurements indicate that vanadium–picolinate complexes are present in treated organs [18,24].

Sakurai and coworkers developed an interesting technique called *in vivo* blood-circulation monitoring–electron spin resonance (BCM-ESR) to monitor the pharmacokinetics of VPA and other insulin-mimetic

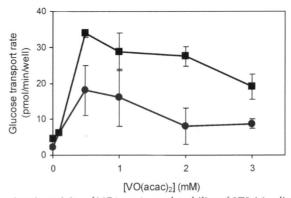

Figure 7.7 *Plot of insulin-mimetic activity of VO(acac)₂ on the ability of 3T3-L1 adipocytes in a culture medium containing 20 μM radioactive 2-deoxy-D-[1-¹⁴C]glucose to take up glucose in the presence (squares) and absence (circles) of 1 mM bovine serum albumin (BSA). The glucose transport rate, pmol min⁻¹ well⁻¹, was measured by exposing the cells to the medium for 5 minutes, washing and analyzing them for their radioactivity content using standard scintillation methods. Adapted from Makinen, M.W. and Brady, M.J., J. Biol. Chem. 2002, 277, 12215–20*

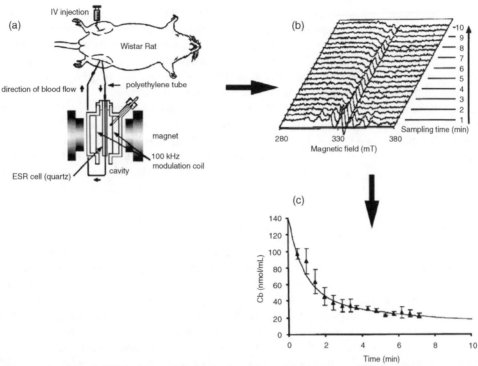

Figure 7.8   *In vivo blood-circulation monitoring using electron spin resonance. (a) Schematic representation of the experimental set up involving the animal and the ESR spectrometer. (b) ESR spectra of a vanadyl VO^{2+} species in blood as a function of time after injecting the Wister rat with a solution containing the insulin-mimic $VO(pic)_2$, VPA. (c) Plot of the concentration of the vanadyl species, C_b, as a function of time in the blood of the rat. Adapted from J. Inorg. Biochem., 78, H. Yasui et al., Metallokinetic Analysis of Disposition of Vanadyl Complexes as Insulin-Mimetics in Rats Using BCM-ESR Method, 185–196. Copyright 2000, with permission from Elsevier*

agents in real time in the blood of a rat [32]. As shown in Figure 7.8a, the investigators connected heparinized silicon tubes to the left femoral artery and vein of a Wister rat, creating a closed loop that would allow the blood of the animal to pass, under the action of the beating heart, through a quartz cell in an ESR spectrometer. Next, VPA was administered by IV bolus into the right femoral artery of the anestiszed rat, and EPR spectra on the blood passing through the EPR cell were collected at 30 second intervals (Figure 7.8b). By carrying out a series of control experiments, the investigators determined that VPA quickly reacts to form two EPR-detectable compounds in blood, labeled α and β, which have EPR parameters (α): $g_\perp = 1.977$, $A_\perp = 84.4 \times 10^{-4}\,\text{cm}^{-1}$, $g_\parallel = 1.943$, $A_\parallel = 170.9 \times 10^{-4}\,\text{cm}^{-1}$ and (β): $g_\perp = 1.980$, $A_\perp = 78.8 \times 10^{-4}\,\text{cm}^{-1}$ and $g_\parallel = 1.946$, $A_\parallel = 164.5 \times 10^{-4}\,\text{cm}^{-1}$, and that these compounds are relatively stable in blood with time. They also found that these same two species are observed in the circulating blood of the rat given a dose of VPA, but they disappear from blood with half lives $t_{1/2} = 0.4 \pm 0.2$ (α) and 8.2 ± 1.1 minutes (β) (Figure 7.8c). The investigators suggested that the disappearance of the compounds with time in the circulating blood of the rat is due to their distribution to tissue with the indicated half lives, followed by absorption by bone, liver and kidney prior to their ultimate elimination in the urine.

Hydrogen peroxide, H_2O_2, a strong oxidizing agent, contains two acidic protons that can be removed, allowing it to act as a bidentate dianionic ligand (hard base) toward V^{+5} and other metal ions [33]. Shown in

Figure 7.1 is [bpV(phen)]$^-$, a seven-coordinate V^{+5} complex with an oxo-ligand, two bidentate peroxo-ligands and the well-known metal-chelating agent 1,10-phenanthroline (phen). The compound, which can be made by reacting sodium vanadate in water with hydrogen peroxide and phen, is a potent insulin-mimetic agent [7,37]. While the peroxovanadates show promise as antidiabetic agents, the compounds have poor oral bioavailability and are not very stable, especially to acid, suggesting that they may be difficult to formulate and that they would rapidly decompose in the stomach in oral administration.

In addition to the compounds shown in Figure 7.1, many other vanadium compounds are being evaluated for their antidiabetic properties. The carrier ligands in these agents exhibit wide structural diversity, ranging from substituted β-diketonates, maltolates and picolinates [11–13,18,38], acyclic tetradentates [39,40] and cyclic tetradentate anionic porphyrins [12,13] to analogs of vitamin B$_{12}$ [41]. While the bulk of the research in this area has focused on vanadium, metal complexes of Co, Cr, Mn, Zn, Se, Mo and W have also been reported to have antidiabetic properties [18].

7.1.4 Anticancer effects of vanadium compounds

In addition to exhibiting antidiabetic properties, some vanadium complexes have anticancer effects, a versatility that makes them similar to gold complexes that are active against more than one disease state (see Chapter 6). The anticancer effects of vanadium were first reported by Köpf-Maier and Köpf, who described the antitumor activity of vanadocene dichloride, [(η^5-Cp)$_2$VCl$_2$] (Figure 7.9), the vanadium analog of the earlier-discussed anticancer agent titanocene dichloride (Figure 5.15) [42]. Vanadocene dichloride, which contains vanadium in the formal oxidation state (+4), exhibits high antitumor activity against Ehrlich ascites tumor as well as other types of cancer with less toxicity and fewer side effects than cisplatin. Vanadocene dichloride is probably transported by transferrin to the cell, where, like titanocene dichloride, it is believed to exert its antitumor effects by binding to DNA [42,43].

The bis-peroxovanadium compound, [bpV(phen)]$^-$ (Figure 7.1), and analogues containing substituted 1,10-phenanthrolines, also exhibit antineoplastic activity [44]. Examination of the cytotoxicity of a series of

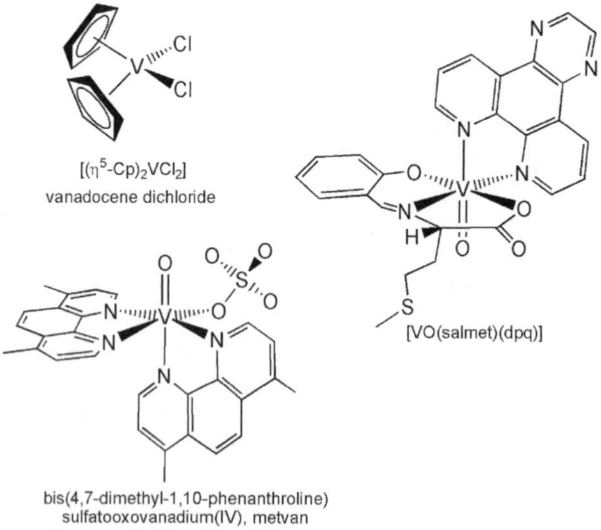

[(η^5-Cp)$_2$VCl$_2$]
vanadocene dichloride

[VO(salmet)(dpq)]

bis(4,7-dimethyl-1,10-phenanthroline)
sulfatooxovanadium(IV), metvan

Figure 7.9 *Structures of vanadocene dichloride and metvan, which have antitumor properties, and [VO(salmet) (dpq)], which may be used as a sensitizer in photodynamic therapy (PDT). The metvan isomer with the Δ absolute configuration is shown*

bis-peroxovanadium compounds against a panel of 28 different cell lines showed that some analogs have IC_{50} values in the submicromolar range for a 48 hour exposure. The biological target for the vanadium complexes appears to be the phosphatase enzyme, Cdc25A, which is responsible for removing phosphate residues from tyrosine, threonine and serine groups of proteins that are required for cell growth. Since Cdc25A is overexpressed in several human cancers, and *in vitro* studies showed that the vanadium complexes inhibit the enzyme, Cdc25A may be the target for the metal complexes in the cell [44]. Interestingly, [bpV-(phen)]$^-$ can be activated by light, and the reducing agents NADPH, GSH and the amino acid cysteine, to cleave closed circular pBR322 DNA [42]. Analysis using agarose gel electrophoresis shows that [bpV-(phen)]$^-$ is a single-strand DNA cleavage agent exhibiting some base and/or sequence preference. The complex [VO(H$_2$O)$_2$(phen)]$^{2+}$, which contains vanadyl, has also been reported to cleave closed circular ColE1 DNA in the presence of H$_2$O$_2$ by a mechanism that likely involves the generation of hydroxyl radical, HO$^\bullet$ [46].

Shown in Figure 7.9 is the compound bis(4,7-dimethyl-1,10-phenanthroline)sulfatooxovanadium(IV), metvan, which looks promising in preclinical studies as a potential anticancer drug [47–49]. The compound induces apoptosis in human leukemia, multiple myeloma and solid tumor cells derived from breast cancer and is highly effective against cisplatin-resistant ovarian and testicular cell lines. Metvan shows favorable pharmacokinetics in mice and does not cause acute or sub-acute toxicity at the dose levels tested (12.5–50 mg kg^{-1}). The compound has an octahedral structure with the oxo- and coordinated sulfate ligands occupying *cis*-positions of the octahedron and the two substituted 1,10-phenanthroline ligands occupying the remaining four coordination sites. Metvan exists as a racemic mixture, the Δ enantiomer of which (see Box 5.3), is shown in Figure 7.9. While the molecular mechanism of action is not known, metvan appears to produce ROS as well as reduce the concentration of GSH in cells, either of which has the potential to induce apoptosis.

In addition to exhibiting antidiabetic properties, VO(acac)$_2$ (Figure 7.1), blocks the normal cell cycle of hepatocellular liver carcinoma, HepG2, cells at the G1 phase [50]. While the mechanism of cytotoxicity appears to be related to VO(acac)$_2$ inhibiting the phosphorylation of the protein *retinoblastoma tumor suppressor*, the compound has been shown to cleave purified closed circular pA1 DNA (Figure 7.10) [51]. Interestingly, the DNA cleavage reaction, which involves the conversion form I → form II → form III DNA (see Box 3.4), appears to take place by simply mixing the complex with DNA, but dissolved oxygen gas may be required. The investigators showed that hydroxyl radical, HO$^\bullet$, which is produced in the reaction, is likely the DNA damaging agent and that binding of the metal complex to the phosphate residues of DNA may be important in the cleavage process.

The use of light to destroy tumors containing a photosensitizer is called *photodynamic therapy, PDT* (see Box 9.1). Chakravarty and coworkers showed that oxovanadium(IV) complexes containing Schiff

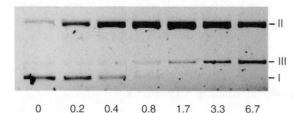

0	0.2	0.4	0.8	1.7	3.3	6.7

Figure 7.10 *Agarose gel of closed circular pA1 DNA cleaved by VO(acac)$_2$. The forms of DNA – form I, closed circular; form II, nicked circular; form III, linear DNA – are indicated at the right of the gel and the ratio of VO(acac)$_2$ to DNA base pairs is indicated at the bottom of the gel. The negative image of the ethidium-stained gel is shown. Reprinted from J. Inorg. Biochem. 103, N. Butenko et al., DNA Cleavage Activity of V(IV)O(acac)2 and Derivatives, 622–32. Copyright 2009, with permission from Elsevier*

bases formed between salicylaldehyde and various amino acids and having a coordinated phenanthroline-type ligand may be useful as photosensitizers in PDT [52]. The complex [VO(salmet)(dpq)] (Figure 7.9), which contains the Schiff base formed by the reaction of the amino acid methionine with salicylaldehyde, salmet and dipyrido[3,2-d:2′,3′-f]quinoxaline, dpq, binds to calf thymus DNA via a groove or DNA surface with $K_b = 2.7$ $(\pm 0.6) \times 10^5\,M^{-1}$. The complex exhibits *d-d* bands associated with the $3d^1$ electronic configuration of V^{+4} at 848 nm (ε, $30\,M^{-1}\,cm^{-1}$) and 701 nm (ε, $40\,M^{-1}\,cm^{-1}$). When the complex is bound to closed circular pUC19 DNA and irradiated with red light (>750 nm) into the *d-d* transitions, singlet oxygen, $^1O_2$, is produced, which facilitates the cleavage of form I DNA, converting it into nicked circular form II DNA. The investigators pointed out that since light with a wavelength greater than 750 nm is highly penetrating in tissue, and vanadium complexes are less toxic that other metal-based alternatives, [VO(salmet)(dpq)] and its analogues may be useful as sensitizers in PDT.

7.2   Role of copper and other metal ions in Alzheimer's disease

Alzheimer disease, AD, a form of dementia, is a neurodegenerative disease diagnosed in individuals that are mainly over the age of 65 [53]. People with AD progressively lose their memory, senses and bodily functions, and die within approximately 10 years of being diagnosed with the disease. During the latter stages of AD, patients have difficulty speaking and eating, they become incontinent and in general they lose the ability to respond to their environment. Since there is no cure for AD, late-stage patients are treated with various drugs in a palliative manner so that they maintain some quality of life. AD also creates a significant physical and emotional burden on the caregiver, usually a family member, who is witness to the slow, irreversible decline of a loved one. In 2008 there were about 5 million AD patients in the United States, making it the sixth leading cause of death, surpassing diabetes. The number of AD cases is expected to double in the next 20 years.

While the biological cause of AD is unknown, one hypothesis, called the *amyloid cascade hypothesis*, is that AD is caused by the formation of *plaques* and *neurofibrillary tangles* mainly in the temporal lobes of the brain, which are important for memory [54–63]. The plaques, which form in the extracellular space between neurons, are mainly composed of a peptide 39–43 amino acids long, termed the *Aβ peptide*, that has been cleaved from the extracellular portion of a transmembrane protein, called the *amyloid precursor protein, APP*. Analysis of the plaque shows that it contains many copies of the Aβ peptide in an insoluble oligomeric β-sheet aggregate with high amounts of Cu^{+2}, Zn^{+2} and Fe^{+3}. While it was initially thought that Aβ was simply a metal-binding peptide, it has been suggested that the plaque behaves more like a metalloprotein, which, with the incorporated Cu^{+2} ions and possibly also Fe^{+3}, may be capable of facilitating the production of ROS, which damages cellular components and kills neurons [54]. An additional hallmark of AD is the presence of *neurofibrillary tangles*, which form inside the neuron and consist of an insoluble hyperphosphorlyated aggregate of a protein called *tau*, which is important in microtubule assembly. While both the plaques and the tangles appear to be toxic to neurons, it is unclear if they cause AD or if they are the product of some biological process that causes the disease.

7.2.1   Metal chelating agents for treating Alzheimer's disease

Currently, a large number of existing drugs, amino acids, peptides, proteins and other chemical agents are in clinical trials in the United States as potential new agents for treating AD [64]. Since it is known that the amount of copper in the brain increases with age, and especially in patients with AD, an initial idea for treating the disease was to use chelating agents which could bind to and remove Cu^{+2} and other metal ions in the plaque, thereby breaking up the structure of the plaque into its less toxic components. This approach, called

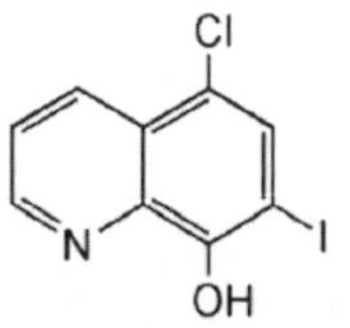

Figure 7.11 *Structures of some chelating agents being examined for the treatment of Alzheimer disease*

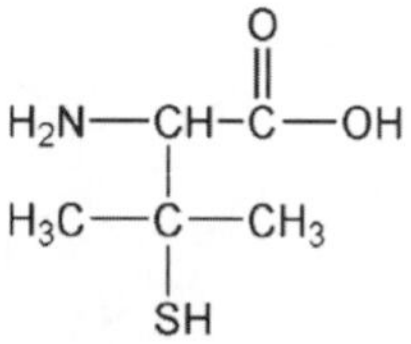

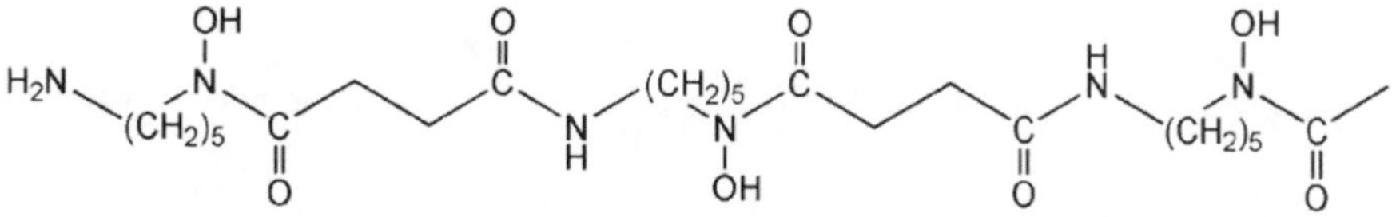

chelation therapy, has been used in medicine for many years to detoxify individuals with *heavy metal poisoning*, who, through occupational exposure or by other means, have high amounts of Pb^{+2}, Hg^{+2} and so on in their body. The agent of choice in chelation therapy for this purpose is calcium-disodium EDTA (Figure 1.26), but since EDTA is a multivalent anion at neutral pH and cannot cross the blood–brain barrier (BBB) (see Section 8.2.1) it is of little use for chelation therapy in the brain.

Shown in Figure 7.11 are chelating agents that are in or have completed clinical trials for treatment of AD. The compound desferrioxamine, DFO, has three hydroxamic acid functional groups, which in their deprotonated, hydroximate, forms have high affinity for Fe^{+3} [54]. Desferrioxamine, which is used to reduce iron overload produced by blood transfusions given to people with *Thalassaemia*, has been shown to reduce the rate of decline in daily living activities of AD patients. The drug D-penicillamine is a chelating agent that contains both hard (N,O) and soft (S) donors and is effective in removing copper (as either as Cu^{+2} or Cu^{+}) from the body. Penicillamine, which is used to reduce the high levels of copper in patients with Wilson disease, was shown to have no or only a modest effect on AD [54].

Clioquinol, CQ, 5-chloro-7-iodo-8-hydroxyquinoline, was until the late 1960s approved for the treatment of human *intestinal amebiasis*, a disease caused by a parasite. However, CQ produced optic atrophy, mainly in the Japanese population, possibly by complexing with the cobalt ion in vitamin B_{12}, and for this reason the use of the drug was discontinued [65]. Clinical trails with CQ in Alzheimer patients indicate that the orally-administered compound has potential for treating AD and related neurological disorders, although some questions concerning the details of the trials have been raised in the literature [66–69]. It is well known that 8-hydroxy quinoline binds to a metal ion via the pyridine nitrogen and the phenolate oxygen (deprotonated phenol), which is evident in the structures of the Cu^{+2} and Zn^{+2} complexes of CQ shown in Figure 7.12 [70]. As can be seen from the figure, the Cu^{+2} complex is square planar, with the donor atoms of the CQ ligands in the *trans* geometry, while the Zn^{+2} complex is five-coordinate, having a distorted trigonal bipyramidal geometry with a water molecule occupying one of the coordination sites [70]. The stability constants of CQ toward Cu^{+2} and Zn^{+2}, measured by spectroscopic and polarographic (electrochemical) methods in a medium that simulates the fluid found in the brain, are 1.2×10^{10} and $7.0 \times 10^8\,M^{-2}$ for $[Cu(CQ)_2]$ and $[Zn(CQ)_2]$ respectively, showing that CQ has a greater affinity for Cu^{+2} [65]. Since the stability constant of Cu^{+2} found in the plaque appears to be higher than that of $[Cu(CQ)_2]$, it has been suggested that rather than acting as a chelating agent that breaks up the plaque, CQ may serve as an *ionophore* for transporting 'free' Cu^{+2} ions in the blood to the brain and eventually into the neuron [54,65,71]. Supporting this concept is the observation that treatment with CQ *increases* the levels of both copper and zinc in the brain, and studies

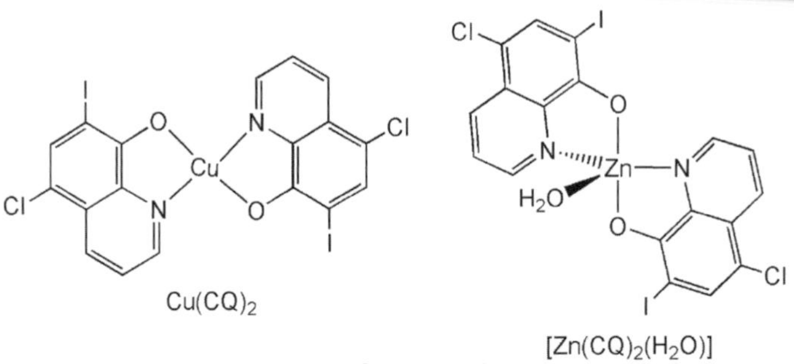

Figure 7.12 *Structures of the Cu^{+2} and Zn^{+2} complexes of clioquinol, CQ*

using a yeast model system show that the levels of copper inside the yeast cell *increase* if CQ is added to a culture medium that contains Cu^{+2} [72]. Since patients with AD seem to have faulty homeostasis of copper in the brain – that is, an inordinate amount of copper is found in the plaque *outside* the neuron – CQ may be simply a vehicle for delivering Cu^{+2} to neurons from which the ion has been depleted, with Cu^{+2}, once inside the neuron, inhibiting the cellular processes responsible for the production of the Aβ peptide from APP. Interestingly, while CQ can facilitate the transport of Cu^{+2} into yeast cells in culture, it is unable to transfer Zn^{+2} into the cell under similar conditions [72]. However, the results are very different for human prostate cancer cells that have been exposed to Zn^{+2} and CQ, which take up the metal ion from the medium, depositing it inside lysosomes, causing lysosome rupture and apoptotic cell death [73]. Despite the promising clinical results of CQ for treating AD, the exact mechanism by which the agent affects the metal ion distribution within the brain and is able to slow the progression of the disease remains unclear.

7.2.2   Complexes of the amyloid beta (Aβ) peptide

The peptide at the center of discussion of AD is the Aβ peptide. The main problem in studying this peptide is that it readily oligomerizes to form higher-order structures in solution, which may be facilitated by trace amounts of metal ions such as Cu^{+2}, and the final oligomerized product is insoluble in the medium [54,74]. In general, chemical processes in which a monomer self-associates to form larger structures are inherently difficult to study and the general approach used by many investigators is to drive the equilibrium in favor of the monomeric state, where the system is relatively simple. Since oligomerization processes are concentration-dependent, this means working at dilute concentrations, which also makes it more difficult to obtain useful information using any physical technique; that is, the concentration of solute is low. If the substance is soluble in the monomeric state but higher-order structures are insoluble, equilibrium in the conventional sense is lost because not all of the components are in the same phase. Adding to the challenge in studying the binding of metal to Aβ is that the buffers often employed in these experiments contain components that can bind to the metal ion, which may shift equilibria in favor of buffer–metal interactions or otherwise produce species that are not relevant to the biochemical system. In the fully-oligomerized end product, called a *fibril*, the structure is insoluble in aqueous media, which makes it difficult to study with techniques such as high-resolution NMR. However, using a technique called *solid-state NMR*, some of the morphology of the fibril can be uncovered, but many structural details are lacking [75]. Divalent copper, which has a $3d^9$ electronic configuration and one unpaired electron ($S = \frac{1}{2}$), complexed to monomeric and aggregated Aβ, can be studied with EPR, but the yield of detailed structural information through the measured values of g and hyperfine coupling constants, A, is not extensive [76]. Since Zn^{+2} has an electronic configuration of $3d^{10}$, which is diamagnetic ($S = 0$), its binding to monomeric Aβ can be studied using high-resolution NMR, which has yielded structural detail [77].

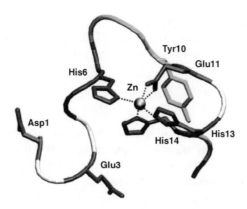

$A\beta$ = NH_2-DAEFRHDSGYEVHHQKLVFFAEDVGSNKGAIIGLMVGGVVIA
          10   20   30   40

Figure 7.13 *Solution NMR structure of a segment (residues 1–16) of the amyloid beta (Aβ) peptide coordinated to Zn^{+2} (PDB 1ZE9). While Aβ peptides found in plaques are 39–43 amino acids long, the sequence of residues 1–42 is shown at the bottom of the figure. Adapted with kind permission from Springer Science + Business Media: Eur. Biophys. J., X-Ray Absorption and Diffraction Studies of the Metal Binding Sites in Amyloid β-Peptide, 37, 2008, 257–263, V. Streltsov, Fig. 1*

Shown in Figure 7.13 is the solution structure of the Zn^{+2} complex of the Aβ peptide (residues 1–16), determined using ^1H NMR, and the amino acid sequence of the Aβ peptide itself (residues 1–42) [74,77]. As is evident from the structure of the zinc complex, the Zn^{+2} ion is five-coordinate, having bound imidazole residues, His_6 (N-δ), His_{13} (N-ε), His_4 (N-δ) and a bidentate carboxylate of Glu_{11}. While there is general agreement on the histidine residues as peptide donors for Zn^{+2}, the fourth donor in the full-length peptide can be from any of the side chains other than Glu_{11}, and Asp_1 is commonly found in complexes with Aβ. The reported equilibrium binding constant for the Zn^{+2}-Aβ complex is in the range 10^6 to $5 \times 10^4 M^{-1}$ [59].

The structure of Cu^{+2}-Aβ has been studied by many investigators and a recent review and analysis of the results of these studies suggests that the binding models shown in Figure 7.14 are possible [59]. In Figure 7.14a, the Cu^{+2} ion is ligated by N-ε of His_6, His_{13} or His_{14}, the amino group on the N-terminus of the

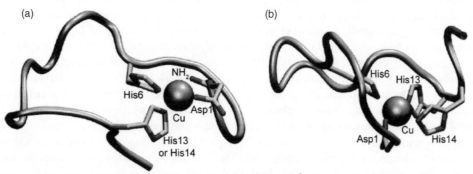

Figure 7.14 *Models of two possible coordination spheres of the Cu^{+2}-Aβ complex. From P. Faller and C. Hureau, Bioinorganic Chemistry of Copper and Zinc Ions Coordinated to Amyloid-β Peptide, Dalton Trans., 2009, 1080–1094. Reproduced by permission of the Royal Society of Chemistry*

peptide and a carboxylate oxygen of Asp_1. In the second model, Figure 7.14b, the donors are N-ε of His_6, His_{13} and His_{14}, and a carboxylate oxygen of Asp_1. The reported binding constant for the Cu^{+2}-Aβ complex varies widely with the conditions of the study and values in the range 10^7–10^{11} M^{-1} have been reported, with one as high as $\sim 10^{18}$ M^{-1} [71].

7.2.3 Radical production in AD

One of the aspects of AD is that there is oxidative stress (damage) to the brain [78]. Since the brain utilizes $\sim 20\%$ of the oxygen that enters the body, there are ample amounts of the dissolved oxygen gas for generating ROS that can react with and modify biological molecules in the brain. Supporting this is the observation that patients with AD have damage to a broad range of biomolecules, including free carbonyls, lipid peroxidation adducts, nitrated proteins and DNA oxidation products [57]. At the heart of the radical production mechanism is Cu^{+2} (and possibly Fe^{+3}), which if reduced to Cu^+ can react with molecular oxygen to generate ROS by a series of well-known reactions (7.5)–(7.9).

$$Cu^{+2} \rightarrow Cu^+ \text{ (metal reduction)} \tag{7.5}$$

$$Cu^+ + O_2 \rightarrow Cu^{+2} + O_2^{-\bullet} \text{ (formation of superoxide)} \tag{7.6}$$

$$2\,O_2^{-\bullet} + 2H^+ \rightarrow H_2O_2 + O_2 \text{ (formation of } H_2O_2) \tag{7.7}$$

$$Cu^+ + H_2O_2 \rightarrow Cu^{+2} + HO^\bullet + OH^- \text{ (Fenton chemistry)} \tag{7.8}$$

$$O_2^{-\bullet} + H_2O_2 \rightarrow HO^\bullet + OH^- + O_2 \text{ (Haber-Weiss reaction)} \tag{7.9}$$

$$R\text{-}H + HO^\bullet \rightarrow R^\bullet + H_2O \text{ (H atom abstraction)} \tag{7.10}$$

These reactions are most often written in connection with Fe^{+3}/Fe^{+2} but they also apply to Cu^{+2}/Cu^+. The first step in the metal-mediated process is the reduction of the metal ion, which can occur with a biologically-common reducing agent such as ascorbate [79] or can even be carried out by Aβ itself. For example, there is evidence that methionine, Met 35, of Aβ can be oxidized by Cu^{+2}, which produces Cu^+ and oxidized Met (methionine sulfoxide) as products [57]. Once Cu^{+2} is reduced to Cu^+, the latter can react with molecular oxygen to produce the anion radical, superoxide, $O_2^{-\bullet}$ and Cu^{+2}(7.6). Since there is an abundance of the metalloenzyme *superoxide dismutase* in the cell, two superoxide radical anions can be converted by the enzyme into hydrogen peroxide and oxygen, according to (7.7). While both superoxide and hydrogen peroxide can directly react with and modify biological molecules, H_2O_2 can react with reduced copper to produce hydroxyl radical, $HO^{\bullet -}$ this is called *Fenton chemistry* (7.8) – and superoxide can react with hydrogen peroxide to also produce hydroxyl radical in the *Haber–Weiss reaction* (7.9). Because hydroxyl radical can abstract a *hydrogen atom* from a wide variety of target molecules in the cell, including proteins, DNA, lipids and carbohydrates, to form water (7.10), the radical is much more reactive than either superoxide or hydrogen peroxide. Since abstraction of a hydrogen atom from a biological target transfers the radical originally on $HO^\bullet$ to the target, producing an unstable product, the modified biomolecule usually undergoes a series of reactions that ultimately result in a change in its chemical structure. Depending on the exact nature of the chemical modification and the number of molecules affected, this type of chemical damage usually results in impairment of some biochemical function, causing the cell to die. Since the plaque is the site of radical generation and is outside the cell, it is not clear how ROS produced in the vicinity of the plaque cause damage to the cell, especially since the most damaging radical, $HO^\bullet$, cannot migrate very far before it collides with and modifies a target. This observation is supported by a recent study which showed that while both the monomeric and fibril forms of Cu-Aβ produce H_2O_2 in the presence of ascorbate, hydroxyl radicals released in the reaction do not escape into solution but rather react with and damage $A\beta$ itself, with the damage occurring at the histidine and methionine residues of the peptide [79]. However, it is possible that a soluble oligomeric form of Cu-$A\beta$ could break off from the plaque and attach itself to the surface of a neuron, thus causing damage to the lipids in the membrane, which could kill the cell.

Clearly, considerable progress has been made on understanding the biochemical origins of Alzheimer disease, but many questions, including the central one of the role of plaque formation in the pathogenesis of the disease, remain unanswered. However, the rapid pace with which the scientific literature is unfolding on the subject suggests that the answer to this, and to the more important question of how to treat the disease, is forthcoming.

7.3 Copper in Wilson's and Menkes diseases

Wilson disease is a relatively rare genetic disorder (1 in 30 000 births) that is characterized by high levels of copper in tissue, especially in the liver and brain, which affects the function of these organs [80–82]. Patients with Wilson disease display reduced hepatic ability, cirrhosis and/or neurological or psychiatric problems that can be Parkinson-like in nature. Biochemically, the disease is caused by a mutation in the ATP7B protein (Box 3.3), which causes the copper levels in blood to be higher than normal. There are two strategies for treating Wilson disease. One approach is to give the patient agents that bind copper and cause the ion to be eliminated from the body as a metal chelate; the other is to administer Zn^{+2} ion, for example zinc acetate, which is relatively nontoxic and induces the production of the natural chelating protein metallothionein, MT (Box 2.1), in intestinal cells [83,84]. When these cells become necrotic and are sloughed off, the complexed copper is eliminated from the body.

Compounds used for treating Wilson disease via chelation are D-penicillamine (Figure 7.11), triethylenetetramine (trientine) and ammonium tetrathiomolybdate (TTM) (Figure 7.15). As described above, D-penicillamine, which is being studied in connection with Alzheimer disease, strongly binds copper, which it probably eliminates from the body as Cu^+. However, D-penicillamine is relatively toxic, which may in part be related to its ability to generate ROS via a Cu^{+2} to Cu^+ reduction process [80,85]. Trientine (Figure 7.15), which is also used to treat Wilson disease, is a well-known metal-chelating agent similar to the acyclic tetradentate ligand 232tet, shown in Figure 1.27. Trientine strongly binds Cu^{+2}, as well as other metal ions.

The copper-lowering ability of TTM was discovered by accident in the 1940s when it was noticed that cows and sheep grazing in certain pastures in Australia and New Zealand developed copper-deficiency syndrome [86]. Upon further investigation it was found that the soils in the pastures were rich in molybdenum and it was suspected that the grazing animals were converting the molybdenum in their food supply into a compound that blocked the absorption of copper. While molybdenum salts fed directly to cows and sheep caused copper-deficiency syndrome, the same treatment involving rats did not induce the syndrome. Eventually it was realized that sheep and cows, which are ruminants and produce high amounts of sulfide, S^{2-}, in their stomach, were converting the ingested molybdenum into tetrathiomolybdate, $[MoS_4]^{2-}$, and that

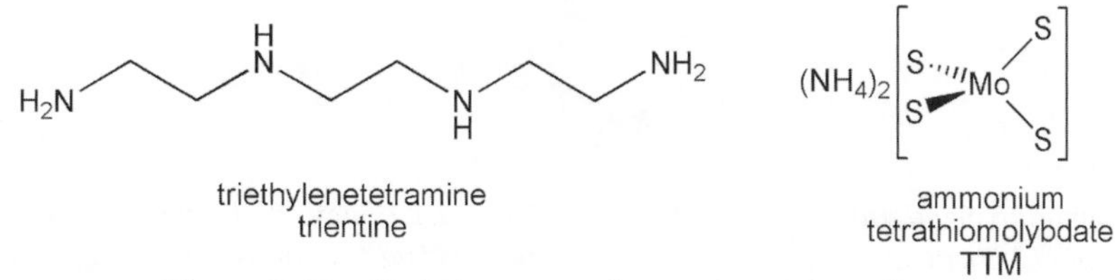

Figure 7.15 *Chelating agents for treating Wilson disease*

the complex anion was binding copper and blocking it from being absorbed into the animals. Since TTM proved to be a potent binding ligand for copper, and it exhibited low toxicity, Brewer and coworkers explored the use of the compound for treating Wilson disease [81]. Ammonium tetrathiomolybdate is currently in the latter stages of the drug approval process in the United States for treatment of the neurological effects of Wilson disease.

The molybdenum ion in TTM is in the $+6$ oxidation state, having the electronic rare-gas configuration [Kr]. The ion has tetrahedral geometry, with both σ- and π-type bonds formed between the metal ion and the attached sulfide ions. The red color of TTM is due to a strong transition at $\sim$467 nm in the visible region of the spectrum, which is likely due to $d \leftarrow p$ charge transfer transitions from S^{-2} to Mo^{+6} [87]. The chelating ability of TTM is due to the presence of the lone pairs of electrons on the bound sulfide ions, which coordinate copper, especially Cu^+, with high affinity. The copper-binding geometry is based on the structure of the tetrahedron, which has six edges (the octahedron has 12 edges (Figure 1.1)) that allow up to six Cu^+ ions to bind to TTM to form multinuclear clusters. However, the most common species appear to have $Cu:Mo$ stoichiometries of 3 and 4 [88]. Shown in Figure 7.16 is the calculated structure of $[(CH_3SCu)_4MoS_4]^{2-}$, in which four Cu^+-SCH_3 units are bound to four of the six edges of the tetrahedral $[MoS_4]^{2-}$ ion [88]. When TTM is administered in therapy it is thought that Cu^{+2} is reduced to Cu^+ and that the latter is complexed to TTM with glutathione, GSH, producing three-coordiante Cu^+ in the cluster. The simple thiolate, CH_3S^-, simulates GS^- in the structure shown in Figure 7.16.

Orally-administered TTM reaches the blood, where it complexes 'free copper', $\sim$10% of the total copper, which is not tightly bound to the protein – *ceruloplasmin* – making the copper ion unavailable for cellular absorption [89]. Tetrathiomolybdate also reaches the liver, where it can remove some of the copper that is weakly bound to metallothionein (MT has strong and weak Cu sites), and upon interaction with GSH and the help of glutathione-linked transporters, transfer the copper to the kidney for elimination in the urine.

While TTM appears to be an excellent compound for treating patients with Wilson disease, it may also be useful for treating cancer and other afflictions [89]. Ammonium tetrathiomolybdate is currently in phase II clinical trials for the treatment of advanced kidney cancer, hormone refractory prostate cancer and malignant mesothelioma, underscoring the importance of copper metabolism in many disease states.

Menkes disease, also called *kinky hair disease*, is a progressive neurodegenerative genetic disorder that affects about 1 in 200 000 births [80,82]. The biological origin of Menkes disease is a mutation in the ATP7A gene which forces copper to remain in the cells in the intestinal tract, blocking its passage to the blood for distribution throughout the body. Lack of copper affects a number of copper-containing enzymes and

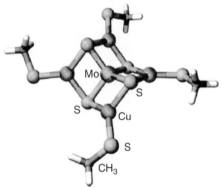

Figure 7.16   *Energy-minimized structure of $[(CH_3SCu)_4MoS_4]^{2-}$ determined by density functional theory. Adapted with permission from L. Zhang et al., Tracing Copper Thiomolybdate Complexes in a Prospective Treatment for Wilson's Disease, Biochemistry, 48, 891–97. Copyright 2009 American Chemical Society*

proteins, including lysyl oxidase, tyrosinase, cytochrome oxidase, dopamine β-hydroxylase, superoxide dismutase and dopamine oxidase, which are important for the normal structure and function of bone, skin, hair, blood vessels and the nervous system. Children with Menkes exhibit delayed development, kinky and brittle hair, seizures and mental retardation, and if they are not treated early for the disease they usually die before the age of three.

In contrast to Wilson disease, where excess copper is removed from the body with a chelating agent, the therapeutic approach for treating Menkes disease involves giving the patient a nontoxic copper complex, which can supply copper ions to the biological systems in the body that require them for function. Since the copper–histidine complex, which is currently in phase II clinical trials, is the main complex found in blood serum, it can be used to treat patients with Menkes disease.

The Cu^{+2}-bis-histidine complex is easily synthesized by mixing aquated $CuCl_2$ with two equivalents of histidine and adjusting the pH $\sim$7. Although the complex is easy to prepare, the fact that histidine is a tridentate ligand with three good copper donor groups – that is, the amino and imidazole functions and the carboxylate, all of which have protonated and deprotonated forms – and Cu^{+2} forms five-coordinate square pyramidal and six-coordinate complexes in water, characterization of Cu^{+2}-His species in aqueous solution is challenging. Since the electronic configuration of Cu^{+2} is $3d^9$, the metal ion is a classic Jahn–Teller system (see Chapter 1), which for a six-coordinate complex is usually manifested by displacement of two axial ligands farther away from the metal ion than the other four, producing a tetragonally-distorted complex.

The distribution of species in the Cu^{+2}-histidine system was determined by Kruck and Sarkar [90], and later by Mesu *et al.* [91], the latter investigators using a variety of physical techniques to show that the main species in water at pH 7.4 (25 °C) is [Cu(His)$_2$], $\sim$80%, with a smaller amount of [Cu(His)$_2$(H$_2$O)], $\sim$20%. As shown in Figure 7.17, both of these complexes are octahedral, having the all-*trans* geometry for the two His ligands, but in [Cu(His)$_2$(H$_2$O)] a water molecule has replaced one of the ligated carboxylates. A single crystal x-ray analysis of a complex isolated from an aqueous solution containing a copper–histidine of 1 : 2 at pH 7.4 shows that the bis-histidine complex is five-coordinate, square pyramidal, with the imidazole group of one of the histidine ligands not bound to the copper ion [92,93].

For treatment of Menkes disease, the copper–histidine complex is prepared in normal saline solution and injected, usually subcutaneously, into the patient [80]. While the stability constant for the formation of [Cu (His)$_2$] is relatively high, log $\beta_2 = 18.5$, when the complex reaches the blood, one of the histidine ligands is displaced and the resulting mono-histidine-aqua complex enters into an equilibrium with the copper-binding site on human serum albumin, forming a ternary complex, HSA-Cu-His [94]. In this way, both HSA and His act as copper buffers to regulate the amount of 'exchangeable' copper in blood, which is approximately 10% of the copper that is not tightly bound to the main copper transport protein, ceruloplasmin.

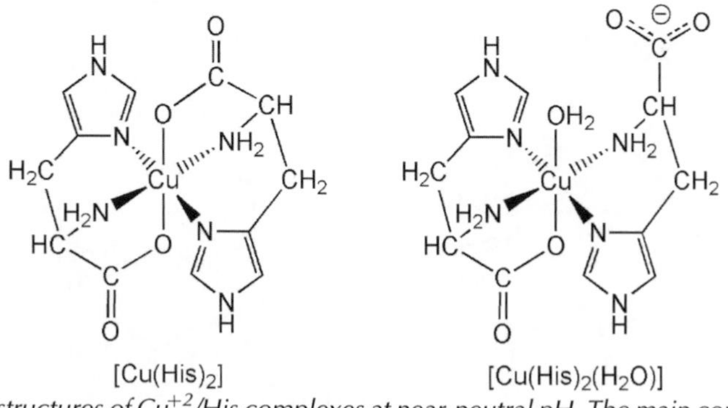

$[Cu(His)_2]$ $[Cu(His)_2(H_2O)]$

Figure 7.17 *Proposed structures of Cu^{+2}/His complexes at near-neutral pH. The main complex present, pH 7.4, is [Cu(His)$_2$], with a smaller amount of [Cu(His)$_2$(H$_2$O)]*

Treatment with copper–histidine early in life prevents serious complications associated with Menkes disease, mainly because copper is made available during the critical period of development of the central nervous system. While the number of patients is not large, those that have been diagnosed early with the disease and have been treated with copper–histidine have survived well beyond the early death expectancy, $\sim$3 years, and while they are not totally normal, they have manageable deficiencies [80].

7.4   Zinc–bicyclam: a chemokine receptor antagonist

In the search for drugs to combat human immunodeficiency virus, HIV, the virus that causes AIDS, De Clercq *et al.* examined the well-known macrocyclic chelate compound cyclam (Figure 7.18) [95–98]. The researchers tested several commercially-available samples of cyclam and found that one of the samples showed anti-HIV activity against HIV replication in MT-4 cells in culture. By analyzing the active sample using HPLC and other techniques, they found that the material contained 1–2% of an impurity, and after isolating and characterizing the impurity they discovered that it was the bicyclam compound, with two cyclam units connected to each other through a carbon–carbon bond, JM1657 (Figure 7.18). After showing that JM1657 was indeed the anti-HIV active compound in the mixture, they attempted to synthesize the compound, but lack of success prompted them to make other, more readily accessible bicyclams, one of which was the aromatic bridged compound containing the linker 1,4-phenylene-bis(methylene), AMD3100. This compound proved to be very active against two strains of the HIV virus at concentrations in the 0.005 μg ml^{-1} range and remarkably it was not toxic to host cells at concentrations of $\sim$500 μg ml^{-1}, which made the selectivity of the compound for the virus one of the highest ever recorded. The compound was entered into clinical trials for treatment of AIDS but due to significant cardiac side effects it failed in phase II and was withdrawn from further development [99].

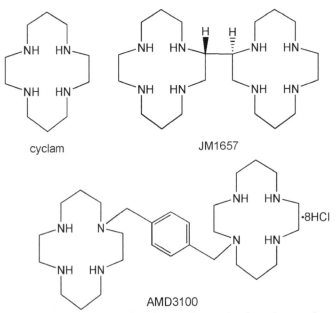

Figure 7.18   *Structures of cyclam and two of its bicyclam analogs*

An interesting finding in the clinical trials with AMD3100 was that the white blood cell (WBC) counts of subjects increased dramatically and reached a peak $\sim$6 hours after IV administration of the bicyclam, an observation that ultimately allowed AMD3100 to gain a new lease of life as a pharmaceutical agent. After determining that the compound was releasing CD34 + stem cells, which are found in bone marrow, into the blood, AMD3100 was entered into clinical trials as a new agent for labializing *hematopoietic stem cells*, and in 2008 the United States FDA approved AMD3100, named Mozobil and plerixafor, for treatment of patients with *non-Hodgkin's lymphoma* and *multiple myeloma* undergoing bone marrow transplants. The compound, which is intravenously administered prior to the transplant, mobilizes stem cells to the blood, whereupon the cells are collected and, after the transplant is completed, returned to the patient to facilitate grafting of the new marrow to the bone.

Since cyclam is a strong metal-binding ligand, early efforts were made to determine the role, if any, that naturally-occurring metal ions play in the mechanism of action of AMD3100. If a link to metal ions could be established, AMD3100 could be similar to the clinically-used anticancer drug *bleomycin*, which although not administered to the patient as a metal complex, is considered a *prodrug* in that it reacts with iron in the biological system to form the active complex [100]. The bleomycin–iron complex is thought to facilitate a metal-centered redox reaction that destroys genomic DNA, which ultimately causes the death of the tumor cell. While there is no evidence that AMD3100 works by a similar redox mechanism, it appears that the compound may be reacting with Zn^{+2} ion in blood to form $[Zn_2(AMD3100)]^{4+}$, which is the active form of the compound in the body.

The concentration of Zn^{+2} in blood is $\sim$19 μM, and since the stability constant of the zinc–cyclam is $\log K \sim 15$, AMD3100, which is similar in structure to cyclam, would be expected to be complexed by Zn^{+2} in blood [101–103]. Cyclam has pK_as of 11.6, 10.6, <2, <2, implying that at physiological pH the free ligand is diprotonated, $[H_2cyclam]^{2+}$ [104]. When cyclam binds to a metal ion, which is initially through the deprotonated amine nitrogen atoms, the protons on the protonated amines are easily displaced and the metal ion becomes coordinated to all four nitrogen atoms. Competing with Zn^{+2} for AMD3100 in blood is Cu^{+2}. Although the concentration of the latter ion is lower than that of Zn^{+2} in blood, the binding constant for $[Cu(cyclam)]^{+2}$ is about *11 orders of magnitude higher* than the corresponding Zn^{+2} complex, showing that the formation of the copper complex is greatly thermodynamically favored over that of the zinc complex. In order to determine which metal ion, Zn^{+2} or Cu^{+2}, becomes bound to the drug when AMD3100 is given to a patient, Paisey and Sadler carried out a series of kinetic studies using NMR and UV–visible spectroscopies under conditions simulating those found in blood [103]. They found that although Zn^{+2} ion has the lower binding constant toward cyclam, the kinetics of zinc binding to the macrocyclic ligand are much faster than those of copper binding, showing that the initially-formed complex in blood is $[Zn(cyclam)]^{2+}$. However, as expected this complex slowly reacts with Cu^{+2} to produce the thermodynamically-stable product [Cu (cyclam)]$^{2+}$, with a second-order rate constant of $\sim$6 $\times$ 10$^{-9} M^{-1} s^{-1}$ (0.5 cacodylate buffer, pH 7.4, 310 K). The study suggested that although the free ligand, AMD3100, is administered to the patient, it is probably a *prodrug* for $[Zn_2(AMD3100)]^{4+}$, which is the active compound in treatment.

The biological target for AMD3100 is the *trans*-membrane protein CXCR4 which is a receptor protein on the surface of leukocytes, cells in the immune system responsible for defending the body against infectious diseases [105–109]. The CXCR4 receptor molecule and the small protein that binds to it, SDF-1α (stromal cell-derived factor-1α), which is called a *chemokine*, control the migration and tissue-targeting of leukocytes as well as the metastatic spread and survival of different types of cancer cells. When SDF-1α is bound to CXCR4, hematopoietic stem cells remain bound in bone marrow, but if the SDF-1 is blocked from binding to its receptor, as would happen if AMD3100 were bound to the receptor site, the stem cells are released to the blood. Since the CXCR4 protein is also found on the surface of T-lymphocytes, it is one of the receptors utilized by the gp-120 protein on the surface of HIV to recognize and infect the T-cell, which explains the connection between the anti-HIV and stem-cell-mobilization activities of AMD3100 [106].

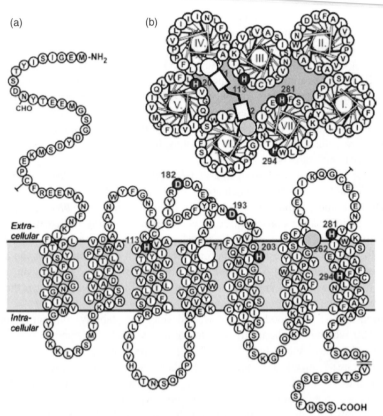

(a) (b)

Extra-
cellular

Intra-
cellular

Figure 7.19 *Serpentine (a) and helical wheel diagram (b) of the CXCR4 receptor. Shown are the locations of Asp₁₇₁ (white circle) and Asp₂₆₂ (gray circle), which are important binding sites for AMD3100 and its metal complexes (indicated by connected rectangles in (b)). The binding amino acid residues are located near the extracellular surface of the membrane. Adapted with permission from L.O. Gerlach et al., Metal Ion Enhanced Binding of AMD3100 to Asp262 in the CXCR4 Receptor, Biochemistry, 42, 710–717. Copyright 2003 American Chemical Society*

The serpentine (side view) and helical wheel diagram (top view) of the CXCR4 protein imbedded in the cell membrane are shown in Figure 7.19. The protein consists of seven *trans*-membrane α-helical segments, with the key helices being IV and VI, which contain the residues Asp$_{171}$ and Asp$_{262}$. These residues have carboxylate groups directed into a pocket that is important for the binding of AMD3100 and its metal complexes (Figure 7.19b). Studies have shown that binding of metal ions to AMD3100 enhances the affinity of the drug for the CXCR4 receptor site and improves the anti-HIV activity in the order: $(Zn^{+2})_2 >$ AMD3100 $(Ni^{+2})_2 > (Cu^{+2})_2 \gg (Co^{+3})_2 \gg (Pd^{+2})_2$ [103,109]. A striking feature of the series is that the Pd^{+2} complex, which can only bind the cyclam ligand in a square planar configuration, is inactive, but Zn^{+2}, which can easily adopt a variety of different stereochemistries (Zn^{+2} has zero CFSE), is the most active complex. In order to determine the role that ligand conformation and metal coordination geometry may play in the interaction of $[Zn_2(AMD3100)]^{4+}$ with the CXCR4 receptor, Sadler and coworkers used [^{1}H, ^{15}N] and [^{1}H, ^{13}C] HSQC NMR spectroscopy to study the structure of the Zn^{+2} complex in the presence of acetate ion, which is a good model for the carboxylate groups of Asp$_{171}$ and Asp$_{262}$ in the CXCR4 receptor site. In the absence of acetate ion, the major conformations of the cyclam ligand in $[Zn_2(AMD3100)]^{4+}$ are *trans*-I (R,S,R,S) and *trans*-III (S,S,R,R), where the R,S notation denotes the absolute configurations of the asymmetric nitrogen atoms in the

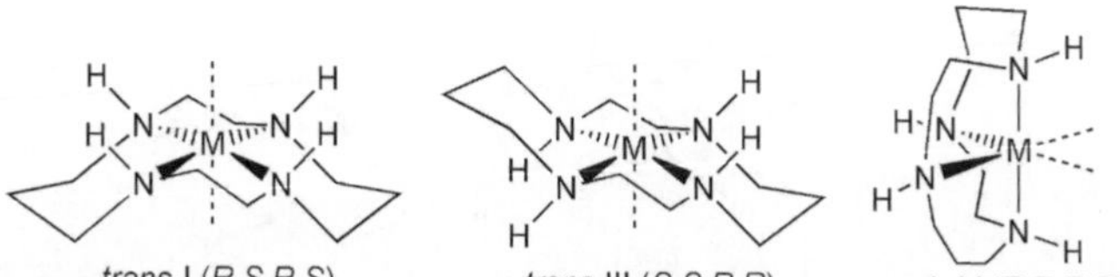

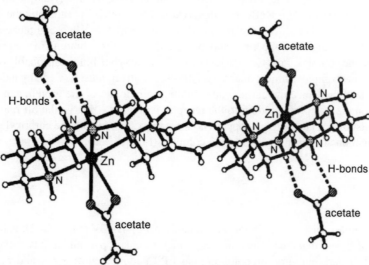

Figure 7.20   *Some possible isomers of cyclam bound to a metal ion. The absolute configurations of the asymmetric nitrogen atoms in the complexes are denoted by R and S*

complex (Figure 7.20). When acetate ion is incrementally added to a solution containing $[Zn_2(AMD3100)]^{4+}$, one of the zinc–cyclam groups of the complex rearranges to the *cis*-V conformation (Figure 7.20), and an acetate ligand becomes bound in a bidentate fashion to two *cis* coordination sites of the Zn^{+2} ion, producing, as the major isomer in solution, *cis*-V-*trans*-I. Interestingly, the *solid* isolated from a solution containing $[Zn_2(AMD3100)]^{4+}$ and acetate ion has *both* of its cyclam ligands in the *cis*-V configuration, with bidentate acetates coordinated to the Zn^{+2} ions in the complex and double hydrogen bonds from a second acetate helping to stabilize the *cis*-V configuration (Figure 7.21). Since the solid state and solution structures are different, the free-energy difference between the cyclam isomers may not be very large, and since the binding constant of $[Zn_2(AMD3100)]^{4+}$ toward the CXCR4 receptor is substantial, $\sim10^8\,M^{-1}$, there may be sufficient binding energy available to modify ligand configurations about the metal ions of $[Zn_2(AMD3100)]^{4+}$ when the compound is bound to the protein [105].

In an effort to minimize the number of cyclam isomers that can form and generate structures with high affinity for the CXCR4 receptor, chemically-modified bicyclam ligands with restricted conformations have been made [111,112]. The bicyclam (a) in Figure 7.22, which forms a complex with Zn^{+2} with the *trans*

Figure 7.21   *X-ray crystal structure of $[Zn_2(AMD3100)(OAc)_2](OAc)_2 \cdot 2CH_3OH$ (CH_3OH not shown). On each of the zinc–cyclam units, which have the cis-V configuration, one of the acetate groups acts as a bidentate ligand to Zn^{+2}, coordinated through the two oxygen atoms, and a second acetate is doubly hydrogen bonded to two amine hydrogen atoms of the cyclam ligand. Reprinted with permission from X. Liang et al., Structure and Dynamics of Metallomacrocycles: Recognition of Zinc Xylyl-Bicyclam by an HIV Coreceptor, J. Am. Chem. Soc, 124, 9105–12. Copyright 2002 American Chemical Society*

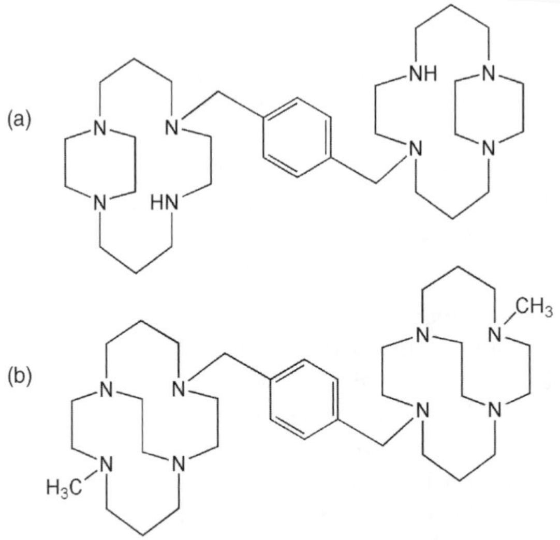

Figure 7.22 *Structures of two conformationally-restricted bicyclam ligands*

configuration, exhibits anti-HIV activity comparable to $[Zn_2(AMD3100)]^{4+}$, but the free ligand is *three orders of magnitude less active* as an anti-HIV agent, which underscores the role of the metal ion in the interaction between the drug and receptor. The bicyclam (b) in Figure 7.22 has a bis-methylene bridge between two ring nitrogen atoms that forces the ligand to adopt the *cis*-V configuration (Figure 7.20) [112]. When compared to $[Cu_2(AMD3100)]^{4+}$, which has a *trans* square planar geometry for the coordinated copper ions, the copper complex of the *cis*-V bicyclam ligand, Figure 7.22(b), was found to only slowly dissociate from the CXCR4 receptor and was about 10 times more effective than $[Cu_2(AMD3100)]^{4+}$ in blocking the 4R strain of HIV-1 from attacking MT-4 cells. The difference in anti-HIV activity of these two complexes suggests that the coordination geometry around the copper ion may be important for direct binding of the protein carboxylates to the copper ions in the complex. For $[Cu_2(AMD3100)]^{4+}$, carboxylate binding probably occurs via the axial coordination sites of the complex, which, due to the Jahn–Teller effect, must be weak ligand-binding sites. The investigators suggested that since the carboxylates in the receptor site of the protein can only bind to axial coordination sites on the complex, binding to the receptor is weaker – that is, the off rate is large – and the biological actively is lower than with the copper-*cis*-V complex, with which the carboxylates of the receptor can bind to coordination sites in the plane of the complex (Figure 7.20).

Problems

1. The vanadyl complex, bis(ethylmaltolato)oxovanadium(IV), BEOV (Figure 7.1), is in clinical trials for the treatment of type 2 diabetes. The complex, which is square pyramidal, has a magnetic moment of $\mu_{eff} = 1.72\,BM$, is dark blue-gray in color and has EPR parameters of: g_x, g_y and g_z, 1.988, 1.976 and 1.935, respectively, and A_x, A_y and A_z, 60.5, 60.0 and $170.0 \times 10^{-4}\,cm^{-1}$, respectively.

   a. Using Table 1.1, which gives the relative energies of the *d*-orbitals, give the *d*-orbital splitting pattern for a square pyramidal complex with five identical ligands.

   b. Both the EPR parameters and the value of the magnetic moment indicate that the crystal field of BEOV has lower symmetry than the idealized square pyramidal geometry given for the orbitals in Table 1.1.

Indicate the approximate *d*-orbital splitting pattern for BEOV and write the electronic occupancy, for example d_{xy}^1 and so on, for the V^{+4} ion in the complex.

c. How many *d-d* electronic transitions are theoretically possible for BEOV? Write the possible transitions for the single *d*-electron in the complex, for example $d_{xz} \leftarrow d_{xy}$ and so on, in order of increasing energy.

2.   Calculate the *wavelength* of radiation required for the resonance condition for a Q-band EPR instrument with $H = 1.25\,T$ and $g = 2$, $c = 3 \times 10^8\,\mathrm{cm\,s^{-1}}$.

3.   In a phase I clinical trial, a patient was given a 25 mg oral dose of a vanadium-containing insulin-mimetic agent and blood samples were collected at various times immediately following administration. After analyzing the blood samples for their vanadium content using AAS, the following pharmacokinetic data were collected: $T_{max} = 3.5$ hours, $C_{max} = 38.5\,\mu\mathrm{g\,ml^{-1}}$ and $t_{1/2} = 45.1$ hours for the clearance of vanadium from blood.

   a. From the information given, sketch the concentration versus time curve for the agent, being sure to label the axes and the various parameters given.
   b. Estimate the area under curve, AUC, from the information given.

4.   The bis-peroxovanadium compound, $[\mathrm{bpV(phen)}]^-$, can be chemically and photochemically activated to cut closed circular DNA in a random single-strand cleavage fashion. Sketch the set of curves that indicate how the relative concentrations of the three forms of DNA, form I (closed circular), form II (nicked circular) and form III (linear), change with time.

5.   Enantiomeric recognition may be important for the interaction of vanadium complexes exhibiting anticancer activity with chiral receptor sites in the body.

   a. Draw the mirror-image isomer of the antitumor active vanadyl complex, metvan, shown in Figure 7.6.
   b. Sketch the diastereomer of $[\mathrm{VO(salmet)(dpq)}]$, which has the *S* absolute configuration of the chiral carbon atom but the enantiomeric configuration of the chelate rings.

6.   Give the molecular orbital configuration and bond order for the following:

   a. O_2, molecular oxygen
   b. $O_2^{-\bullet}$, superoxide
   c. O_2^{-2}, peroxide.

7.   One of the approaches for treating Wilson disease is to administer simple Zn^{+2} salts, for example zinc acetate, to the patient, which is believed to induce the production of metallothionein, MT.

   a. Since MT induced by treatment with Zn^{+2} is likely to produce Zn_7MT, use crystal field arguments to briefly explain why Cu^{+2} may be able to displace Zn^{+2} from metal-binding sites on MT.
   b. There is evidence that the reaction of Cu^{+2} with apo-MT produces Cu^+-MT. Briefly describe how this reduction may occur.

8.   It is known that AMD3100 forms complexes with Zn^{+2}, Ni^{+2}, Cu^{+2}, Co^{+3}, Pd^{+2}.

   a. Assuming AMD3100 binds two metal ions, indicate the net charge on a complex formed with each of the metal ions listed above. If additional ligands are necessary to complete the coordination sphere, assume that they are uncharged; for example, water.
   b. Assuming AMD3100 binds only one metal ion or is not bound to any metal ion, what is likely to be the net charge on the compound under physiological conditions?
   c. Which of the metal ions listed above would be most appropriate for a potential pharmaceutical preparation containing metallated AMD3100? Briefly explain your choice.

d. Aside from providing a rigid linker between the two cyclam units, speculate on another possible role that the xylyl group may play when AMD3100 is bound to the CXCR4 receptor on the surface of the cell.

9. Metal ions are implicated in Alzheimer disease, AD. On the sequence below of the amyloid beta ($A\beta$) peptide, the major peptide found in an AD plaque, indicate the residues that are thought to be bound to Cu^{+2}, $3d^9$, in the plaque and the approximate coordination geometry around the metal ion in the Cu^{+2} complex.

$$A\beta = NH_2\text{-DAEFRHDSHYEVHHQKLVFFAEDVGSNKGAIIGLMVGGVVIA}$$
$$\phantom{A\beta = NH_2\text{-DAEFRH}}10203040$$

References

1. Lyonnet, B., Martz, M., and Martin, E. (1899) L'emploi Thérapeutique des derives du vanadium. *La Presse Médicale*, **32**, 191–192.
2. Tolman, E.L., Barris, E., Burns, M. *et al.* (1979) Effects of vanadium on glucose metabolism *in vitro*. *Life Sciences*, **25**, 1159–1164.
3. Dubyak, G.R. and Kleinzeller, A. (1980) The insulin-mimetic effects of vanadate in isolated rat adipocytes. Dissociation from effects of vanadate as a $(Na^+\text{-}K^+)ATPase$ inhibitor. *The Journal of Biological Chemistry*, **255**, 5306–5312.
4. Shechter, Y. and Karlish, S.J. (1980) Insulin-like stimulation of glucose oxidation in rat adipocytes by vanadyl (IV) ions. *Nature*, **284**, 556–558.
5. Heyliger, C.E., Tahiliani, A.G., and McNeill, J.H. (1985) Effect of vanadate on elevated blood glucose and depressed cardiac performance of diabetic rats. *Science*, **227**, 1474–1477.
6. Poucheret, P., Verma, S., Grynpas, M.D., and McNeill, J.H. (1998) Vanadium and diabetes. *Molecular and Cellular Biochemistry*, **188**, 73–80.
7. Thompson, K.H. and Orvig, C. (2004) Vanadium compounds in the treatment of diabetes. *Metal Ions in Biological Systems*, **41**, 221–252.
8. Thompson, K.H., McNeill, J.H., and Orvig, C. (1999) Vanadium compounds as insulin mimics. *Chemical Reviews*, **99**, 2561–2571.
9. Thompson, K.H. and Orvig, C. (2006) Vanadium in diabetes: 100 years from phase 0 to phase I. *Journal of Inorganic Biochemistry*, **100**, 1925–1935.
10. Thompson, K.H., Lichter, J., LeBel, C. *et al.* (2009) Vanadium treatment of type 2 diabetes: A view to the future. *Journal of Inorganic Biochemistry*, **103**, 554–558.
11. Crans, D.C. (2000) Chemistry and insulin-like properties of vanadium(IV) and vanadium(V) compounds. *Journal of Inorganic Biochemistry*, **80**, 123–131.
12. Sakurai, H., Yoshikawa, Y., and Yasui, H. (2008) Current state for the development of metallopharmaceuticals and anti-diabetic metal complexes. *Chemical Society Reviews*, **37**, 2383–2392.
13. Hiromura, M. and Sakurai, H. (2008) Action mechanism of insulin-mimetic vanadyl-allixin complex. *Chemistry & Biodiversity*, **5**, 1615–1621.
14. Mukherjee, B., Patra, B., Mahapatra, S. *et al.* (2004) Vanadium-an element of atypical biological significance. *Toxicology Letters*, **150**, 135–143.
15. Aureliano, M. and Crans, D.C. (2008) Decavanadate $[V_{10}O_{28}]^{6-}$ and oxovanadates: Oxometallates with many biological activities. *Journal of Inorganic Biochemistry*, **103**, 536–546.
16. Wüthrich, K. and Connick, R.E. (1968) Nuclear magnetic resonance studies of the coordination of vanadyl complexes in solution and the rate of elimination of coordinated water molecules. *Inorganic Chemistry*, **7**, 1377–1388.
17. Sasagawa, T., Yoshikawa, Y., Kawabe, K. *et al.* (2002) Bis(6-ethylpicolinato)oxovanadium(IV) complex with normoglycemic activity in KK-A^y mice. *Journal of Inorganic Biochemistry*, **88**, 108–112.
18. Sakurai, H. (2002) A new concept: The use of vanadium complexes in the treatment of diabetes mellitus. *The Chemical Record*, **2**, 237–248.

19. McNeill, J.H., Yuen, V.G., Hoveyda, H.R., and Orvig, C. (1992) Bis(maltolate)oxovanadium(IV) is a potent insulin mimic. *Journal of Medicinal Chemistry*, **35**, 1489–1491.

20. Thompson, K.H. *et al.* (2003) Preparation and characterization of vanadyl complexes with bidentate maltol-type ligands; *in vivo* compairson of anti-diabetic therapeutic potential. *Journal of Biological Inorganic Chemistry: JBIC: A Publication of the Society of Biological Inorganic Chemistry*, **8**, 66–74.

21. Caravan, P. *et al.* (1995) Reaction chemistry of BMOV, bis(maltolate)oxovanadium(IV) – A potent insulin mimetic agent. *Journal of the American Chemical Society*, **117**, 12759–12770.

22. Liboiron, B.D., Thompson, K.H., Hanson, G.R. *et al.* (2005) New insights into the interactions of serum proteins with bis(maltolate)oxovanadium(IV): transport and biotransformation of insulin-enhancing vanadium pharmaceuticals. *Journal of the American Chemical Society*, **127**, 5104–5115.

23. Kiss, E., Fábián, I., and Kiss, T. (2002) Kinetics of ligand substitution reactions in the oxovandium(IV)-maltol system. *Inorganica Chimica Acta*, **340**, 114–118.

24. Kiss, T., Kiss, E., Garribba, E., and Sakurai, H. (2000) Speciation of insulin-memetic VO(IV)-containing drugs in blood serum. *Journal of Inorganic Biochemistry*, **80**, 65–73.

25. Sanna, D., Garribba, E., and Micera, G. (2009) Interaction of VO^{2+} ion with human serum transferin and albmin. *Journal of Inorganic Biochemistry*, **103**, 648–655.

26. Bordbar, A.-K., Creagh, A.L., Mohammadi, F. *et al.* (2009) Calorimetric studies of the interaction between the insulin-enhancing drug candiate bis(maltolate)oxovanadium(IV) (BMOV) and human serum apo-transferrin. *Journal of Inorganic Biochemistry*, **103**, 643–647.

27. Fugono, J., Yasui, H., and Sakurai, H. (2001) Pharmacokinetic study on gastrointestinal absorption of insulinomimetic vanadyl complexes in rats by ESR spectroscopy. *The Journal of Pharmacy and Pharmacology*, **53**, 1247–1255.

28. Zhang, S.-Q., Zhong, X.-Y., Chen, G.-H. *et al.* (2008) The anti-diabetic effects and pharmacokinetic profiles of Bis(maltolato)oxovanadium in non-diabetic and diabetic rats. *The Journal of Pharmacy and Pharmacology*, **60**, 99–105.

29. Mehdi, M.Z., Pandey, S.K., Théberge, J.F., and Srivastava, A.K. (2006) Insulin signal mimicry as a mechanism for the insulin-like effects of vanadium. *Cell Biochemistry and Biophysics*, **44**, 73–81.

30. Drago, R.S. (1992) *Physical Methods for Chemists*, 2nd edn, Saunders College Publishing, Fort Worth, TX.

31. Haken, H. and Wolf, H.C. (2004) Molecular physics and elements of quantum chemistry, *Introduction to Theory and Experiments*, 2nd edn, Springer, Berlin, Germany.

32. Atkins, P. (1997) *Physical Chemistry*, 6th edn, W. H. Freeman and Company, New York, NY.

33. Makinen, M.W. and Brady, M.J. (2002) Structural origins of the insulin-mimetic activity of Bis(acetylacetonato) oxovanadium(IV). *The Journal of Biological Chemistry*, **277**, 12215–12220.

34. Sakurai, H., Fujii, K., Watanabe, H., and Tamura, H. (1995) Orally active and long-term acting insulin-mimetic vanadyl complex: bis(picolinato)oxovanadium(IV). *Biochemical and Biophysical Research Communications*, **214**, 1095–1101.

35. Yasui, H., Takechi, K., and Sakurai, H. (2000) Metallokinetic analysis of disposition of vanadyl complexes as insulin-mimetics in rats using BCM-ESR method. *Journal of Inorganic Biochemistry*, **78**, 185–196.

36. Butler, A., Clague, M.J., and Meister, G.E. (1994) Vanadium peroxide complexes. *Chemical Reviews*, **94**, 625–638.

37. Posner, B.I. *et al.* (1994) Peroxovanadium compounds. *The Journal of Biological Chemistry*, **269**, 4596–4604.

38. Sheela, A., Roopan, S.M., and Vijayaraghavan, R. (2008) New diketone based vanadium complexes as insulin mimics. *European Journal of Medicinal Chemistry*, **43**, 2206–2210.

39. Faneca, H. *et al.* (2008) Vanadium compounds as therapeutic agents: Some chemical and biochemical studies. *Journal of Inorganic Biochemistry*, **103**, 601–608.

40. Xie, M., Xu, G., Li, L. *et al.* (2007) *In vivo* insulin-mimetic activity of [N, N'-1,3-propyl-bis(salicyladimine)] oxovanadium(IV). *European Journal of Medicinal Chemistry*, **42**, 817–822.

41. Mukherjee, R. *et al.* (2008) Vanadium-vitamin B_{12} bioconjugates as potential therapeutics for treating diabetes. *Chemical Communications*, 3783–3785.

42. Köpf-Maier, P. and Köpf, H. (1987) Non-platinum group metal antitumor agents: History, current status, and perspectives. *Chemical Reviews*, **87**, 1137–1152.

43. Du, H., Xiang, J., Zhang, Y. *et al.* (2008) Binding of V^{IV} to human transferrin: Potential relevance to anticancer activity of vanadocene dichloride. *Journal of Inorganic Biochemistry*, **102**, 146–149.

44. Scrivens, P.J., Alaoui-Jamali, M.A., Giannini, G. *et al.* (2003) Cdc25A-inhibitory properties and antineoplastic activity of bisperoxovanadium analogues. *Molecular Cancer Therapeutics*, **2**, 1053–1059.
45. Hiort, C., Goodisman, J., and Dabrowiak, J.C. (1996) Cleavage of DNA by the insulin-mimetic compound, $NH_4[VO-(O_2)_2(phen)]$. *Biochemistry*, **35**, 12354–12362.
46. Sakurai, H., Tamura, H., and Okatani, K. (1995) Mechanism for a new antitumor vaandium complex: Hydroxyl radical-dependent DNA cleavage by 1,10-phenanthroline-vanadyl complex in the presence of hydrogen peroxide. *Biochemical and Biophysical Research Communications*, **206**, 133–137.
47. D'Cruz, O.J. and Uckun, F.M. (2002) Metvan: A novel oxovanadium(IV) complex with broad spectrum anticancer activity. *Expert Opinion on Investigational Drugs*, **11**, 1829–1836.
48. Dong, Y., Narla, R.K., Sudbeck, E., and Uckun, F.M. (2000) Synthesis, X-Ray structure, and anti-leukemic activity of oxovanadium(IV) complexes. *Journal of Inorganic Biochemistry*, **78**, 321–330.
49. Narla, R.K., Chen, C.L., Dong, Y., and Uckun, F.M. (2001) *In vivo* antitumor activity of Bis(4,7-dimethyl-1,10-phenanthroline)sulfatooxovanadium(IV)(METVAN) [$VO(SO_4)(Me_2$-phen$)_2$]. *Clinical Cancer Research*, **7**, 2124–2133.
50. Fu, Y., Wang, Q., Yang, X.G. *et al.* (2008) Vanadyl bisacetylacetonate induced G1/S cell cycle arrest via high intensity ERK phosphorylation in HepG2 cells. *Journal of Biological Inorganic Chemistry: JBIC: A Publication of the Society of Biological Inorganic Chemistry*, **13**, 1001–1009.
51. Butenko, N. *et al.* (2009) DNA cleavage activity of $V^{IV}O(acac)_2$ and derivatives. *Journal of Inorganic Biochemistry*, **103**, 622–632.
52. Sasmal, P.K., Patra, A.K., Nethaji, M., and Chakravarty, A.R. (2007) DNA cleavage by new oxovanadium(IV) complexes of n-salicylidene amino acids and phenanthroline bases in the photodynamic therapy window. *Inorganic Chemistry*, **46**, 11112–11121.
53. http://www.alz.org (03/17/09).
54. Bush, A.I. (2008) Drug development based on the metals hypothesis of Alzheimer's disease. *Journal of Alzheimer's Disease*, **15**, 223–240.
55. Bush, A.I., Masters, C.L., and Tanzi, R.E. (2003) Copper, β-amyloid, and Alzheimer's disease: Tapping a sensitive connection. *Proceedings of the National Academy of Sciences of the United States of America*, **100**, 11193–11194.
56. Barnham, K.j. and Busch, A.I. (2008) Metals in Alzheimer's and Parkinson's diseases. *Current Opinion in Chemical Biology*, **12**, 222–228.
57. Smith, D.G., Cappai, R., and Barnham, K.J. (2007) The redox chemistry of the Alzheimer's disease amyloid β peptide. *Biochimica et Biophysica Acta*, **1768**, 1976–1990.
58. Donnelly, P.S., Xiao, Z., and Wedd, A.G. (2007) Copper and Alzheimer's disease. *Current Opinion in Chemical Biology*, **11**, 128–133.
59. Faller, P. and Hureau, C. (2009) Bioinorganic chemistry of copper and zinc ions coordinated to amyloid-β peptide. *Dalton Transactions*, 1080–1094.
60. Gaggelli, E., Kozlowski, H., Valensin, D., and Valensin, G. (2006) Copper homeostasis and neurodegenerative disorders (Alzheimer's, prion, and Parkinson's diseases and amyotrophic lateral sclerosis). *Chemical Reviews*, **106**, 1995–2044.
61. Drago, D., Bolognin, S., and Zatta, P. (2008) Role of metal ions in the A-beta oligomerization in Alzheimer's disease and in other neurological disorders. *Current Alzheimer Research*, **5**, 500–507.
62. Liu, G., Garrett, M.R., Men, P. *et al.* (2005) Nanoparticle and other metal chelation therapeutics in Alzheimer disease. *Biochimica et Biophysica Acta*, **1741**, 246–252.
63. Drago, D. *et al.* (2008) Potential pathogenic role of β-Amyloid 1-42-aluminum complex in Alzheimer's disease. *The International Journal of Biochemistry & Cell Biology*, **40**, 731–746.
64. http://clinicaltrials.gov/ (03/17/09).
65. Ferrada, E., Arancibia, V., Loeb, B., and Norambuena, E. (2007) Stoichiometry and conditional stability contstants of Cu(II) or Zn(II) clioquinol complexes; implications for Alzheimer's and Huntington's disease therapy. *NeuorToxicol*, **28**, 445–449.
66. Ritchie, C.W. *et al.* (2003) Meta-protein attenuation with iodochlorohydroxyquin (Clioquinol) targeting a-beta amyloid deposition and toxicity in Alzheimer's disease: A pilot phase 2 clinical trial. *Archives of Neurology*, **60**, 1685–1691.

67. Lannfelt, L. *et al.* (2008) Safety, efficacy, and biomarker findings of PBT2 in targeting Aβ as a modifying therapy for Alzheimer's disease: A phase IIa, double blind, randomised, placebo-controlled trail. *Lancet Neurology*, **7**, 779–786.

68. Jenagaratnam, L. and McShane, R. (2006) Clioquinol for the treatment of Alzheimer's disease. *Cochrane Database of Systematic Reviews*, CD005380.

69. Sampson, E., Jenagaratnam, L., and McShane, R. (2008) Metal protein attenuating compounds for the treatment of Alzheimer's disease. *Cochrane Database of Systematic Reviews*, CD005380.

70. Di Vaira, M., Bazzicalupi, C., Orioli, P. *et al.* (2004) Clioquinol, a drug for Alzheimer's disease specifically interfering with brain metal metabolism: Structural characterization of its zinc(II) and copper(II) complexes. *Inorganic Chemistry*, **43**, 3795–3797.

71. Atwood, C.S., Scarpa, R.C., Huang, X. *et al.* (2000) Characterization of copper interactions with Alzheimer amyloid beta peptides: Identification of an attomolar-affinity copper binding site on amyloid beta 1–42. *Journal of Neurochemistry*, **75**, 1219–1233.

72. Treiber, C. *et al.* (2004) Clioquinol mediates copper uptake and counteracts copper efflux activities of the amyloid precursor protein of Alzheimer's disease. *The Journal of Biological Chemistry*, **279**, 51958–51964.

73. Yu, H., Zhou, Y., Lind, S.E., and Ding, W.Q. (2009) Clioquinol targets zinc to lysosomes in human cancer cells. *The Biochemical Journal*, **417**, 133–139.

74. Streltsov, V. (2008) X-ray absorption and diffraction studies of the metal binding sites in amyloid β-peptide. *European Biophysics Journal*, **37**, 257–263.

75. Petkova, A.T., Leapman, R.D., Guo, Z. *et al.* (2005) Self-propagating, molecular-level polymorphism in Alzheimer's β-amyloid fibrils. *Science*, **307**, 262–265.

76. Syme, C.D., Nadal, R.C., Rigby, S.E., and Viles, J.H. (2004) Copper binding to the amyloid-β (Aβ) peptide associated with Alzheimer's disease. *The Journal of Biological Chemistry*, **279**, 18169–18177.

77. Zirah, S. *et al.* (2006) Structural changes of region 1–16 of the Alzheimer disease amyloid β-peptide upon zinc binding and *in vitro* aging. *The Journal of Biological Chemistry*, **281**, 2151–2161.

78. Butterfield, A.D., Reed, T., Newman, S.F., and Sultana, R. (2007) Roles of amyloid β-peptide-associated oxidative stress and brain protein modifications in the pathogenesis of Alzheimer's disease and mild cognitive impairment. *Free Radical Biology and Medicine*, **43**, 658–677.

79. Nadal, R.C., Rigby, S.E., and Viles, J.H. (2008) Amyloid β-Cu^{+2} complexes in both monomeric and fibrillar forms do not generate H_2O_2 catalytically but quench hydroxyl radicals. *Biochemistry*, **47**, 11653–11664.

80. Sarkar, B. (1999) Treatment of Wilson and Menkes diseases. *Chemical Reviews*, **99**, 2535–2544.

81. Brewer, G.J. *et al.* (1994) Treatment of Wilson's disease with ammonium tetrathiomolybdate. *Archives of Neurology*, **51**, 54553.

82. Daniel, K.G., Harbach, R.H., Guida, W.C., and Dou, Q.P. (2004) Copper storage diseases: Menkes, Wilson's, and cancer. *Frontiers in Bioscience*, **9**, 2652–2662.

83. Hoogenraad, T.U. (2006) Paradigm shift in treatment of Wilson's disease: Zinc therapy now treatment of choice. *Brain & Development*, **28**, 141–146.

84. Jabłonska-Kaszewska, I., Dabrowska, E., Drobinska-Jurowiecka, A., and Falkiewicz, B. (2003) Treatment of Wilson's disease. *Medical Science Monitor: International Medical Journal of Experimental and Clinical Research*, **9** (Suppl 3), 5–8.

85. Gupte, A. and Mumper, R.J. (2007) An investigation into copper catalyzed D-penicillamine oxidation and subsequent hydrogen peroxide generation. *Journal of Inorganic Biochemistry*, **101**, 594–602.

86. Dick, A.T. and Bull, L.B. (1945) Some preliminary observations of the effect of molybdenum on copper metabolism in herbivorous animals. *Australian Veterinary Journal*, **21**, 70–72.

87. Shin, B.-K. and Han, J. (2008) Structure and physical properties of copper thiomolybdate complex, $(n-Bu_4N)_3$ $[MoS_4Cu_3Cl_4]$. *Bulletin of the Korean Chemical Society*, **29**, 2299–2302.

88. Zhang, L., Lichtmannegger, J., Summer, K.H. *et al.* (2009) Tracing copper thiomolybdate complexes in a prospective treatment for Wilson's disease. *Biochemistry*, **48**, 891–897.

89. Brewer, G.J. (2009) The use of copper-lowering therapy with tetrathiomolybdate in medicine. *Expert Opinion on Investigational Drugs*, **18**, 89–97.

90. Kruck, T.P. and Sarkar, B. (1973) Structure of the species in the copper(II)-L-histidine system. *Canadian Journal of Chemistry*, **51**, 3563–3571.
91. Mesu, J.G. *et al.* (2006) New insights into the coordination chemistry and molecular structure of copper(II) histidine complexes in aqueous solutions. *Inorganic Chemistry*, **45**, 1960–1971.
92. Deschamps, P., Kulkarni, P.P., and Sarkar, B. (2004) X-ray structure of physological copper(II)-Bis(L-histidinato) complex. *Inorganic Chemistry*, **43**, 3338–3340.
93. Deschamps, P., Kulkarni, P.P., Gautam-Basak, M., and Sarkar, B. (2005) The saga of copper(II)-L-histidine. *Coordination Chemistry Reviews*, **249**, 895–909.
94. Tabata, M. and Sarkar, B. (1985) Kinetic mechanism of copper(II) transfer between the native sequence peptide representing the copper(II)-transport site of human serum albumin and L-histidine. *Canadian Journal of Chemistry*, **63**, 3111–3116.
95. De Clercq, E. (2005) Potential clinical applications of the CXCR4 antagonist bicyclam AMD3100. *Mini-Reviews in Medicinal Chemistry*, **5**, 805–824.
96. De Clercq, E. (2003) The bicyclam AMD3100 story. *Nature Reviews. Drug Discovery*, **2**, 581–587.
97. De Clercq, E. (2001) Inhibition of HIV infection by CXCR4 and CCR5 chemokine receptor antagonists. *Antiviral Chemistry & Chemotherapy*, **12** (Suppl 1), 19–31.
98. De Clercq, E. *et al.* (1992) Potent and selective inhibition of human immunodeficiency virus (HIV)-1 and HIV-2 replication by a class of bicyclams interacting with a viral uncoating event. *Proceedings of the National Academy of Sciences of the United States of America*, **89**, 5286–5290.
99. Scozzafava, A., Mastrolorenzo, A., and Supuran, C.T. (2002) Non-peptidic chemokine receptors antogenists as emerging anti-HIV agents. *Journal of Enzyme Inhibition and Medicinal Chemistry*, **17**, 69–76.
100. Hecht, S.M. (2000) Bleomycin: New perspectives on the mechanism of action. *Journal of Natural Products*, **63**, 158–168.
101. Liang, X. and Sadler, P.J. (2004) Cyclam complexes and their applications in medicine. *Chemical Society Reviews*, **33**, 246–266.
102. Liang, X. *et al.* (2002) Structure and dynamics of metallomacrocycles: recognition of zinc xylyl-bicyclam by an HIV coreceptor. *Journal of the American Chemical Society*, **124**, 9105–9112.
103. Paisey, S.J. and Sadler, P.J. (2004) Anti-viral cyclam macrocycles: Rapid zinc uptake at physiological pH. *Chemical Communications*, 306–307.
104. Röper, J.R. and Elias, H. (1992) Kinetic studies of nickel(II) and copper(II) complexes with N_4 macrocycles of the cyclam type. 1. Kinetics and mechanism of complex formation with different N-methylated 1,4,8,11-tetraazacyclotetradecanes. *Inorganic Chemistry*, **31**, 1202–1210.
105. Gerlach, L.O. *et al.* (2003) Metal ion enhanced binding of AMD3100 to Asp in the CXCR4 receptor. *Biochemistry*, **42**, 710–717.
106. Choi, W.-T. *et al.* (2005) Unique ligand binding sites on CXCR4 probed by a chemical biology approach: implications for the design of selective human immunodeficiency virus type 1 inhibitors. *Journal of Virology*, **79**, 15398–15404.
107. Hatse, S. *et al.* (2001) Mutation of Asp[171] and Asp[262] of the chemokine receptor CXCR4 impairs its coreceptor function for human immunodeficiency virus-1 entry and abrogates the antagonistic activity of AMD3100. *Molecular Pharmacology*, **60**, 164–173.
108. Rosenkilde, M.M. *et al.* (2007) Molecular mechanism of action of monocyclam versus bicyclam non-peptide antagonists in the CXCR4 chemokine receptor. *The Journal of Biological Chemistry*, **282**, 27354–27365.
109. Esté, J.A. *et al.* (1999) Activity of different bicyclam derivatives against human immunodeficiency virus depends on their interaction with the CXCR4 chemokine receptor. *Molecular Pharmacology*, **55**, 67–73.
110. Gerlach, L.O., Skerlj, R.T., Bridger, G.J., and Schwartz, T.W. (2001) Molecular interactions of cyclam and bicyclam non-peptide antagonists with the CXCR4 chemokine receptor. *The Journal of Biological Chemistry*, **276**, 14153–14160.
111. Valks, G.C. *et al.* (2006) Configurationally restricted bismacrocyclic CXCR4 receptor antagonists. *Journal of Medicinal Chemistry*, **49**, 6162–6165.
112. Khan, A. *et al.* (2009) Binding optimization through coordination chemistry: CXCR4 chemokine receptor antagonists from ultrarigid metal complexes. *Journal of the American Chemical Society*, **131**, 3416–3417.

Further reading

Housecroft, C.E. and Sharpe, A.G. (2007) *Inorganic Chemsitry*, 3rd edn, Pearson Education Limited, Edinburgh Gate, UK.

Huheey, J.E., Keiter, E.A., and Keiter, R.L. (1993) Inorganic chemistry, *Principles of Structure and Reactivity*, 4th edn, Benjamin-Cummings Publishing Co., San Francisco, CA.

Mathews, C.K., van Holde, K.E., and Ahern, K.G. (2000) *Biochemistry*, 3rd edn, Benjamin Cummings Publishing Co., San Francisco, CA.

Rehder, D. (2008) *Bioinorganic Vanadium Chemistry*, John Wiley & Sons Ltd., Chichester, UK.

8

Metal Complexes for Diagnosing Disease

The previous chapters have focused on metal complexes for treating disease. These compounds mainly express their biological activity by losing one or more bound ligand and reacting directly with a target molecule in the cell. If the target was critical for the function of the cell, and if the cell could not in some way mitigate the effects of chemical modification, the cell entered into apoptosis and died. This behavior sharply contrasts with metal compounds used for diagnosing disease; for these agents, lack of chemical reactivity is the rule, and if an interaction takes place between the diagnostic agent and a biological target in the cell, it is generally brief and nondisruptive, and causes little harm to the cell. Since agents used for imaging can be detected in tissue with an analytical technique, an image of their location in the body can be created that is useful for diagnosing disease.

Metal complexes for detecting disease fall into two areas of diagnostic medicine, the oldest of which is *diagnostic nuclear medicine* [1–10]. In diagnostic nuclear medicine, often simply called *nuclear medicine*, a patient is given a small amount of a radioactive metal complex that is not harmful to cells. Since diseased and healthy tissues often do not absorb substances from their environment in the same way or at the same rate, a camera that is sensitive to the emitted radiation can be used to capture an image of the location of the radioactive compound in the body, which is helpful to the physician for determining which treatment options would be most beneficial to the patient.

In addition to radioactive agents for imaging tissue, nonradioactive metal complexes are used as contrast agents in *magnetic resonance imaging, MRI* [11–16]. In MRI, a patient is placed in a magnetic field and the relaxation properties of water molecules in their body are measured using nuclear magnetic resonance, NMR. Since water molecules in healthy tissue have different NMR properties than water in diseased tissue, MRI can be used to produce a three-dimensional image based on the relaxation rates of water molecules in different regions of the body. In order to enhance the quality of the image by accentuating the borders or contrast between normal and diseased tissue detected by MRI, paramagnetic metal compounds called *contrast agents* are given to the patient. Since these compounds greatly change the relaxation properties of water molecules in their vicinity, they improve the contrast of the image, which ultimately provides better information on which more accurate diagnoses can be made.

In addition to diagnosing disease, certain radioactive complexes, which emit strong radiation, can be used to control severe pain associated with bone metastases, and others are available for treating leukemia [17,18]. A sample of these agents will also be presented and discussed in this chapter.

Metals in Medicine   James C. Dabrowiak
© 2009 John Wiley & Sons, Ltd

8.1   Technetium in diagnostic nuclear medicine

Technetium, element 43, is a 'synthetic' element that was discovered in 1937 in a segment of molybdenum foil that had been used as a target for high-energy deuterons and protons in a cyclotron [10]. While a small amount of naturally-occurring technetium exists in uranium and molybdenum ores due to fission processes, the most stable isotope of technetium, ^{98}Tc, has a half life of 4.2 million years, which is much shorter than the period since the earth was formed, when the elements were created. In spite of the relatively late discovery of the element, technetium chemistry has been actively explored, especially since it was learned that one of its isotopes, ^{99m}Tc, has excellent imaging properties for application in nuclear medicine.

As shown in Figure 8.1, the decay of ^{99}Mo produces ^{99}Tc and ^{99m}Tc, the latter of which decays with a first-order half life of 6 hours to ^{99}Tc, with the release of a 140.5 keV γ-ray. Since the emitted γ-ray is too weak to cause damage to the cell, but energetic enough to be easily detected by scintillation, it is in the ideal range for imaging tissue in the body. An additional attractive feature of ^{99m}Tc is that its disappearance half life is 6 hours, which allows time to prepare a compound containing the isotope, administer it to the patient and capture images of the γ-rays released from the agent while it is in the body. The fact the half life of ^{99m}Tc is relatively short means that the radioactivity administered to the body decreases rapidly with time, producing a radiation exposure equivalent to that obtained from a single normal x-ray. As shown in Figure 8.1, the decay product of ^{99m}Tc, which is ^{99}Tc, is also radioactive. While this isotope has a long half life, 2.14×10^5 years, the fact that it is a weak β-emitter (electron), that its concentration is low and that it is cleared from the body suggests that it is not a risk factor in diagnostic nuclear medicine.

A general procedure for synthesizing a ^{99m}Tc compound used in diagnostic nuclear medicine is outlined in Figure 8.2. The molybdenum-99, as the tetrahedral anion molybdate, ^{99}MoO$_4{}^{-2}$, produced in a nuclear reactor is preloaded on to the top of a small chromatography column packed with alumina (aluminum oxide, Al$_2$O$_3$) that is inside a compact, lead-lined container. The resulting device, called a *Tc 99m generator*, is periodically shipped to hospitals and clinics carrying out diagnostic nuclear medicine; when the amount of ^{99m}Tc falls below a usable level, the generator is returned to the supplier for recharging. On the day that a ^{99m}Tc complex is needed for imaging, ^{99m}TcO$_4{}^-$, and a small amount of the daughter, ^{99}TcO$_4{}^-$, which are *pertechnetate* ions on the top of the column in the generator, are separated from ^{99}MoO$_4{}^{-2}$, also on the top of the column, by elution with normal saline solution (150 mM NaCl). Since the surfaces of alumina have exposed cationic sites to which anionic pertechnetate and molybdate are bound, the chloride ions in the saline solution passing down the column in the *eluent* easily displace the monoanionic pertechnetate from the alumina surface, but molybdate, because it has two negative charges, moves very slowly or not at all with the eluent. While the concentration

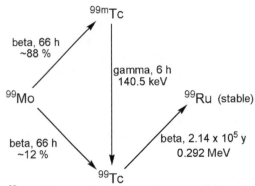

Figure 8.1   *Decay scheme for ^{99}Mo, showing the nature and energy of the radiation released and the first-order decay half life, $t_{1/2}$*

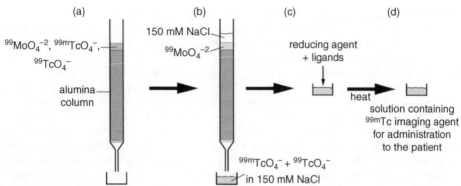

Figure 8.2 *General procedure for preparing an imaging agent containing ^{99m}Tc. While the apparatus used in a hospital setting for isolating $^{99m}TcO_4{}^-$, pertechnetate, called a Tc 99m generator, is compact and different than the conventional 'gravity' column shown, the figure provides the basic essentials of how $^{99m}TcO_4{}^-$ is isolated and used to make a ^{99m}Tc imaging agent. (a) An alumina column with adsorbed $^{99}MoO_4{}^{-2}$, $^{99m}TcO_4{}^-$ and $^{99}TcO_4{}^-$ at the top of the column; (b) the pertechnetate ions, $^{99m}TcO_4{}^-$ and $^{99}TcO_4{}^-$, are separated from the permolybdate ion, $^{99}MoO_4{}^{-2}$, by eluting with normal saline solution (150 mM NaCl); (c) both $^{99m}TcO_4{}^-$ and $^{99}TcO_4{}^-$ are reduced using $SnCl_2$ (stannous chloride), and reduced technetium species are reacted with the ligands to form an imaging complex; (d) a solution containing the ^{99m}Tc imaging agent is administered to the patient*

of pertechnetate in the eluent is very low, $\sim 10^{-10}\,M$, the solution contains enough ^{99m}Tc, which can be converted into an imaging agent, for diagnostic purposes.

Since the pertechnetate ion contains technetium in the $+7$ oxidation state bound to four oxo (O^{-2}) ligands, the metal ion in the complex is relatively unreactive and is not suitable for complexation with ligands that are required for delivering ^{99m}Tc to specific sites in the body. In order to enhance the chemical reactivity of technetium in the eluent toward the ligands, the pertechnetate is treated with a strong reducing agent, usually stannous chloride, $SnCl_2$, which reduces the Tc^{+7} to a lower oxidation state, Tc^+–Tc^{+5}, in the process converting Sn^{+2} to Sn^{+4} (Figure 8.2). In a hospital setting, this is typically done by adding a specified volume of the eluent containing $^{99m}TcO_4{}^-$ obtained from the generator to a 'kit' that contains large excesses (10^5-fold) of the ligand or ligands to be bound to the technetium ion, plus stannous chloride. After adding the solution containing the pertechnetate to the kit with the reducing agent and the ligands, the resulting solution is heated for a period of time to ensure that reduction and complexation of the technetium are complete and, since all solutions are sterile, a specified volume of the resulting solution is administered by IV injection to the patient. After waiting for the radiolabeled compound to reach its intended target site in the body, the area of interest in the patient is imaged using *single photon emission computed tomography, SPECT*.

Box 8.1 Single photon emission computed tomography (SPECT)

In diagnostic nuclear medicine, a radioactive compound is given to the patient and an image of the radiation in their body is obtained using a technique called *single photon emission computed tomography, SPECT* [10,19,20]. After allowing for the radioactive compound to localize in the tissue of the patient, the emitted γ-radiation is detected using a *gamma camera* (Figure 8.3), which takes a number of 2D images of the radiation from different perspectives. When all of these images are analyzed using a procedure called *tomography*, a 3D image of the radiation in the body is obtained. The heart of the gamma camera is a sodium iodide crystal, which contains a small amount of thallium iodide.

When an iodide ion in this crystal is impacted by a γ-ray, an electron is momentarily ejected from the iodide ion to produce an iodine atom. Although this electron 'roams' the crystal for a time, it eventually recombines with the iodine atom to regenerate iodide ion, but the capture results in the release of a photon of light. The process of converting the γ-ray to a photon of light that can be measured is called *scintillation*.

An important feature of the gamma camera, which allows it to resolve the spatial distribution of γ-rays being released from the tissue in the patient, is the *collimator*. This part of the camera is a lead plate that contains a large number of small holes. Since lead resists the transmission of γ-rays, only those rays that are emitted from tissue directly below (in line with) the holes in the plate can reach the sodium iodide crystal and cause a scintillation. By placing a number of photomultiplier tubes (PMTs), which are able to detect photons of light, on the side of the sodium iodide crystal opposite the collimator, the spatial location and intensity of the scintillation events in the crystal can be determined. When the signals from the PMTs are collected and entered into a computer program, the position and intensity of the γ-rays reaching the crystal can be reconstructed to form a 2D image called a *tomogram*. If the camera is placed in other positions relative to the patient and additional tomograms in different planes are collected, the resulting set of images can be use to construct a 3D image of the target tissue – a process called *tomography* – which can be used by the physician to determine the health of the tissue and what medical action, if any, might be needed.

A major challenge in characterizing technetium imaging agents at the *tracer level*, the concentration of complex used in diagnosis, is that the concentration is too low for many of the usual characterization techniques: NMR, IR, x-ray crystallography and so on. The approach taken over the years has been to work with and characterize compounds of ^{99}Tc, which can be obtained in 10–20 mg quantities, using a variety of physical techniques and to compare its retention time using an HPLC instrument equipped with a γ-detector to the retention time of the analogous ^{99m}Tc compound which has been produced at the tracer level [2]. If the retention times for the ^{99}Tc and ^{99m}Tc compounds match, it is assumed that the structures of the compounds are the same.

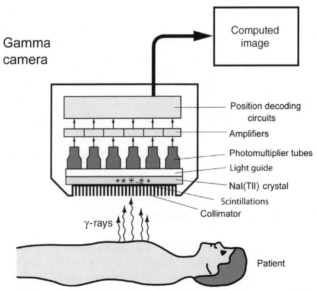

Figure 8.3 *A gamma camera is used to carry out single photon emission computed tomography, SPECT, on a patient who has received a ^{99m}Tc imaging agent. Reprinted with permission from the Federation of American Scientists. http://www.fas.org/irp/imint/docs/rst/Intro/img003.gif*

8.1.1   Clinically used ^{99m}Tc imaging agents

Figure 8.4 shows a few of the many ^{99m}Tc compounds that are used in diagnostic nuclear medicine. One of these, *Cardiolite*, or *99mTc-sestamibi*, is used as a heart-imaging agent [2,10]. The compound is made by adding a pertechnetate solution obtained from a generator to a kit that contains stannous chloride and the tetrahedral Cu^+ complex of the isonitrile ligand, methoxyisobutylisonirtile, BIMI; that is, $[Cu(BIMI)_4]BF_4$. Since the carbon atom of the isonitrile is a soft base and Cu^+ is a soft acid, complexation of the ligand to Cu^+ offers a means of stabilizing the somewhat reactive isonitrile in the kit until it is needed for complexation to Tc. When pertechnetate is added to the kit, the Tc^{+7} ion is reduced to Tc^+, which, because it is a soft acid, is complexed and stabilized by BIMI, forming the octahedral complex $[Tc(BIMI)_6]^+$ (Figure 8.4). Because the CN group of the isonitrile ligand

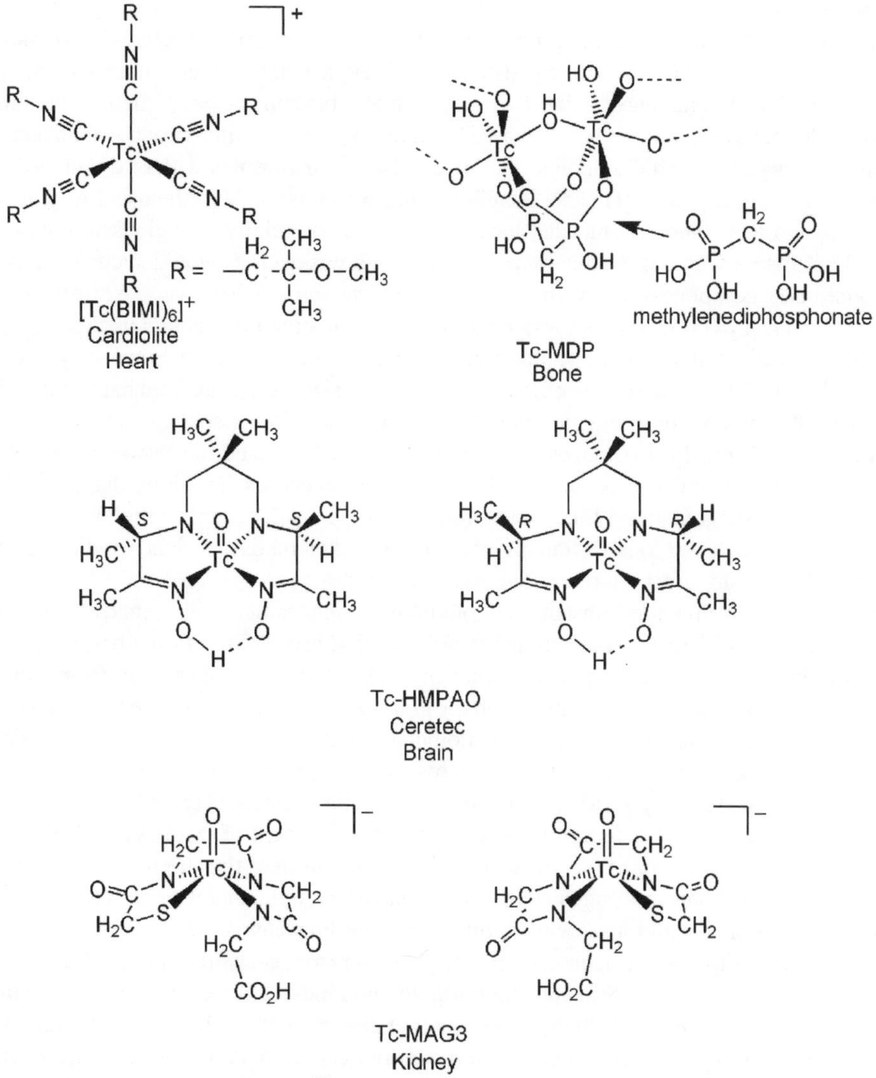

Figure 8.4   *Structures of commonly-used ^{99m}Tc imaging agents. The brain and kidney imaging agents exist as mirror-image isomers; that is, enantiomers*

is similar in electronic structure to carbon monoxide, CO, the bonding between the Tc^+ ion and the isonitrile is mainly covalent in nature, with both σ- and π-bonds being formed (see Figure 5.4).

Group I cations (K^+, Cs^+ and Rb^+) accumulate in the muscle tissue of the heart called the *myocardium*. Since the 'pump' that allows these cations to enter heart cells also permits Cardiolite to enter, the imaging agent can be used to detect and image healthy heart tissue. After a heart attack, heart tissue undergoes reversible damage called *myocardial ischemia*, caused by the obstruction of the coronary artery, which prevents oxygenated blood from reaching cells. If the affected tissue remains without oxygen for more than ~ 1 hour, it becomes *infarcted*, which means that the cells become necrotic and die. Because Cardiolite localizes in healthy heart tissue, imaging a patient who has suffered a heart attack with Cardiolite provides valuable information on which parts of the heart have been damaged by the attack and what treatment may be necessary [10]. In addition to being a useful imaging agent for the heart, Cardiolite can also be used to detect cancerous tissue in the breast and thyroid.

Bone is a complex structure, containing the mineral hydroxyapatite, $[Ca_{10}(PO_4)_6(OH)_2]$, and the protein collagen. The basic unit of collagen is a fiber that consists of a triple helix of three α-helical polypeptide chains that are ~ 1000 amino acids long. In bone, these wire cable-like triple helices are covalently crosslinked to each other to form a higher-order structure that has embedded hydroxyapatite. As shown in Figure 8.4, the technetium imaging agent, Tc-MDP, contains Tc^{+4} in a polynuclear complex with bridging methylenediphosphonate, medronic acid, ligands [21]. This imaging agent, which is made in the usual way, involving a kit containing the ligand and stannous chloride, localizes in bone, especially in regions where new growth is taking place. The new growth may be due to accelerated bone turnover, as would occur with the repair of a break in the bone or the new bone response that accompanies cancerous bone lesions. While the mechanism by which the agent localizes in bone is not known, the bifunctional nature of the methylenediphosphonate ligand suggests that one part of the ligand could be attached to Ca^{+2} in hydroxyapatite of bone, while the other part of the ligand is bound to $^{99m}Tc^{+4}$. Since the oxygen atoms of the MDP ligand are hard bases and both Tc^{+4} and Ca^{+2} are hard acids, this type of interaction is consistent with the HSAB concept.

Shown in Figure 8.5 are SPECT images with ^{99m}Tc-MDP of the mandible (lower jaw) and cranium of patients who are suffering from *hyperparathyroidism*, a disease that causes the overproduction of the hormone, PTHi, that leads to bone degradation [22]. As shown in Figure 8.5, patients with a high level of PTHi in their blood exhibit a greater degree of localization of ^{99m}Tc-MDP in skeletal tissue, clearly indicating the extent to which hyperparathyroidism affects normal bone structure.

The brain is protected from many substances in blood by a complex set of barriers called the *blood–brain barrier, BBB*. The *endothelial* cells that line the capillaries that bring blood to the brain are tightly joined together so that there are no channels or pores between them [10]. Separating these cells from brain tissue are a basement membrane and a fatty tissue lining called the *glial foot*, which also must be traversed by a diffusing agent. If an agent is successful in entering the endothelial cells – a process called *influx* – through some transport mechanism, it is most often *effluxed* back across the membrane by *p-glycoprotein*, which is capable of effluxing a variety of different types of small molecules from the cell [23,24] (see Box 3.3). Thus, whether a substance gains entry to the brain tissue depends on its relative rates of influx and efflux involving the endothelial cells, as well as its ability to traverse hydrophobic barriers that lie beyond the capillary wall. Molecules that are successful in crossing the BBB usually have low molecular weights – for example, proteins are prevented from crossing – and low, or most often no molecular charge.

The structure of the brain-imaging agent *Ceretec, Tc99m exametazime*, which contains Tc^{+5} coordinated to the tetradentate oxime ligand, *R,R, S,S*, hexamethylpropylene-amine oxime (HMPAO), is shown in Figure 8.4. The free oxime ligand has a number of hydrophobic methyl groups on its periphery and four relatively acidic protons which, when the ligand is complexed to a metal ion, can be lost. The protons are the two alcohol protons of the oxime and two protons of the secondary amine. However, when the ligand is bound to Tc^{+5} in Ceretec, only three of these protons are lost (one proton remains bound and is equally shared with the two

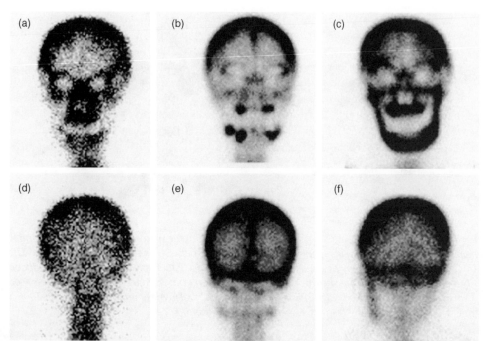

Figure 8.5   *SPECT images using ^{99m}Tc-MDP of the mandible (lower jaw) and cranium of a patient having secondary hyperparathyroidism (SHPT). (a)-(c), anterior (front) and (d)-(f) posterior (back) views. (a), (d) control group; (b), (e) patient with SHPT with a concentration of parathyroid hormone (PTHi) in the blood of ≤ 1000 pg mL^{-1} and (c), (f) patient with SHPT with PTHi > 1000 pg mL^{-1}. From Fig 1 p 57 of Boasquevisque, E. et al., ^{99m}Tc-MDP Bone Uptake in Secondary Hyperparathyroidism: Comparison of the Mandible, Cranium, Radius, and Femur. Oral Radiol., 2008, 24, 55–58*

oxime oxygens), making the net charge on the coordinated ligand in the complex (-3). Since the remaining group attached to the technetium ion is a doubly-negatively-charged oxo ligand, which is a vestige of one of the four original oxo ligands attached to pertechnetate, the complex has a net charge of zero. Ceretec, which possesses the $[TcO]^{3+}$ core, has a five-coordinate square pyramidal geometry, with the metal ion and the oxo group above the four nitrogen donor plane of the ligand (Figure 8.4).

The kit that is used to make Ceretec contains the racemic form of the ligand, with half of the molecules having the R,R configuration of the two asymmetric carbon atoms and the other half having the S,S configuration about these two stereocenters. The ligand isomer with R,S stereocenters, called the *meso* isomer, is not present in the kit used for preparing Ceretec [25]. Thus, the synthesis of Ceretec produces two ^{99m}Tc compounds that are mirror images of each other – that is, enantiomers – both of which are shown in Figure 8.4.

In an effort to determine if one optical isomer of Ceretec is selectively taken up by brain tissue, Ding and coworkers synthesized the pure forms of both enantiomers of HMPAO and made the corresponding ^{99m}Tc complexes [25]. They found that the imaging agent containing the S,S enantiomer had higher accumulation and was retained longer in the brain of a rat than either the ^{99m}Tc complex with the R,R enantiomer or the racimate, clearly showing that chiral recognition is important in determining how the agent distributes to tissue [26].

Ceretec is used for diagnosing various brain disorders including stroke, vascular disease and dementia, and an example of how it is used as a diagnostic tool for Alzheimer disease, AD, is summarized in the images shown in Figure 8.6 [27,28]. In this study, normal elderly patients and those with mild AD were administered Ceretec and images of their brains were captured using SPECT (Figure 8.6). Since Ceretec is distributed throughout the

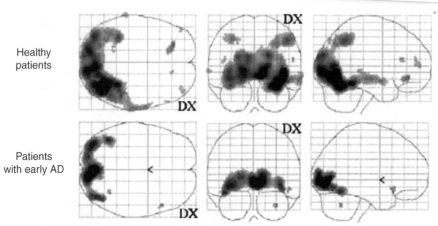

Healthy patients

Patients with early AD

Figure 8.6 *Three SPECT images using the ^{99m}Tc imaging agent Ceretec of normal elderly patients and those with early Alzheimer disease, AD. The images shown are the differences between the regional cerebral blood flow (rCBF) detected by Ceretec during a memory provocation minus rCBF at rest. Patients with early AD have reduced blood to the brain. Three views of the brain are shown. With kind permission from Springer Science + Business Media: Eur. J. Nucl. Mol. Imag., Memory-Provoked rCBF-SPECT as a Diagnostic Tool in Alzheimer's Disease? 33, 2006, 73–80, T. Sundström, E. Elgh, A. Larsson, B. Näsman, L. Nyberg, K.Å, Riklund, Fig 1*

brain by the blood, and blood flow in various regions (referred to as *regional cerebral blood flow, rCBF*) is believed to be a measure of brain activity, the amount of the imaging agent present in various parts of the brain may be used to determine regional brain activity. In the study, the patients who were given Ceretec were initially placed in a resting condition, which was followed by memory-provocation condition. While different memory provocations were used, one was to show the patient a series of faces with names, give them a distracting task and then ask them to match a face with the correct name. The data, which were analyzed by taking the difference between SPECT images in the memory-provocation state and in the resting state, show that patients with early AD have significantly less blood flow (brain activity) in the left parietal cortex, as measured by the relative amounts of Ceretec in the region, than do healthy individuals (Figure 8.6). The researchers were optimistic that this diagnostic approach could be used to determine the early stages of AD in patients.

The kidney is responsible for eliminating water-soluble substances in the blood from the body [10]. The *nephron*, the functional unit of the kidney, is composed of the *glomerulus*, which is connected to a multi-segmented *tubule*, and together they retain essential small molecules and ions in the blood and pass unwanted substances to the urine for elimination. Early in the development of technetium compounds as imaging agents it was discovered that complexes that are negatively charged have prolonged retention times in the kidney and thus they may be useful for imaging the organ. One compound, discovered by Fritzberg *et al.*, that is currently in widespread use for imaging the kidney is Tc-MAG3 (Figure 8.4) [29].

The compound can be made using a kit which contains the *S*-benzoyl-protected tripeptide ligand, stannous chloride and tartaric acid, which when heated in solution with ^{99m}TcO$_4$$^-$ produces Tc-MAG3. Due to the fact that the ligand is not symmetric, and the geometry of the complex is square pyramidal, two mirror-image isomers – that is, enantiomers – are possible (Figure 8.4). Both of these isomers have been separated and evaluated but they show only small differences in plasma clearance and renal transit [30,31]. While the ligand has a total of five acidic protons – three amide, one carboxylic acid and one thiol proton – only four are released to solution when the ligand is bound to the $[TcO]^{+3}$ core at neutral pH, giving the complex a net (-1) charge. Evidence suggests that the negative charge on Tc-MAG3 may be the basis for the compound entering the cell via *organic anion transporter 1, OAT1* [32,33].

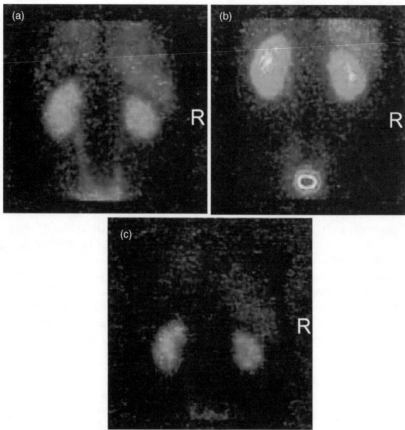

Figure 8.7 *Three SPECT images of the kidneys of a patient using the imaging agent Tc-MAG3. (a) Image of kidneys before administration of the drug candidate, Seliciclib; (b) image three days after administration; (c) image 14 days after administration. The imaging agent shows that the renal function of the right kidney (R) is poorer than that of the left and that drug-induced impairment of renal function, lighter shades of gray on (b), is recovered in 14 days. Reprinted by permission from Macmillan Publishers Ltd: C. Benson et al., A Phase I Trial of the Selective Oral Cyclin-Dependent Kinase Inhibitor Seliciclib (CYC202; R-Roscovitine), Administered Twice Daily for 7 Days Every 21 Days, Br. J. Cancer, 96, 29–37. Copyright 2007*

Figure 8.7 is an example of how Tc-MAG3 can be used to evaluate renal function. In a recent phase I clinical trial, patients were given the drug candidate *seliciclib*, which is an orally-administered agent with antitumor activity against a broad range of cancer cell lines [34]. Since one patient taking seliciclib in the trial showed abnormal kidney function, which appeared to be reversible, the kidneys of the patient were imaged with Tc-MAG3. As shown in Figure 8.7a, Tc-MAG3 indicates that while there is an asymmetry in kidney function, with the right kidney functioning less well than the left, both kidneys were performing normally before administration of seliciclib. However, three days after administration of the drug candidate, both kidneys showed reduced renal function, as evidenced by enhanced localization (retention) of Tc-MAG3 in each kidney Figure 8.7b), but14 days after administration, normal kidney function was returned (Figure 8.7c). This study shows the diagnostic power of nuclear medicine as a means of uncovering the side effects of a potential new drug for the treatment of cancer.

Figure 8.8   *Schematic of a ^{99m}Tc compound with an appended molecule for targeting specific sites in the body*

8.1.2   Technetium imaging agents for selective targeting

The huge success of imaging agents containing ^{99m}Tc has stimulated worldwide efforts to develop new technetium agents that can target different tissue types at the cellular level. Unlike compounds which owe their biodistribution to size, charge and lipophilicity, attaching (conjugating) a targeting vector to a metal chelate region capable of binding technetium (Figure 8.8) raises the possibility that ^{99m}Tc agents can be designed to target specific receptor sites on the cell surfaces. This has led to the synthesis and evaluation of a large number of compounds with appended sugars, peptides, steroids, proteins and other biologically-important groups capable of cellular targeting [1–9,36,37]. Consider the imaging agent [TcO]depreotide, shown in Figure 8.9, which contains a short peptide of the cyclic peptide hormone somatostatin, conjugated through a linker to a metal-chelating, MAG3-type ligand, which can bind ^{99m}Tc. Since somatostatin regulates cell growth, and cancer cells have more somatostatin receptors (SSTRs) on their surface than normal cells, [TcO]depreotide is a useful imaging agent for tumors.

The imaging agent is made by adding pertechnetate obtained from a Tc 99m generator to a kit containing the ligand stannous chloride and other additives. As shown in Figure 8.9, the receptor-binding portion of the ligand in [TcO]depreotide contains many chiral carbon atoms that properly align groups on the peptide to interact

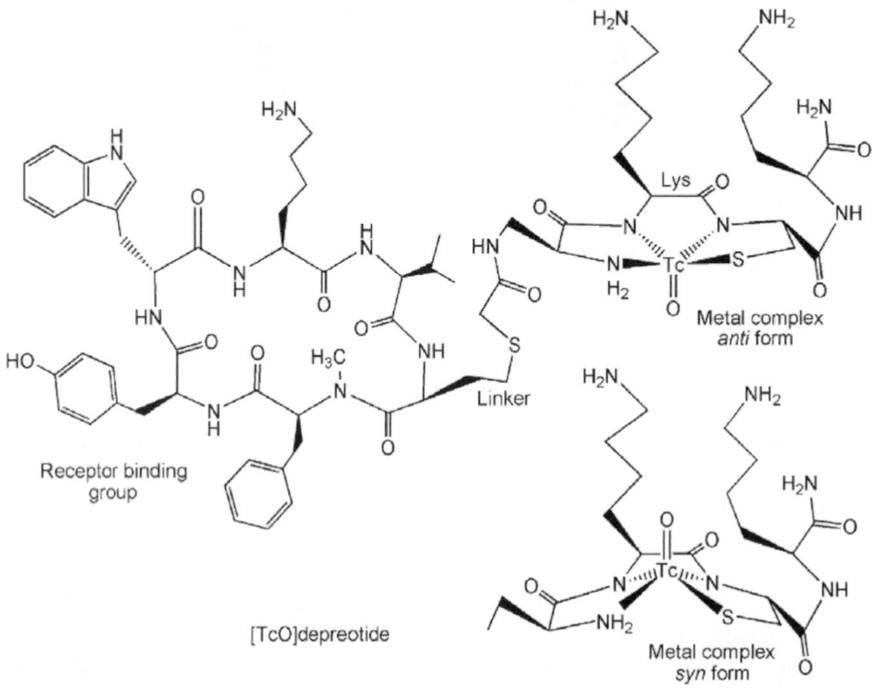

Figure 8.9   *Structure of [TcO]depreotide, showing the receptor-binding group, the linker and syn and anti diastereomers associated with the metal-binding region. The syn and anti isomers are referenced to the relative dispositions of the oxo and substituent on the α-carbon atom of lysine (Lys)*

with sites on the receptor protein on the surface of the cell. The metal-binding region of the ligand is similar to the N_3S donor peptide of MAG3, but unlike the latter, which is made up of optically-inactive glycine residues, the metal-binding peptide of [TcO]depreotide has amino acid residues that have asymmetric carbon atoms. Since the reaction using the kit produces a square pyramidal complex of Tc^{+5}, which has a TcO^{+3} core, the stereochemistry associated with the metal-binding region results in two isomers that are differentiated from each other by their relative arrangements of the oxo ligand and the lysine side chain of the N_3S donor peptide (Figure 8.9). The *syn* diastereomer has the oxo ligand on the same 'side' as the lysine side chain, while the *anti* diastereomer has the oxo ligand opposite the substituent on the α-carbon of Lys. Recent studies by Cyr *et al.* [37], using HPLC, NMR and circular dichroism, showed that the ratio of the *syn* form to the *anti* form in the metallation reaction is ~9, indicating that for stereochemical reasons the *syn* isomer is highly favored. These investigators also showed that the binding affinity of *syn* isomer is about six times greater than that of the *anti* isomer toward the somatostatin receptor, and that the former isomer also has greater absorption by the tumor.

The main use of Tc-depreotide is for imaging lung cancer; a patient with squamous cell carcinoma in the right lung imaged with the compound is shown in Figure 8.10 [38]. The patient was part of a study to determine whether Tc-depreotide could also be used to detect the spread of cancer (metastasis) from the lung to other

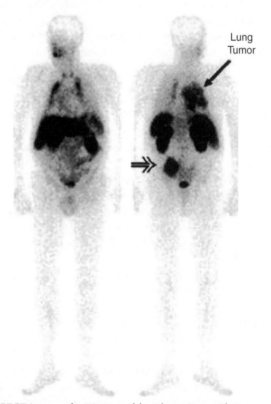

Figure 8.10 *Whole-body SPECT image of a 70-year-old male patient with squamous cell carcinoma in the right lung, imaged using ^{99m}Tc-depreotide. That the primary tumor in the lung has spread to bone is evident from the intense image of the left sacro-iliac, double-headed arrow. With kind permission from Springer Science + Business Media: Eur. J. Nucl. Med. Mol. Imaging, ^{99m}Tc-Depreotide Scintigraphy of Bone Lesions in Patients with Lung Cancer, 31, 2004, 1399–1404, E. Mena, V. Camacho, M. Estorch, J. Fuertes, A. Flotats, I. Carrió, Fig 4*

parts of the body, in this case the bone. Since the metastasized tumor will also have somatostatin receptors, if Tc-depreotide could localize to the region, it could be used to detect malignancies in bone. As shown in Figure 8.10, the patient has a massive tumor in the right lung, which, as is shown by Tc-depreotide, has spread to the left sacro-iliac joint. The study also used a ^{99m}Tc bone-imaging agent and other techniques to show that the cancer in the lung had indeed spread to the sacro-iliac joint and that malignancies are also present in the ribs adjacent to the diseased lung.

8.1.3 Innovations in synthesis, the $[Tc(CO)_3]^+$ core

Since the earliest days of the use of ^{99m}Tc in diagnostic nuclear medicine, one of the challenges has been to identify ligands that can securely bind the technetium ion, but at the same time do not produce coordination isomers on the technetium ion, which might lead to different biological distributions of the agent in the body. While the effects of coordination geometry on the metal center are not important for large structures – for example, a protein antibody conjugated to a ^{99m}Tc ion – most agents used in diagnostic nuclear medicine are small, which means that structural and/or charge features associated with the metal center are important for biodistribution. This is clearly the case with Tc-depreotide, for which changes in the location of the oxo ligand (*syn* vs. *anti*) affect the imaging properties of the agent [37]. An important advance in diagnostic nuclear medicine has been the development of a new type of technetium core, which consists of three facially-coordinated carbon monoxide molecules bound to Tc^+; that is, $[Tc(CO)_3]^+$ [39,40]. Since this core limits isomerization associated with metal centers and is available in kit form, Isolink, it is a useful way to quickly label a variety of different types of targeting vector with ^{99m}Tc and not create isomerization at the metal center.

As shown in Figure 8.11a, the key precursor compound for the attachment of the $[Tc(CO)_3]^+$ core to another molecule is the octahedral triaqua tricarbonyl complex, *fac*-$[Tc(H_2O)_3(CO)_3]^+$. This complex can be made by the reaction of pertechnetate obtained from a Tc 99m generator with the potassium salt of borano hydrogencarbonate, which serves as both a reducing agent and a source of CO in the reaction. While the mechanism of formation of *fac*-$[Tc(H_2O)_3(CO)_3]^+$ is unknown, the compound has three ligands that can easily be displaced – the water molecules – and three ligands that are essentially inert to substitution – the carbon monoxide groups. In this sense the compound is similar to the ruthenium arenes discussed in Box 5.1, in which half of the compound is organometallic – the $Tc(CO)_3$ part – and half is a coordination compound – the $Tc(H_2O)_3$ part. If the compound is reacted with a tridentate ligand that has three good donor groups capable of occupying a face of the octahedron, the water molecules, which are connected to the Tc ion through salt-like bonds (crystal field), are readily displaced and the $[Tc(CO)_3]^+$ core becomes bound to the ligand. Since the electronic configuration of Tc^+ is $[Kr]4d^6$, and the ion is in the second-row transition metal series, the crystal field stabilization energy, Δ, for the complex is substantial, which along with the tridentate nature of an attached ligand produces a stable adduct between the $[Tc(CO)_3]^+$ core and the ligand, with no opportunities for geometric isomerism.

An example of how the $[Tc(CO)_3]^+$ core can be used to construct an imaging agent is shown in Figure 8.11b [41,42]. The small peptide fMLF, which is capable of binding to the formyl peptide receptor on white blood cells (neutrophils), was modified on its carboxylic acid end by the incorporation of bis-quinoline tertiary amine tridentate ligand via a linker. In order to demonstrate the versatility of the metal tricarbonyl core, and to help characterize the complex formed between *fac*-$[Tc(H_2O)_3(CO)_3]^+$ and the modified peptide, the investigators synthesized *fac*-$[Re(H_2O)_3(CO)_3]^+$, which since rhenium is directly below technetium in the periodic chart, is iso-electronic and iso-structural with the analogous technetium complex. However, unlike ^{99m}Tc, for which syntheses are carried out in a highly-dilute so-called *tracer-level* environment ($10^{-10} M$), rhenium is available in large quantities – grams amounts – and its naturally-occurring isotopes are not radioactive. By synthesizing and characterizing the rhenium tricarbonyl adduct of the

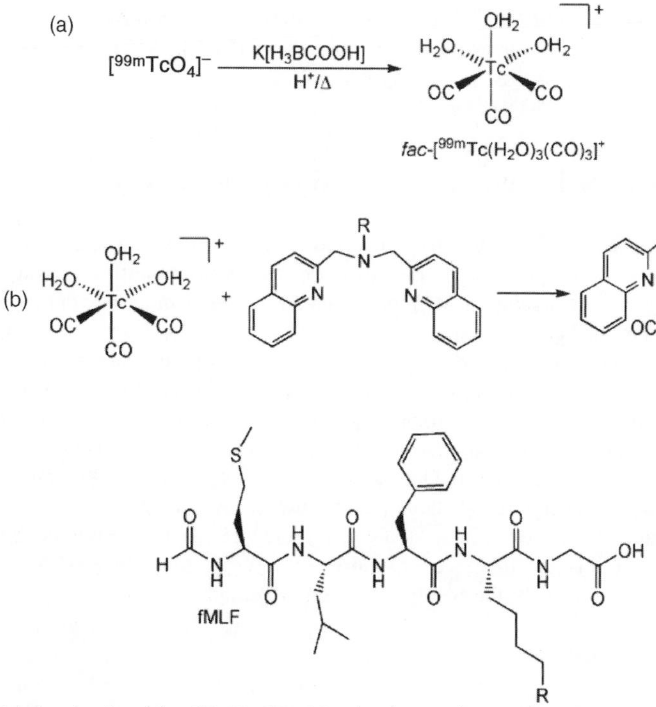

Figure 8.11   *(a) Synthesis of fac-[Tc(H₂O)₃(CO)₃]⁺ ; (b) synthesis of an imaging agent with the [Tc(CO)₃]⁺ core attached via a linker to the fMLF peptide capable of recognizing the formyl peptide receptor (FPR) on white blood cells (neutrophils)*

modified peptide, the investigators were able to use HPLC to show that the compounds formed in the reactions of the peptide with the technetium and the rhenium triaqua tricarbonyl complexes are the same.

A useful feature of rhenium complexes, especially those containing aromatic ligands with nitrogen donor atoms, is that they have properties that allow them to be detected in cells with fluorescence microscopy. Since the chemistries of Tc and Re are similar, the investigators pointed out that making molecules with the same targeting vector attached to either $[Tc(CO)_3]^+$ or $[Re(CO)_3]^+$ provides a way to correlate the low-resolution images obtained with ^{99m}Tc in living subjects using SPECT with cellular- and sub-cellular-resolution images of the corresponding Re analog in isolated cells using fluorescence microscopy [41,42].

8.2   Metal compounds as contrast agents for MRI

The discovery that *nuclear magnetic resonance, NMR,* could be used to detect differences in the properties of water molecules in diseased and normal tissue quickly propelled this premier physical technique from the purely chemical realm into medicine [11–16,44,45]. By enlarging the NMR magnet, which normally accepted a small-diameter glass tube containing a chemical sample, to a size large enough to accept the human body, magnetic resonance imaging (MRI) instruments were created, which could detect subtle differences between water molecules located in fluids and in soft tissue of the body. While the technique could detect differences between tissue types, it was discovered that if the patient was given a paramagnetic metal complex prior to MRI, the definition, or *contrast*, of the resulting image was greatly enhanced. This effect was caused by the paramagnetic metal complex acting as a tiny magnet, which induced NMR changes in water molecules that could be detected by MRI. Since these initial discoveries, many compounds containing gadolinium, manganese and iron have been approved for use as *contrast agents*, *CAs*, in MRI, and many other image-enhancing agents are in various stages of testing and development [11–16].

Box 8.2 Magnetic resonance imaging, MRI, and contrast agents

The basis for diagnostic imaging with MRI is directly connected to the theory of nuclear magnetic resonance, NMR [46,47]. While conventional NMR spectroscopy primarily uses the chemical shift of protons as a means of characterizing the structures of molecules in solution, MRI focuses on the relaxation properties of the protons on water molecules located in soft tissues of the body.

The proton has a *nuclear spin quantum number* (sometimes referred to simply as nuclear spin), I, of 1/2, indicating that there is one unpaired nuclear spin. Since the proton is spinning on its axis, the unpaired spin generates a *nuclear magnetic moment* perpendicular to the spin. When the proton is placed in a magnetic field, quantum mechanics requires that only discrete orientations of the spin moment, specified by the *nuclear spin angular momentum quantum number*, m_I, relative to the applied magnetic field, are allowed. One orientation with $m_I = +1/2$ has the spin moment aligned with the applied field and is lowest in energy (most stable), while another orientation with $m_I = -1/2$ has the spin moment opposing the applied field and is highest in energy (less stable) (Figure 8.12). If there is no applied field, there is no preferred orientation and there is no energy difference between the states (no splitting). In this case, both m_I states are said to be *degenerate in energy*. Experimentally, the energy separation between the two states in Joules, J, ΔE, is given by (8.1):

$$\Delta E = \gamma \hbar H \tag{8.1}$$

where γ is the magnetogyric ratio, a constant for a given nucleus with units of $2\pi\,s^{-1}\,T^{-1}$; $\hbar$ is Planck's constant, h, divided by 2π; and H is the strength of the applied magnetic field in Teslas, T.

The strength of the magnetic field used in MRI is approximately 1 T, which is about an order of magnitude weaker than most magnetic fields employed in conventional NMR spectroscopy. For a proton placed in a magnetic field of this strength, the value of ΔE for the splitting between the two levels is very small, corresponding to electromagnetic radiation in the radio frequency range, in terms of wavenumbers, $\sim 10^{-3}$ cm^{-1}. By comparison, the splitting between the d-orbitals for a transition metal complex, measured by the crystal field splitting parameter, Δ, is $\sim 10^4$ cm^{-1}, which is seven orders of magnitude larger than the energy separation of the states in Figure 8.12. If a large number of water molecules are placed in a magnetic field, not all of their protons will have their nuclear magnetic moments aligned in the manner corresponding to the lowest-energy, $m_I = +1/2$, state. This is because at the temperatures at which most NMR, as well as MRI, measurements are made, there is enough thermal energy available – about 200 cm^{-1} at room temperature – that both levels are almost equally populated, but there is a slight excess of nuclei in the lower, $m_I = +1/2$, state.

The expression that addresses the number of nuclei to be found in each state is the *Boltzmann distribution equation*, which is given by (8.2).

$$\frac{N(-1/2)}{N(+1/2)} = e^{-\frac{\Delta E}{kT}} \tag{8.2}$$

In this expression, $N(-1/2)$ and $N(+1/2)$ are the number of nuclei in the $m_I = -1/2$ and $m_I = +1/2$ states, respectively; ΔE is the separation between the two levels in cm^{-1}; k is the Boltzmann constant, $k = 0.695$ cm^{-1} K^{-1}; and T is the temperature in degrees Kelvin, K. From (8.2) it is evident that at room temperature (20 °C), the ratio $N(-1/2)/N(+1/2)$ is 0.9999954, showing that for an ensemble of nuclei there are slightly more in the lower, more stable, state than in the upper state. NMR spectroscopy takes advantage of this slight population difference between the two states by irradiating the sample while it is in the magnetic field with electromagnetic radiation that exactly corresponds to ΔE, a condition called *resonance*. This is done by aligning a radio frequency (RF) field perpendicular to the main magnetic field, which results in the promotion of the nuclei in the lower state to the upper state. If the RF field remains on for a period of time, the population in the two states will become equal and the system is said to reach *saturation*. After the RF field is turned off, the population of nuclei begins immediately to *relax* to the original equilibrium distribution, the rate of which is characterized by some rate constant, k. Since the transition from the $m_I = -1/2$ level to the $m_I = +1/2$ state emits radiation, the rate of *relaxation* can be directly monitored by a detector – *receiver coil* – sensitive to RF radiation that is located at right angles (90°) to both the magnetic and the RF fields. The data accumulated by the detector as a function of time after the RF field is turned off can be used to measure the rate and the lifetime of the relaxation process.

One of the relaxation parameters measured in NMR is the *longitudinal* or *spin lattice relaxation time*, T_1, which is the *time constant*, $1/k$, where k is the observed rate constant, associated with the relaxation of the spin population at saturation to its thermal equilibrium. This parameter is mainly affected by fluctuations of neighboring molecules and by paramagnetic substances with large magnetic moments, which generate fluctuating magnetic fields at positions of the relaxing nuclei. Typical values of T_1 for liquids are in the range 10^{-4}–10 seconds, and for solids in the range 10^{-2}–10^3 seconds.

A second relaxation parameter measured in NMR is the *transverse or spin–spin relaxation time*, T_2. This relaxation time constant, which cannot be described using the quantum-mechanical model in Figure 8.12, is a measure of the rate of loss of phase coherence among the spins in the plane perpendicular to the applied field. In the classical model of the NMR experiment, an ensemble of excited nuclei have their spins precessing in unison (coherence) in the applied field, but local fields acting on different nuclei in the ensemble and/or interactions between spins can destroy the coherence of the population. This effect, which involves changes in the net magnetization in the plane perpendicular to the applied field, is described by the transverse or spin–spin relaxation time constant, T_2, which, for liquids, is similar in magnitude to that of T_1. A third relaxation parameter, T_2^*, is caused by the effects of inhomogeneities in the applied field on T_2.

MRI parallels the NMR experiment outlined above except that the water molecules are in the body and the constants, T_1, T_2 and other parameters, are determined in three dimensions within the tissue of the patient [44]. As shown in Figure 8.13, the patient is positioned on a surface that can be moved into the bore of a large electromagnet, which produces the main magnetic field. At a right angle to the magnetic field is an RF field and at right angles to both the main magnetic field and the RF field is a receiver coil that captures the radiation emitted from the spin population.

The spatial resolution of the relaxation parameters is carried out using three *gradient magnets* at right angles to one another, which, when turned on, produce linear field gradients in specific directions; that is, the magnetic field changes along the gradient. The activation of these electromagnets creates the annoying sounds heard during an MRI scan. By producing gradients in the field in different directions, the relaxation properties for water molecules along the gradients within the tissue of the patient can be obtained. When the data are analyzed with a computer, the relaxation parameters within many small-volume elements, called *voxels*, are calculated and after attaching a color (or a shade of gray) to a specific value of the relaxation parameter, an image based on the colors is created, showing the relaxation properties of water within the voxels. Depending on the size of the field of view (the size of the region in the body over which relaxation data are collected) and other parameters, spatial resolution on the order of a few millimeters can be obtained in a typical MRI scan.

Since the contrast of the image, and thus its usefulness for diagnosis, critically depends on the differences in relaxation parameters between various regions in tissue, chemical compounds called *contrast agents*, *CAs*, which dramatically affect the relaxation properties of water, have been developed. Since these compounds are nontoxic and distribute to only certain types of fluid and tissue in the body, they can be given to the patient immediately before an MRI scan. If a CA has access to a particular type of tissue, it will dramatically change the relaxation properties of water molecules located in the tissue, leading to greater differences in the relaxation properties (contrast) between tissues with and without the CA.

Many CAs presently in use contain the lanthanide gadolinium, which, in its stable trivalent oxidation state, Gd^{+3}, is highly paramagnetic with the electronic configuration $[Xe]4f^7$ and $S = 7/2$. CAs containing Gd^{+3} mainly affect the spin lattice relaxation time, T_1, by two mechanisms. The first involves the direct exchange of a water molecule that is bound to the metal ion with water molecules that are free in fluids and tissue; the other is the result of water molecules passing through the local magnetic field produced by the Gd^{+3} ion; that is, no direct binding to the metal ion occurs. In addition to compounds containing Gd^{+3}, small nanometer-sized particles containing iron have been developed as CAs. These compounds, which have the spin moments of many iron centers aligned in the same direction, produce strong local magnetic fields that mainly shorten the transverse or spin–spin relaxation time, T_2, of water molecules passing through the field.

The effectiveness of a CA in decreasing T_1 and T_2 of bulk water depends on a property of the CA, called the *relaxivity*, r_i which, along with the concentration of the CA, affects relaxation properties of water according to (8.3).

$$R_{i(obs)} = \frac{1}{T_{i(obs)}} = \frac{1}{T_{i(diam)}} + r_i C; \quad i = 1 \quad \text{or} \quad 2 \qquad (8.3)$$

In this expression, $i = 1$ is the effect of the CA on T_1, and $i = 2$ is its effect on T_2 [44]. The quantities $R_{i(obs)}$ and $1/T_{i(obs)}$ are the global relaxation rate constants of water molecules in the system (s^{-1}); $T_{i(diam)}$ is the relaxation time of water before addition of the paramagnetic CA (s); C is the concentration of the paramagnetic ion in the CA $(mmol^{-1} L)$; and r_i is the *relaxivity* $(s^{-1} mmol^{-1} L)$. Experimentally, the relaxivity of a CA, which is also dependent on the strength of the applied magnetic field (generally stated in Tesla units, T) and the temperature of the solution, is determined by measuring $R_{i(obs)}$ at different values of C and, from a plot of $R_{i(obs)}$ versus C, calculating the slope, which is r_i.

The work of Solomon, Bloembergen and Morgan pointed out that certain properties of the paramagnetic metal complex affect the magnitude of r [44]. One property is the number of exchangeable water molecules bound to the metal ion, denoted by q. Another is the time constant for the exchange of a bound water molecule with one in solvent, indicated by τ_M, where $\tau_M = 1/k_{ex}$. If q is large and τ_M is small (k_{ex} is large), many water molecules in solvent will be allowed to directly contact, 'visit', the paramagnetic metal ion during the experiment, which will have the effect of *decreasing* the relaxation time of water observed in the experiment. The tumbling (rotational) rate of the CA in solution, called the *rotational correlation time*, denoted by the time constant, τ_R, where $\tau_R = 1/k_{rot}$, also affects the relaxivity, r, of the CA. If τ_R is large (k_{rot} is small), as would be the case for a large, slowly-rotating CA, the 'contact time' between the local magnetic field produced by the CA and a water molecule in solvent is relatively long, which also increases r of the CA. Much of the current effort in the MRI/CA field is focused on maximizing r by changing the structure of the CA in various ways to affect one or more of the above parameters.

8.2.1 Clinically used MRI contrast agents

Shown in Figure 8.14 are three of the many CAs that are currently used in MRI [43,44,47]. Two of the compounds, $[Gd(DTPA)]^{2-}$ (Magnevist) and $[Gd(DOTA)]^{-}$ (Dotarem), contain Gd^{+3}, which has the electronic configuration $[Xe]4f^7$. The f-level of the atom can accept a maximum of 14 electrons, and since the level is half filled for Gd^{+3}, the total spin quantum number of the ground state of the ion is $S = 7/2$, making Gd^{+3} complexes highly paramagnetic. Since the 4f-orbitals are buried deep within the atom and experience negligible perturbations from the attached ligands, there is no or negligible crystal field stabilization

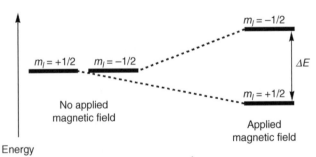

Figure 8.12   *A proton with a nuclear spin quantum number, $I = \frac{1}{2}$ indicating the presence of one unpaired nuclear spin. The spinning proton generates a magnetic moment which has two possible orientations in an applied magnetic field, which are described by the nuclear spin angular momentum quantum number m_I. The splitting between the $m_I = +1/2$ and $m_I = -1/2$ states in the presence of the magnetic field, ΔE, is proportional to the strength of the applied magnetic field, H_o*

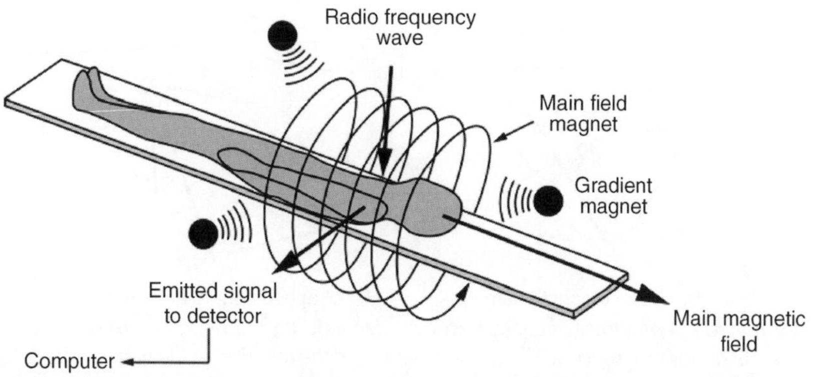

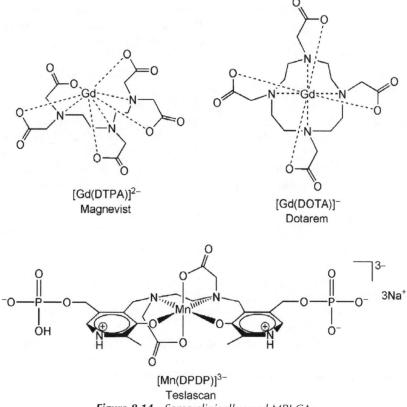

Figure 8.13 *Schematic diagram of a patient in a magnetic resonance imaging (MRI) instrument. The main magnetic field is created by an electromagnet and the field in many specific planes within the tissue of the patient is modified by gradient electromagnets. After analysis using a computer program, a 3D `image' of the relaxation lifetimes of water molecules in the tissue of the patient is obtained*

energy and the number and geometry of attached ligands is governed largely by the size of the cation and steric/electronic effects produced by the ligands. Like other lanthanides, the most common coordination number for Gd^{+3} is 9, with the ligands being arranged in either a tricapped trigonal prismatic or a monocapped square antiprismatic geometry (Figure 8.15). In the case of Magnevist and Dotarem, the attached carboxylate–amine

$[Gd(DTPA)]^{2-}$
Magnevist

$[Gd(DOTA)]^{-}$
Dotarem

$[Mn(DPDP)]^{3-}$
Teslascan

Figure 8.14 *Some clinically-used MRI CAs*

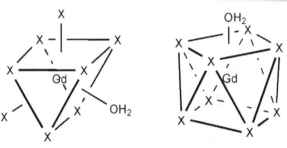

Tricapped trigonal prism   Monocapped square antiprism

Figure 8.15 *Coordination geometries of Gd^{+3} MRI CAs. Depending on the structure of the octadentate ligand, which contributes eight donor groups to Gd^{+3}, denoted by X, and the location of the water molecule, each geometry can exist in a number of isomeric forms. The monocapped square antiprism is a cube in which two opposite faces have been rotated by 45° relative to each other*

ligand is octadentate, which leaves one site available for the coordination of a water molecule. Since attaching a polydentate ligand to either a tricapped trigonal prism or a monocapped square antiprism can be done in a number of different ways, the possibility of geometric and optical isomers exists, which complicates the solution characterization of the compounds [43]. The stability constant for Dotarem is log $K = 25.2$ (25 °C), which is about three orders of magnitude higher than the stability constant of Magnevist. While the ligands of both compounds form five-membered chelate rings with the Gd^{+3} ion, Dotarem contains a macrocyclic ring, which, due to the *macrocyclic effect*, imparts greater stability to the Gd^{+3} complex than the *acyclic* ligand of Magnevist. The exchange rate constant of the water molecule for both Magnevist and Dotarem is $k_{ex} = 10^6\,s^{-1}$, showing that water exchange is very fast [44]. This feature of the complexes allows a large number of water molecules in solvent to directly contact the paramagnetic Gd^{+3} ion during the relatively long data collection time of the NMR experiment.

While Gd^{+3} complexes are very useful as CAs for MRI, the free aquated ion, $[Gd(H_2O)_9]^{3+}$, is toxic to cells and causes nephrogenic systemic fibrosis (NSF), which is a disabling and potentially fatal disorder that causes fibrosis to skin, joints, eyes and internal organs [48–50]. An important property of the Gd-based CAs is that their retention time in the body is relatively short. Since they are given by IV injection, they quickly distribute in the blood and extracellular fluids, and due to their anionic charge they are rapidly eliminated through the renal system, with short half lives, $t_{1/2} < 2$ hours. However, if the renal system is impaired, as would be the case with patients on dialysis, and if the stability of the CA is relatively low, as with compounds containing acyclic ligands like Magnevist, the risk exposure to NSF is increased, and for these patients other CAs or diagnostic approaches are used [48–50]. Recent studies also show that Gd CAs that have acyclic ligands are less stable in human serum than their macrocyclic counterparts [51].

Magnevist, a CA that primarily affects T_1, with $r_1 = 3.3\,mM^{-1}\,s^{-1}$ (1.5 T, 37 °C, water), was the first Gd^{+3} CA to be approved for clinical use in MRI [43]. While CAs do not normally cross the blood–brain barrier (BBB), if there is damage to the BBB, as would occur with a tumor in the brain, the CA can 'leak' inside and affect the relaxation rates of water protons in the brain. Shown in Figure 8.16 is a patient with a brain tumor who was scanned using MRI with and without the presence of Magnevist as the CA. As is evident from the figure, the tumor is not visible in the scan without Magnevist, but after administration of the CA the tumor is clearly visible as a bright mass in the MRI scan.

Dotarem (Figure 8.14), also mainly a T_1 relaxation CA with $r_1 = 2.9\,mM^{-1}\,s^{-1}$ and $r_2 = 3.2\,mM^{-1}\,s^{-1}$ at 1.5 T at 37 °C, has many different uses, but one is imaging the blood pool of the patient [43]. Dotarem is often used to detect coronary heart disease and can enhance the MR image of the carotid artery of patients that have been given a single injection of the CA (Figure 8.17).

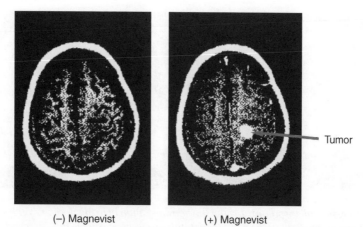

(–) Magnevist   (+) Magnevist

Figure 8.16   *Patient with a brain tumor imaged using MRI using a T_1-weighted scan. Scan before (left) and after (right) patient was given Magnevist. The tumor is opaque in the pre-contrast image but it appears as a bright mass in the MR image in the presence of Magnevist. With kind permission from Springer Science + Business Media: Handb. Exp. Pharmacol. Contrast Agents: Magnetic Resonance, 185, 2008, 135–65, C. Burtea, S. Laurent, E.L. Vander and R. N. Muller, Fig 6*

The third compound shown in Figure 8.14, $[Mn(DPDP)]^{3-}$, mangafodipir trisodium, Teslascan, which is in clinical use as a CA, contains the transition metal ion Mn^{+2}. The complex has a distorted octahedral geometry, the Mn^{+2} ion, $[Ar]3d^5$, is high-spin with $S = 5/2$, and in aqueous solution r_1 of $[Mn(DPDP)]^{3-}$ is 2.8 mM^{-1} s^{-1} and r_2 is 3.7 mM^{-1} s^{-1} at 0.47 T and 40 °C [52]. The hexadentate ligand in $[Mn(DPDP)]^{3-}$, which is an analog

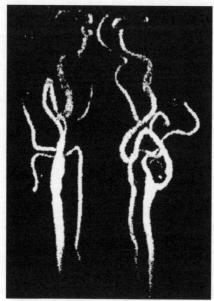

Figure 8.17   *MRI of vessels in the neck (carotid artery) of a healthy individual after a bolus injection of Dotarem. With kind permission from Springer Science + Business Media: Handb. Exp. Pharmacol. Contrast Agents: Magnetic Resonance, 185, 2008, 135–65, C. Burtea, S. Laurent, E.L. Vander and R.N. Muller, Fig 7*

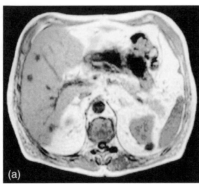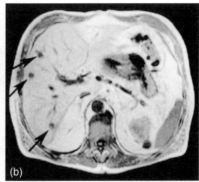

Figure 8.18 T_1-weighted MR image of cancer (arrows) in the liver of a patient: (a) without the administration of the CA [Mn(DPDP)]$^{3-}$, mangafodipir trisodium; (b) after the administration of mangafodipir. With kind permission from Springer Science + Business Media: Eur. Radiol., Hepatobiliary Contrast Agents for Contrast-Enhanced MRI of the Liver: Properties, Clinical Development and Applications, 14, 2004, 559–578, P. Reimer, G. Schneider and W. Schima, Fig 1

of vitamin B$_6$, has two tertiary amine nitrogen donors, two carboxylates and two phenolate oxygen atoms coordinated to the Mn^{+2} ion [53]. Due to the dispositions of the two carboxylate chelate 'arms' on the octahedron, the compound exists as mirror-image, Λ/Δ, isomers, both of which are present in the crystal structure of the complex in equal amounts; that is, the compound crystallizes as a racimate [53].

Divalent aquated manganese was the first simple CA to be studied in connection with MRI, but since the ion is toxic, the ligand for [Mn(DPDP)]$^{3-}$ was synthesized to lower its toxicity and enhance its uptake into tissue [54]. Divalent manganese is an essential ion for cellular function, but chronic (continuous, long-term) exposure to simple manganese salts can cause *manganism*, which produces effects similar to Parkinson disease, and acute (short-term) exposure of high concentrations of the ion may cause hepatic failure and cardiac toxicity [55–57].

Mangafodipir trisodium is given by IV injection and, as shown in Figure 8.18, is useful for detecting cancer in the liver using MRI [58]. The phosphate groups of the compound appear to be quickly removed in blood and, due to the presence of Zn^{+2} in blood, which has a higher binding constant toward the metabolized ligand than Mn^{+2}, log K = 18.95 (Zn^{+2}), 15.1 (Mn^{+2}) at 25 °C, some Mn^{+2} may be displaced and transferred to proteins, which carry the ion to liver cells, where it is internalized [52,58,59]. Once inside liver cells, the ion has much greater relaxivity than when it is free in solution. After injection, the high levels of manganese in serum quickly drop to detection levels, with the ion being eliminated in the urine and feces.

8.2.2 Contrast agents in development

The success of Gd^{+3} complexes as CAs for MRI stimulated the search for new agents with high relaxivities, altered delivery characteristics and structures that could be biologically modified to enhance relaxivity *in vivo*. One way to enhance relaxivity of a CA is to slow the time that it takes to move the magnetic field of the paramagnetic compound through space, which is related to the *rotational correlation time*, τ_R, of the complex (see Box 8.2). If the complex slowly rotates – that is, the compound has a large τ_R – the contact time of a water molecule in the magnetic field of the CA increases, which increases the effect on the relaxation rate of the water protons. While a number of different strategies have been used to increase τ_R, including attaching Gd^{+3} to antibodies, dendrimers and other large structures [44], one interesting example is the CA recently reported by Meade and coworkers, which has seven Gd^{+3} ions attached to β-cyclodextrin, a molecule that occurs naturally in starch (Figure 8.19) [60]. The CA, which can be made by the reaction of seven equivalents of an activated

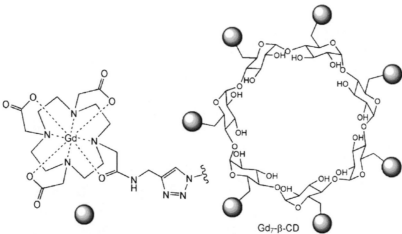

Figure 8.19 *Structure of a multimeric CA. The β-cyclodextrin molecule has seven Gd-DOTA-type complexes (sphere) attached via a triazole linker to form Gd$_7$-β-CD. Based on Figure 1, p. 6662 of Song, Y., Kohlmeir, E.K., Meade, T.J. (2008) Synthesis of multimeric MR contrast agents for cellular imaging. J. Am. Chem. Soc., 130, 6662–6663*

Gd-DOTA type complex with a modified cyclodextrin, produces Gd$_7$-β-CD, with MW = 5480, which rotates more slowly than the low-MW mononuclear material. The relaxivity of Gd$_7$-β-CD in water is $r_1 = 12.20 \pm 0.54 \, \text{m}M^{-1} \, \text{s}^{-1}$ per Gd^{+3} ion, which is much larger than the simple mononuclear Gd-DOTA precursor complex $r_1 = 3.21$ ($T = 1.41$, 37 °C), showing that Gd$_7$-β-CD is a more effective relaxation agent for water.

Labeling cells with MRI CAs allows the possibility of detecting the movements of single cells or clusters of cells within the body using MRI [15,61]. When NIH-3T3 fibroblast cells growing in culture were exposed to Gd$_7$-β-CD, the cells absorbed the compound to a much greater extent than a mononuclear Gd^{+3} complex [60]. This was determined by isolating the cells using centrifugation to produce a cell pellet, which was subjected to analysis for its Gd content with ICP-MS. When the cells containing the CA are subjected to MRI, they appear bright white in a T_1-weighted scan, indicating that Gd$_7$-β-CD could indeed be useful for labeling cells and following their movements in the body.

Some gadolinium compounds that are being investigated as new CAs for MRI take advantage of biological molecules that serve as transport vehicles in the blood. For example, human serum albumin, HSA, has a number of binding sites for fatty acids, steroids, drugs and other small molecules, and since it circulates in the blood, it could be used to carry a CA in order to image the blood pool. Shown in Figure 8.20 is a novel blood-pool CA, MS-325, containing Gd^{+3}, that is currently in phase III clinical trials as a potential new agent for detecting blockages in arteries using MRI [62]. The complex, which is a trivalent anion, has an appended hydrophobic diphenyl-cyclohexyl group, which allows it to bind to a hydrophobic binding site, site II, on HSA. A binding study using ultrafitration showed that a total of four molecules of MS-325 can bind to HSA at pH 7.4 in a stepwise manner. As indicated in Figure 8.21, the spin–lattice relaxation time, T_1, of water is significantly reduced when MS-325 is in the presence of HSA. By measuring relaxation rates for the analogous Gd compound without an exchangeable water molecule and by making other measurements, the researchers were able to show that the reduction in T_1 is mainly due to an increase in the rotational correlation time, $\tau_R = 10.1 \pm$ ns bound versus 115 ps free, when MS-325 is bound to the 67 kDa HSA protein [62]. This type of *receptor-induced magnetization enhancement* (*RIME*) is an often-used strategy to enhance the relaxivity of a CA by attaching it to a naturally-occurring slower-rotating larger molecule, which, in this case, is an equilibrium, noncovalent interaction with HSA.

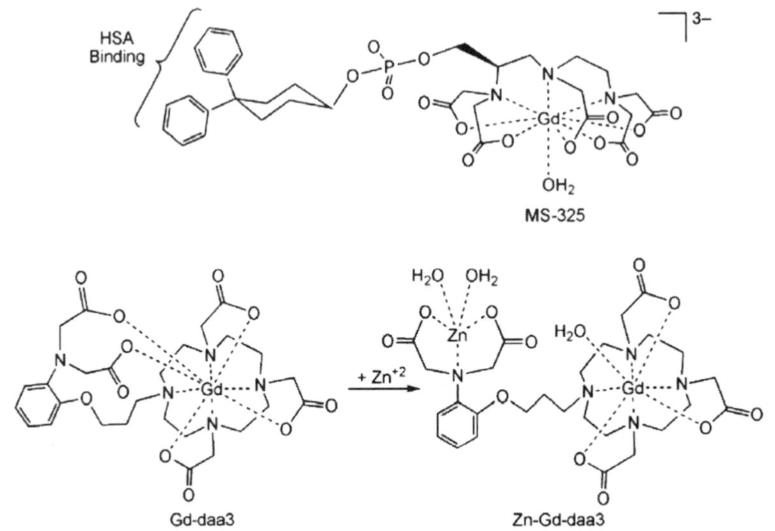

Figure 8.20   *Potential new CAs for MRI*

In addition to enhancing the relaxivity of a gadolinium CA by changing its rotational correlation time, some CAs can be activated by environmental conditions to change the number of exchangeable water molecules, usually from no exchangeable waters, for which $q = 0$, to a complex with one exchangeable water molecule, where $q = 1$. These agents can be activated by pH, hydrolysis by enzymes, changes in the concentration of dissolved oxygen and other factors [44]. Divalent zinc plays a critical role in the structure and reactivity of many metalloproteins and enzymes and has been implicated in Alzheimer and other diseases [63]. Recently, Meade and coworkers described the synthesis and properties of an unusual Gd^{+3} CA with a relaxivity that depends on the concentration of Zn^{+2} ion in solution [64]. As shown in Figure 8.20, the compound, Gd-daa3, has Gd^{+3} bound to a DOTA type macrocyclic framework (Figure 8.14) and an appended aminodiacetate. In the absence of Zn^{+2}, the chelating arms of the appended acetates coordinate to the Gd^{+3} ion, which along with the remaining donors produces nine-coordinate Gd^{+3} Gd-daa3. However, in the presence of Zn^{+2}, the chelate arms are displaced from Gd^{+3} to Zn^{+2}, thus allowing the Gd^{+3} ion to bind water and produce Zn-Gd-daa3. By studying the analogous Tb^{+3} complex, Zn-Tb-daa3, the fluorescence of which is indicative of the number of metal-bound water molecules, the investigators showed that Zn-Gd-daa3 very likely has only one exchangeable water molecule (Figure 8.20), but the ligand occupying the remaining, ninth, coordination site was not specified. Since the relaxivity of Zn-Gd-daa3, $r_1 = 5.07 \, \text{mM}^{-1}\text{s}^{-1}$, is more than twice that of Gd-daa3, $r_1 = 2.33 \, \text{mM}^{-1}\text{s}^{-1}$ (1.41 T, 37 °C), the presence of the exchangeable water molecule significantly increases the relaxivity of the agent.

The compound Gd-daa3 is much more selective for Zn^{+2} than for two similar ions, Ca^{+2} and Mg^{+2}, which are more biologically abundant than Zn^{+2}. As shown in Figure 8.22, there is no significant change in the relaxivity of water induced by Gd-daa3 in the presence of Mg^{+2} and Ca^{+2}, even when there is an excess of these ions in solution [64]. The binding constant of Zn^{+2} toward Gd-daa3 was found to be $4.2 \times 10^3 \, M^{-1}$ and *in vitro* magnetic resonance images showed that concentrations of Zn^{+2} as low as 100 μM can be detected with the CA. Since the intended goal of the CA is to detect Zn^{+2} in the brain during neurotransmission, which results in zinc levels of ~300 μM, Gd-daa3 has the potential to monitor zinc levels in the brain under these conditions. In addition to Gd-daa3, a zinc-detecting Mn^{+3} porphyrin, with a porphyrin ligand similar in structure to that shown in Figure 1.27, has also been reported [65]. Interestingly, in this case, binding of Zn^{+2} to ligands appended to the periphery of the Mn-porphyrin caused a *decrease* in r_1 of the CA.

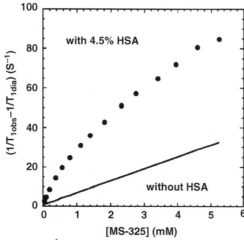

Figure 8.21 *Plot of* $(1/T_{1obs} - 1/T_{1dia})(s^{-1})$ *versus the concentration, in mM, of the* Gd^{+3} *CA, MS-325, in the presence and absence of 4.5% human serum albumin, HSA, the approximate concentration of the protein found in blood. Adapted with permission from P. Caravan et al., The Interaction of MS-325 with Human Serum Albumin and its Effect on Proton Relaxation Rates, J. Am. Chem. Soc., 124, 3152–3162. Copyright 2002 American Chemical Society*

8.2.3 Iron oxide particles

Magnevist, Dotarem and mangafodipir trisodium (Figure 8.14) are a few of the clinically-approved CAs for MRI. These compounds, which contain a single paramagnetic metal ion each, mainly affect the spin–lattice relaxation time of water and are used to obtain so-called T_1-*weighted* MRI scans. A second class of CA in wide

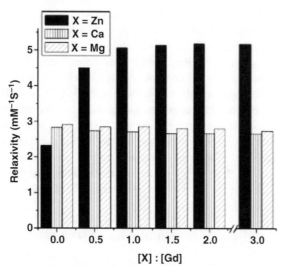

Figure 8.22 *Relaxivity of Gd-daa3 at 1.41 T, 310 K in the presence of XCl_2, where $X = Zn^{+2}$, Ca^{+2} or Mg^{+2}. All solutions were prepared in 100 μM KCl/100 mM Hepes buffer at pH 7.4. Reproduced with permission from J.L. Major et al., The Synthesis and In Vitro Testing of a Zinc-Activated MRI Contrast Agent, Proc. Natl. Acad. Sci., 104, 13881–86. Copyright 2007 National Academy of Sciences, U.S.A*

clinical use consists of nanometer-sized particles of iron oxide, an example of which is the well-known mixed iron-oxide magnetite, Fe_2O_3/Fe_3O_4 [43,44,66]. While magnetite shortens both T_1 and T_2 of water, its greatest effect is on the *spin–spin relaxation time*, T_2. Due to the way in which the MRI data are analyzed, magnetite produces an MRI image that is darker than the image without the addition of the CA and for this reason magnetite is referred to as a *negative CA*.

Magnetite is an inorganic compound that contains high-spin Fe^{+2} ($S = 2$) and high-spin Fe^{+3} ($S = 5/2$) ions in an oxide lattice. Due to coupling between the net spins of the individual iron ions in the oxide lattice, the bulk magnetic properties of magnetite depend on the physical size of the particle. For magnetite that is greater than $\sim$5 nm in size, the large number of adjacent spins in the particle can couple in the presence of an external field to give the particle a net magnetic moment. However, if the particle is less than $\sim$5 nm in size, there are fewer spins that can couple and, since the stabilization of the coupled state depends on the number of coupling spins (an entropy effect), small particles of magnetite have no net magnetic moment in an applied field. Due to the random orientation of the spins of individual iron ions in the lattice, the particles are paramagnetic [67]. In tissue, the magnetic properties of the particles produce local inhomogeneities in the magnetic field, which causes water molecules diffusing through the field to lose their phase coherence, which affects T_2.

In order to make magnetite particles useful as CAs for MRI, they are coated with biocompatible substances such as citrate, silanes, synthetic polymers, starch and so on, and when the coated particles are put into water they form colloidal suspensions [66,67]. If the diameter of the coated particle is greater than 50 nm, the particle is called *super paramagnetic iron oxide, SPIO*, and if it is less than 50 nm it is called *ultrasmall super paramagnetic iron oxide, USPIO*. Very large SPIOs, in the 300 nm to 3.5 μm range, are not injected into the blood but rather are given orally to enhance contrast of the gastrointestional (GI) tract in MRI. Smaller SPIOs in the range 60 nm to 150 nm, which are given by IV injection, are mainly used to image the blood pool, liver and spleen. Following injection, these particles are quickly removed (within minutes) from circulation by the reticuloendothelial system, mainly by the Kupffer macrophage cells in the liver. Since tumor nodules do not have Kupffer cells, they do not take up SPIOs and as a result the contrast between normal tissue-containing SPOIs and the tumor is enhanced. An example of this can be seen in Fig. 8.23 which shows negative contrast of a tumor in the liver of a patient imaged with magnetic resonance using the clinically approved contrast agent, Resovist (Ferucarbotran), which is an iron oxide particle coated with carboxydextran [68]. As is evident from the figure, the tumor appears white while normal tissue, which contains Resovist, is dark in shade.

Because of their very small size, the USPIOs take a much longer time to clear the body, mainly because they cross the capillary wall and are taken up by lymph nodes and bone marrow. These agents show promise for identifying cancerous lymph nodes in a number of metastatic cancers, and since they are taken up by macrophage cells they can be used to detect inflamed regions in the body, as would occur in the brain following a stroke, near an atherosclerotic plaque or as the result of rejection after an organ transplant [66]. In addition to using USPIOs to enhance the image of tissue in the body, it is also possible to label individual cells with USPIOs by incubating the cells with the particles. For example, stem cells that have absorbed the particles by endocytosis can be introduced into the body and the locations of the cells tracked over extended periods of time using MRI. By attaching DNA oligonucleotides that can hybridize to specific sequences of DNA or RNA in the cell, the particles may also prove to be useful for targeting genes for specific diseases at the molecular level [66].

8.3   Radionuclides for palliative care and cancer treatment

Malignant tumors that accompany advanced stages of prostate, breast and lung cancer often metastasize to bone, causing painful bone cancer [18,69–72]. Since this type of pain is especially difficult to treat with existing analgesics, patients with progressive disease usually require palliation of bone pain in order to maintain some quality of life. A group of compounds that show promise for palliative treatment of bone

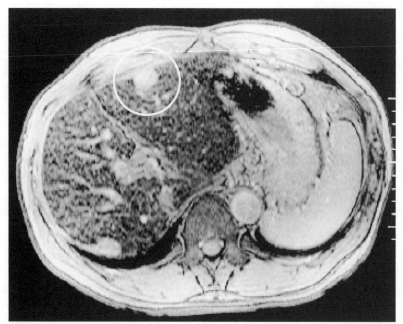

Figure 8.23 *Magnetic resonance image (T2*weighted) of the liver of a 58-year man with heptocellular carcinoma and type B-type cirrhosis in the presence of the superparamagnetic iron oxide contrast agent, Resovist. The tumor appears bright (circled) relative to the normal darker tissue of the liver. From Fig 2 of: Kim, T.; Murakami, T.; Hori, M.; Onishi, H.; Tomoda, K.; Nakamura, H. Effect of Superparamagnetic Iron Oxide on Tumor-to-liver Contrast at T2*-Weighted Gradient-Echo MRI: Comparison Between 3.0 T and 1.5 T MR Systems. J. Magn. Reson. Imaging, 2009, 29, 595–600*

metastases are radioactive complexes that have an attached ligand that can target bone. These compounds generally consist of a readily-available radionuclide that has a relatively short half life and emits β-radiation of $\sim$2 MeV. Since this energy radiation is *ionizing radiation*, which can ionize water and produce very reactive hydroxyl radical (HO$^\bullet$), cell death, which occurs within $\sim$5 mm of the site of radical generation, is caused by hydroxyl radicals chemically modifying important biomolecules on the surface of or inside the cell. Bone metastases often have an osteoblastic response (cells responsible for bone growth), which leads to uptake of the radioactive compound by normal cells adjacent to the tumor colony. When this occurs, the healthy cells, as well as nearby tumor cells, become exposed to numerous β-rays from the emitting radionuclides, which kills some of them, temporally mitigating the pain associated with the cancer.

Nuclides used in the palliation of pain associated with bone metastases are two isotopes of rhenium, ^{186}Re and ^{188}Re, and an isotope of yttrium, ^{90}Y [18,69–72]. Rhenium-186 emits a 1.07 MeV β-particle, which produces damage via ionizing radiation to proximally-located cells. This radionuclide also releases a 137 keV γ-ray, which while not harmful to cells, can be used to determine the location of the radiolabeled compound in the body using SPECT. The radionuclide, which has a short half life of 89.3 hours, is produced by neutron capture by ^{185}Re in a nuclear reactor and can be supplied as the perrhenate anion, $[^{186}\text{ReO}_4]^-$, to a hospital or clinic for pain treatment. For palliation of bone metastases, the perrhenate is reduced with stannous chloride and the reduction product reacted with an excess of 1-hydroxy-ethylenediamine-diphosphonic acid, HEDP (Figure 8.24), to produce ^{186}Re-HEDP. The resulting complex, which is approved for use in the European Union for palliation of pain, localizes in regions of bone metastases. The agent, which is given to the patient by IV injection, has a clearance half life from blood of $\sim$40 hours, with most of the radiation being eliminated in the urine [71]. The rapid clearance means that doses of the agent for controlling pain can be given as often as at

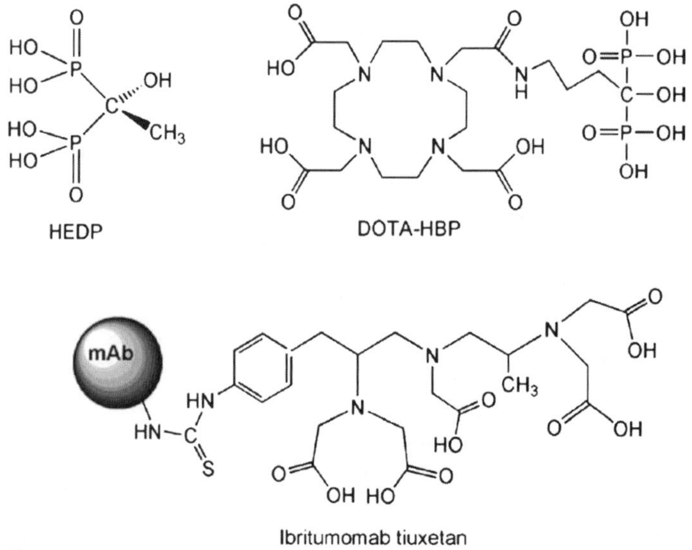

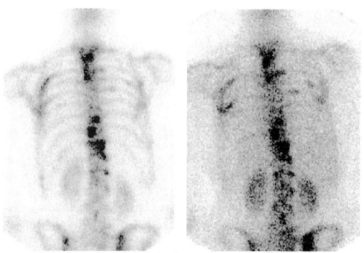

Figure 8.24   *Ligands which when complexed to radionuclides can be used for palliative treatment of pain associated with bone metastases to improve the quality of life of the patient (HEDP and DOTA-HBP) and to treat non-Hodgkin lymphoma (ibritumomab tiuxetan)*

~200 hour intervals. That the complex accumulates in bone is clearly seen from the SPECT images in Figure 8.25, which show a patient with breast cancer that has metastasized to the skeletal regions [71]. The patient was given ^{186}Re-HEDP and, for comparison, the known bone-imaging agent ^{99m}Tc-HDP, a complex similar to Tc-MDP (shown in Figure 8.4). Both compounds show the same number and locations of bone metastases in the patient.

Rhenium-188 emits a 2.12 MeV β-particle, as well as a 155 keV γ-ray. The nuclide has a short half life of 17 hours and can be produced by neutron radiation in a nuclear reactor or on-site in a hospital or clinic with a

Figure 8.25   *A patient with breast cancer and skeletal metastases. Left, ^{99m}Tc-HDP SPECT image 3 hours after injection; right, ^{186}Re-HEDP SPECT image 3 hours after injection. With kind permission from Springer Science + Business Media: Eur. J. Med. Mol. Imaging, 186Re-HEDP for Metastatic Bone Pain in Breast Cancer Patients, 31, 2004, S162–170, M.G. Lam, J.M. de Klerk and P.P. van Rijk, Fig 1*

$^{188}W/^{188}Re$ generator similar to that used for the generation of ^{99m}Tc, Figure 8.2. The product from the reactor, sodium perrhenate, $Na[^{188}ReO_4]$, can be reduced with stannous chloride and the reduction product reacted with HEDP to produce the agent, which when injected provides relief of pain from bone metastases [18,70].

Yttrium-90, element 39, is produced by high-purity separation from strontium-90, a fission product of uranium in a nuclear reactor. The most common oxidation state of yttrium is $3+$ and the trichloride, $^{90}YCl_3$, is available for applications in nuclear medicine. Yttrium-90 is a pure β-emitter with a maximum energy of 2.28 MeV and a half life of 64.1 hours, making it a useful radionuclide for palliation of pain in bone metastases. Ogawa and coworkers recently reported that the DOTA-based conjugate with a bisphosphonate, DOTA-HBP (Figure 8.24), reacts with $^{90}YCl_3$ to produce ^{90}Y-DOTA-HBP [72]. In this complex, the macrocyclic amine/acetate framework of the ligand is thought to be bound to the $^{90}Y^{+3}$ ion, while the bisphosphonate portion is free to interact with the hydroxyapatite matrix of bone. While ^{90}Y-DOTA-HBP has a lower accumulation in bone than ^{90}Y-citrate, the clearance of the former compound from blood and almost all soft tissue is faster than the citrate analog and is accompanied by a decrease in the level of unnecessary radiation compared to ^{90}Y-citrate.

In addition to alleviating pain, an yttrium-90 complex of the antibody–chelate conjugate, ibritumomab tiuxetan, called Zevalin (Figure 8.24), was approved by the United States FDA in 2002 for the treatment of non-Hodgkin lymphoma (NHL), the fifth most common form of cancer in the United States [73–77]. Zevalin, which is a member of a class of agents for *radioimmunotherapy*, has a monoclonal antibody from a mouse conjugated to a chelating pentaacetate ligand which binds the radionuclide. Since the attached antibody has high specificity for the CD20 antigen located on the surface of B-lymphocytes, the drug delivers the radionuclide to the cell with high specificity, where the emitted β-radiation produces hydroxyl radicals that kill the cell. Yttrium-90 does not emit a gamma ray, which would allow easy visualization of the drug in the body, but if ^{90}Y is substituted with ^{111}In (indium), which is a commonly available gamma emitter, the location of ^{111}In-ibritumomab tiuxetan can easily be determined using SPECT. The drug is available in a kit which contains the antibody–chelate conjugate, to which is added $^{90}YCl_3$ prior to IV injection of the solution into the patient. Zevalin is very effective in late-stage patients that have failed different types of chemotherapy and, because of the presence of the antibody, the drug produces fewer side effects than other forms of chemotherapy.

Problems

1. Of the ligands shown below, briefly explain which would be best suited to binding to fac-$[^{99m}Tc(H_2O)_3(CO)_3]^+$, the 'core' that is often used in synthesizing a ^{99m}Tc complex used in diagnostic nuclear medicine:

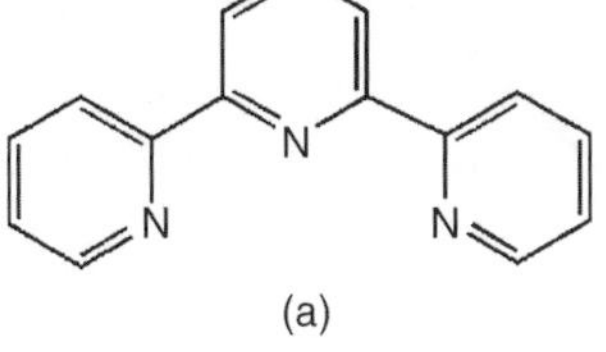

(a)

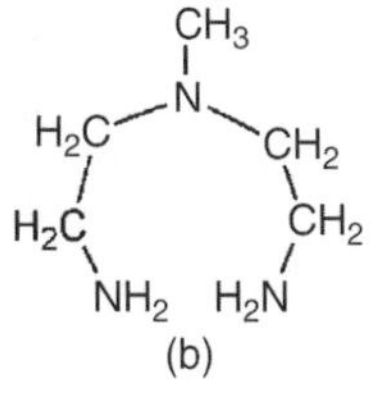

(b)

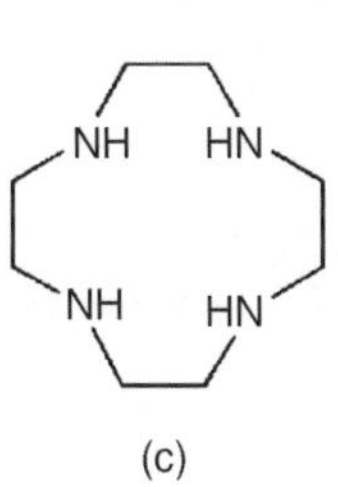

(c)

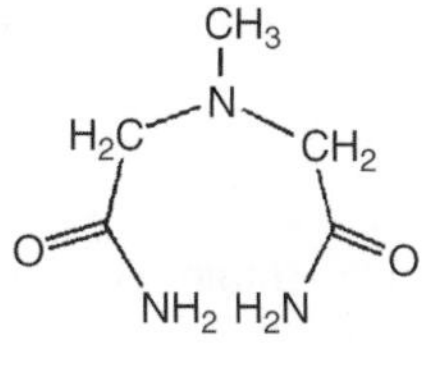

(d)

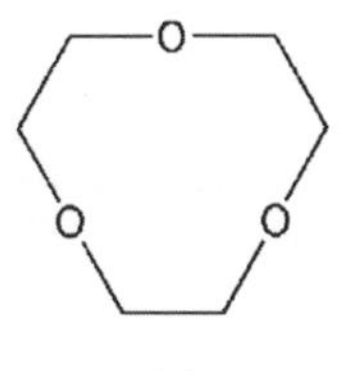

(e)

2. Determine the following, associated with the decay of ^{99m}Tc to ^{99}Tc:

    a. The length of time required for a sample of ^{99m}Tc to decay to 60% of its initial value.
    b. The wavelength of a photon released in the decay. Speed of light, $c = 3.0 \times 10^8 \, m \, s^{-1}$; $1 \, eV = 1.602 \times 10^{-19} \, J$; $h = 6.626 \times 10^{-34} \, J \, s$.

3. The tripeptide ligand, L, an analog of MAG3, reacts with $[^{99}TcO_4]^-$, with the loss of the thiol proton, in the presence of $SnCl_2$ to give two anionic compounds, 'A' and 'B', with the formula $[^{99}TcOL]^-$. Circular dichroism measurements show that both compounds are optically active and HPLC shows that both have different retention times. Sketch the structures of the complexes.

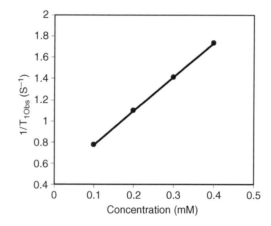

4. The compound ^{99m}Tc-MDP (Figure 8.4), which contains a bisphosphonate ligand, is thought to be an effective imaging agent for bone because it can bind to Ca^{+2} ions in hydroxyapatite, which is part of bone. Propose structures for the following:

    a. A dinuclear complex with a bridging MDP ligand, one Tc^{+4} and one Ca^{+2}.
    b. A trinuclear complex with a bridging MDP ligand, two Tc^{+4} ions and one Ca^{+2} ion.

5. Shown below is a plot of the reciprocal of the spin–lattice relaxation time T_1 versus concentration for a CA containing Gd^{+3} in water at 37 °C at 1.5 T. If the T_1 of water without the CA is 2.2 seconds, determine the relaxivity, r_1, of the CA from the plot.

6. A gadolinium CA has a relaxivity of $5.2 \, mM^{-1} \, s^{-1}$. When the concentration of the CA in blood is 150 μM the measured value of T_1 is 657 ms. Determine the value of T_1 for water (in seconds) in blood in the *absence* of the CA.

7. In NMR, if the ratio $N(-1/2)/N(+1/2) = 0.9999977$ for a proton at 25 °C, calculate ΔE in terms of wave numbers (cm^{-1}) between the $m_I = -1/2$ and $m_I = +1/2$ states, $k = 0.695$ cm^{-1} K^{-1}.

8. The stepwise stability (binding) constants for the first two steps in the interaction of the gadolinium CA MS-325 with human serum albumin, HSA, at 37 °C at pH 7.4 are $11.0 \times 10^3\,M^{-1}$ and $0.84 \times 10^3\,M^{-1}$ for K_1 and K_2, respectively. Calculate β_2 and ΔG associated with β_2 for MS-325 binding to HSA. $R = 8.314$ J K mol^{-1}.

References

1. Banerjee, S.R., Maresca, K.P., Francesconi, L. *et al.* (2005) New directions in the coordination chemistry of ^{99m}Tc: a reflection on technetium core structures and a strategy for new chelate design. *Nuclear Medicine and Biology*, **32**, 1–20.
2. Dilworth, J.R. and Parrott, S.J. (1998) The biomedical chemistry of technetium and rhenium. *Chemical Society Reviews*, 43–55.
3. Bowen, M.L. and Orvig, C. (2008) 99m-technetium carbohydrate conjugates as potential agents in molecular imaging. *Chemical Communications*, 5077–5091.
4. Zolle, I. (ed.) (2007) *Preparation and Quality Control in Nuclear Medicine*, Springer, Berlin.
5. Jurisson, S.S. and Lydon, J.D. (1999) Potential technetium small molecule radiopharmaceuticals. *Chemical Reviews*, **99**, 2205–2218.
6. Liu, S. and Edwards, D.S. (1999) 99m-Tc-labeled small peptides as diagnostic radiopharmaceuticals. *Chemical Reviews*, **99**, 2235–2268.
7. Ginj, M. and Maecke, H.R. (2004) Radiometallo-labeled peptides in tumor diagnosis and therapy. *Metal Ions in Biological Systems*, **42**, 109–142.
8. Imam, S.K. (2005) Molecular nuclear imaging: The radiopharmaceuticals (Review). *Cancer Biother Radio*, **20**, 163–172.
9. Jurisson, S., Cutler, C., and Smith, S.V. (2008) Radiometal complexes: Characterization and relevant *in vitro* studies. *Quarterly Journal of Nuclear Medicine and Molecular Imaging*, **52**, 222–234.
10. Pinkerton, T.C., Desilets, C.P., Hoch, D.J. *et al.* (1985) Bioinorganic activity of technetium radiopharmaceuticals. *Journal of Chemical Education*, **62**, 965–973.
11. Bottrill, M., Kwok, L., and Long, N.J. (2006) Lanthanides in magnetic resonance imaging. *Chemical Society Reviews*, **35**, 557–571.
12. Hermann, P., Kotek, J., Kubíček, V., and Lukes, I. (2008) Gadolinum(III) complexes as MRI contrasts agents: Ligand design and properties of the complexes. *Dalton Transactions*, 3027–3047.
13. Storr, T., Thompson, K.H., and Orvig, C. (2006) Design of targeting ligands in medicinal inorganic chemistry. *Chemical Society Reviews*, **35**, 534–544.
14. Caravan, P. (2006) Strategies for increasing the sensitivity of gadolinium based MRI contrast agents. *Chemical Society Reviews*, **35**, 512–523.
15. Aime, S., Barge, A., Cabella, C. *et al.* (2004) Targeting cells with MR imaging probes based on paramagnetic Gd(III) chelates. *Current Pharmaceutical Biotechnology*, **5**, 509–518.
16. Allen, M.J. and Meade, T.J. (2004) Magnetic resonance contrast agents for medical and molecular imaging. *Metal Ions in Biological Systems*, **42**, 1–38.
17. Fricker, S.P. (2006) The therapeutic application of lanthanides. *Chemical Society Reviews*, **35**, 524–533.
18. Lungu, V., Niculae, D., Bouziotis, P. *et al.* (2007) Radiolabeled phosphonates for bone metastases therapy. *Journal of Radioanalytical and Nuclear Chemistry*, **273**, 663–667.
19. http://en.wikipedia.org/wiki/Tomography.
20. http://en.wikipedia.org/wiki/Gamma_camera.
21. Libson, K., Deutsch, E., and Barnett, B.L. (1980) Structural characterization of a ^{99}Tc-Diposphonate complex. Implications for the chemistry of ^{99m}Tc skeletal imaging agents. *Journal of the American Chemical Society*, **102**, 2476–2478.

22. Boasquevisque, E. *et al.* (2008) ^{99m}Tc-MDP bone uptake in secondary hyperparathyroidism: comparison of the mandible, cranium, radius, and femur. *Oral Radiology*, **24**, 55–58.

23. Seelig, A. (1998) A general pattern for substrate recognition by P-gylcolprotein. *European Journal of Biochemistry*, **251**, 252–261.

24. Seelig, A. (2007) The role of size and charge for blood brain barrier permeation of drugs and fatty acids. *Journal of Molecular Neuroscience*, **33**, 32–41.

25. Vanderghinste, D., Eeckhoudt, M.V., Terwinghe, C. *et al.* (2003) An efficient HPLC method for the analysis of isomeric purity of technetium-99m-exametazime and identity confirmation using LC-MS. *Journal of Pharmaceutical and Biomedical Analysis*, **32**, 679–685.

26. Ding, H.-Y., Huang, Y.F., Tzeng, C.-C. *et al.* (1999) Synthesis of D,D-HMPAO and Tc-L,L-HMPAO and their comparison of chemical and biological properties. *Bioorganic & Medicinal Chemistry Letters*, **9**, 3199–3202.

27. Neirinckx, R.D. *et al.* (1987) Technetium-99m d,l-HM-PAO: A new radiopharmaceutical for SPECT imaging of regional cerebral blood perfusion. *Journal of Nuclear Medicine*, **28**, 191–202.

28. Sundström, T., Elgh, E., Larsson, A. *et al.* (2006) Memory-provoked rCBF-SPECT as a diagnostic tool in Alzheimer's disease? *European Journal of Nuclear Medicine and Molecular Imaging*, **33**, 73–80.

29. Fritzberg, A.R., Kasina, S., Eshima, D., and Johnson, D.L. (1986) Synthesis and biological evaluation of technetium-99m MAG3 as a hippuran replacement. *Journal of Nuclear Medicine*, **27**, 111–116.

30. Verbruggen, A., Bormans, G., Cleyhens, B. *et al.* (1989) Separation of the enantiomers of technetium-99m-MAG$_3$ and their renal excretion in baboons and a volunteer. *Nuklearmedizin*, **25**, 436–439.

31. Lipowska, M., Hanson, L., Xu, X. *et al.* (2002) New N$_3$S donor ligand small ligand peptide analogues of the N-Mercaptoacetyl-glycylglycylglycine ligand in the clinically used Tc-99m renal imaging agent: Evidence for unusual amide oxygen coordination by two new ligands. *Inorganic Chemistry*, **41**, 3032–3041.

32. Shikano, N., Kanai, Y., Kawai, K. *et al.* (2004) Transport of 99m Tc-MAG3 via rat renal organic anion transporter 1. *Journal of Nuclear Medicine*, **45**, 80–85.

33. Blaufox, M.D. (2004) Transport of ^{99m}Tc-MAG3 via rat renal organic anion. *Journal of Nuclear Medicine*, **45**, 86–88.

34. Benson, C. *et al.* (2007) A phase I trial of the selective oral cyclin-dependent kinase inhibitor seliciclib (CYC202; *R*-Roscovitine), administered twice daily for 7 days every 21 days. *British Journal of Cancer*, **96**, 29–37.

35. Reubi, J.C. and Maecke, H.R. (2008) Peptide-based probes for cancer imaging. *Journal of Nuclear Medicine*, **49**, 1735–1738.

36. Cortez-Retamozo, V. *et al.* (2008) ^{99m}Tc-labeled nanobodies: A new type of targeted probes for imaging antigen expression. *Current Radiopharmaceuticals*, **1**, 37–41.

37. Cyr, J.E. (2007) Isolation, characterization, and biological evaluation of *Syn* and *Anti* diastereomers of [^{99m}Tc] technetium depreotide: A somatostatin receptor binding tumor imaging agent. *Journal of Medicinal Chemistry*, **50**, 4295–4303.

38. Mena, E., Camacho, V., Estorch, M. *et al.* (2004) ^{99m}Tc-depreotide scintigraphy of bone lesions in patients with lung cancer. *European Journal of Nuclear Medicine and Molecular Imaging*, **31**, 1399–1404.

39. Alberto, R., Schibli, R., Egli, A., and Schubiger, A.P. (1998) A novel organometallic aqua complex of technetium for the labeling of biomolecules: Synthesis of [^{99m}Tc(OH$_2$)$_3$(CO)$_3$]$^+$ from [^{99m}TcO$_4$] in aqueous solution and its reaction with a bifunctional ligand. *Journal of the American Chemical Society*, **120**, 7987–7988.

40. Alberto, R., Ortner, K., Wheatly, N. *et al.* (2001) Synthesis and properties of boranocarbonate: A convenient *in situ* CO source for the aqueous preparation of [^{99m}Tc(OH$_2$)$_3$(CO)$_3$]$^+$. *Journal of the American Chemical Society*, **123**, 3135–3136.

41. Stephenson, K.A. *et al.* (2004) Bridging the gap between in vitro and *in vivo* imaging: isostructural Re and ^{99m}Tc complexes for correlating fluorescence and radioimaging studies. *Journal of the American Chemical Society*, **126**, 8598–8599.

42. Bartholomä, M., Valliant, J.V., Maresca, K.P. *et al.* (2009) Singel amino acid chelates (SAAC): A stragety for the design of technetium and rhodium radiopharmaceuticals. *Chemical Communications*, 493–512.

43. Burtea, C., Laurent, S., Vander, E.L., and Muller, R.N. (2008) Contrast agents: Magnetic resonance. *Handbook of Experimental Pharmacology*, **185** (Pt 1), 135–165.

44. Krause, W. (2002) Contrast agents I. Magnetic resonance imaging, *Top. Curr. Chem.*, **221**, Springer, Berlin, Germany.

45. Drago, R.S. (1992) *Physical Methods for Chemists*, 2nd edn, Saunders College Publishing.

46. Harken, H. and Wolf, H.C. (1994) Molecular physics and elements of qunatum chemistry, *Introduction to Experiments and Theory*, Springer-Verlag, Berlin, Heidleberg, Germany.

47. Werner, E.J., Datta, A., Jocher, C.J., and Raymond, K.N. (2008) High-relaxivity MRI contrast agents: where coordination chemistry meets medical imaging. *Angewandte Chemie-International Edition*, **47**, 8568–8580.

48. Penfield, J.G. and Reilly, R.F. (2008) Nephrogenic systemic fibrosis risk: is there a difference between gadolinium-based contrast agents? *Seminars in Dialysis*, **21**, 129–134.

49. ten Dam, M.A. and Wetzels, J.F. (2008) Toxicity of contrast media: an update. *Netherlands Journal of Medicine*, **66**, 416–422.

50. Abraham, J.L. and Thakral, C. (2008) Tissue distribution and kinetics of gadolinium and nephrogenic systemic fibrosis. *European Journal of Radiology*, **66**, 200–207.

51. Frenzel, T., Lengsfeld, P., Schimer, H. *et al.* (2008) Stability of gadolinium-based magnetic resonance imaging contrast agents in human serum at 37 degrees C. *Investigative Radiology*, **43**, 817–828.

52. Rockledge, S.M., Cacheris, W.P., Quay, S.C. *et al.* (1989) Manganese(II) N, N'-dipyridoxylethylenediamine-N,N'-diacetate 5,5'-Bis(phosphate). Synthesis and characterization of a paramagnetic chelate for magnetic resonance imaging enhancement. *Inorganic Chemistry*, **28**, 477–485.

53. Tirkkonen, B. *et al.* (1997) Physicochemical characterization of mangafodipir sodium. *Acta Radiologica*, **38**, 780–789.

54. Mendonça-Dias, M.H., Gaggelli, E., and Lauterbur, P.C. (1983) Paramagnetic contrast agents in nuclear magnetic resonance medical imaging. *Seminars in Nuclear Medicine*, **13**, 364–376.

55. Silva, A. and Bock, N.A. (2008) Manganese-enhanced MRI: An exceptional tool in translational neuroimaging. *Schizophrenia Bulletin*, **34**, 595–604.

56. Chandra, S.V. and Shukla, G.S. (1976) Role of iron deficiency in inducing susceptibility to manganese toxicity. *Archives of Toxicology*, **35**, 319–323.

57. Wolf, G.L. and Baum, L. (1983) Cardiovascular toxicity and tissue proton T1 response to manganese injection in the dog and rabbit. *American Journal of Roentgenology*, **141**, 193–197.

58. Reimer, P., Schneider, G., and Schima, W. (2004) Hepatobiliary contrast agents for contrast-enhanced MRI of the liver: Properties, clinical development and applications. *European Radiology*, **14**, 559–578.

59. Toft, K.G., Hustvedt, S.O., Grant, D. *et al.* (1997) Metabolism and pharmacokinetics of MnDPDP in man. *Acta Radiologica*, **38**, 677–689.

60. Song, Y., Kohlmeir, E.K., and Meade, T.J. (2008) Synthesis of multimeric MR contrast agents for cellular imaging. *Journal of the American Chemical Society*, **130**, 6662–6663.

61. Arbab, A.S., Liu, W., and Frank, J.A. (2006) Cellular magnetic resonance imaging: Current status and future prospects. *Expert Review of Medical Devices*, **3**, 427–439.

62. Caravan, P. *et al.* (2002) The interaction of MS-325 with human serum albumin and its effect on proton relaxation rates. *Journal of the American Chemical Society*, **124**, 3152–3162.

63. Bush, A.I. (2008) Drug development based on the metal hypothesis of Alzheimer's disease. *Journal of Alzheimer's Disease*, **15**, 223–240.

64. Major, J.L., Parigi, G., Luchinat, C., and Meade, T.J. (2007) The synthesis and *in vitro* testing of a Zinc-activated MRI contrast agent. *Proceedings of the National Academy of Sciences of the United States of America*, **104**, 13881–13886.

65. Zhang, X., Lovejoy, K.S., Jasanoff, A., and Lippard, S.J. (2007) Water soluble porphyrins as a dual-function molecular imaging platform fro MRI and fluorescence zinc sensing. *Proceedings of the National Academy of Sciences of the United States of America*, **104**, 10780–10785.

66. LaConte, L., Nitin, N., and Bao, G. (2005) Magnetic nanoparticle probes. *Nanotoday*, May, 32–38.

67. Ferrucci, J.T. and Stark, D.D. (1990) Iron oxide-enhanced MR imaging of the liver and spleen: Review of the first 5 years. *American Journal of Roentgenology*, **155**, 943–950.

68. Kim, T. *et al.* (2009) Effect of Superparamagnetic Iron Oxide on Tumor-to-liver Contrast at T_2^*-Weighted Gradient-Echo MRI: Comparison Between 3.0 T and 1.5 T MR Systems. *Journal of Magnetic Resonance Imaging*, **29**, 595–600.

69. Ferro-Flores, G. and Arteaga de Murphy, C. (2008) Pharmacokinetics and dosimetry of [188]Re-pharmaceuticals. *Advanced Drug Delivery Reviews*, **60**, 1389–1401.

70. Liepe, K., Kropp, J., Runge, R., and Kotzerke, J. (2003) Therapeutic efficiency of rhenium-188-HEDP in human prostate cancer skeletal metastases. *British Journal of Cancer*, **89**, 625–629.

71. Lam, M.G., de Klerk, J.M., and van Rijk, P.P. (2004) [186]Re-HEDP for metastatic bone pain in breast cancer patients. *European Journal of Nuclear Medicine and Molecular Imaging*, (Suppl 1), S162–S170.

72. Ogawa, K. *et al.* (2009) Development of [^{90}Y]DOTA-conjugated bisphosphonate for treatment of painful bone metastases. *Nuclear Medicine and Biology*, **36**, 129–135.

73. Assié, K., Dieudonné, A., Gardin, I. *et al.* (2008) Comparison between 2D and 3D dosimetry protocols in ^{90}Y-Ibritumomab tiuxetan radioimmunotherapy of patients with non-Hodgkin's lymphoma. *Cancer Biotherapy & Radiopharmaceuticals*, **23**, 53–64.

74. Sgouros, G. (2008) Molecular radiotherapy: Survey and current status. *Cancer Biotherapy & Radiopharmaceuticals*, **23**, 531–540.

75. Aarts, F., Bleichrodt, R.P., Oyen, W.J., and Boerman, O.C. (2008) Intracavity radioimmunotherapy to treat solid tumors. *Cancer Biotherapy & Radiopharmaceuticals*, **23**, 92–107.

76. Nijsen, J.F., Krijger, G.C., and van Het Schip, A.D. (2007) The bright future of radionuclides for cancer therapy. *Anti-Cancer Agents in Medicinal Chemistry*, **7**, 271–290.

77. Chapuy, B., Hohloch, K., and Trümper, L. (2007) Yttrium 90 ibritumomab tiuxetan (Zevalin[(R)]): A new bullet in the fight against malignant lymphoma? *Biotechnology Journal*, **2**, 1435–1443.

Further reading

Huheey, J.E., Keiter, E.A., and Keiter, R.L. (1993) Inorganic chemistry, *Principles of Structure and Reactivity*, 4th edn, Benjamin-Cummings Publishing Co., San Francisco.

Levitt, M.H. (2001) Spin dynamics, *Basics of Nuclear Magnetic Resonance*, John Wiley & Sons Ltd, Chichester, UK.

9

Nanomedicine

Most drugs in use today are not very specific for the diseases they are intended to treat. Commonly-used drugs have relatively narrow *therapeutic windows*, suggesting that their efficacies are due to relatively small differences in their concentration in healthy and diseased tissue in the body. In an effort to enhance the specificity of drugs and minimize their side effects, science is examining the potential of nanometer-sized engineered particles to act as delivery vehicles for drugs [1–5]. Since nanomaterials have a variety of dimensions, shapes and properties (Figure 9.1), and they can be taken into the cell by endocytotic mechanisms, loading nanoparticles with a cytotoxic agent or attaching drugs to their surface may be a means of delivering a high 'payload' of a pharmacologically-active agent to a specific site in the body. If the particle is also equipped with a cell-recognizing feature such as an antibody, the resulting assembly becomes a highly specific system for delivering chemotherapeutic agents to certain types of tissue. In addition to treating disease, nanoparticles have the potential for detecting disease. For example, if a nanoparticle has many attached paramagnetic ions and is equipped with an antibody that can bind to an antigen on the surface of certain types of cancer cell, the resulting nanosystem could be used as a high-relaxivity MRI contrast agent (CA) for detecting cancer in the body [6–8].

While the application of nanoscience and nanotechnology to medicine, often called *nanomedicine*, promises to revolutionize many aspects of medical science, engineered nano-sized particles are totally alien to the biological environment and as such they may pose serious health risks to humans, especially if they are introduced directly into the body in the form of therapeutic drugs or diagnostic agents [9]. Thus, in addition to describing a few of the exciting new advances in the field of nanomedicine that involve inorganic chemistry, it will be important to address some of the issues facing investigators who are attempting to evaluate the health risks of nanomaterials, and to briefly summarize what has been learned concerning the compatibility of some of these agents with living systems.

9.1 Nanoscience for treating disease

9.1.1 Single walled carbon nanotubes

Single-walled carbon nanotubes, SWNTs, which are related to the well-known spherical compound fullerene, C_{60}, are small-diameter ($\sim$1 nm) long tubular structures made up of many carbon atoms joined to each other in a hexagonal array [10]. The *aspect ratio* of SWNT, which is the ratio of the length of the particle to its diameter,

Metals in Medicine James C. Dabrowiak
© 2009 John Wiley & Sons, Ltd

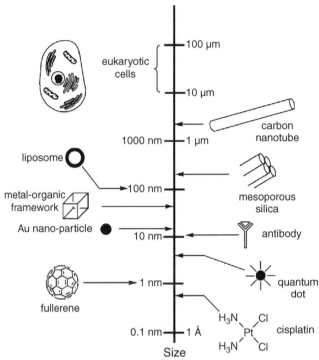

Figure 9.1 *The scale of some particles and substances encountered in nanomedicine*

varies, but it can be large, $> \sim 10^3$, showing that the particles have a rod-like structure. To visualize the general shape of an SWNT, consider hexagonal-holed (chicken wire) fencing, in which the vertices of the hexagons are carbon atoms. If a long section of fencing were rolled and sealed to form a small-diameter 'tube', it would look like an SWNT. Single-walled carbon nanotubes can be made using arc discharge, chemical vapor deposition or by other methods, and procedures are available to obtain lengths that are relatively monodisperse (all SWNTs in the sample have approximately the same length).

Using SWNTs, Lippard, Dai and their coworkers [11] synthesized a nanoparticle that is capable of delivering many toxic platinum complexes to cells. The particle, which the investigators called a *longboat delivery system*, consisted of an SWNT with ~ 65 molecules of a Pt^{+4} compound attached (Figure 9.2). The platinum compound, which is the 'cargo part' of the nanoparticle, was produced by oxidizing cisplatin using hydrogen peroxide in ethanol, which resulted in a six-coordinate Pt^{+4} complex with the cisplatin ligands in the plane and ethoxide and hydroxo (OH) ligands *trans* to each other on the axis of the compound. This oxidation is similar to that described in connection with the synthesis of the anticancer Pt^{+4} complex satraplatin, outlined in Box 4.4. Since the bound hydroxide ligand is nucleophilic, it was reacted with succinic anhydride to produce the compound c,c,t-$[Pt(NH_3)_2Cl_2(OEt)(O_2CCH_2CH_2CO_2H)]$ (shown in Figure 9.2a). This complex, which has antitumor properties, is believed to be reduced in the body to cisplatin. In order to make the delivery system, the researchers synthesized a long organic linker consisting of a polyethylene glycol (PEG) chain 45 units long appended to two 16-carbon alkyl chains (Figure 9.2a). The polyethylene glycol chain imparted water solubility to the final product, while two 16-carbon alkyl chains, because they strongly associated with SWNT through hydrophobic interactions, allowed for the connection of the linker to the carbon nanotube. After allowing the linker to bind to the SWNT, the resulting product was reacted with c,c,t-$[Pt(NH_3)_2Cl_2(OEt)(O_2CCH_2CH_2CO_2H)]$ in the presence of a carbodiimide coupling reagent that formed an amide linkage between the free amino group on the end of the PEG portion

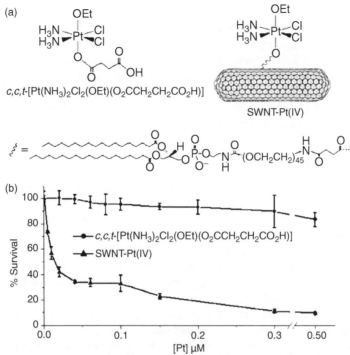

Figure 9.2   *(a) c,c,t-[Pt(NH₃)₂Cl₂(OEt)(O₂CCH₂CH₂CO₂H)] and the SWNT-tethered conjugate, SWNT-Pt(IV). (b) Cytotoxicity of free c,c,t-[Pt(NH₃)₂Cl₂(OEt)(O₂CCH₂CH₂CO₂H)] and SWNT-Pt(IV) in NTera-2 cells. Adapted with permission from R.P. Feazell et al., Soluble Single-Walled Carbon Nanotubes as Longboat Delivery Systems for Platinum(IV) Anticancer Drug Design, J. Am. Chem. Soc. 129, 8438–8439. Copyright 2007 American Chemical Society*

of the linker and the free carboxyl group on the Pt^{+4} complex. The final product, the longboat delivery system (SWNT-P(IV)), was purified by dialysis (Figure 9.2a). The investigators used atomic absorption spectroscopy (AAS) to determine that each SWNT carried an average of 65 tethered Pt^{+4} complexes.

The cytotoxicity of the longboat delivery system and, as a control, the untethered Pt^{+4} complex against the testicular carcinoma cell line, NTera-2, was determined using a standard cell assay [11]. After exposing the cells to various concentrations of agent for four days and determining the number of surviving cells at the end of this period, the longboat delivery system was found to be much more toxic *on a per-platinum basis* than the untethered Pt^{+4} complex, which had negligible cytotoxicity (Figure 9.2b).

In addition to the cytotoxicity studies, the investigators also determined the cellular location of the longboat delivery system inside NTera-2 cells [11]. This was done by attaching a fluorescent molecule to the linker region of the delivery system and determining the locations of fluorescence in the cell using microscopy. They found that the fluorophore localized in *endosomes*, small ∼2 μm acidic vesicles inside the cell, supporting earlier studies that suggested that SWNT enter cells through a mechanism called *endocytosis*, involving the protein *clathrin* [12]. In this process, the SWNT binds to a 'pit' or depression on the surface of the cell, where it is enveloped into the cell as a clathrin-coated vesicle. These clathrin-coated vesicles eventually merge with other vesicles to form an endosome, which releases the contents to *lysosomes* and ultimately to the cytoplasm for distribution in the cell.

In a continuation of these studies, Dai, Lippard and their coworkers [13] replaced the ethoxide ligand of the Pt^{+4} complex shown in Figure 9.2a with a carboxylate group that has an attached folic acid residue. Since folic acid receptors are produced in large numbers on the surface of many different types of cancer cell – for example, ovarian, endometrial, breast, lung, renal and colon cancer – covalently attaching folic acid to the Pt^{+4} complex produces a longboat delivery system that can recognize folic acid receptors on the surface of cells. The study showed that NTera-2 cells with a large number of folic acid receptors on their surface were more easily killed by the delivery system than the same cells with the normal number of receptors, indicating that folic acid enhances the cytotoxicity of the delivery system. The investigators also used a fluorescent antibody that was able to identify a GG, 1,2-intrastrand platinum crosslink on DNA to show that a longboat delivery system does indeed produce the same crosslink on genomic DNA as cisplatin.

9.1.2   Metal–organic frameworks

The term *metal–organic frameworks* or *MOFs*, which are sometimes called *coordination polymers*, refers to compounds that have a crystal lattice made up of metal complexes containing organic ligands [14]. In order to form a MOF, a simple metal complex with some regular geometry, for example octahedral coordination, is first formed in solution by reaction with an organic ligand. The ligand is carefully chosen so that it has two or more good metal donor groups, usually carboxylates, but not all of the groups on the ligand can bind to the same metal ion. This ensures that some of the donor groups on the ligand are available to form *bridges* to other metal ions, which produces a lattice with a *multinuclear* network. The formation of the network begins when an initially-formed *mononuclear* complex self-associates to form a larger, *multinuclear* structure, for example a di-, tri- or tetranuclear complex. Under the proper conditions, these multinuclear complexes may also self-associate to form even larger structures. Which product forms depends in a complicated way on the solubility of materials, the nature of the metal ion and ligands, and other experimental conditions.

A wide variety of three-dimensional lattices are formed by MOFs, but all contain *voids* or *pores*. If a pore is large enough to accept a small molecule such as a drug, the process of moving the small molecule into the pore is called *adsorption* and the small molecule is considered the *guest*, while the lattice framework is referred to as the *host*.

Recently, Horcajada *et al.* [15] studied the ability of two MOFs to adsorb the pain-relieving drug ibuprofen. The MOFs, denoted as MIL-100 and MIL-101 in Figure 9.3, were made by reacting the simple water-soluble chromium salt $[Cr(H_2O)_6](NO_3)_3$, with either 1,4-benzenedicarboxylic acid (1,4-BDC), which led to the formation of MIL-101, or 1,3,5-benzenetricarboxylic acid (1,3,5-BTC), which formed MIL-100, in the presence of hydrofluoric acid in water at high temperature (220 °C). Due to the fact that the temperature of the reaction was above the boiling point of water and thus under normal conditions the water would simply boil away as steam, the reaction, which is called a *hydrothermal reaction*, was carried out in a thick-walled metal container called a 'bomb', which was able to contain the high pressures developed in the reaction. While the mechanistic details of the reaction are not known, the synthesis of MIL-100 and MIL-101 probably begins with the formation of a *trinuclear* Cr^{+3} complex, which has three octahedral Cr^{+3} ions joined together in a cyclic structure (Figure 9.3). This structure, referred to by the investigators as a 'trimer of chromium octahedra', has bridging carboxylates ligands and an oxo ligand (O^{2-}, di-deprotonated water) which connects three octahedral Cr^{+3} ions. The investigators proposed that this structure associates through a ligand exchange process to form a tetramer, indicated as 'T' in Figure 9.3, and that 'T' ultimately further associates to form MIL-100 and MIL-101, which crystallize from the reaction medium as $\sim$60 nm crystalline particles. The joining of the 'T' structures to each other to form the crystal lattice produces two different-sized holes or pores in the framework, which are capable of adsorbing small guest molecules. The sizes of the pores are 2.5–3.4 nm, with MIL-101 having the larger pores.

Ibuprofen is a well-known nonprescription drug with potent anti-inflammatory and analgesic (pain-killing) properties. Since ibuprofen is especially useful for relieving the symptoms of arthritis, finding formulations of the drug that would allow it to be slowly released in the body over a long period of time would be beneficial to

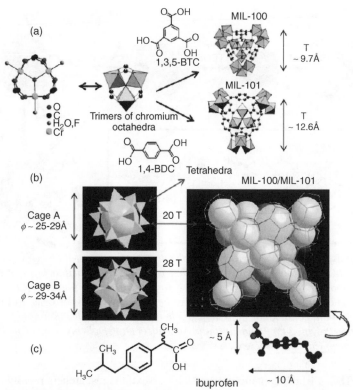

Figure 9.3 *(a) Schematic 3D representation of tetrahedral (T) built up from trimers of chromium octahedra and 1,4-benzenedicarboxylate groups or 1,3,5-benzenetricarboxylate groups in MIL-101 and MIL-100, respectively. (b) Schematic 3D representation of the mobil-39 (MTN) zeotype architecture of MIL-100 and MIL-101; left: smaller A cages (spheres formed by 20 tetrahedra) and larger B cages (spheres formed by 28 tetrahedra); right: a unit cell with lines connecting the tetrahedral centers. (c) Chemical structure of ibuprofen and its approximate molecular dimensions. From P. Horcajada et al., Metal-Organic Frameworks as Efficient Materials for Drug Delivery, Angew. Chem. Int. Ed. 2006, 45, 5974–78. Copyright Wiley-VCH Verlag GmbH & Co. KGaA. Reproduced with permission*

patients suffering from arthritis. As shown at the bottom of Figure 9.3, the molecular size of ibuprofen is smaller than the pores (cages sizes) of either MIL-100 or MIL-101. In order to see whether ibuprofen could be adsorbed into either MOF, the investigators exposed hexane solutions of the drug to MIL-100 and MIL-101. They found that both MOFs adsorbed the drug, with MIL-101 adsorbing 140% of its own weight of ibuprofen. After obtaining the drug-loaded MOFs, the researchers also studied the release of the drug into an aqueous medium that simulates body fluids. They found that while some drug is released within hours by both MOFs, complete removal of the drug from the particles required ~6 days, suggesting that MOFs of this type could be useful for the slow release of ibuprofen in pharmaceutical preparations such as a transdermal patch. Although Cr^{+3} is quite toxic to mammalian cells, the investigators noted that the molecular lattice need not be built around Cr^{+3} octahedra, and that Fe^{+3}, which is far less toxic than Cr^{+3} and readily forms octahedra, would be a useful alternative for constructing MOFs for drug storage and release under biological conditions.

Because of their success as anticancer agents, platinum complexes are being directly incorporated into nanostructures. An interesting example involves the complex *c,c,t*-diamminedichlorodisuccinatoplatinum (IV), which is a prodrug for cisplatin (Figure 9.4). Lin and coworkers [16] dissolved the platinum complex in water, adjusted the pH to 5.5, which deprotonates the two free carboxyl groups of the compound, and added the

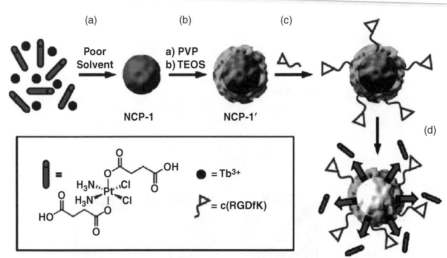

Figure 9.4   *(a) Synthesis of a nanoscale coordination polymer (NCP) from c,c,t-[Pt(NH$_3$)$_2$Cl$_2$(O$_2$CCH$_2$CH$_2$CO$_2$H)$_2$] and the lanthanide, TbCl$_3$, to form NCP-1. (b) Coating of NCP-1 with a thin layer of silica, where PVP is polyvinylpyrrollidone and TEOS is tetraethylorthosilicate, to give the coated particle NCP-1'. (c) Attachment of a small cyclic peptide, c(RGDfK), which has high affinity for angiogenic cancer cells, to the surface of NCP-1'. (d) Release of toxic platinum compounds from the particle inside cancer cells. Reprinted from W.J. Rieter et al., Nanoscale Coordination Polymers for Platinum-Based Anticancer Drug Delivery, J. Am. Chem. Soc. 130, 11584–5. Copyright 2008 American Chemical Society*

simple terbium salt TbCl$_3$, which contains aquated Tb^{+3}. Next, the researchers rapidly added ethanol to the stirred solution, which dramatically lowered the dielectric constant of the medium (the dielectric constant of ethanol is much lower than that of water), causing a solid to precipitate from solution. The isolated solid, which contained terbium and the platinum complex, existed as relatively uniform-sized nanoparticles with a diameter of 58.3 ± 11.3 nm, but studies showed that it was amorphous in structure; that is, not crystalline. Since the solid readily dissociated in water, the researchers coated the surface of the particles with a thin layer of silica (2–7 nm thick), which slowed the attack of water on the particle. The coating was done using a *grafting* procedure, in which SiO$_2$ was chemically bonded to sites on the surface of the particle. In the final step, the investigators covalently grafted a small peptide to the silica surface of the particle. Since the peptide has a high affinity for a protein on the surface of HT-29 angiogenic cancer cells, the final particle expectedly exhibited higher toxicity – lower IC$_{50}$ values – than cisplatin toward this cell line. The researchers pointed out that this general approach could be used to incorporate not just cytotoxic drugs but also imaging agents, for which slow release of active compounds and site-specific delivery are important [16].

9.1.3   Mesoporous silica

In 1992, researchers at the Mobil Research and Development Corporation reported the synthesis and characterization of a highly-ordered silica structure that was made in a template reaction involving surfactants and silicate ions (Figure 9.5) [17]. The new material, called MCM-41 (Mobil Crystalline Material-41) was made by combining the surfactant hexamethyltrimethylammonium ion, C$_{16}$H$_{33}$(CH$_3$)$_3$N$^+$OH/Cl, with tetramethylammonium silicate, [(CH$_3$)$_4$N]$_4$SiO$_4$. In solution, the surfactant molecules align themselves to form a rod-like micelle with the *hydrophobic* alkyl chains of the surfactant pointed inward and the *hydrophilic* positively-charged *head group* pointed out toward solvent. Interestingly, these individual micelles align themselves to produce larger bundles, which are stacked in a hexagonal pattern, called a *hexagonal closest-*

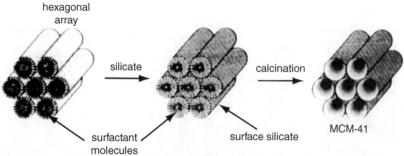

hexagonal
array

silicate

calcination

surfactant
molecules

surface silicate

MCM-41

Figure 9.5 *Schematic of the synthesis of MCM-41 mesoporous silica through a liquid crystal templating mecha-nism. Hexagonal arrays of cylindrical micelles form with the polar groups of the surfactants (light gray) to the outside. Silicate species (dark gray) occupy the spaces between the cylinders and in the final step (calcination) the micelles are removed at high temperature, leaving hollow cylinders of MCM-41. Reprinted in part by permission from Macmillan Publishers Ltd: C.T. Kresge et al. Ordered Mesoporous Molecular Sieves Synthesized by a Liquid-Crystal Template Mechanism, Nature, 359, 710–712. Copyright 1992*

packed array. By heating the solution for a period of time, the silicate anions arrange themselves on the surface of the individual micelles in the bundles and react with each other to form a glass-like matrix, $(SiO_2)_n$, which coats each micelle and joins the entire structure together in a rigid form. Once the SiO_2 matrix is in place, the material is heated to a very high temperature, 540 °C, which 'burns off' the surfactant organic molecules, releasing them to the vapor phase, but the silica matrix remains intact. The end result is a rigid mesoporous silica nanoparticle, MSN, with pore sizes >2 nm, that has the appearance of a stack of irrigation tubes (Figure 9.5). Close inspection of the tubes reveals that they are not perfect cylinders but rather are hexagonally-shaped tubes; that is, the pore opening is a hexagon and not a circle.

Researchers have found ways to change the particle and pore size of MSNs, as well as to modify their surface by attaching a wide range of organic molecules. The nanomaterials are characterized using a variety of methods, one of which is x-ray powder diffraction, which mainly measures the distance between repeat units in the solids, in this case the center-to-center distance between the tubes. Solid-state ^{29}Si NMR is also used to characterize mesoporous silica. Although ^{29}Si has the same nuclear spin as the proton, $I = \frac{1}{2}$, the fact that the silicon atoms are in a rigid solid in MCM-41and not rapidly tumbling in solution leads to broadened NMR resonance lines. However, the chemical shifts of the broad resonances still provide useful information on the nature of groups attached to the silicon atoms in MCM-41.

A useful property of mesoporous silica is that it has a very large surface area, $\sim 1000 \, m^2$ per gram weight of material. Surface areas for these and similar nanomaterials are determined using *BET N_2 gas adsorption* ('B', 'E' and 'T' are the first initials of the surnames of the researchers who discovered the technique). This produces an *isotherm* showing the number of N_2 molecules adsorbed on to the surface of the solid as a function of the applied relative pressure of the gas. Since the N_2 molecules form a *monolayer* on all of the surfaces of the particle (the N_2 molecules are in van der Waals contact with each), the total surface area per gram, S_{BET}, can be calculated from the size and molecular weight of N_2 and the experimentally-determined adsorption isotherm.

The interesting property of mesoporous materials like MCM-41 is that, since their pores are large enough to adsorb small molecules, they can be used to deliver drugs in chemotherapy. An example of how this might work was published by Zink and Tamanoi and their coworkers [18], who showed that the anticancer drug camptothecin is adsorbed by modified mesoporous silica nanoparticles, and that the drug-loaded particles can inhibit the growth of a number of different cancer cell lines.

Camptothecin is a quinoline alkaloid natural product that blocks the ability of the enzyme *topoisomerase* to change the degree of supercoiling of DNA that is necessary for transcription and translation. A major drawback of the drug is that it has low solubility in water, and while other water-soluble analogs have been identified, they

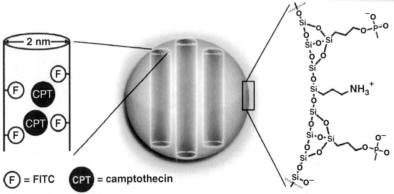

$\textcircled{F}$ = FITC $\textbf{CPT}$ = camptothecin

Figure 9.6 *Schematic representation of camptothecin (CPT)-loaded fluorescent mesoporous silica spherical nanoparticles. The 2 nm diameter pores of the particle were derivatized with a fluorescent dye, F (FITC), and filled with CPT. The surfaces of the nanoparticles were modified with trihydroxysilylpropyl methylphosphonate to reduce particle–particle interactions in solution. From J. Lu et al., Mesoporous Silica Nanoparticles as a Delivery System for Hydrophobic Anticancer Drugs, Small, 2007, 3, 1341–46. Copyright Wiley-VCH Verlag GmbH & Co. KGaA. Reproduced with permission*

are less effective than camptothecin in attacking cancer. In order to study the feasibility of using mesoporous silica as a carrier for camptothecin, Zink and Tamanoi and their coworkers dissolved the drug in dimethylsulfoxide and applied the solution to derivatized nanoparticles (Figure 9.6), which adsorbed ~80 nmol of camptothecin per 50 mg of particle [18]. They showed that the drug-loaded particles were highly toxic to colon, pancreatic and stomach cancer cell lines and, through the use of fluorescence microscopy, that the nanoparticles entered the cells and localized in their hydrophobic regions. The authors suggested that MSNs may be a useful way to deliver hydrophobic drugs in cancer chemotherapy, and while no metallo-drugs were examined in the study, it would appear that they would behave in a manner similar to that of camptothecin and be adsorbed and released by mesoporous silica.

9.1.4 Encapsulation

Liposomes, which are nano-sized spherical vesicles in the 100–200 nm range (Figure 9.1), have been studied for many years as drug-delivery vehicles for treating cancer [19]. A couple of the attractive features of liposomes as drug carriers are that they are compatible with the biological system – that is, their shell is made up a lipid bilayer – and their interiors are large, allowing them to hold a high concentration of drug. Since liposomes are relatively stable in blood, they can circulate to a tumor, where, because there are few lymphatic vessels to remove them, they get caught and ultimately release their drug load. Both cisplatin and carboplatin have been incorporated into liposomes to produce nano-sized structures, and while neither formulation has yet been approved for clinical use, carboplatin nanocapsules exhibit cytotoxicities against cancer cells in culture that are three orders of magnitude higher than that of non-encapsulated carboplatin [20,21].

The Swiss physician Paracelsus, who is considered the father of modern toxicology, pointed out that the dose of a substance is the important factor in determining whether or not it is a poison, or in a modernized version of what he actually said, '*It is the dose that makes the poison.*' This could not be more true than for arsenic trioxide (As_2O_3, ATO), which has been used as a poison throughout history and is now making a comeback as an agent for treating cancer [22,23]. ATO for injection, which is called Trisenox, is approved for use in the United States for the treatment of acute promyelocytic leukemia (APL), a cancer of the blood and bone marrow. While the mechanism of action of ATO is not known, a recent study shows that both it and the

Au$^+$-containing drug auranofin block the metabolism of selenium in lung cancer cells and that inhibition of the enzyme thioredoxin reductase, TrxR, may be involved [23]. As was pointed out earlier (see Box 4.3), TrxR has a selenocysteine residue in its active site and since As$^+$ is a soft acid and the selenate ion is a soft base, ATO would be expected to have a high affinity for this site in the protein, which could be important in the mechanism of action of the compound.

An interesting strategy for encapsulating ATO in a slow-release formulation into liposomes so that it could be used for treating cancer was described by O'Halloran and coworkers (Figure 9.7) [24]. At physiological pH, ATO exists mainly as its triprotonated from, H_3AsO_3, which, since it is neutral, rapidly diffuses across the lipid bilayer. If it is initially placed inside a liposome, it leaks out of the vesicle with a half life of $t_{1/2} \approx 50$ min (4 °C, pH 7.2), which is too short to be an effective time-release system for ATO in chemotherapy. In order to produce formulations that would slowly release ATO, the researchers formed precipitated, insoluble ATO by combining it with a transition metal ion inside the liposome. This was done by dissolving a transition metal acetate, for example $Ni(OAc)_2$, in solution, forming the liposome in the same solution and then separating the liposomes, which had trapped some metal ions, from the remainder of the solution. Next, they added the transition metal ion-loaded liposomes to a solution containing ATO, which at neutral pH is mainly $As(OH)_3$ (Figure 9.7a). When the neutral ATO moved inside the liposome, it immediately reacted with the aquated metal acetate inside, with the loss of two protons, to produce an insoluble precipitate, which in the case of $Ni(OAc)_2$ is an Ni^{+2} arsenite. Due to the release of two protons from $As(OH)_3$, acetic acid is also produced in the reaction, which crosses the lipid bilayer and effluxes to the medium. This arsenite-loaded liposome, indicated as Lip(Ni, As), is stable at neutral pH, but if the pH of the medium is lowered, for example to pH ~5, protons will leak inside the liposome, react with the arsenite and release ATO, which leaks out across the bilayer of the liposome.

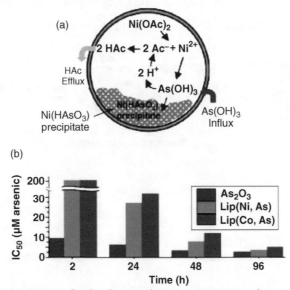

Figure 9.7 *(a) Schematic of $As(OH)_3$ influx loading mechanism into a 100 nm liposome containing $Ni(OAc)_2$. The driving force for the loading is the formation of an insoluble precipitate of Ni^{+2} arsenite inside the liposome and the efflux of acetic acid, HAc. (b) Cytotoxicity studies against SU-DHL-4-human lymphoma cells. The order of the bars, left to right, is the order of the listing of substances (inset, top to bottom) inside the liposome. Adapted with permission from H. Chen et al., Lipid Encapsulation of Arsenic Trioxide Attenuates Cytotoxicity and Allows for Controlled Anticancer Drug Relaese, J. Am. Chem. Soc. 128, 13348–49. Copyright 2006 American Chemical Society*

The researchers evaluated the cytotoxicity of As_2O_3, Lip(Ni, As) and Lip(Co, As) against SU-DHL-4-human lymphoma cells (Figure 9.7b). They found that at short incubation times with the cells, As_2O_3 is highly cytotoxic – that is, it exhibits a low IC_{50} value – but Lip(Ni, As) and Lip(Co, As) are not very toxic – that is, they have high IC_{50} values. However, as the incubation time is increased, the arsenite-liposomes became as toxic as non-liposome-protected As_2O_3. The authors suggested that the increased cytotoxicity with longer incubation times for the arsenite-liposomes was due to the slow release of the ATO from the liposome, which killed the cells. They also speculated that uptake of liposomal arsenite could make its way to enodsomes, which, because they are acidic, could facilitate the release of toxic ATO inside the cell.

The native iron-storage protein ferritin (Ft) could potentially be used as a delivery vehicle for drugs in cancer chemotherapy. Since there are receptor sites for Ft on cancer cells and the protein is taken up by the cells through endocytosis, loading Ft with drug molecules could be a way to deliver high loads of cytotoxic agents to cells.

Ft is found mainly in the liver, bone morrow and spleen, and its function is to store iron until it is needed by cells (see Box 5.2). The protein is made up of 24 subunits, which self-assemble, forming a spherical shell-like structure containing a core of $\sim$4500 high-spin (S = 5/2) Fe^{+3} ions in a complex phosphate matrix. Since the inside diameter of the shell is $\sim$8 nm, apoferritin (AFt), the protein without the iron core, is capable of accepting a variety of different species, ranging in size from small drug molecules to nano-sized particles.

Guo and coworkers [25] outlined methods for encapsulating the anticancer drugs cisplatin and carboplatin into AFt to produce drug-loaded transferrin that could be useful in cancer chemotherapy (Figure 9.8). One approach for encapsulating the drugs was to take advantage of the assembly–disassembly of AFt, which is controlled by pH (Figure 9.8a). It is known that in acidic media, pH $\sim$2, AFt dissociates into its monomeric units, and if the pH is adjusted to neutral, the system will reassemble into AFt. When this process was carried out in a solution containing either cisplatin (2 mM) or carboplatin (30 mM) and AFt (10 µM), some of the drug molecules in solution become encapsulated in AFt. Next, the investigators exhaustively dialyzed the solutions

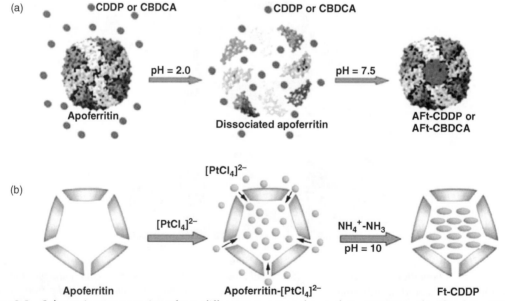

Figure 9.8 *Schematic representation of two different ways to make cisplatin (CDDP) and carboplatin (CBDCA) loaded into the central cavity of apoferritin: (a) the unfolding–refolding method; (b) the in situ method. From Z. Yang et al., Encapsulation of Platinum Anticancer Drugs by Apoferritin, Chem. Commun. 2007, 3453–55. Reproduced by permission of the Royal Society of Chemistry*

containing AFt-cisplatin and AFt-carboplatin, which removed drug that was not encapsulated in AFt, and determined that each AFt trapped only two molecules of cisplatin and five molecules of carboplatin.

In an effort to increase the number of encapsulated drug molecules inside AFt, Guo and coworkers [25] synthesized cisplatin inside AFt (Figure 9.8b). This was done by adding a concentrated solution of $K_2[PtCl_4]$ to a solution of AFt. Because there are small pores/channels between the monomeric proteins that form the shell of AFt, the platinum complex is allowed to diffuse into AFt. Next, they added an NH_4^+-NH_3 buffer (0.3 M, pH 10), which reacted with $[PtCl_4]^{2-}$ to form both cisplatin and transplatin, as confirmed using $[^1H$-$^{15}N]$ HSQC NMR and ^{15}N-labeled ammonia. Unlike iodide ion, which is used to direct ammonia molecules into *cis* sites for the synthesis of cisplatin (Chapter 1), chloride ion is not an especially strong *trans*-directing ligand. This explains why both cisplatin *and* transplatin were produced in the reaction of ammonia with $[PtCl_4]^{2-}$. While this encapsulation method trapped about 15 cisplatin and 15 transplatin molecules inside AFt, the resulting drug-loaded AFt was only slightly more toxic toward rat pheochromocytoma (PC12) cells than cisplatin.

9.1.5 Gold nanoparticles

Photodynamic therapy uses light, a sensitizer and tissue oxygen to treat cancer. Since many of the presently-used sensitizers for PDT require 24 hours or more to reach maximum accumulation in the tumor, Fei and Burda and their coworkers made an unusual gold nanoparticle–phthalocyanine conjugate which localizes in the tumor in less than 2 hours [26]. As shown in Figure 9.9, the conjugate consists of a spherical metallic gold (Au^0) nanoparticle with attached polyethylene glycol molecules and many six-coordinate Si^{+4}-phthalocyanine complexes (Pc 4). The Au nanoparticles were made by reducing the square planar Au^{+3} complex, $HAuCl_4$, with sodium borohydride, $NaBH_4$, in aqueous media in the presence of surfactants. Next, in order to impart good water-solubility to the conjugate, PEG with a terminal thiol group was attached to the gold by mixing the gold particles with PEG-SH. While the attachment chemistry is not known with certainty, it likely involves the formation of atomic or molecular hydrogen adsorbed on the gold surface and the formation of two electron Au-S bonds [27–29].

Phthalocyanine (Pc), which has a macrocyclic structure similar to that of a porphyrin, is an excellent sensitizer for PDT, but, since it is hydrophobic, it has very low water-solubility. In order to attach phthalocyanine to the gold nanoparticle and make a system that would release Pc at a tumor site, the investigators made the Si^{+4} complex of Pc and attached two axial ligands to the bound silicon ion to produce

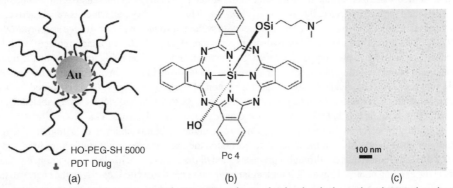

HO-PEG-SH 5000
PDT Drug

(a)

Pc 4

(b)

100 nm

(c)

Figure 9.9 *(a) 5 nm diameter Au-nanoparticle (Au NP) with attached polyethylene glycol (PEGylated) and attached photodynamic therapy drugs (sensitizer). (b) Structure of Pc 4. (c) Transmission electron micrograph (TEM) image of the conjugate, PEGylated Au NP-Pc 4. Adapted with permission from Y. Cheng et al., Highly Efficient Drug Delivery with Gold Nanoparticle Vectors for In Vivo Photodynamic Therapy of Cancer, J. Am. Chem. Soc. 130, 10643–10647. Copyright 2008 American Chemical Society*

the six-coordinate silicon complex Pc 4 (Figure 9.9). In the final step, Pc 4 was adsorbed on to the surface of the PEGylated gold nanoparticle in a nonaqueous solvent, most likely through hydrophobic interactions between PEG and Pc 4, to produce the conjugate. A transmission electron micrograph (TEM), a physical technique for visualizing nano-sized objects, showed that the diameter of the gold particle in the conjugate is ∼5 nm, and other studies showed that the *hydrodynamic diameter* of the particle, which includes the PEG and Pc 4 molecules on the surface, is much larger, ∼32 nm.

In order to test the conjugate as a sensitizer for PDT, it was injected into tumor-bearing mice, and the fluorescence of the Pc 4 as a function of time after injection was measured with an image-capture device. As is evident in Figure 9.10, the fluorescence due to the Pc 4 can easily be detected in the tumor in very short times, <2 hours, indicating that the conjugate quickly makes it way to the tumor. Moreover, since the fluorescence is quenched (does not appear) when Pc 4 is bond in the conjugate, the presence of fluorescence in the mouse implies that Pc 4 has been dissociated from the carrier nanoparticle and is possibly localized in the lipid membrane of the cells. Irradiation of the tumor with light (>500 nm) caused it to shrink in size in one week, indicating that conjugate is a useful delivery system for the sensitizer in PDT.

Box 9.1 Photodynamic therapy

Photodynamic therapy (PDT) is a clinically-used cancer treatment that utilizes a sensitizing agent, biological molecular oxygen and light to generate reactive chemical species that kill cancer cells [30–32]. In PDT, the patient is administered a nontoxic sensitizer molecule either by direct application of the sensitizer to an exposed tumor or by intravenous injection of the compound, which results in its localization in a tumor in an internal organ. Since the therapy involves irradiation of the cancer with a high-intensity light source, treatable tumors are mainly on the exposed portions of the body (skin), but fiber optics have made it possible to irradiate cancers in internal areas of the body; for example, lung, throat, colon and so on.

Porphyrins are the most-used sensitizing agents in PDT. These macrocyclic structures are naturally-occurring, nontoxic and have relatively strong absorptions in the red and near-infrared region of the spectrum that can be stimulated by light (Figure 9.11). In PDT, light from an external source is directed at the sensitizing agent that has become localized in diseased tissue. If the energy of the photons in the irradiation source corresponds to an absorption band of the sensitizer, a transition between electronic states of the sensitizer can occur, causing the sensitizer to become *photoexcited* (Figure 9.12). Since there is molecular oxygen in and around cells, the collision of an oxygen molecule with the photoexcited sensitizer results in the transfer of the stored photoenergy in the sensitizer to molecular oxygen, producing an excited form of oxygen called *singlet oxygen*, $^1O_2$. The singlet oxygen, which is very reactive, can either directly attack macromolecules in the cell or it can react with water to produce *reactive oxygen species, ROSs*, such as superoxide, $O_2^{-\bullet}$, hydroxyl radical, $HO^\bullet$, or hydrogen peroxide, H_2O_2, which can attack molecules in the cell. If enough damage is caused by singlet oxygen and/or ROSs, the tissue in the region being irradiated by the light will die.

The sequence of photoexcitation and energy transfer processes involved in PDT, in the form of a *Jablonski diagram*, is shown in Figure 9.12b. The diagram shows that if a photon of the proper energy, *hv*, is absorbed by a sensitizer molecule, the sensitizer is converted from its ground *singlet* state, S_0, to an excited singlet state, labeled S_1, where the non-italicized letter 'S' indicates that the states involved have no unpaired electrons. This process, which is very fast, $\sim 10^{-15}$ s, promotes an electron from a bonding MO to a higher vibrational level of a higher electronic state (a nonbonding or antibonding MO) of the molecule. Normally, the excited state is not stable very long and it decays within $\sim 10^{-8}$ s to the ground state, S_0, through a process called *fluorescence*, which releases a photon, *hv'*, that is lower in energy than the exciting photon. However, if molecular oxygen is near the excited sensitizer, the energy in the excited state can be efficiently transferred to molecular oxygen by a process called *intermolecular intersystem crossing*, which involves the triplet state in the sensitizer, indicated by 'T' (which has two unpaired electrons), and the singlet state of molecular oxygen, denoted by 'S' (which has no unpaired electrons). As is shown in the diagram, the conversion of the excited S_1 state of the sensitizer to the T state and finally to the S state of O_2 is energetically 'downhill' and thus thermodynamically favored.

Simple molecular orbital theory applied to homonuclear diatomic molecules shows that molecular oxygen, O_2, in its ground state is *paramagnetic* and has two unpaired electrons. In this case the spin quantum number, $S = \frac{1}{2} + \frac{1}{2} = 1$, and the multiplicity of the state, which is given by $2S + 1$, is 3, or the state is said to be a 'triplet' state; that is, $^3O_2$. This is because there is one electron in each of the two antibonding π-type (π^*) molecular orbitals that has the same energy; that is, the electronic configuration is $\pi^*(2p_y)^1$, $\pi^*(2p_z)^1$, which corresponds to the triplet, $S = 1$, state of O_2. At higher energy, $95\,kJ\,mol^{-1}$ above the triplet state, is the singlet ($S = 0$) state. This state has both electrons in the same π^* MO with the spins paired; that is, $\pi^*(2p_y)^2$, $\pi^*(2p_z)^0$ or the equivalent configuration, $\pi^*(2p_y)^0$, $\pi^*(2p_z)^2$. Since the singlet state of oxygen is lower in energy than the T state of the sensitizer, the collision of an oxygen molecule with an excited sensitizer molecule can result in the transfer of energy from the T state of the sensitizer to the S state of oxygen.

Singlet oxygen is chemically quite reactive and can either attack biological molecules near the sensitizer or can react with water to produce ROSs which attack cellular targets. The location of the damage in the cell depends on the nature of the sensitizer, but proteins and lipids in the nuclear and outer membrane, and mitochondria (usually not DNA) are most often damaged in PDT. If the rate of damage to important biomolecules exceeds the rate at which the cell can produce new biomolecules or repair damaged molecules, the cell will enter into apoptosis and die.

An important advantage of PDT over conventional forms of chemotherapy is that since light is the activating agent, the area of irradiation in the body can be carefully controlled, which minimizes the damage caused to surrounding healthy tissue. A drawback of the therapy is that the wavelengths of radiation used can only penetrate a relatively short distance into tissue, $\sim 1\,cm$, and thus the tumor must be accessible and relatively small in order for PDT to be effective.

An essential element of PDT is molecular oxygen. Since tumors contain rapidly-dividing cells which have a high demand for oxygen, and a *hypoxic* environment can limit the effectiveness of PDT, efforts are being made to develop photoactivated metal complexes that can kill cells in the absence of molecular oxygen. These complexes, which contain Pt^{+4}, Rh^{+3} and Ru^{+2}, can be photolyzed to produce products that interact with DNA and proteins by a variety of mechanisms which lead to the death of the cell [33].

9.2 Nanomedicine in diagnosing disease

9.2.1 Computed tomography

Computed tomography, CT, or sometimes CAT, is a fast and relatively inexpensive way to diagnose disease. The approach involves the passage of x-rays through the patient, wherein the x-ray source and detector are moved relative to a target area in the body. By collecting a series of images, each from a different direction, and by using a computer algorithm, a *tomogram*, which contains three-dimensional information about the target, can be obtained. Since the scattering of x-rays depends on the atomic number of the element involved in the scattering, with the heaviest elements producing the most scattering, a detector pointed in the direction of the incoming x-ray beam detects fewer x-rays from a heavy element placed in the beam. To the detector, it appears that the beam has been attenuated, or X-rays have been absorbed, when in fact they have been scattered in directions that cannot be captured by the detector. In order to increase the scattering between diseased and normal tissue, patients are often given iodinated organic compounds, which, since iodine is atomic number 53, scatter x-rays quite well. If the iodinated compound localizes in the fluid surrounding diseased tissue, the demarcation or *contrast* between diseased and normal tissue will be enhanced, thereby allowing the physician to arrive at firmer conclusions concerning the state and progression of the disease.

Using standard procedures, Jon and Jeong and their coworkers [34] synthesized gold nanoparticles (GNPs) and attached thiol-polyethylene glycol, PEG-SH, to their surface. Imaging with scanning electron microscopy (SEM) and light-scattering measurements showed that the GNPs had a relatively narrow-diameter distribution centered at $\sim 30\,nm$. By coating the gold nanoparticle with PEG, binding to blood proteins was minimized, which allowed the GNPs to escape rapid removal by the reticuloendothelial system, RES, which is responsible for removing foreign matter from the blood.

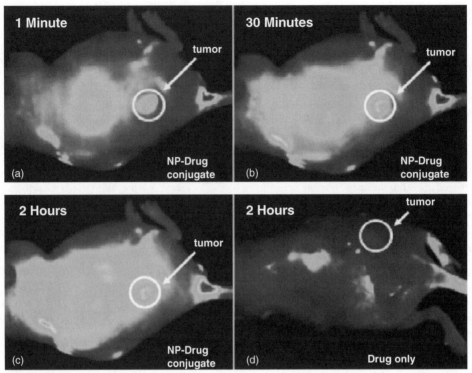

Figure 9.10   *Fluorescence images of a tumor-bearing mouse after being administered the Au NP-Pc 4 conjugate in normal saline solution (a) 1 minute, (b) 30 minutes, (c) 120 minutes after injection and (d) 120 minutes after injection with Pc 4 alone, not conjugated to Au NP. The light regions are due to the fluorescence of Pc 4 in the mouse and the circle shows the location of the tumor. Adapted with permission from Y. Cheng et al., Highly Efficient Drug Delivery with Gold Nanoparticle Vectors for In Vivo Photodynamic Therapy of Cancer, J. Am. Chem. Soc. 130, 10643–10647. Copyright 2008 American Chemical Society*

As is shown in Figure 9.13a, the absorption of X-rays by the GNPs is $\sim$1.9 times greater than the iodine containing organic molecule Ultravist which is currently used as a contrast agent for CT. To show the effectiveness of the GNPs as CAs, the investigators imaged the blood pool of a rat, the result of which is shown in Figure 9.13b. As shown in the figure, the heart and great vessels can easily be distinguished on the GNP-enhanced CT image of the animal, with good contrast. An additional positive aspect of GNPs as imaging tools is that their blood half life is about three times longer than that of Ultravist, which allows health care personnel more flexibility in imaging patients and collecting data that can improve the diagnosis of disease. In other studies the investigators showed that the gold particles could also image liver cancer in rats and that, in cell-culture studies, the particles exhibit low toxicity toward HepG2 hepatocyte cells.

9.2.2   Magnetic resonance imaging

Gadolinium (III) is an ideal ion for enhancing the contrast in magnetic resonance imaging. The ion, which has seven unpaired electrons, $S = 7/2$, and a high magnetic susceptibility, $\mu_{so} = 7.94$ BM, greatly shortens the *longitudinal relaxation* time, T_1 of water. Wilson and coworkers [35] recently described the synthesis and properties of super-paramagnetic gadonanotubes as high-performance MRI CAs. The tubes were made by

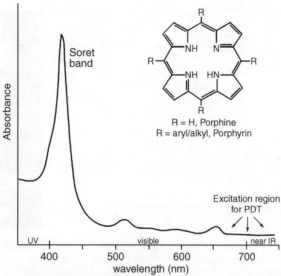

Figure 9.11   *Photodynamic therapy. Structure of prophine, the unsubstituted porphyrin, and a general structure of meso-substituted porphyrin ligands. The absorption spectrum of a typical porphyrin is also shown, with the excitation region for PDT indicated*

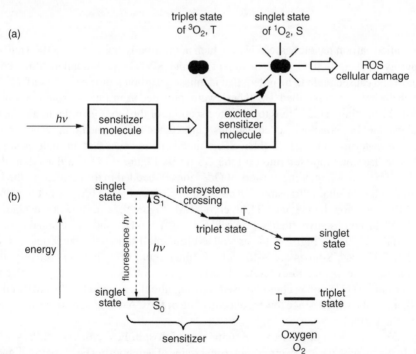

Figure 9.12   *Photodynamic therapy. (a) Schematic of photodynamic therapy, PDT. The process begins with the sensitizer molecule absorbing a photon, which converts normal triplet oxygen into reactive singlet oxygen through direct contact of oxygen with the excited sensitizer. The reactive singlet oxygen reacts with biological targets and also reacts with water to produce reactive oxygen species, ROSs. (b) Modified Jablonski diagram showing the relative energy levels of the sensitizer and molecular oxygen in PDT. The arrows show how the energy in the photon absorbed by the sensitizer is transferred to oxygen to produce reactive singlet oxygen*

(a)

(b)

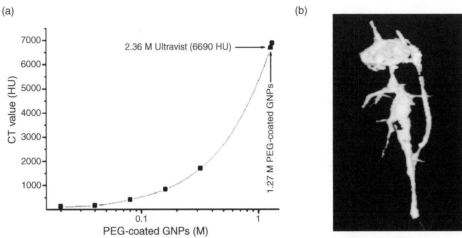

Figure 9.13   *(a) The measurement shows that PEG-coated GNPs are ~1.9 times more effective as CAs than the conventional iodine CA, Ultravist. The effectiveness of the agent is measured in terms of radiodensity Hounsfield units (HU). (b) Three-dimensional in vivo CT angiogram image of the heart and great vessels obtained 10 minutes after injection of PEG-coated GNPs into the tail vein of a Sprague–Dawley rat. Adapted with permission from D. Kim, S. Park, J.H. Lee, Y.Y. Jeong, S. Jon, Antibiofouling Polymer-Coated Gold Nanoparticles as a Contrast Agent for In Vivo Computed Tomography Imaging. J. Am. Chem. Soc. 129, 7661–7665. Copyright 2007 American Chemical Society*

treating single-walled carbon nanotubes, SWNTs, which are normally quite long, >1000 nm, with fluorine, followed by pyrolysis at a 1000 °C. This treatment cut the SWNTs into smaller, ultra-short, nanotubes (20–100 nm long) and caused them to be pitted; that is, missing carbon atoms on their surface. The pyrolized nanotubes, which the investigators called 'US-tubes', were sonicated in an aqueous solution containing $GdCl_3$, which resulted in the binding of Gd^{+3} ions to the pitted sections of the tubes. In order to use the Gd^{+3} tubes in relaxation studies, the investigators added various surfactants to the medium, which produced water suspensions of the material, $Gd^{3+}{}_n$@US-tubes. This moniker indicates that an undetermined number of Gd^{+3} ions (n) have been incorporated into (@) the US-tubes. Figure 9.14a is a depiction of the probable structure of $Gd^{3+}{}_n$@US-tubes, showing clusters of Gd^{+3} ions imbedded in the pit areas of the US-tubes. The figure also shows a high-resolution transmission electron microscopy (HRTEM), revealing clusters of Gd^{+3} ions (Figure 9.14b), and a low-temperature TEM, cryo-TEM (Figure 9.14c), showing that $Gd^{3+}{}_n$@US-tubes are 20–80 nm long. In comparative studies between $Gd^{3+}{}_n$@US-tubes and the commonly-used MIR CA, magnevist, $[Gd(DTPA)(H_2O)]^{2-}$, Figure 8.14, as well as ultra-small superparamagnetic iron oxide particles, it was found that $Gd^{3+}{}_n$@US-tubes are ~40 and ~8 times more effective, respectively, than the other materials in reducing the relaxation rate of water. This is a significant finding and suggests that relatively low concentrations of $Gd^{3+}{}_n$@US-tubes could be used to bring about the same level of MRI enhancement as produced by other agents, which, since lower concentration of the CA would need to be administered, would be beneficial to the patient.

Maghemite, γ-Fe_2O_3, a form of ferric oxide containing high-spin Fe^{3+}, $3d^5$, $S = 5/2$, is a *ferromagnetic* substance [36]. In maghemite the net electron spin from groups of neighboring Fe^{3+} ions is aligned in the same direction (the spins are coupled with each other), which in effect produces a tiny magnet with a high magnetic moment. Since spherical nanometer-sized maghemite particles can be made, these *superparamagnetic iron oxide* (SPIO) nanoparticles have potential as MRI CAs [37]. Motte and coworkers [38] created ultra-small SPIOs by attaching 5-hydroxy-5,5-bis(phosphono)pentanoic acid, HMBP-COOH, to the surface of 10 nm

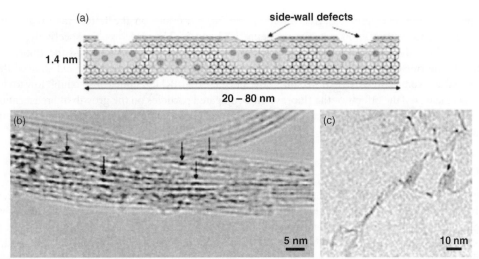

Figure 9.14 *(a) Depiction of a single carbon nanotube, 'US-tube', loaded with hydrated Gd^{+3} ions (filled black circles). (b) High-resolution transmission electron microscopy image of the Gd^{3+}_n@US-tubes showing the Gd^{+3}_n clusters (arrows). (c) Transition electron micrograph of Gd^{3+}_n@US-tubes. From B. Sitharaman et al., Super-paramagnetic Gadonadotubes are High-Performance MRI Contrast Agents, Chem. Commun. 2005, 3915–3917. Reproduced by permission of the Royal Society of Chemistry*

diameter maghemite particles (Figure 9.15). The HMBP ligand, which was bound through the two phosphonate groups to iron ions on the particle, modified the surface of the particle so that it did not aggregate in solution, and provided a free carboxylic acid functional group that could be used for covalent attachment of other groups to the particle. In studying the enhancement effects of the particles on both the *longitudinal* (T_1) and the *transverse* (T_2) proton relaxation times of water as a function of iron concentration,

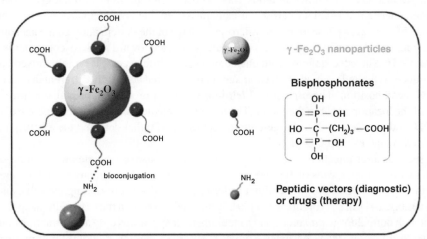

Figure 9.15 *Maghemite nanocrystals with a surface-bound bisphosphonate for magnetic resonance imaging and drug delivery. The free carboxyl group on the particle can be used to link agents for treating and diagnosing disease. From Y. Lalatonne et al., Bis-Phosphonates-Ultra Small Superparamagnetic Iron Oxide Nanoparticles: A Platform Towards Diagnosis and Therapy, Chem. Commun. 2008, 2553–2555. Reproduced by permission of the Royal Society of Chemistry*

they found that while the particles affect both T_1 and T_2, the effect on the latter was greater, with the longitudinal, r_1, and transverse, r_2, relaxivities being 1.83 and 55.7 Fe mM^{-1} s^{-1}, respectively.

In order to show the versatile nature of the particles, the investigators conjugated the fluorescent dye fluorescein to the free carboxylic acid functional group of the modified nanoparticle and showed that each particle bound an average of 590 dye molecules [38]. Since useful MRI CAs must also exhibit low toxicity, the investigators measured the effects of the fluorescein-conjugated particles on the growth of breast cancer cells and found that, although the particles were taken up by the cells through endocytosis, they exhibited low cytotoxicity.

9.2.3 Quantum dots

Quantum dots (QDs) are small nanometer-sized particles composed of semiconductor material which when excited by ambient light, emit light that is nearly monochromatic; that is, a narrow range of wavelengths. Since the wavelength of emitted radiation depends on the size of the particle and the material from which the particle is composed, QDs have unique, tunable, optical properties that make them useful for diagnosing and, potentially, treating disease.

The monoclonal antibody trastuzumab (Herceptin), which is clinically approved for treating breast cancer, targets HER2 receptors that are overexpressed on the surface of breast cancer cells. When the antibody binds to a receptor it is taken into the cell, where it inhibits the production of critical signaling proteins, which eventually causes the death of the cell. Higuchi and coworkers recently described the attachment of a QD to a protein fragment of trastuzumab to produce the conjugate, QT [39]. With the aid of a specially-designed microscope-camera system to detect the conjugate, they tracked in real time the movements of QT from the bloodsteam to the inside of a breast cancer cell growing in a live mouse.

To carry out the study, the researchers first treated the intact trastuzumab antibody with a reducing agent, which reduced the disulfide bonds in the protein to thiols. Since parts of different chains of the antibody are held together by disulfide linkages, this fragmented the protein. Next, they attached a commercially-available PEGylated QD composed of semiconducting CdTe to the antibody fragments using standard coupling reagents and isolated a product, QT, with three antibody fragments per QD. In order to track QT in real time in the mouse, the researchers constructed a microscope-camera system that could capture images at high magnification at the rate of 30 images per second. The microscope stage of the system allowed the anesthetized mouse to lie flat on a transparent window, with the exposed tumor directed toward a microscope pointing upward, as shown in Figure 9.16. Since the stage was attached to a piezoelectric crystal, it rapidly moved back and forth very small distances (vibrated), which allowed images to be captured at slightly different displacements of the stage about its mean position. These *right* and *left* images, when processed, provided a three-dimensional (stereo) view of the location of QT in the tumor. The remaining elements of the system included a laser, which excited the QD, and a camera with a charge-coupled device (CCD) that detected the 800 nm (near-infrared) radiation emitted by the excited QD of QT.

Within 30 seconds after injection of QT into the tail of the mouse, conjugates were observed in a blood vessel, moving with average speeds of 100–600 μm s^{-1}, and after two hours many particles were observed to leak through the vessel and into the extracellular space (extravasation) surrounding the cells, and their speed slowed considerably, 1–4 μm s^{-1} (Figure 9.17) [39]. While their motion in the extracellular space was random and consistent with normal *Brownian* motion in a medium, their speed was only a tiny fraction of that expected in solution due to diffusion. Many particles were also observed moving randomly within a small volume, $\sim$1 μm, for a period of time, and suddenly hopping from the volume to another region near the original volume. The authors suggested that this behavior is due to the existence of a matrix in the extracellular space that they could not directly detect with the camera, and that the QTs initially become trapped in volume elements of the matrix but eventually break free and move to a new, empty, site in the matrix.

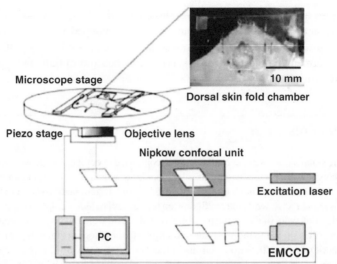

Microscope stage

10 mm

Dorsal skin fold chamber

Piezo stage

Objective lens

Nipkow confocal unit

Excitation laser

PC

EMCCD

Figure 9.16   *Three-dimensional microscopic system for detecting the real-time movement of QD-labeled antibodies in a live mouse. Reprinted with permission from H. Tada et al., In Vivo Real-Time Tracking of Single Quantum Dots Conjugated with Monoclonal Anti-HER2 Antibody in Tumors of Mice, Cancer Res, 2007, 67, 1138–1143, figure 2*

Six hours after injection, many QTs were bound to the membrane of the cells where the HER2 receptor is located and they exhibited motion within a ~500 nm restricted region. The authors suggested that the motion of the particles was due to the HER2 antibody being anchored to a flexible part of the cytoskeleton, such as an actin filament, which allowed the particle to move around some mean position. Eventually, the particles entered cytoplasm and moved along the inner face of the cell membrane at speeds of ~600 nm s^{-1} in what may

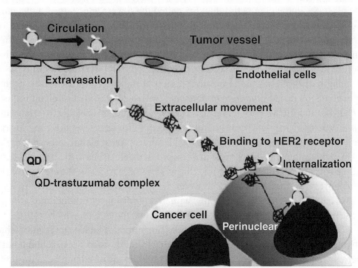

Circulation

Tumor vessel

Extravasation

Endothelial cells

Extracellular movement

Binding to HER2 receptor

QD

Internalization

QD-trastuzumab complex

Cancer cell

Perinuclear

Figure 9.17   *Schematic illustration of the real-time detection of a QD-labeled antibody, QD-trastuzumab, moving from the blood into a cancer cell growing in a live mouse. Reprinted with permission from H. Tada et al., In Vivo Real-Time Tracking of Single Quantum Dots Conjugated with Monoclonal Anti-HER2 Antibody in Tumors of Mice, Cancer Res, 2007, 67, 1138–1143, figure 6*

have been transport by the actomyosin system of vesicles containing the QTs. When the particles reached the *perinuclear space* (nuclear envelope) their directional movement stopped and Brownian motion commenced in a small-diameter, ~ 1 µm, region. The study demonstrated the remarkable ability of QDs with an attached antibody drug for the treatment of breast cancer to serve as tiny beacons of light and report their location in a tumor growing in a live animal.

Box 9.2 Quantum dots

Quantum dots (QDs) are small particles in the 2–10 nm range composed of compounds that are semiconductors [40,41]. Consider the simple inorganic compound cadmium selenide, CdSe, which is a semiconductor. While there are two forms of CdSe in the solid state, both have the Cd^{+2} and Se^{-2} ions in tetrahedral site geometries. The electronic configuration of Cd^{+2} is $[Kr]4d^{10}5s^0$, while the configuration for Se^{-2} is $[Ar]3d^{10}4s^24p^6$. By applying simple molecular orbital theory to a single 'molecule' of CdSe, the 5s orbital on Cd^{+2} can overlap with the 4s orbital on Se^{-2} to produce bonding and antibonding MOs similar to those produced for molecular hydrogen (Figure 9.18a). Since Se^{-2} brings two electrons to the simple MO scheme (the two electrons in the 4s orbital) and Cd^{+2} bring no electrons (the 5s orbital is empty), the two electrons go into the bonding MO, leaving the antibonding MO vacant, which again is like H_2 (Chapter 1).

This is the case for the CdSe 'molecule', but cadmium selenide in the solid crystal is an ordered array of Cd^{+2} and Se^{-2} ions, so there are many diagrams of the type shown in Figure 9.18a. Since the bonding and antibonding orbitals in the crystal are combinations of the bonding and antibonding orbitals for individual pairs (diagrams), there are many states. The bonding states or MOs are spread over a range of energy, and the antibonding states are also spread over a range of energy, with all of the latter being at higher energy (less stable) than all of the former (more stable). These energy ranges are called 'bands', with the lowest-energy band being the *valence band* and the highest band being the *conduction band*. The number of bonding states and the number of antibonding states are the same, and there will be enough electrons to fill half of the total number of states (like an individual CdSe molecule). This way, all bonding states will be filled and all the antibonding states will be empty.

The group of bonding states in the valence band is called the *highest occupied molecular orbitals (HOMO)* and the group of antibonding states in the conduction band is called the *lowest unoccupied molecular orbitals (LUMO)*. The way that semiconductors work is, if an electron in one of the HOMOs, which is in the *valence band* of the semiconductor, is promoted to one of the LUMOs, in the *conduction band*, the electron can, in effect, move over all of the ions in three dimensions in the solid; that is, the solid becomes a conductor of electrical charge.

For large millimeter-sized crystals of CdSe the separation between the HOMOs and LUMOs is $\sim 14\,035$ cm^{-1}, which corresponds to a wavelength of ~ 713 nm in the infrared region of the spectrum. However, as the size of the crystal becomes smaller, a phenomenon called *quantum confinement*, which is a product of quantum mechanics, causes the band gap between HOMO and LUMO to *increase*. There is a famous exercise in quantum mechanics (QM) called the *particle in a box* problem, which, when solved, gives the spacing between the energy levels associated with the π-systems of certain organic molecules [42]. Consider the simplest alkene, ethene (ethylene), which has one π-bond between the two carbon atoms. In the particle in a box problem, electrons in the π-bond of ethene are confined to a line between the two carbon atoms, with the carbons acting as stops or barriers for the electrons at both ends of the line. The solution to the problem reveals that the spacing between energy levels associated with the π-system depends on the distance between the two barriers, the two carbon atoms, with *greater* distances producing *smaller* spacing. That this is in fact the case can be seen by considering conjugated polyene organic molecules, which from the viewpoint of QM are simply longer versions of ethene. For example, a molecule like β-carotene, which is the pigment that gives the carrot its orange color, contains a conjugated 22-carbon-long chain with 11 double bonds. It is found that the spacing between energy levels of β-carotene are much smaller than for comparable levels of ethene because the 'box', which is the distance between the two end carbon atoms in the conjugated chain, is now much bigger for β-carotene than for ethene. The effect of this is that β-carotene has absorptions in the visible region of the spectrum (low energy, small spacing between levels), while ethene has absorptions in the ultraviolet region of the of the spectrum (high energy, large spacing between levels).

Figure 9.18a is the MO diagram for a single Cd-Se bond in cadmium selenide, but in a crystal of CdSe there are many such bonds, which leads to (as was earlier pointed out) bonding and antibonding states spread over a range of energy. Since all of the individual Cd-Se 'molecules' are interconnected in the solid state, the MOs shown in Figure 9.18a are really multicenter MOs, which incorporate the $5s$ (Cd) and $4s$ (Se) wavefunctions for *all* of the ions in the crystal. Placing an electron into this multicentered MO means that it can move throughout the entire structure, which is a *three-dimensional* particle in a box problem. In three dimensions, the solution to the problem is essentially the same as with one dimension, but in three dimensions the 'box' is a cube and not a line. As with the one-dimensional case, a *decrease* in the size of the cube (or a sphere which approximates a cube) will *increase* the spacing between the energy levels. Without presenting the formula that gives the energies of the various states, the spacing between levels does not change much for macroscopic objects. However, as the dimensions of the object approach the low nanometer range, the change in spacing is dramatic, and it become even more pronounced as the dimensions of the object are reduced to those of a typical metal-ligand bond, ~ 0.2 nm.

Figure 9.18b shows how the spacing between the HOMO and LUMO levels for QDs composed of CdSe change as a function of the diameter of the particle. As is evident from the figure, the spacing between the two levels is $15\,380\,\text{cm}^{-1}$ (650 nm, red light) for an 8 nm diameter particle and increases to $22\,220\,\text{cm}^{-1}$ (450 nm, blue light) for a 2 nm particle. This means that if an 8 nm diameter CdSe particle is irradiated so that an electron in the HOMO (valence band) set is promoted to one of the levels in the LUMO set (conduction band), the electron will ultimately return to the HOMO, with the emission of a 450 nm photon making the particle appear blue in color. Changing the nature of the semiconductor material or the size of the particle changes the band gap. This makes it possible to construct QDs capable of emitting a wide variety of different colors and since 'color-coded' particles can be attached to biological macromolecules, QDs have the potential to become important new tools for nanomedicine.

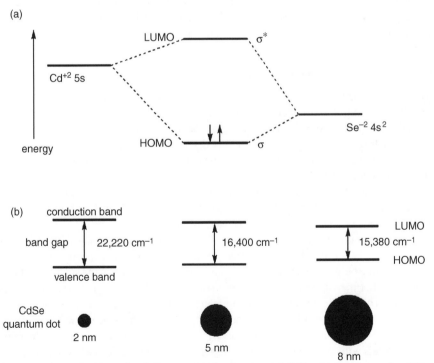

Figure 9.18   (a) Simple molecular orbital diagram for the 'molecule' CdSe. (b) Band gaps for QDs of CdSe as a function of the diameter of the particle

9.3 Potential health risks of nanoparticles

While nanomaterials have certainly captured the imagination of the scientific community, they have also raised many questions about risks they may pose to human health. Most of the nanomaterials that are being explored for the treatment or diagnosis of disease are composed of substances that are totally foreign to the biological system and are not easy degraded in the body. In addition to having an attached or adsorbed active agent such as a drug or paramagnetic ion, the particles are often coated with polyethylene glycol or a similar substance, which although biocompatible, generally decreases the rate at which the particle can be removed from the circulating system. While long circulation times can be beneficial – for imaging agents, for example – they also mean long exposure to the particles and to any toxic effects that they may have. Since the dimensions of the particles are smaller than the typical eukaryotic cell, and they can enter the cell through receptor-mediated endocytosis, the particles can have a dramatic effect on the functioning of the cell.

At first glance it would appear easy to measure the toxicity of these new nanomaterials toward cells in culture and animals, but investigators in this area face a number of obstacles not normally encountered in measuring the toxicities of conventional therapeutic and diagnostic agents. One major impediment is that nano-sized particles do not in most cases dissolve in aqueous media. Sometimes they form colloidal suspensions in water and sometimes they start by being suspended in the medium but over time they slowly settle to the bottom of the container. This and their tendency to agglomerate and self-associate makes quantitative measurements of their toxicities using standard techniques such as cell-culture assays quite challenging.

The goal in this section is to briefly summarize some of the issues faced by toxicologists in measuring the cytotoxicity of nanomaterials and to outline what has been learned about the toxicity of nanoparticles that may someday make their way to medicine. A more detailed presentation of the toxicities of nanomaterials can be found in the excellent articles by Davies [9], Drezek [43], Diabaté [44], Kostarelos [45], Warheit [46–48], Bhatia [49], Yamamoto [50], Huang [51] and their coworkers.

9.3.1 Cell culture assays and nanomaterials

The mainstay in testing new agents for possible use in medicine is the cell-culture assay. In the typical cytotoxicity study involving a toxic agent such as a drug, cells growing on a surface (adherent cells) are exposed to a solution of the drug for some time and, later, the number of live (or dead) cells is determined using an assay. In this case, drug molecules that are randomly dispersed in solution encounter the surface of the cell, and some encounters are productive in the sense that the cell absorbs the drug molecule. Since there is a very large number of drug molecules in the medium above the cells, and only a small fraction of drug molecules are actually removed from solution by the cells, the drug concentration in the medium above the cells remains constant with time. This means that the collision rate between the cell and the drug for a given concentration of drug is constant for the time that the cells are exposed to the agent.

In a cytotoxicity study involving a nanoparticle, the same general approach is used but sometimes the nanomaterial, because it is not dissolved in solution, sediments during the exposure time to the cells. If this happens, the collision rate between the nanoparticles and the cells growing on the surface of the well or culture flask is not constant with time. Since the internalization of the particle into the cell requires a collision, and the collision rate increases with time during the exposure, the internalization rate of the particle into the cell could increases with time. What impact this has had on the many studies that have been published on the cytotoxicity of nanoparticles to date is difficult to assess. However, if sedimentation occurs during the exposure time to the agent and if materials with different sedimentation rates are being studied and compared, the results of the assay could be misleading in that agents that are quite toxic could register low toxicity simply because their sedimentation rates are small.

Another potential problem in determining the cytotoxicity of nanoparticles is that they often self-associate (clump) in solution and/or interact with substances in the culture medium, such as proteins, to produce aggregates. Since aggregation is very likely concentration- and time-dependent and clumps and individual particles could produce different cellular responses, data from this type of heterogeneous system could be nonreproducible and difficult to analyze. This underscores the importance of knowing the form of the nanomaterial *under the conditions of the cell experiment*, which could be quite different from the conditions that were used to characterize the nanomaterial [46,52].

In the typical cytotoxicity study leading to the determination of the value of IC_{50}, which is the concentration of agent needed to reduce the growth of a cell population relative to a control by 50%, the concentration of the agent is plotted on the x-axis versus the percentage survival on the y-axis of a graph. Since many nanoparticles are not dissolved in the medium, the term 'concentration', which implies that solute molecules are randomly distributed in solution, does not apply. Since the toxic entity toward the cell is the particle itself, a metric that addresses the *number* of particles per unit volume applied to the cells would be more appropriate than one looking at the *weight* of particles per unit volume, which is often reported in publications measuring the cytotoxicity of nanoparticles. This is especially important in comparing the toxic effects of different nanomaterials where densities of materials and sizes of particles vary considerably. Also, if one type of particle is modified by attachment of a number of organic molecules to its surface, the weight of an individual particle will increase. This means that identical weights of modified and unmodified particles will not contain the same number of particles, and since the particle is the toxic entity, the cytotoxicity results based on weights could be deceiving. Clearly, what seems to be a simple experiment is in reality quite complicated, which may in part be the reason why the cytotoxicities of nanomaterials reported in the literature seem to span a wide range.

9.3.2 Cytotoxicity of nanoparticles

Some nanoparticles are inherently cytotoxic because they contain metal ions that are toxic. For example, QDs, which are semiconductors containing Cd^{+2}, Pb^{+2}, Hg^{+2} and other heavy metal ions and which are known to be toxic to cells, release heavy metal ions inside the cells. These ions can bind to proteins and other targets and/or can interact with mitochondria to generate ROS, which modify biomolecules and kill the cell [43,49,55]. In the case of QDs made of the semiconductor CdSe, oxidation of selenate ions on the surface of the particle to selenium dioxide, SeO_2, by molecular oxygen inside the cell releases Cd^{+2} ions, but if the surface is coated with a layer of ZnS or the protein albumin, surface oxidation is minimized and the particle is less toxic than the uncoated QD.

Since gold nanoparticles were used to treat rheumatoid arthritis as early as the 1920s, they have a history of being relatively nontoxic to humans [46,56]. Studies on the cellular uptake of these particles show that they enter the cell through receptor-mediated endocytosis, but they are apparently unable to leave the endosome inside the cell. Coating gold nanoparticles with various agents, for example biotin, citrate, cysteine and glucose, and changing their size in the 4–18 nm range has little effect on their nontoxic nature [57]. Superparamagnetic iron oxide particles such as maghemite, γ-Fe_2O_3, which may be useful as a CA for MRI, also appear to be relatively nontoxic to human cells. While the bare particle exhibits some cytotoxicity, toxicity can be reduced by coating it with polyethylene glycol (PEG) [46,58].

Carbon nanoparticles exist in a number of different forms. The small compound fullerene, C_{60}, Figure 9.1 has been reported to be only modestly taken up by murine macrophages and it exhibits relatively low toxicity toward the cells [58]. The compound also exhibits low toxicity toward human monocyte macrophages, but it was reported to enter the cells and accumulate in the lysosomes and cytoplasm along the nuclear membrane, and inside the nucleus of the macrophages [60].

A second form of nano-sized carbon that has potential use in medicine is the single-walled carbon nanotube, SWNT, which most studies have shown is more cytotoxic than C_{60} [43]. Cui *et al.* [61] studied the effects of

SWNTs on human HEK293 cells, finding that carbon nanotubes inhibit cell growth and decrease cell adhesion in a dose- and time-dependent manner, and that they ultimately induce apoptosis. However, if the surfaces of the SWNTs are functionalized or otherwise coated with organic molecules, the resulting modified nanotubes seem to be less toxic than unfunctionalized SWNTs. One hypothesis to explain the toxicity of SWNTs is that iron impurities introduced in the production of the nanotubes could facilitate the production of ROS, which damage the nucleus and mitochondria of cells [43,44]. A second hypothesis is that the SWNTs aggregate and that larger clusters of particles are in some way responsible for the cytotoxic effects of the nanotubes. This hypothesis is supported by the observation that coated SWNTs are less toxic to cells than pristine SWNTs.

Mesoporous silica nanoparticles, MSNs, are also being considered as a new delivery vehicle for drugs. The attractive features of MSNs are that they contain chambers that are large enough to adsorb many small drug molecules and their exteriors can be modified by attaching a wide variety of functional groups. Huang and coworkers showed that MSNs that are modified by attachment of positively-charged groups to the particle are taken into 3T3-L1 mouse embryonic cells and human mesenchymal stem cells, hMSCs, by endocytotic mechanisms that involve clathrin- and actin-dependent pathways [51]. They also showed that the MSNs do not affect the viability, proliferation or differentiation of either cell type, but that changes in the amount of surface charge on the particle may be important for uptake for different cell types.

Asefa, Dabrowiak and their coworkers determined the cytotoxicity of MCM-41, an MSN, two of its 'grafted' analogs and silicon dioxide (nanospheres) toward human neuroblastoma (SK-N-SH) cells [53]. In the study, the researchers determined the number of nanoparticles per gram weight of each of the materials, finding that on a per-particle basis pristine (unmodified) MCM-41 was the most toxic substance studied toward the neuroblastoma cells. The authors suggested that in studies where individual nanoparticles are the toxic entities toward cells, a useful metric for quantitating cytotoxicity is the *number of particles* rather than the *weight* of material applied to the cells.

The biocompatibility of mesoporous silica (MCM-41, SBA-15 and MCF) of particle sizes $\sim$150 nm, $\sim$800 nm and $\sim$4 μm and pore sizes 3 mm, 7 mm and 16 nm, respectively, was recently examined by Kohane and coworkers [62]. In *in vitro* studies with mesothelial cells (cells in the thorax), the MSN showed toxicity at high concentrations, 0.5 mg ml^{-1}, over 72 and 96 hours exposures. If the MSNs were subcutaneously injected (under the skin) into rats, the amount of residual material decreased over a three month period with good biocompatibility. However, if the materials were given by intraperitoneal or intravenous injection to mice, the animals died, probably due to thrombosis (formation of blood clots). The investigators concluded that although local tissue reaction to MSN was benign, the materials cause severe systemic toxicity, but that the toxicity might be mitigated by modification of the nanomaterials.

Nanometer-sized engineered particles exist in a seemingly limitless variety of shapes, sizes and compositions, and since their properties can be tuned for drug delivery, imaging and biosensing, they have the potential to create a wide range of clinically-useful medicines and devices. While work on developing nanomaterials for application in medicine is moving at warp speed, determination of the health risks that these materials pose to humans is preceding at a slow pace, which if not greatly accelerated, may severely limit the development of this exciting new area of science.

Problems

1. Octahedral Pt^{+4} complexes are believed to be reduced in the body to square planar Pt^{+2} compounds.

    a. Write a balanced reaction for the reduction of *c,c,t*-[Pt(NH$_3$)$_2$Cl$_2$(OH)$_2$] to a platinum Pt^{+2} compound by glutathione at pH 7.4 and indicate the structure of the Pt^{+2} product. Glutathione has the following pK_as: 2.12, 3.59, 8.75 and 9.65.
    b. Since the rate of reduction is fast, suggest the mechanism by which the reduction occurs.

2. A metal–organic framework, MOF, crystalline compound is saturated with a water-soluble adsorbed drug. The free drug has an absorption maximum at 360 nm, with $\varepsilon = 8300\,M^{-1}\,cm^{-1}$, and it was determined that the MOF contained 42% by weight of a drug with a molecular weight of 242 g mol^{-1}. In measuring the rate of drug release from the MOF, a student suspended 100 mg of the drug-saturated MOF in 100 ml of water and stirred the solution for 30 minutes, after which time the suspended particles were removed by centrifugation and the absorbance at 360 nm of the solution in a 1 cm path-length cell was determined to be 0.130.

   a. Assuming that drug release is far from equilibrium, calculate the *rate* of drug release from the particles in units of moles of drug per hour (mol hour^{-1}) from the information given.
   b. If the absorbance of the solution after the system reached equilibrium in 72 hours was 1.163, calculate the ratio of adsorbed drug to free drug in the system.

3. Apoferritin is being considered as a potential delivery vehicle for drugs. Consider a solution which contains the apoferritin monomer, $2.4 \times 10^{-4}\,M$, and a drug, 30 mM, which is originally at pH 2.0. When the pH of the solution is raised to 7.4, the individual monomeric subunits associate to form intact apoferririn, which contains 24 subunits in a shell-like structure. If the internal diameter of the apoferritin shell is 8 nm and it is approximately spherical in structure, and there is no active 'recruitment' of drug molecules to the inside of the shell when it forms, calculate the approximate average number of drug molecules that would be trapped inside intact apoferritin. The volume of a sphere is $V = 4/3\pi r^3$.

4. The energy of the singlet state of oxygen is 95 kJ mol^{-1} above the stable triplet state of O_2. Knowing that $E = h\nu$, calculate the wavelength of light in nm that would be needed for the triplet-to-singlet state conversion in a photoexcitation. $h = 6.626 \times 10^{-34}\,J\,s$, $c = 3 \times 10^8\,m\,s^{-1}$.

5. The longitudinal relaxation rate constant, r_1, at 40 °C, 40 MHz, for $Gd^{3+}{}_n$@US-tube samples that are potential magnetic resonance imaging agents was obtained using $(T_{1(obs)})^{-1} = (T_{1(diam)})^{-1} + r_1[Gd^{+3}]$, where $T_{1(obs)}$ and $T_{1(diam)}$ are the relaxation time constants $(1/k)$ of the sample and the matrix (aqueous surfactant solution) in seconds and $[Gd^{+3}]$ is the Gd concentration in mM. Calculate $T_{1(obs)}$ if $[Gd^{+3}] = 0.044\,mM$, $r_1 = 173\,mM^{-1}\,s^{-1}$ and $(T_{1(diam)})^{-1} = 0.25\,s^{-1}$.

6. The crystal structure of a gold nanoparticle with bound *p*-mercaptobenzoic acid (*p*-MBA) molecules with the formula $Au_{102}(p\text{-MBA})_{44}$ has been published [29]. The electronic configuration of Au^0 is $[Xe]4f^{14}5d^{10}6s^1$. The authors of the paper suggested that the reason the cluster is stable is that 44 gold atoms of the cluster contribute their 6s electrons to bonding the 44-*p*-MBA ligands to the cluster, and the remaining 58 Au^0 atoms contribute one electron each to a 'shell', associated with the entire cluster, that can hold a total of 58 electrons. This accounts for the total of 102, 6s, electrons that are associated with the Au^0 atoms in the cluster. If 44 Au atoms react with 44 thiols to form 44 two-electron Au-S bonds, write a balanced chemical reaction for what may be taking place when thiols are attached to the gold surface.

References

1. Whitesides, G.M. (2003) The 'right' size in nanobiotechnology. *Nature Biotechnology*, **21**, 116–1165.
2. Moghimi, S.M. and Kissel, T. (2006) Particulate nanomedicines. *Advanced Drug Delivery Reviews*, **58**, 1451–1455.
3. Cohen, S.M. (2007) New approaches for medicinal applications of bioinorganic chemistry. *Current Opinion in Chemical Biology*, **11**, 115–120.
4. Forrest, M.L. and Kwon, G.S. (2008) Clinical developments in drug delivery nanotechnology. *Advanced Drug Delivery Reviews*, **60**, 861–862.
5. Slowing, I.I., Vivero-Escoto, J.L., Wu, C.-W., and Lin, V.S.-Y. (2008) Mesoporous silica nanoparticles as controlled release drug delivery and gene transfection carriers. *Advanced Drug Delivery Reviews*, **60**, 1278–1288.
6. Bolskar, R.D. (2008) Gadofullerene MRI contrast agents. *Nanomedicine*, **3**, 201–213.

7. Sun, C., Lee, J.S.H., and Zhang, M. (2008) Magnetic nanoparticles in MR imaging and drug delivery. *Advanced Drug Delivery Reviews*, **60**, 1252–1265.

8. Smith, A.M., Duan, H., Mofs, A.M., and Nie, S. (2008) Bioconjugated quantum dots for *in vivo* molecular and cellular imaging. *Advanced Drug Delivery Reviews*, **60**, 1226–1240.

9. Vega-Villa, K.R., Takemoto, J.K., Yáñez, J.A. *et al.* (2008) Clinical toxicities of nanocarrier systems. *Advanced Drug Delivery Reviews*, **60**, 926–938.

10. Belin, T. and Epron, F. (2005) Characterization methods for carbon nanotubes: A review. *Materials Science and Engineering: B*, **119**, 105–118.

11. Feazell, R.P., Nakayama-Ratchford, N., Dai, H., and Lippard, S.J. (2007) Soluble single-walled carbon nanotubes as longboat delivery systems for platinum (IV) anticancer drug design. *Journal of the American Chemical Society*, **129**, 8438–8439.

12. Mousavi, S.A., Malerød, L., Berg, T., and Kjeken, R. (2004) Clathrin-dependent endocytosis. *The Biochemical Journal*, **377**, 1–16.

13. Dhar, S., Liu, Z., Thomale, J. *et al.* (2008) Targeted single-walled carbon nanotube-mediated Pt(IV) prodrug delivery using folate as a homing device. *Journal of the American Chemical Society*, **130**, 11467–11476.

14. James, J.L. (2003) Metal-organic frameworks. *Chemical Society Reviews*, **32**, 276–288.

15. Horcajada, P., Serre, C., Vallet-Regé, M. *et al.* (2006) Meta-organic frameworks as efficient materials for drug delivery. *Angewandte Chemie-International Edition*, **45**, 5974–5978.

16. Rieter, W.J., Pott, K.M., Taylor, K.M.L., and Lin, W. (2008) Nanoscale coordiantion polymers for platinum-based anticancer drug delivery. *Journal of the American Chemical Society*, **130**, 11584–11585.

17. Kresge, C.T., Leonowicz, M.E., Roth, W.J. *et al.* (1992) Ordered mesoporous molecular sieves synthesized by a liquid-crystal template mechanism. *Nature*, **359**, 710–712.

18. Lu, J., Liong, M., Zink, J.I., and Tamanoi, F. (2007) Mesoporous silica nanoparticles as a delivery system for hydrophobic anticancer drugs. *Small*, **3**, 1341–1346.

19. Allen, T.M. and Cullis, P.R. (2004) Drug delivery systems: Entering the mainstream. *Science*, **303**, 1818–1822.

20. Zamboni, W.C. *et al.* (2004) Systemic and tumor disposition of platinum administration of cisplatin or STEALTH liposomal-cisplatin formulations (SPI-077 and SPI-077 B103) in a preclinical tumor model of melonoma. *Cancer Chemotherapy and Pharmacology*, **53**, 329–336.

21. Hamelers, I.H.L., van Loenen, E., Staffhorst, R.W.H.M. *et al.* (2006) Carboplatin nanocapsules: A highly cytotoxic, phospholipid-based formulation of carboplatin. *Molecular Cancer Therapeutics*, **5**, 2007–2012.

22. Litzow, M.R. (2008) Arsenic trioxide. *Expert Opinion on Pharmacotherapy*, **9**, 1773–1785.

23. Talbot, S., Nelson, R., and Self, W.T. (2008) Arsenic trioxide and auranofin inhibit selenoprotein synthesis: Implications for chemotherapy for acute promyelocytic leukemia. *British Journal of Pharmacology*, **154**, 940–948.

24. Chen, H., MacDonald, R.C., Li, S. *et al.* (2006) Lipid encapsulation of arsenic trioxide attenuates cytotoxicity and allows for controlled anticancer drug release. *Journal of the American Chemical Society*, **128**, 13348–13349.

25. Yang, Z., Wang, X., Diao, H. *et al.* (2007) Encapsulation of platinum anticancer drugs by apoferritin. *Chemical Communications*, 3453–3455.

26. Cheng, Y., Samia, A.C., Meyers, J.D. *et al.* (2008) Highly efficient drug delivery with gold nanoparticle vectors for *in vivo* photodynamic therapy of cancer. *Journal of the American Chemical Society*, **130**, 10643–10647.

27. Hansan, M., Bethell, D., and Brust, M. (2002) The fate of sulfur-bound hydrogen on formation of self-assembled thiol monolayers on gold: ^{1}H NMR spectroscoptic evidence from solutions of gold clusters. *Journal of the American Chemical Society*, **124**, 1132–1133.

28. Walter, M. *et al.* (2008) A Unified view of ligand-protected gold clusters as superatom clusters. *Proceedings of the National Academy of Sciences of the United States of America*, **105**, 9157–9162.

29. Jadzinsky, P.D., Calero, G., Ackerson, C.J. *et al.* (2007) Structure of a thiol monolayer-protected gold nanoparticle at 1.1 Å resolution. *Science*, **318**, 430–433.

30. Dougherty, T.J., Gomer, C.J., Henderson, B.W. *et al.* (1998) Photodynamic therapy. *Journal of the National Cancer Institute*, **90**, 889.

31. Kolarova, H., Nevrelova, P., Tomankova, K. *et al.* (2008) Production of reactive oxygen species after photodynamic therapy by porphyrin sensitizers. *General Physiology and Biophysics*, **27**, 101–105.

32. Detty, M.R., Gibson, S.L., and Wagner, S.J. (2004) Current clinical and preclinical photosensitizers for use in photodynamic therapy. *Journal of Medicinal Chemistry*, **47**, 3897–3915.

33. Bednarski, P.J., Mackay, F.S., and Sadler, P.J. (2007) Photoactivatable platinum complexes. *Anti-Cancer Agents in Medicinal Chemistry*, **7**, 75–93.

34. Kim, D., Park, S., Lee, J.H. *et al.* (2007) Antibiofouling polymer-coated gold nanoparticles as a contrast agent for *in vivo* computed tomography imaging. *Journal of the American Chemical Society*, **129**, 7661–7665.

35. Sitharaman, B. *et al.* (2005) Superparamagnetic gadonadotubes are high-performance MRI contrast agents. *Chemical Communications*, 3915–3917.

36. Liu, X.-M., Fu, S.-Y., and Xiao, H.M. (2006) Synthesis of maghemite sub-microspheres by simple solvothermal reduction method. *Journal of Solid State Chemistry*, **179**, 1554–1558.

37. McCarthy, J.R., Kelly, K.A., Sun, E.Y., and Weissleder, R. (2007) Targeted delivery of multifunctional magnetic nanoparticles. *Nanomedicine*, **2**, 153–167.

38. Lalatonne, Y., Paris, C., Serfaty, J.M. *et al.* (2008) Bis-phosphonates-ultra small superparamagnetic iron oxide nanoparticles: A platform towards diagnosis and therapy. *Chemical Communications*, 2553–2555.

39. Tada, H., Higuchi, H., Wanatabe, T., and Ohuchi, N. (2007) *In vivo* real-time tracking of single quantum dots conjugated with monoclonal Anti-HER2 antibody in tumors of mice. *Cancer Research*, **67**, 1138–1143.

40. Michalet, X., Pinaud, F.F., Bentolila, L.A. *et al.* (2005) Quantum dots for live cells, *in vivo* imaging, and diagnostics. *Science*, **307**, 538–544.

41. Alivisatos, A.P. (1996) Semiconductor clusters, nanocrystals, and quantum dots. *Science*, **271**, 933–937.

42. Tinoco, I. Jr., Sauer, K., Wang, J.C., and Puglisi, J.D. (2002) Physical chemistry, *Principles and Applications to Biological Sciences*, 4 edn, Prentice Hall, Upper Saddle River, NJ.

43. Lewinski, N., Colvin, V., and Drezek, R. (2008) Cytotoxicity of nanoparticles. *Small*, **4**, 26–49.

44. Pulskamp, K., Diabaté, S., and Krug, H.F. (2007) Carbon nanotubes show no sign of acute toxicity but induce intracellular reactive oxygen species in dependence on contaminants. *Toxicology Letters*, **168**, 58–74.

45. Lacerda, L., Bianco, A., Prato, M., and Kostarelos, K. (2006) Carbon nanotubes as nanomedicines: from toxicology to pharmacology. *Advanced Drug Delivery Reviews*, **58**, 1460–1770.

46. Warheit, D.B. (2008) How meaningful are the results of nanotoxicity studies in the absence of adequate material characterization? *Toxicological Sciences*, **101**, 183–185.

47. Sayes, C.M., Reed, K.L., and Warheit, D.B. (2007) Assessing toxicity of fine and nanoparticles: Comparing *in vitro* measurements to *in vivo* pulmonary toxicity profiles. *Toxicological Sciences*, **97**, 163–180.

48. Warheit, D.B., Sayes, C.M., Reed, K.L., and Swain, K.A. (2008) Health effects related to nanoparticle exposure: Environmental, health, and saftey considerations for assessing hazards and risks. *Pharmacology & Therapeutics*, **120**, 35–42.

49. Derfus, A.M., Chan, W.C.W., and Bhatia, S.N. (2004) Probing the cytotoxicity of semiconductor quantum dots. *Nano Letters*, **4**, 11–18.

50. Hoshino, A., Fujioka, K., Oku, T. *et al.* (2004) Physicochemical properties and cellular toxicity of nanocrystal quantum dots depends on their surface modification. *Nano Letters*, **4**, 2163–2169.

51. Chung, T.-H., Wu, S.-H., Yao, M. *et al.* (2007) The effect of surface charge on the uptake and biological function of mesoporous silica nanoparticles in 3T3-L1 cells and human mesenchymal stem cells. *Biomaterials*, **28**, 2959–2966.

52. Murdock, R.C., Braydich-Stolle, L., Schrand, A.M. *et al.* (2008) Characterication of nanomaterial dispersion in solution prior to *in vitro* exposure using dynamic light scattering techniques. *Toxicological Sciences*, **101**, 239–253.

53. Di Pasqua, A.J., Sharma, K.K., Shi, Y.-L. *et al.* (2008) Cytotoxity of mesoporous silica nanomaterials. *Journal of Inorganic Biochemistry*, **102**, 1416–1423.

54. Wittmaack, K. (2007) In search of the most relevant parameter for quantifying lung inflammatory response to nanoparticle exposure: Particle number, surface area, or what? *Environmental Health Perspectives*, **115**, 187–194.

55. Tang, M., Wang, M., Xing, T. *et al.* (2008) Mechanisms of unmodified CdSe quantum dot-induced elevation of cytoplasmic calcium levels in primary cultures of rat hippocampal neurons. *Biomaterials*, **29**, 4383–4391.

56. Asseth, J., Haugen, M., and Førre, O. (1998) Rheumatoid arthritis and metal compounds-perspectives on the role of oxygen radical detoxifixcation. *Analyst*, **123**, 2–6.

57. Conner, E.E., Mwamuka, J., Gole, A. *et al.* (2005) Gold naonparticles are taken up by human cells but do not cause acute cytotoxicity. *Small*, **1**, 325–327.

58. Gupta, A.K. and Wells, S. (2004) Surface-modified superparamagnetic nanoparticles for drug delivery: Preparation, characterization, and cytotoxicity studies. *IEEE Transactions on Nanobioscience*, **3**, 66–73.

59. Fiorito, S., Serafino, A., Andreola, F. *et al.* (2006) Toxicity and biocompatibility of carbon nanoparticles. *Journal of Nanoscience and Nanotechnology*, **6**, 591–599.

60. Porter, A.E., Muller, K., Skepper, J. *et al.* (2006) Uptake of C_{60} by human monocyte macrophages, its localization and implications for toxicity: Studies by high resolution electron microscopy and electron tomography. *Acta Biomaterialia*, **2**, 409–419.

61. Cui, D., Tian, F., Ozkan, C.S. *et al.* (2005) Effect of single walled carbon nanotubes on human HEK293 cells. *Toxicology Letters*, **155**, 73–85.

62. Hudson, S.P., Padera, R.F., Langer, R., and Kohane, D.S. (2008) The biocompatibility of mesoporous silicates. *Biomaterials*, **29**, 4045–4055.

Further reading

Huheey, J.E., Keiter, E.A., and Keiter, R.L. (1993) Inorganic chemistry, *Principles of Structure and Reactivity*, 4th edn, Benjamin-Cummings Publishing Co., San Francisco, CA.

Matthews, C.K., van Holde, K.E., and Ahern, K.G. (2000) *Biochemistry*, 3rd edn, Addison Wesley Longman, San Francisco, CA.

Engel, T., Drobny, G., and Reid, P. (2008) *Physical Chemistry for the Life Sciences*, 1st edn, Prentice Hall, Upper Saddle River, NJ.

Index

This index was prepared by Neil Manley.

The Periodic Table

Legend:

0.98	— Pauling electronegativity
3	— Atomic number
Li	— Element
6.941	— Atomic weight (^{12}C)

1	2.20	2	
H		He	
1.008		4.003	

Group 1	Group 2	Group : 3	4	5	6	7 d transition elements	8	9	10	11	12	Group 13	Group 14	Group 15	Group 16	Group 17	Group 18
3 0.98 Li 6.941	4 1.57 Be 9.012											5 2.04 B 10.811	6 2.55 C 12.011	7 3.04 N 14.007	8 3.44 O 15.999	9 3.98 F 18.998	10 Ne 20.179
11 0.93 Na 22.990	12 1.31 Mg 24.305											13 1.61 Al 26.98	14 1.90 Si 28.086	15 2.19 P 30.974	16 2.58 S 32.064	17 3.16 Cl 35.453	18 Ar 39.948
19 0.82 K 39.102	20 1.00 Ca 40.08	21 Sc 44.956	22 1.54 Ti 47.90	23 V 50.941	24 Cr 51.996	25 Mn 54.938	26 Fe 55.847	27 Co 58.933	28 Ni 58.71	29 Cu 63.546	30 Zn 65.37	31 1.81 Ga 69.72	32 2.01 Ge 72.59	33 2.18 As 74.922	34 2.55 Se 78.96	35 2.96 Br 79.909	36 Kr 83.80
37 0.82 Rb 85.47	38 0.95 Sr 87.62	39 Y 88.906	40 Zr 91.22	41 Nb 92.906	42 Mo 95.94	43 Tc (99)	44 Ru 101.07	45 Rh 102.91	46 Pd 106.4	47 Ag 107.87	48 Cd 112.40	49 1.78 In 114.82	50 1.96 Sn 118.69	51 2.05 Sb 121.75	52 2.10 Te 127.60	53 2.66 I 126.90	54 Xe 131.30
55 0.79 Cs 132.91	56 0.89 Ba 137.34	57 La 138.91	72 Hf 178.49	73 Ta 180.95	74 W 183.85	75 Re 186.2	76 Os 190.2	77 Ir 192.22	78 Pt 195.09	79 Au 196.97	80 Hg 200.59	81 2.04 Tl 204.37	82 2.32 Pb 207.19	83 2.02 Bi 208.98	84 2.10 Po (210)	85 At (210)	86 Rn (222)
87 Fr (223)	88 Ra 226.025	89 Ac 227.0	104 Rf (261)	105 Db (262)	106 Sg (263)	107 Bh	108 Hs	109 Mt	110 Uun	111 Uuu	112 Unb						

58 Ce 140.12	59 Pr 140.91	60 Nd 144.24	61 Pm (147)	62 Sm 150.35	63 Eu 151.96	64 Gd 157.25	65 Tb 158.92	66 Dy 162.50	67 Ho 164.93	68 Er 167.26	69 Tm 168.93	70 Yb 173.04	71 Lu 174.97
90 Th 232.04	91 Pa (231)	92 U 238.03	93 Np (237)	94 Pu (242)	95 Am (243)	96 Cm (247)	97 Bk (247)	98 Cf (249)	99 Es (254)	100 Fm (253)	101 Md (253)	102 No (256)	103 Lw (260)